Complex Analysis and Potential Theory

Volume 55

CRM PROCEEDINGS & LECTURE NOTES

Centre de Recherches Mathématiques
Montréal

Complex Analysis and Potential Theory

André Boivin
Javad Mashreghi
Editors

The Centre de Recherches Mathématiques (CRM) of the Université de Montréal was created in 1968 to promote research in pure and applied mathematics and related disciplines. Among its activities are special theme years, summer schools, workshops, postdoctoral programs, and publishing. The CRM is supported by the Université de Montréal, the Province of Québec (FQRNT), and the Natural Sciences and Engineering Research Council of Canada. It is affiliated with the Institut des Sciences Mathématiques (ISM) of Montréal. The CRM may be reached on the Web at www.crm.math.ca.

American Mathematical Society
Providence, Rhode Island USA

The production of this volume was supported in part by the Fonds pour la Formation de Chercheurs et l'Aide à la Recherche (Fonds FCAR) and the Natural Sciences and Engineering Research Council of Canada (NSERC).

2000 *Mathematics Subject Classification.* Primary 30-XX, 31-XX, 32-XX.

Photo of Paul Gauthier courtesy of M. Peter Krasselt. Photo of Kohur N. GowriSankaran courtesy of Kohur N. GowriSankaran.

Library of Congress Cataloging-in-Publication Data
ISBN-13 978-0-8218-9173-5

This volume was submitted to the American Mathematical Society
in camera ready form by the Centre de Recherches Mathématiques.
Visit the AMS home page at `http://www.ams.org/`

10 9 8 7 6 5 4 3 2 1   17 16 15 14 13 12

Contents

Preface

A four-day international conference entitled "Complex Analysis and Potential Theory" was held in Montréal on June 20–23, 2011 at the Centre de recherches mathématiques (CRM). More than 50 mathematicians from 15 countries participated in the conference. The program consisted of 24 invited presentations and 14 contributed talks. The lectures given by some of the most established specialists in the fields covered a large range of topics: Geometric Function Theory; Approximation Theory; Operator Theory; Function Spaces; Classical, Abstract and Probabilistic Potential Theory; Pluripotential Theory; Harmonic Analysis; Orthogonal Polynomials; Complex Dynamics; and many others. The conference has provided means to bring together specialists, young researchers and graduate students from both Complex Analysis and Potential Theory to foster further cooperation and exchange of ideas and techniques to find new research perspectives.

The Conference was held to honour the important contributions to mathematics and the mathematical community of two influential analysts from Montréal: Professor Kohur N. GowriSankaran (McGill University) and Professor Paul M. Gauthier (Université de Montréal), both of whom have had long and distinguished careers in Montréal extending over four decades. Short biographies are provided below.

This proceedings contains 24 surveys and research articles from the participants and their co-authors. These papers are dedicated to Professors GowriSankaran and Gauthier. From potential theory on trees to approximation on Riemann surfaces, from universality to inner and outer functions and the disc algebra, from branching processes to harmonic extension and capacities, from harmonic mappings and the Harnack principle to integration formulae in several complex variables and the Hartogs phenomenon, from fine harmonicity and plurisubharmonic functions to the binomial identity and the Riemann hypothesis, and more, the diversity of subjects reflects the wide range of research interests of the two honored guests at this conference.

André Boivin
Javad Mashreghi

Paul M. Gauthier

Kohur N. GowriSankaran

Paul M. Gauthier

Paul is Franco-American. That is, his parents were French Canadians and he was born in the U.S. (in 1940). In 1967, Paul completed his military service in the National Guard and his doctoral studies at Wayne State University, under the direction of Wladimir Seidel. Seidel was a student of Constantin Carathéodory and was one of the founders of the theory of cluster sets and also of inner functions.

In 1967, Paul returned to the land of his ancestors, Canada, to join Université de Montréal. He was quite active in the Canadian Mathematical Society, where he served as Vice-president from 1993 to 1995. He was twice director of CMS summer research institutes and was scientific director for a meeting of the CMS. For many years, he was on the human rights committees of the CMS as well as the AMS and in both he was chair of the committees at some time. He was also scientific director for a Séminaire de mathématiques supérieures at Université de Montréal.

Paul has visited numerous institutions in his career. In particular, this list includes University of Maryland, 1972–1973; Wayne State University, May 1973; Universität Konstanz, fall 1974; Université de Paris, spring 1975; Steklov Institute, 1981–1982, 1989–1990; Bar Ilan University, May 1987; Beijing University, 1996–1997, fall 2003; Universität Oldenburg, summer 2003; Universität Potsdam, November 2003, summer 2007, spring 2010; Centre de Recerca Matemàtica, Barcelona, winter–spring, 2007; Jammu University, September 2010. He lectured also in Armenia, Austria, Bulgaria, China, Cyprus, Denmark, Finland, France, Gaza, Germany, Georgia, Greece, Hong Kong, Iran, India, Ireland, Israel, Italy, Japan, South Korea, Malaysia, Mexico, Morocco, Poland, Romania, Russia, Saudi Arabia, Soviet Union, Spain, Switzerland, Taiwan, Tunisia, Turkey, Ukraine, United Arab Emirates, United Kingdom, United States, Uzbekistan, Vietnam and Yugoslavia.

Paul has given plenary lectures at the Canadian Mathematical Society meeting in 1983, twice at the Colloque des mathématiques du Québec, at the Irish Mathematical Society meeting in 1994, at the Saudi Arabia Mathematical Society meeting in 2002, and in the Iranian Mathematical Society meeting in 2005.

Paul has more than 130 publications (recorded in Zentralblatt). In particular, he has translated four books, and several important articles, from Russian to English. At the Ph.D. level, Paul has had 11 students who completed and 17 descendants. He directed many students at the master's level and several postdoctoral students. In recent years, he has also directed four undergraduate research students, all of whom have published the results of their work.

Since retirement in 2008, Paul has been adjunct professor at Université de Montréal and membre régulier of the Centre de recherches mathématiques. At the time of this writing he has one postdoctoral student. In the past, Paul's presentations have been in complex analysis and potential theory, but in 2012, Paul also gave a talk at a meeting on algebraic geometry and an invited talk on the Riemann hypothesis, in a special session of the AMS on number theory.

Kohur N. GowriSankaran

Already in his twenties while at the Tata Institute of Fundamental Research (TIFR) Bombay he demonstrated his skill and vision. He was chosen to occupy for about a year before he left for the States in 1967, the position of the Secretary of the powerful Library Committee of the TIFR. This Committee was responsible for policy decisions of acquisition, hiring for Library, etc and had four other members who were all professor at the Institute. Though he was a junior member, he brought in policies for Library acquisition, particularly journals, in the face of tremendous foreign exchange problems that India faced at that time. He laid the groundwork for the future.

Soon after his arrival at McGill University in 1968, he and a few other colleagues were mainly responsible in establishing standards that are consistent with those practiced at some of the best universities, for the Preliminary Examinations for doctoral students.

During the next several years, he held positions in several Departmental and University committees. However, under the Chairmanship of Peter Russell, he agreed to take up the responsibility to chair the Committee on Graduate Affairs and become the Graduate Program Director in 1988, a position he held for four years. It is under his leadership that the graduate registration that had sunk to a low of about 30 in the fall of 1988 was brought up to over hundred in just a couple of years.

Once again, this time under the chairmanship of Georg Schmidt, he took up the position of Undergraduate Program Director and did it for three years during the period 1994-1998. In this role he established several important policies including cooperation with other departments which is very critical for Mathematics departments.

It is in 1994 he realized that there are in general very few mathematicians who participate at policy making and governance at the senior level of the University. He decided to make his presence felt at that level. He started off by getting elected to Senate of McGill University and got elected to the powerful Senate Steering Committee (1999–2010). He served four continuous elected terms on the Board of Governors and during some of those years was on the Executive Committee of the Board. He chaired the University Committee on Scholarships and Student Aid from 1999–2002. He was also chosen to become the President of the McGill Association of University Teachers (MAUT) for a term and he was on the Executive Committee from 2001–2004. The indirect benefit of increasing the profile of Mathematics from these activities is enormous. There are very few mathematicians who take up this kind of challenge unlike for example chemists, exceptions in the recent years in Canada (to the best of our knowledge) being Luc Vinet (Recteur at Université de Montréal) and Eddy Campbell (Provost at Memorial University and now President at The University of New Brunswick).

He was invited by the Dean of the Faculty of Science in 1999 to chair the Department. Despite being heavily involved in University Committees, he accepted the challenge and chaired the Department for six years until June 1, 2005. Among his many accomplishments, he made altogether almost 20 new faculty appointments during those six years. He brought enormous talent, both at the junior and senior level, to set the standard and ensure a very bright future for the Department for the coming twenty to thirty years.

He has been a member of the Canadian Mathematical Society for many years. He was named to a couple of committees in the eighties. He participated very vigorously as a chair during the last six years. In particular, he worked hard to make sure the meeting of the CMS held in Montréal in December of 2004 was a huge success. In the process he also raised several thousand dollars for CMS in conjunction with the meeting held in Montréal.

All along he never for any length of time forgot the raison d'être: Mathematics and research. He continues to be very active in research and publishes in some of the leading journals (Potential Analysis, American Journal of Mathematics, etc.). He has been invited to a large number of subject related conferences in several countries like France, Germany, Czech Republic, Romania, Tunisia, etc. and has given over six plenary talks just in the past decade. A total of nine students successfully completed their Ph.D. degrees under his supervision. That is one of the highest numbers in the entire history of the department, matched by a couple of other colleagues. Besides those his contribution to research has been in another dimension too. He was successful in getting funding directly from NATO (not an institutional process) to hold an Advanced Research Workshop in France (at Château de Bonas) in 1993. This was supplemented by an additional grant by the French Government! He has organized many special sessions at meetings of the American Mathematical Society as well as one under the banner of ACFAS (Association francophone pour le savoir).

Centre de Recherches Mathématiques
CRM Proceedings and Lecture Notes
Volume **55**, 2012

A Self-Contained Proof of the Strong-Type Capacitary Inequality for the Dirichlet Space

Omar El-Fallah, Karim Kellay, Javad Mashreghi, and Thomas Ransford

Abstract. In this expository article, we give a self-contained proof of the strong-type capacitary inequality for functions in the Dirichlet space. We also show that such functions possess boundary limits in exponentially tangential approach regions at almost every point on the unit circle.

1. Introduction

Let f be a function holomorphic in the unit disk $\mathbb{D}$. The *Dirichlet integral* of f is defined by

$$\mathcal{D}(f) := \frac{1}{\pi}\int_{\mathbb{D}} |f'(z)|^2 \, \mathrm{d}A(z),$$

where $\mathrm{d}A$ denotes area measure. The *Dirichlet space* is the vector space of all f holomorphic in $\mathbb{D}$ for which $\mathcal{D}(f) < \infty$.

An easy calculation shows that, if $f(z) = \sum_{k\geq 0} a_k z^k$, then $\mathcal{D}(f) = \sum_{k\geq 1} k|a_k|^2$. It follows that $\mathcal{D}$ is a subspace of the Hardy space H^2, and that it is a Hilbert space with respect to the norm $\|\cdot\|_{\mathcal{D}}$ given by

$$\|f\|_{\mathcal{D}}^2 := \|f\|_{H^2}^2 + \mathcal{D}(f) = \sum_{k\geq 0}(k+1)|a_k|^2.$$

We denote the corresponding inner product by $\langle \cdot, \cdot \rangle_{\mathcal{D}}$.

It is well known that, if $f \in H^2$, then the nontangential limit $f^*(\zeta)$ exists at almost every point ζ of the unit circle $\mathbb{T}$, and $f^* \in L^2(\mathbb{T})$. Since $\mathcal{D}$ is contained in H^2, it obviously inherits the same property. But in fact much more is true: if $f \in \mathcal{D}$, then

- the nontangential limit $f^*(\zeta)$ exists for all $\zeta \in \mathbb{T}$ outside a set of logarithmic capacity zero;

2010 *Mathematics Subject Classification.* Primary 31C25; Secondary 31A15.

Key words and phrases. Dirichlet space, capacity, strong-type inequality, tangential approach region.

First author supported by UMV-Agdal, CNRST and AI:MA/03.

Second author supported by ANR Dynop and Egide Volubulis.

Third author supported by NSERC and Le Fonds Racine.

Fourth author supported by NSERC and the Canada research chairs program.

This is the final form of the paper.

- the boundary limit function f^* satisfies the capacitary strong-type inequality
$$\int_0^\infty c(|f^*| > t) t\,\mathrm{d}t \le A\|f\|_{\mathcal{D}}^2,$$
where c denotes logarithmic capacity, and A is an absolute constant;
- for all $\zeta \in \mathbb{T}$ outside a set of Lebesgue measure zero, $f(z) \to f^*(\zeta)$ as $z \to \zeta$ in each region
$$|z-\zeta| < \kappa\Big(\log\frac{1}{1-|z|}\Big)^{-1} \qquad (\kappa > 0).$$

Our aim in this expository article is to give self-contained proofs of these results. In particular, we assume no prior knowledge of capacity. Our approach follows an idea of A. Borichev, which appears to lead to simpler proofs than the ones commonly found in the literature.

2. Capacity

Throughout this section, we fix a compact metric space (X, d) and a continuous decreasing function $K\colon (0,\infty) \to [0,\infty)$. The function K is called a *kernel*. We extend the definition of K to 0 by defining $K(0) := \lim_{t\to 0^+} K(t)$. It may well happen that $K(0) = \infty$, and in fact this is the case for most interesting kernels, though we do not insist upon it. However, in order to avoid trivialities, we do assume that $K \not\equiv 0$.

Definition 2.1. Let μ be a finite positive Borel measure on X. Its *K-potential* is the function $K\mu\colon X \to [0,\infty]$ defined by
$$K\mu(x) := \int_X K\big(d(x,y)\big)\,\mathrm{d}\mu(y) \qquad (x \in X).$$
The *K-energy* of μ is defined as
$$I_K(\mu) := \int K\mu\,\mathrm{d}\mu = \iint K\big(d(x,y)\big)\,\mathrm{d}\mu(x)\,\mathrm{d}\mu(y).$$

Clearly $I_K(\mu) \in [0,\infty]$. In fact it is not hard to see that $I_K(\mu) > 0$. However, it is quite possible that $I_K(\mu) = \infty$.

Definition 2.2. Let F be a compact subset of X. We write $\mathcal{P}(F)$ for the set of Borel probability measures on F. The *K-capacity* of F is defined by
$$c_K(F) := 1/\inf\{I_K(\mu) : \mu \in \mathcal{P}(F)\}. \tag{2.1}$$

In particular $c_K(F) = 0$ if and only if there is no measure $\mu \in \mathcal{P}(F)$ with $I_K(\mu) < \infty$. The following result gives some basic properties of capacity as a set function.

Theorem 2.3. (i) $c_K(\varnothing) = 0$.
(ii) *If $F_1 \subset F_2$, then $c_K(F_1) \le c_K(F_2)$.*
(iii) $c_K(F_1 \cup \cdots \cup F_n) \le c_K(F_1) + \cdots + c_K(F_n)$.

PROOF. Parts (i) and (ii) are obvious. For (iii), set $F := \bigcup_{k=1}^n F_k$. Let $\mu \in \mathcal{P}(F)$. For each k, let μ_k be the measure given by $\mu_k(S) := \mu\big(S \cap F_k \setminus (F_1 \cup \cdots \cup F_{k-1})\big)$. Then $\mu = \sum_{k=1}^n \mu_k$, so
$$I_K(\mu) \ge \sum_{k=1}^n I_K(\mu_k).$$

On the other hand, each measure μ_k is supported on F_k, so, by the definition of capacity and a simple scaling argument, $I_K(\mu_k)c_K(F_k) \geq \mu_k(F_k)^2$. Hence, using Schwarz's inequality, we have

$$1 = \mu(F) = \sum_{k=1}^{n} \mu_k(F_k) \leq \sum_{k=1}^{n} \sqrt{I_K(\mu_k)c_K(F_k)} \leq \Big(\sum_{k=1}^{n} I_K(\mu_k)\Big)^{1/2} \Big(\sum_{k=1}^{n} c_K(F_k)\Big)^{1/2}.$$

Combining these estimates, we obtain $I_K(\mu)\sum_{k=1}^{n} c_K(F_k) \geq 1$. As this holds for all $\mu \in \mathcal{P}(F)$, we deduce that $\sum_{k=1}^{n} c_K(F_k) \geq c_K(F)$, as required. □

Next, here is a simple upper bound for capacity in terms of diameter.

Theorem 2.4. *If F is a compact subset of X, then $c_K(F) \leq 1/K\big(\operatorname{diam}(F)\big)$.*

PROOF. If $\mu \in \mathcal{P}(F)$, then

$$I_K(\mu) \geq \iint K\big(\operatorname{diam}(F)\big)\,\mathrm{d}\mu(x)\,\mathrm{d}\mu(y) = K\big(\operatorname{diam}(F)\big).$$

It follows that $c_K(F) \leq 1/K\big(\operatorname{diam}(F)\big)$. □

Corollary 2.5. *For every compact set F we have $c_K(F) < \infty$.*

PROOF. By assumption $K \not\equiv 0$, so there exists $d_0 > 0$ such that $K(d_0) > 0$. A compact set F can be covered by finitely many compact sets $F_1, \dots, F_n$ of diameter at most d_0, so $c_K(F) \leq c_K(F_1) + \cdots + c_K(F_n) \leq n/K(d_0) < \infty$. □

The next result is sometimes expressed by saying that capacity is upper semicontinuous.

Theorem 2.6. *Let $(F_n)_{n\geq 1}$ be a decreasing sequence of compact subsets of X, and let $F := \cap_n F_n$. Then $c_K(F) = \lim_n c_K(F_n)$.*

The proof makes use of the notion of weak*-convergence of measures. Recall that a sequence of finite measures (μ_n) on X is weak*-convergent to μ if $\int g\,\mathrm{d}\mu_n \to \int g\,\mathrm{d}\mu$ for every continuous function g on X.

Lemma 2.7. *Let (μ_n) be a sequence in $\mathcal{P}(X)$ that is weak*-convergent to $\mu \in \mathcal{P}(X)$. Then $\liminf_{n\to\infty} I_K(\mu_n) \geq I_K(\mu)$.*

PROOF. We first claim that, if $f\colon X \times X \to \mathbb{R}$ is a continuous function, then $\iint f\,\mathrm{d}\mu_n\,\mathrm{d}\mu_n \to \iint f\,\mathrm{d}\mu\,\mathrm{d}\mu$ as $n \to \infty$. Indeed, this is clearly true if f has the form $f(x,y) = g(x)h(y)$, and, using the Stone–Weierstrass theorem, a general continuous f may be uniformly approximated by finite sums of functions of this special form.

If $K(0) < \infty$, then we can apply the claim with $f(x,y) := K\big(d(x,y)\big)$, and deduce that $I_K(\mu_n) \to I_K(\mu)$. For the general case, consider $K_T(t) := \min\{K(t), T\}$. By what we have just proved, $I_{K_T}(\mu) = \lim_{n\to\infty} I_{K_T}(\mu_n)$. Clearly $I_{K_T}(\mu_n) \leq I_K(\mu_n)$ for all n, and so $I_{K_T}(\mu) \leq \liminf_{n\to\infty} I_K(\mu_n)$. Also, by the monotone convergence theorem $I_{K_T}(\mu) \to I_K(\mu)$ as $T \to \infty$. Hence $I_K(\mu) \leq \liminf_{n\to\infty} I_K(\mu_n)$, as required. □

PROOF OF THEOREM 2.6. By Theorem 2.3(ii), the sequence $c_K(F_n)$ decreases and $c_K(F_n) \geq c_K(F)$ for all n. We need to show that $\lim_{n\to\infty} c_K(F_n) \leq c_K(F)$.

For each n, pick $\mu_n \in \mathcal{P}(F_n)$ such that $I_K(\mu_n) < 1/c_K(F_n) + 1/n$. By the Banach–Alaoglu theorem, there exists a subsequence of the μ_n which is weak*-convergent to some $\mu \in \mathcal{P}(X)$. Relabeling, if necessary, we may as well suppose that this is the whole sequence. Note that μ is supported in F_n for each n, so $\mu \in \mathcal{P}(F)$ and $I_K(\mu) \geq 1/c_K(F)$. On the other hand, by Lemma 2.7, we have $I_K(\mu) \leq \liminf_{n\to\infty} I_K(\mu_n) \leq \lim_{n\to\infty} 1/c_K(F_n)$. It follows that $\lim_{n\to\infty} c_K(F_n) \leq c_K(F)$. □

Now we extend the definition of capacity to noncompact sets.

Definition 2.8. Let E be an arbitrary subset of X.

- The *inner K-capacity* of E is defined by

$$c_K(E) := \sup\{c_K(F) : F \subset E, F \text{ compact}\}.$$

- The *outer K-capacity* of E is defined by

$$c_K^*(E) := \inf\{c_K(U) : U \supset E, U \text{ open in } X\}.$$

Evidently $c_K(E) \leq c_K^*(E)$. A set E is called *capacitable* if $c_K(E) = c_K^*(E)$. It is clear that open sets are capacitable, and Theorem 2.6 shows that compact sets are capacitable too. In fact, it can be shown that every Borel subset of X is capacitable (Choquet's theorem), but we do not need that result here.

The next theorem summarizes some basic properties of c_K^*.

Theorem 2.9. (i) *$c_K^*(\varnothing) = 0$.*
(ii) *If $E_1 \subset E_2$, then $c_K^*(E_1) \leq c_K^*(E_2)$.*
(iii) *$c_K^*(\bigcup_{k\geq 1} E_k) \leq \sum_{k\geq 1} c_K^*(E_k)$.*

In particular, the outer capacity of set does not change if we adjoin to it a set of outer capacity zero. Also, a countable union of sets of outer capacity zero is still of outer capacity zero.

Proof. Parts (i) and (ii) are obvious.

For part (iii), we consider first the case when each E_k is an open set U_k. Let F be a compact subset of $\bigcup_k U_k$. By compactness, there exists $n \geq 1$ such that $F \subset U_1 \cup \cdots \cup U_n$. We can write $F = F_1 \cup \cdots \cup F_n$, where each F_k is a compact subset of U_k. Then, using Theorem 2.3(iii), we have

$$c_K(F) \leq \sum_{k=1}^{n} c_K(F_k) \leq \sum_{k=1}^{n} c_K(U_k) \leq \sum_{k\geq 1} c_K(U_k).$$

As this holds for each such F, we obtain $c_K(\bigcup_k U_k) \leq \sum_k c_K(U_k)$, proving the result in this case.

For the general case, we may suppose that $c_K^*(E_k) < \infty$ for all k, otherwise there is nothing to prove. Let $\epsilon > 0$, and, for each k, let U_k be an open set such that $U_k \supset E_k$ and $c(U_k) < c^*(E_k) + \epsilon/2^k$. Then $\bigcup_k U_k$ is an open set containing $\bigcup_k E_k$, and, by what we have already proved,

$$c_K\left(\bigcup_k U_k\right) \leq \sum_k c_K(U_k) \leq \sum_k c_K^*(E_k) + \epsilon.$$

Thus $c_K^*(\bigcup_k E_k) \leq \sum_k c_K^*(E_k) + \epsilon$. This holds for all $\epsilon > 0$, whence the result. □

Next, we consider the important class of measures for which the infimum is attained in (2.1).

Definition 2.10. Let F be a compact subset of X such that $c_K(F) > 0$. An *equilibrium measure* for F is a measure $\nu \in \mathcal{P}(F)$ such that $I_K(\mu) \geq I_K(\nu)$ for all $\mu \in \mathcal{P}(F)$.

Clearly, if ν is an equilibrium measure for F, then $c_K(F) = 1/I_K(\nu)$.

Theorem 2.11. *Let F be a compact subset of X with $c_K(F) > 0$. Then F has an equilibrium measure.*

PROOF. For each $n \geq 1$, choose $\mu_n \in \mathcal{P}(F)$ such that $I_K(\mu_n) < 1/c_K(F)+1/n$. A subsequence of the μ_n converges weak* to some $\nu \in \mathcal{P}(F)$. By Lemma 2.7, this ν must satisfy $I_K(\nu) \leq 1/c_K(F)$, so it is an equilibrium measure for F. □

For many kernels K, it turns out that the equilibrium measure is unique. We shall not pursue this here. We do however need the following version of a theorem of Frostman, sometimes called the fundamental theorem of potential theory, which details some special properties of the potential of an equilibrium measure. We write $\operatorname{supp}\mu$ for the (closed) support of a measure μ.

Theorem 2.12. *Let F be a compact subset of X such that $c_K(F) > 0$. Let ν be an equilibrium measure for F. Then:*

(i) *$K\nu(x) \leq I_K(\nu)$ for all $x \in \operatorname{supp}\nu$,*
(ii) *$K\nu(x) \geq I_K(\nu)$ for all $x \in F \setminus E$, where $c_K^*(E) = 0$.*

Lemma 2.13. *If μ is a finite positive Borel measure on X, then its potential $K\mu$ is a lower semicontinuous function on X.*

PROOF. Let $x_n \to x_0$ in X. By Fatou's lemma, we have

$$\liminf_{n\to\infty} \int K\big(d(x_n, y)\big)\, \mathrm{d}\mu(y) \geq \int K\big(d(x_0, y)\big)\, \mathrm{d}\mu(y).$$

In other words $\liminf_{n\to\infty} K\mu(x_n) \geq K\mu(x_0)$, as required. □

PROOF OF THEOREM 2.12. We first claim that, if $\mu \in \mathcal{P}(F)$ and $I_K(\mu) < \infty$, then $\int K\nu\, \mathrm{d}\mu \geq I_K(\nu)$. In showing this, we may as well suppose that $\int K\nu\, \mathrm{d}\mu < \infty$, otherwise there is nothing to prove. For each $t \in (0, 1)$ consider the measure $\mu_t := t\mu + (1-t)\nu$. A simple calculation shows that

$$I_K(\mu_t) = I_K(\nu) + 2t\Big(\int K\nu\, \mathrm{d}\mu - I_K(\nu)\Big) + t^2\Big(I_K(\mu) + I_K(\nu) - 2\int K\nu\, \mathrm{d}\mu\Big).$$

Also, since $\mu_t \in \mathcal{P}(F)$, it must satisfy $I_K(\mu_t) \geq I_K(\nu)$ for all $t \in (0, 1)$. It follows that $\int K\nu\, \mathrm{d}\mu \geq I_K(\nu)$, as claimed.

(i) Suppose, if possible, that $K\nu(x) > I_K(\nu)$ for some $x \in \operatorname{supp}\nu$. As $K\nu$ is lower semicontinuous, there exists a neighborhood N of x such that $K\nu > I_K(\nu)$ on N. As $x \in \operatorname{supp}\nu$, we must have $\nu(N) > 0$, and hence

$$\int_N K\nu\, \mathrm{d}\nu > I_K(\nu)\nu(N).$$

Also, applying the claim at the beginning of the proof to the measure $\mu(S) := \nu(S \setminus N)/\nu(X \setminus N)$, we have

$$\int_{X\setminus N} K\nu\, \mathrm{d}\nu \geq I_K(\nu)\nu(X \setminus N).$$

Adding together the two inequalities, we obtain

$$\int K\nu \, d\nu > I_K(\nu),$$

which is clearly a contradiction. Hence $K\nu(x) \leq I_K(\nu)$ for all $x \in \operatorname{supp} \nu$.

(ii) For each $n \geq 1$, set

$$E_n := \{x \in F : K\nu(x) \leq I_K(\nu) - 1/n\}.$$

As $K\nu$ is lower semicontinuous, E_n is compact. Suppose, if possible, that $c_K(E_n) > 0$ for some n. Then E_n possesses an equilibrium measure ν_n. By the claim at the beginning of the proof, we have

$$\int K\nu \, d\nu_n \geq I_K(\nu).$$

On the other hand, since $K\nu \leq I_K(\nu) - 1/n$ on E_n, we have

$$\int K\nu \, d\nu_n \leq I_K(\nu) - 1/n.$$

This is clearly a contradiction. Thus, in fact, $c_K(E_n) = 0$ for all n. Setting $E := \bigcup_{n \geq 1} E_n$, we therefore have $c^*(E) = 0$ and $K\nu(x) \geq I_K(\nu)$ for all $x \in F \setminus E$. □

Though the notion of capacity has been developed in some generality, we shall be interested mostly in the following special case.

Definition 2.14. Let $X = \mathbb{T}$ with the metric $d(z, w) := |z - w|$, and let $K(t) := \log^+(2/t)$. The corresponding capacity is called *logarithmic capacity*. We shall denote it simply by $c(\cdot)$.

The choice of 2 in $\log^+(2/t)$ is largely for convenience, since $\mathbb{T}$ has diameter 2. It could be replaced by any other constant $A \geq 2$, and the resulting capacity c_A would be equivalent in the sense that $1/c_A(F) - 1/c(F) = \log(A/2)$ for all F.

Another possibility, often seen in the literature, is to take $K(t) = \log(1/t)$. This gives a nonpositive kernel, but one can nevertheless define

$$\tilde{c}(F) := \exp\Big(- \inf\{I_K(\mu) : \mu \in \mathcal{P}(F)\} \Big).$$

The capacities c and $\tilde{c}$ are related via the formula $1/c = \log(2/\tilde{c})$. In particular, $\tilde{c}(F) = 0$ if and only if $c(F) = 0$.

We shall need two results specific to logarithmic capacity on the circle. The first is an estimate in terms of Lebesgue measure. We write $|E|$ for arc-length measure of E.

Theorem 2.15. *Let E be a Borel subset of $\mathbb{T}$ with $|E| > 0$. Then*

$$c(E) \geq \frac{1}{\log(2\pi e/|E|)}.$$

Proof. Let $K(t) := \log^+(2/t)$. Let F be a compact subset of $\mathbb{T}$ with $|F| > 0$. Let μ be the probability measure on F defined by $\mu(S) := |F \cap S|/|F|$. Then

$$K\mu(z) = \frac{1}{|F|} \int_{e^{i\theta} \in F} \log \frac{2}{|z - e^{i\theta}|} \, d\theta \qquad (z \in \mathbb{T}).$$

For a fixed value of $z \in \mathbb{T}$, the integral is increased by replacing F with an arc in $\mathbb{T}$ of the same length $|F|$, centred at z. Thus

$$K\mu(z) \leq \frac{1}{|F|}\int_{-|F|/2}^{|F|/2} \log\frac{2}{|2\sin(\theta/2)|}\,\mathrm{d}\theta \leq \frac{1}{|F|}\int_{-|F|/2}^{|F|/2} \log\frac{\pi}{|\theta|}\,\mathrm{d}\theta$$
$$= \frac{2\pi}{|F|}\int_{0}^{|F|/2\pi} \log(1/t)\,\mathrm{d}t = \log(2\pi\mathrm{e}/|F|).$$

It follows that $I_K(\mu) \leq \log(2\pi e/|F|)$, and therefore $c_K(F) \geq 1/\log(2\pi e/|F|)$. This proves the result for compact sets.

The result for a general Borel set E now follows easily using the inner regularity of Lebesgue measure, namely the fact that $|E| = \sup\{|F| : F \subset E, F \text{ compact}\}$. □

It is also possible to give an upper bound for the logarithmic capacity of an arc in terms of its measure. However, no such bound is possible for general closed subsets of $\mathbb{T}$. Indeed, there exist Cantor-type sets E such that $|E| = 0$ but $c(E) > 0$.

The second result that we need is sometimes called the maximum principle for potentials.

Theorem 2.16. *Let $X = \mathbb{T}$ with the metric $d(z,w) := |z-w|$, and let $K\colon (0,\infty) \to [0,\infty)$ be a decreasing convex function. If μ is a finite positive Borel measure on $\mathbb{T}$ and $K\mu \leq M$ on $\operatorname{supp}\mu$, then $K\mu \leq M$ on $\mathbb{T}$.*

In particular, this result applies if $K(t) := \log^+(2/t)$.

PROOF. Let I be a connected component of $\mathbb{T}\setminus\operatorname{supp}\mu$, say $I = (\mathrm{e}^{\mathrm{i}\alpha}, \mathrm{e}^{\mathrm{i}\beta})$, where $0 < \beta - \alpha \leq 2\pi$. We shall prove that $K\mu \leq M$ on I.

If $\alpha \leq \gamma \leq \beta$, then

$$K\mu(\mathrm{e}^{\mathrm{i}\gamma}) = \int K(|\mathrm{e}^{\mathrm{i}\theta} - \mathrm{e}^{\mathrm{i}\gamma}|)\,\mathrm{d}\mu(\mathrm{e}^{\mathrm{i}\theta}) = \int_{[\beta,\alpha+2\pi]} K\left(2\sin\frac{\theta-\gamma}{2}\right)\mathrm{d}\mu(\theta).$$

Now sin is concave on $[0,\pi]$, and K is convex and decreasing on $[0,\infty)$. Hence, for each $\theta \in [\beta, \alpha+2\pi]$, the function $\gamma \mapsto K\big(2\sin((\theta-\gamma)/2)\big)$ is convex on $[\alpha,\beta]$ (perhaps infinite at the endpoints). Integrating with respect to μ, we deduce that $\gamma \mapsto K\mu(\mathrm{e}^{\mathrm{i}\gamma})$ is convex on $[\alpha,\beta]$. In particular, $K\mu(\mathrm{e}^{\mathrm{i}\gamma}) \leq \max\{K\mu(\mathrm{e}^{\mathrm{i}\alpha}), K\mu(\mathrm{e}^{\mathrm{i}\beta})\}$ for all $\gamma \in (\alpha,\beta)$. Now $K\mu \leq M$ at $\mathrm{e}^{\mathrm{i}\alpha}, \mathrm{e}^{\mathrm{i}\beta}$, because both these points belong to $\operatorname{supp}\mu$. Therefore $K\mu(\mathrm{e}^{\mathrm{i}\gamma}) \leq M$ for all $\gamma \in (\alpha,\beta)$. In other words $K\mu \leq M$ on I, which is what we wanted to prove. □

Combining this result with Theorem 2.12, we obtain the following important corollary.

Corollary 2.17. *Let $X = \mathbb{T}$ with the metric $d(z,w) := |z-w|$, and let $K\colon (0,\infty) \to [0,\infty)$ be a decreasing convex function. Let F be a compact subset of $\mathbb{T}$ such that $c_K(F) > 0$. Let ν be an equilibrium measure for F. Then:*

(i) *$K\nu(x) \leq I_K(\nu)$ for all $x \in \mathbb{T}$,*
(ii) *$K\nu(x) = I_K(\nu)$ for all $x \in F \setminus E$, where $c_K^*(E) = 0$.*

3. Cauchy transform

The rest of this article is based upon a representation formula for functions in $\mathcal{D}$ in terms of the Cauchy transform, which we shall establish in this section. We begin with some notation.

Definition 3.1. Let $\mathbb{A} := \{z \in \mathbb{C} : 1 < |z| < 2\}$. We write $L^2(\mathbb{A})$ for the Hilbert space of measurable functions $g\colon \mathbb{A} \to \mathbb{C}$ such that

$$\|g\|^2_{L^2(\mathbb{A})} := \frac{1}{\pi}\int_{\mathbb{A}} |g(w)|^2 \, \mathrm{d}A(w) < \infty.$$

Given $g \in L^2(\mathbb{A})$, we define its *Cauchy transform* $\mathcal{C}g\colon \mathbb{D} \to \mathbb{C}$ by

$$\mathcal{C}g(z) := \frac{1}{\pi}\int_{\mathbb{A}} \frac{g(w)}{w-z}\, \mathrm{d}A(w) \qquad (z \in \mathbb{D}).$$

Using Schwarz's inequality, we see that $\mathcal{C}g(z)$ is well-defined for all $z \in \mathbb{D}$, and that $\mathcal{C}g$ is holomorphic in $\mathbb{D}$. In fact, as the next result shows, it belongs to the Dirichlet space.

Theorem 3.2. *If $g \in L^2(\mathbb{A})$, then $\mathcal{C}g \in \mathcal{D}$ and $\|\mathcal{C}g\|_{\mathcal{D}} \le \sqrt{\frac{3}{2}}\|g\|_{L^2(\mathbb{A})}$.*

PROOF. Let $g \in L^2(\mathbb{A})$. For each $z \in \mathbb{D}$, we have

$$\mathcal{C}g(z) = \sum_{k\ge 0}\Big(\frac{1}{\pi}\int_{\mathbb{A}} \frac{g(w)}{w^{k+1}}\, \mathrm{d}A(w)\Big) z^k = \sum_{k\ge 0}\langle g, \phi_k\rangle_{L^2(\mathbb{A})} z^k, \tag{3.1}$$

where $\phi_k(w) := 1/\overline{w}^{k+1}$, and $\langle \cdot, \cdot\rangle_{L^2(\mathbb{A})}$ denotes the inner product on $L^2(\mathbb{A})$. Hence

$$\|\mathcal{C}g\|^2_{\mathcal{D}} = \sum_{k\ge 0}(k+1)|\langle g, \phi_k\rangle_{L^2(\mathbb{A})}|^2.$$

Now $(\phi_k)_{k\ge 0}$ is an orthogonal sequence in $L^2(\mathbb{A})$, so, by Bessel's inequality,

$$\sum_{k\ge 0}(k+1)|\langle g, \phi_k\rangle_{L^2(\mathbb{A})}|^2 \le B\|g\|^2_{L^2(\mathbb{A})},$$

where $B := \sup_{k\ge 0}(k+1)\|\phi_k\|^2_{L^2(\mathbb{A})}$. A calculation gives

$$\|\phi_k\|^2_{L^2(\mathbb{A})} = \frac{1}{\pi}\int_{\mathbb{A}} \frac{1}{|w|^{2k+2}}\, \mathrm{d}A(w) = 2\int_1^2 \frac{dr}{r^{2k+1}} = \begin{cases} \log 4, & k = 0, \\ (1-4^{-k})/k, & k \ge 1. \end{cases}$$

Therefore $B = \frac{3}{2}$ and the result follows. (Note that the constant $\sqrt{\frac{3}{2}}$ in the theorem is sharp: just take $g(w) = \phi_1(w) = 1/\overline{w}^2$.) □

In fact the Cauchy transform maps $L^2(\mathbb{A})$ *onto* $\mathcal{D}$. This is the next result, which may be viewed as a representation theorem for the Dirichlet space.

Theorem 3.3. *Given $f \in \mathcal{D}$, there exists $g \in L^2(\mathbb{A})$ such that $f = \mathcal{C}g$ and $\|g\|_{L^2(\mathbb{A})} \le \|f\|_{\mathcal{D}}$.*

PROOF. Let $f \in \mathcal{D}$, say $f(z) = \sum_{k\ge 0} a_k z^k$. Let $\phi_k(w) := 1/\overline{w}^{k+1}$, as in the previous proof, and consider $g := \sum_{k\ge 0}(a_k/\|\phi_k\|^2_{L^2(\mathbb{A})})\phi_k$. As (ϕ_k) is an orthogonal sequence in $L^2(\mathbb{A})$, and as

$$\sum_{k\ge 0}|a_k|^2/\|\phi_k\|^2_{L^2(\mathbb{A})} \le \sum_{k\ge 0}(k+1)|a_k|^2 = \|f\|^2_{\mathcal{D}} < \infty,$$

the series defining g converges in $L^2(\mathbb{A})$, and $\|g\|_{L^2(\mathbb{A})} \le \|f\|_{\mathcal{D}}$. Furthermore, by (3.1), we have

$$\mathcal{C}\phi_k(z) = \sum_{j\ge 0}\langle \phi_k, \phi_j\rangle_{L^2(\mathbb{A})} z^j = \|\phi_k\|^2_{L^2(\mathbb{A})} z^k \qquad (z \in \mathbb{D}, k \ge 0),$$

and therefore

$$\mathcal{C}g(z) = \sum_{k\geq 0}(a_k/\|\phi_k\|^2_{L^2(\mathbb{A})})\mathcal{C}\phi_k(z) = \sum_{k\geq 0} a_k z^k = f(z) \qquad (z \in \mathbb{D}).$$

This completes the proof. □

4. Beurling's theorem

Recall that $c(E)$ denotes the logarithmic capacity of $E \subset \mathbb{T}$, and that $c^*(E)$ is the corresponding outer capacity. Our main goal in this section is to prove the following theorem of Beurling.

Theorem 4.1. *Let $f \in \mathcal{D}$. Then there exists $E \subset \mathbb{T}$ with $c^*(E) = 0$ such that, if $\zeta \in \mathbb{T} \setminus E$, then $f^*(\zeta) := \lim_{r\to 1^-} f(r\zeta)$ exists, and $f(z) \to f^*(\zeta)$ as $z \to \zeta$ inside each region $|z - \zeta| < \kappa(1 - |z|)$ $(\kappa > 0)$.*

A property is said to hold *quasi-everywhere* (q.e.) on $\mathbb{T}$, if it holds everywhere on $\mathbb{T} \setminus E$ where $c^*(E) = 0$. Thus Beurling's theorem can be summarized by saying that each $f \in \mathcal{D}$ has nontangential limits quasi-everywhere on $\mathbb{T}$.

For the proof of Beurling's theorem, it is helpful to extend the Cauchy transform to the unit circle as follows.

Definition 4.2. Given $g \in L^2(\mathbb{A})$, we define $\widetilde{\mathcal{C}}g\colon \mathbb{T} \to [0, \infty]$ by

$$\widetilde{\mathcal{C}}g(\zeta) := \frac{1}{\pi}\int_{\mathbb{A}} \frac{|g(w)|}{|w-\zeta|}\,\mathrm{d}A(w) \qquad (\zeta \in \mathbb{T}).$$

If $\zeta \in \mathbb{T}$ and $\widetilde{\mathcal{C}}g(\zeta) < \infty$, then we set

$$\mathcal{C}g(\zeta) := \frac{1}{\pi}\int_{\mathbb{A}} \frac{g(w)}{w-\zeta}\,\mathrm{d}A(w).$$

Theorem 4.3. *Let $g \in L^2(\mathbb{A})$. Then $\widetilde{\mathcal{C}}g$ is lower semicontinuous on $\mathbb{T}$.*

Proof. Let $\zeta_n \to \zeta_0$ in $\mathbb{T}$. By Fatou's lemma,

$$\liminf_{n\to\infty} \frac{1}{\pi}\int_{\mathbb{A}} \frac{|g(w)|}{|w-\zeta_n|}\,\mathrm{d}A(w) \geq \frac{1}{\pi}\int_{\mathbb{A}} \frac{|g(w)|}{|w-\zeta_0|}\,\mathrm{d}A(w).$$

In other words, $\liminf_{n\to\infty} \widetilde{\mathcal{C}}g(\zeta_n) \geq \widetilde{\mathcal{C}}g(\zeta_0)$, as required. □

The next theorem shows that $\widetilde{\mathcal{C}}g$ plays the role of a sort of maximal function (for more on this theme, see, e.g., [26, p. 332]).

Theorem 4.4. *Let $g \in L^2(\mathbb{A})$, let $\zeta \in \mathbb{T}$, and suppose that $\widetilde{\mathcal{C}}g(\zeta) < \infty$. Then*

$$|\mathcal{C}g(z)| \leq \Big(1 + \frac{|z-\zeta|}{1-|z|}\Big)\widetilde{\mathcal{C}}g(\zeta) \qquad (z \in \mathbb{D}), \tag{4.1}$$

and $\mathcal{C}g(z) \to \mathcal{C}g(\zeta)$ as $z \to \zeta$ in each region $|z-\zeta| < \kappa(1-|z|)$.

Proof. For $z \in \mathbb{D}$, we have

$$|\mathcal{C}g(z)| \leq \frac{1}{\pi}\int_{\mathbb{A}} \frac{|g(w)|}{|w-z|}\,\mathrm{d}A(w) \leq \sup_{w\in\mathbb{A}} \frac{|w-\zeta|}{|w-z|}\widetilde{\mathcal{C}}g(\zeta).$$

Now, if $z \in \mathbb{D}$ and $w \in \mathbb{A}$, then

$$\frac{|w-\zeta|}{|w-z|} \leq \frac{|w-z|+|z-\zeta|}{|w-z|} \leq 1 + \frac{|z-\zeta|}{|w-z|} \leq 1 + \frac{|z-\zeta|}{1-|z|}.$$

This gives (4.1).

Turning now to the second part of the theorem, for $\delta > 0$ let us define $g_\delta(w) := g(w)1_{\{1<|w|<1+\delta\}}$. Then, for $z \in \mathbb{D}$, we have

$$|\mathcal{C}g(z) - \mathcal{C}g(\zeta)| \le |\mathcal{C}(g - g_\delta)(z) - \mathcal{C}(g - g_\delta)(\zeta)| + |\mathcal{C}g_\delta(z)| + |\mathcal{C}g_\delta(\zeta)|.$$

Now $\mathcal{C}(g - g_\delta)$ is holomorphic on a neighborhood of $\overline{\mathbb{D}}$, so in particular

$$\lim_{\substack{z\to\zeta\\ z\in\mathbb{D}}} \mathcal{C}(g - g_\delta)(z) = \mathcal{C}(g - g_\delta)(\zeta).$$

Also, by (4.1),

$$|\mathcal{C}g_\delta(z)| \le \Bigl(1 + \frac{|z-\zeta|}{1-|z|}\Bigr)\widetilde{\mathcal{C}}g_\delta(\zeta),$$

and clearly $|\mathcal{C}g_\delta(\zeta)| \le \widetilde{\mathcal{C}}g_\delta(\zeta)$. Putting these facts together, we deduce that, for each $\kappa > 0$,

$$\limsup_{\substack{z\to\zeta\\ |z-\zeta|\le\kappa(1-|z|)}} |\mathcal{C}g(z) - \mathcal{C}g(\zeta)| \le (\kappa + 2)\widetilde{\mathcal{C}}g_\delta(\zeta).$$

This inequality holds for each $\delta > 0$. Since $\widetilde{\mathcal{C}}g(\zeta) < \infty$, the dominated convergence theorem implies that $\widetilde{\mathcal{C}}g_\delta(\zeta) \to 0$ as $\delta \to 0$. Hence, for each $\kappa > 0$,

$$\limsup_{\substack{z\to\zeta\\ |z-\zeta|\le\kappa(1-|z|)}} |\mathcal{C}g(z) - \mathcal{C}g(\zeta)| = 0.$$

This gives the result. □

Theorem 4.4 begs the question: on how large a set can $\widetilde{\mathcal{C}}g(\zeta)$ be infinite? This is where capacity enters the picture.

Theorem 4.5. *Let $g \in L^2(\mathbb{A})$. Then*

$$c(\widetilde{\mathcal{C}}g > t) \le A\|g\|^2_{L^2(\mathbb{A})}/t^2 \qquad (t > 0),$$

where A is an absolute constant.

The key to the proof of this theorem is the following elementary lemma.

Lemma 4.6. *There exists $B > 0$ such that, for all $\zeta_1, \zeta_2 \in \mathbb{T}$,*

$$\frac{1}{\pi}\int_{\mathbb{A}} \frac{dA(w)}{|w-\zeta_1|\,|w-\zeta_2|} \le 2\log\frac{B}{|\zeta_1 - \zeta_2|}.$$

PROOF. Making the change of variable $z := (w - \zeta_1)/(\zeta_2 - \zeta_1)$, we obtain

$$\int_{1<|w|<2} \frac{dA(w)}{|w-\zeta_1|\,|w-\zeta_2|} = \int_{1<|\zeta_1+(\zeta_2-\zeta_1)z|<2} \frac{dA(z)}{|z|\,|z-1|} \le \int_{|z|<4/|\zeta_2-\zeta_1|} \frac{dA(z)}{|z|\,|z-1|}.$$

Now

$$\int_{|z|<2} \frac{dA(z)}{|z|\,|z-1|} \le \int_{|z|<2} \Bigl(\frac{1}{|z|} + \frac{1}{|z-1|}\Bigr)\,dA(z) \le \int_{|z|<2} \frac{dA(z)}{|z|} + \int_{|z|<3} \frac{dA(z)}{|z|} = 4\pi + 6\pi.$$

Also, for $R > 2$,

$$\int_{2\le|z|<R} \frac{\mathrm{d}A(z)}{|z|\,|z-1|} \le 2\pi \int_2^R \frac{\mathrm{d}r}{r-1} \le 2\pi \log R.$$

If we put all these inequalities together, then we obtain

$$\frac{1}{\pi}\int_{\mathbb{A}} \frac{dA(w)}{|w-\zeta_1|\,|w-\zeta_2|} \le 10 + 2\log\frac{4}{|\zeta_2-\zeta_1|}.$$

This gives the required inequality with $B = 4\mathrm{e}^5$. □

PROOF OF THEOREM 4.5. For the time being, we work with the capacity c_K, where K is the kernel $K(t) := \log^+(B/t)$, and B is the constant in Lemma 4.6. At the end of the proof we shall make the necessary adjustments to obtain the standard logarithmic capacity c.

Let $t > 0$ and let F be a compact subset of $\{\zeta \in \mathbb{T} : \widetilde{\mathcal{C}}g(\zeta) > t\}$. Let μ be a Borel probability measure on F. Then, by Fubini's theorem and Schwarz's inequality, we have

$$\begin{aligned} t \le \int_F \widetilde{\mathcal{C}}g(\zeta)\,\mathrm{d}\mu(\zeta) &= \frac{1}{\pi}\int_{\mathbb{A}}\int_F \frac{|g(w)|}{|w-\zeta|}\,\mathrm{d}\mu(\zeta)\,\mathrm{d}A(w) \\ &\le \|g\|_{L^2(\mathbb{A})}\Bigl(\frac{1}{\pi}\int_{\mathbb{A}}\Bigl(\int_F \frac{1}{|w-\zeta|}\,\mathrm{d}\mu(\zeta)\Bigr)^2\mathrm{d}A(w)\Bigr)^{1/2}. \end{aligned}$$

Now, using Lemma 4.6, we have

$$\begin{aligned} \frac{1}{\pi}\int_{\mathbb{A}}\Bigl(\int_F \frac{1}{|w-\zeta|}\,\mathrm{d}\mu(\zeta)\Bigr)^2\mathrm{d}A(w) &= \int_F\int_F \frac{1}{\pi}\int_{\mathbb{A}} \frac{dA(w)}{|w-\zeta_1|\,|w-\zeta_2|}\,\mathrm{d}\mu(\zeta_1)\,\mathrm{d}\mu(\zeta_2) \\ &\le \int_F\int_F 2\log\frac{B}{|\zeta_1-\zeta_2|}\,\mathrm{d}\mu(\zeta_1)\,\mathrm{d}\mu(\zeta_2) = 2I_K(\mu). \end{aligned}$$

If we put all this together, then we obtain $t \le \|g\|_{L^2(\mathbb{A})}|(2I_K(\mu)|)^{1/2}$, whence

$$1/I_K(\mu) \le 2\|g\|^2_{L^2(\mathbb{A})}/t^2.$$

As this holds for each $\mu \in \mathcal{P}(F)$, it follows that

$$c_K(F) \le 2\|g\|^2_{L^2(\mathbb{A})}/t^2.$$

And as this holds for each compact F in $\{\widetilde{\mathcal{C}}g > t\}$, we get

$$c_K(\widetilde{\mathcal{C}}g > t) \le 2\|g\|^2_{L^2(\mathbb{A})}/t^2.$$

It remains to relate c_K to c. For this, note that the difference between their respective kernels is $\log B - \log 2 = \log(B/2)$, and so $1/c_K(F) - 1/c(F) = \log(B/2)$ for all compact sets F. Hence

$$c(F)/c_K(F) = 1 + \log(B/2)c(F) \le 1 + \log(B/2)c(\mathbb{T}).$$

It follows that $c(\widetilde{\mathcal{C}}g > t) \le A\|g\|^2_{L^2(\mathbb{A})}/t^2$, where $A := 2\bigl(1 + \log(B/2)c(\mathbb{T})\bigr)$. □

Corollary 4.7. *If $g \in L^2(\mathbb{A})$, then $\widetilde{\mathcal{C}}g < \infty$ quasi-everywhere on $\mathbb{T}$.*

PROOF. Since $\widetilde{\mathcal{C}}g$ is lower semicontinuous, for each $t > 0$ the set $\{\zeta \in \mathbb{T} : \widetilde{\mathcal{C}}g(\zeta) > t\}$ is open in $\mathbb{T}$. Thus $c^*(\widetilde{\mathcal{C}}g = \infty) \le c(\widetilde{\mathcal{C}}g > t) \le A\|g\|^2_{L^2(\mathbb{A})}/t^2$ for all $t > 0$. Letting $t \to \infty$, we get $c^*(\widetilde{\mathcal{C}}g = \infty) = 0$. □

Finally, we are in a position to prove Beurling's theorem.

PROOF OF THEOREM 4.1. Let $f \in \mathcal{D}$. From Theorem 3.3 we know that $f = \mathcal{C}g$ for some $g \in L^2(\mathbb{A})$. By Theorem 4.4, $\mathcal{C}g$ has nontangential limits at every $\zeta \in \mathbb{T}$ for which $\widetilde{\mathcal{C}}g(\zeta) < \infty$. And by Corollary 4.7, we have $\widetilde{\mathcal{C}}g(\zeta) < \infty$ quasi-everywhere. The proof is complete. □

5. Weak-type and strong-type inequalities

The proof of Beurling's theorem above yields a lot more information about the boundary values of functions in $\mathcal{D}$. In particular, the following important result is more or less an immediate consequence.

Theorem 5.1 (Capacitary weak-type inequality). *Let $f \in \mathcal{D}$. Then*

$$c^*(|f^*| > t) \leq A\|f\|_{\mathcal{D}}^2/t^2 \qquad (t > 0), \tag{5.1}$$

where A is an absolute constant.

PROOF. Using Theorem 3.3, we can find a function $g \in L^2(\mathbb{A})$ such that $f = \mathcal{C}g$ and $\|g\|_{L^2(\mathbb{A})} \leq \|f\|_{\mathcal{D}}$. By Theorem 4.3 the set $\{\widetilde{\mathcal{C}}g > t\}$ is open in $\mathbb{T}$, and by Theorem 4.5 we have $c(\widetilde{\mathcal{C}}g > t) \leq A\|g\|_{L^2(\mathbb{A})}^2/t^2$, where A is an absolute constant. Finally, $|f^*| = |\mathcal{C}g| \leq \widetilde{\mathcal{C}}g$ q.e. on $\mathbb{T}$. Assembling the pieces, we obtain (5.1). □

Recall that $|\cdot|$ denotes Lebesgue measure on $\mathbb{T}$, and that it is related to logarithmic capacity via Theorem 2.15.

Corollary 5.2. *Let $f \in \mathcal{D}$. Then*

$$|\{|f^*| > t\}| \leq A\mathrm{e}^{-Bt^2/\|f\|_{\mathcal{D}}^2} \qquad (t > 0),$$

where $A, B > 0$ are absolute constants.

PROOF. Combine Theorems 5.1 and 2.15. □

Corollary 5.3. *Let $f \in \mathcal{D}$. Then* $\exp(|f^*|^2) \in L^1(\mathbb{T})$.

PROOF. Assume first that $\|f\|_{\mathcal{D}}^2 < B$, where B is the constant in Corollary 5.2. Then

$$\begin{aligned}\int_{\mathbb{T}} (\mathrm{e}^{|f^*(\mathrm{e}^{\mathrm{i}\theta})|^2} - 1)\,\mathrm{d}\theta &= \int_{\mathbb{T}} \int_0^{|f^*(\mathrm{e}^{\mathrm{i}\theta})|} 2t\mathrm{e}^{t^2}\,\mathrm{d}t\,\mathrm{d}\theta = \int_{t=0}^{\infty} \int_{\{|f^*|>t\}} 2t\mathrm{e}^{t^2}\,\mathrm{d}\theta\,\mathrm{d}t \\ &= \int_{t=0}^{\infty} 2t\mathrm{e}^{t^2}|\{|f^*| > t\}|\,\mathrm{d}t \leq \int_{t=0}^{\infty} 2At\mathrm{e}^{t^2}e^{-Bt^2/\|f\|_{\mathcal{D}}^2}\,\mathrm{d}t < \infty.\end{aligned}$$

For the general case, write $f = p + g$, where p is a polynomial and $\|g\|_{\mathcal{D}}^2 < B/2$. Then $|f^*|^2 \leq 2|p|^2 + 2|g^*|^2$ q.e. on $\mathbb{T}$, whence

$$\int_{\mathbb{T}} \mathrm{e}^{|f^*(\mathrm{e}^{\mathrm{i}\theta})|^2}\,\mathrm{d}\theta \leq \mathrm{e}^{2\|p\|_\infty^2} \int_{\mathbb{T}} \mathrm{e}^{2|g^*(\mathrm{e}^{\mathrm{i}\theta})|^2}\,\mathrm{d}\theta < \infty. \qquad \square$$

This result is close to optimal. If $p > 2$ and $f = \big(\log\big(1/(1-z)\big)\big)^{1/p}$, then $f \in \mathcal{D}$ but $\exp(|f^*|^p) \notin L^1(\mathbb{T})$.

The weak-type inequality, Theorem 5.1, can be strengthened a little further. This extra strength turns out to be vital for certain applications, such as the characterization of multipliers.

Theorem 5.4 (Capacitary strong-type inequality). *Let $f \in \mathcal{D}$. Then*

$$\int_0^\infty c^*(|f^*| > t)\,\mathrm{d}t^2 \le A\|f\|_{\mathcal{D}}^2, \tag{5.2}$$

where A is an absolute constant.

PROOF. Once again, it will be convenient to work with the capacity c_K, where $K(t) := \log^+(B/t)$ and B is the constant in Lemma 4.6. We shall prove that, if $g \in L^2(\mathbb{A})$, then

$$\int_0^\infty c_K(\widetilde{\mathcal{C}}g > t)\,\mathrm{d}t^2 \le 32\|g\|_{L^2(\mathbb{A})}^2. \tag{5.3}$$

If so, then, since c is majorized by a multiple of c_K, the same inequality holds with c_K by c, and 32 replaced by another absolute constant. We can then finish off the proof by using the representation theorem for $\mathcal{D}$, exactly as in the proof of Theorem 5.1. Thus, it remains to prove (5.3).

Let $n \ge 1$ and, for $k = -n, \dots, n$, let F_k be a compact subset of $\{\widetilde{\mathcal{C}}g > 2^k\}$. Set $\mu := \sum_{k=-n}^n 2^k c_K(F_k)\nu_k$, where ν_k is an equilibrium measure for F_k. By Fubini's theorem and Schwarz's inequality, we have

$$\begin{aligned}\sum_{k=-n}^n 2^{2k} c_K(F_k) &\le \int_{\mathbb{T}} \widetilde{\mathcal{C}}g(\zeta)\,\mathrm{d}\mu(\zeta) = \frac{1}{\pi}\int_{\mathbb{A}}\int_{\mathbb{T}} \frac{|g(w)|}{|w-\zeta|}\,\mathrm{d}\mu(\zeta)\,\mathrm{d}A(w)\\ &\le \|g\|_{L^2(\mathbb{A})}\Big(\frac{1}{\pi}\int_{\mathbb{A}}\Big(\int_{\mathbb{T}}\frac{1}{|w-\zeta|}\,\mathrm{d}\mu(\zeta)\Big)^2\,\mathrm{d}A(w)\Big)^{1/2}.\end{aligned}$$

Using Lemma 4.6, we have

$$\begin{aligned}\frac{1}{\pi}\int_{\mathbb{A}}\Big(\int_{\mathbb{T}}\frac{1}{|w-\zeta|}\,\mathrm{d}\mu(\zeta)\Big)^2\,\mathrm{d}A(w) &= \int_{\mathbb{T}}\int_{\mathbb{T}}\frac{1}{\pi}\int_{\mathbb{A}}\frac{dA(w)}{|w-\zeta_1|\,|w-\zeta_2|}\,\mathrm{d}\mu(\zeta_1)\,\mathrm{d}\mu(\zeta_2)\\ &\le \int_{\mathbb{T}}\int_{\mathbb{T}} 2\log\frac{B}{|\zeta_1-\zeta_2|}\,\mathrm{d}\mu(\zeta_1)\,\mathrm{d}\mu(\zeta_2) = 2I_K(\mu).\end{aligned}$$

Now, by Corollary 2.17, we have $K\nu_k \le I_K(\nu_k)$ on $\mathbb{T}$. It follows that $\int K\nu_j\,\mathrm{d}\nu_k = \int K\nu_k\,\mathrm{d}\nu_j \le I(\nu_k)$ for all j, k. Therefore

$$\begin{aligned}I_K(\mu) &= \sum_{k=-n}^n\sum_{j=-n}^n 2^j c_K(F_j)2^k c_K(F_k)\int_{\mathbb{T}} K\nu_j\,\mathrm{d}\nu_k\\ &\le \sum_{k=-n}^n\sum_{j=-n}^n 2^j c_K(F_j)2^k c_K(F_k)\min\{I_K(\nu_j), I_K(\nu_k)\}\\ &= \sum_{k=-n}^n\sum_{j=-n}^n 2^j 2^k \min\{c_K(F_j), c_K(F_k)\}\\ &\le 2\sum_{k=-n}^n\sum_{j=-n}^k 2^j 2^k c_K(F_k)\\ &\le 4\sum_{k=-n}^n 2^{2k} c_K(F_k).\end{aligned}$$

If we put all this together, then we obtain

$$\sum_{k=-n}^{n} 2^{2k} c_K(F_k) \le \|g\|_{L^2(\mathbb{A})} \Big(8 \sum_{k=-n}^{n} 2^{2k} c_K(F_k)\Big)^{1/2},$$

whence

$$\sum_{k=-n}^{n} 2^{2k} c_K(F_k) \le 8\|g\|^2_{L^2(\mathbb{A})}.$$

As this holds for all such compact sets F_k, we get

$$\sum_{k=-n}^{n} 2^{2k} c_K(\widetilde{\mathcal{C}}g > 2^k) \le 8\|g\|^2_{L^2(\mathbb{A})}.$$

Finally, letting $n \to \infty$, we obtain

$$\sum_{k=-\infty}^{\infty} 2^{2k} c_K(\widetilde{\mathcal{C}}g > 2^k) \le 8\|g\|^2_{L^2(\mathbb{A})},$$

and this implies (5.3). □

6. Tangential approach regions

In this section we shall consider limits along certain exponentially tangential approach regions. Our aim is to prove the following result, due to Nagel, Rudin and Shapiro [28].

Theorem 6.1. *Let $f \in \mathcal{D}$. Then, for a.e. $\zeta \in \mathbb{T}$, we have $f(z) \to f^*(\zeta)$ as $z \to \zeta$ in each region*

$$|z - \zeta| < \kappa\Big(\log \frac{1}{1-|z|}\Big)^{-1} \qquad (\kappa > 0).$$

The proof is based on the representation formula of Theorem 3.3, and on a slightly more sophisticated version of the estimate (4.1). In order to state the estimate, we need to recall the notion of maximal function.

Given $h \in L^1(\mathbb{T})$, its *maximal function* $Mh\colon \mathbb{T} \to [0,\infty]$ is defined by

$$(Mh)(\zeta) := \sup_{\delta>0} \frac{1}{2\delta} \int_{|\theta - \arg\zeta|<\delta} |h(e^{i\theta})|\, d\theta.$$

Mh is lower semicontinuous, and it satisfies the weak-type inequality

$$|\{Mh > t\}| \le A\|h\|_{L^1(\mathbb{T})}/t \qquad (t > 0), \tag{6.1}$$

where A is an absolute constant. For more details, we refer to [32, §7].

Lemma 6.2. *Let $g \in L^2(\mathbb{A})$ and let $\zeta \in \mathbb{T}$. Then*

$$|\mathcal{C}g(z)| \le 2\widetilde{\mathcal{C}}g(\zeta) + 2\Big((Mh_g)(\zeta)|z-\zeta| \log\Big(\frac{3}{1-|z|}\Big)\Big)^{1/2} \qquad (z \in \mathbb{D}), \tag{6.2}$$

where

$$h_g(\zeta) := \int_1^2 |g(r\zeta)|^2 r\, dr \qquad (\zeta \in \mathbb{T}).$$

Proof. Fix $z \in \mathbb{D}$. Clearly

$$|\mathcal{C}g(z)| \leq \frac{1}{\pi}\int_{\mathbb{A}} \frac{|g(w)|}{|w-z|}\,\mathrm{d}A(w).$$

We shall estimate the right-hand side by splitting the domain of integration into two regions, $\mathbb{A}_1$ and $\mathbb{A}_2$, say. Let $\delta := |z-\zeta|$ and set $\mathbb{A}_1 := \mathbb{A} \cap \{w : |w-\zeta| \geq 2\delta\}$. For w in this region, we have

$$\left|\frac{w-\zeta}{w-z}\right| \leq \frac{|w-\zeta|}{|w-\zeta|-\delta} \leq 2,$$

and consequently

$$\frac{1}{\pi}\int_{\mathbb{A}_1} \frac{|g(w)|}{|w-z|}\,\mathrm{d}A(w) \leq \frac{1}{\pi}\int_{\mathbb{A}} \frac{2|g(w)|}{|w-\zeta|}\,\mathrm{d}A(w) = 2\widetilde{\mathcal{C}}g(\zeta).$$

In the complementary region $\mathbb{A}_2$, we apply Schwarz's inequality:

$$\frac{1}{\pi}\int_{\mathbb{A}_2} \frac{|g(w)|}{|w-z|}\,\mathrm{d}A(w) \leq \left(\frac{1}{\pi}\int_{\mathbb{A}_2} |g(w)|^2\,\mathrm{d}A(w)\right)^{1/2}\left(\frac{1}{\pi}\int_{\mathbb{A}_2} \frac{1}{|w-z|^2}\,\mathrm{d}A(w)\right)^{1/2}.$$

Now $\mathbb{A}_2$ is contained in the sector $\{1 < r < 2, |\theta - \arg\zeta| < \pi\delta\}$. Hence

$$\frac{1}{\pi}\int_{\mathbb{A}_2} |g(w)|^2\,\mathrm{d}A(w) \leq \frac{1}{\pi}\int_{|\theta-\arg\zeta|<\pi\delta} h_g(\mathrm{e}^{\mathrm{i}\theta})\,\mathrm{d}\theta \leq 2\delta(Mh_g)(\zeta).$$

Also, we have

$$\frac{1}{\pi}\int_{\mathbb{A}_2} \frac{1}{|w-z|^2}\,\mathrm{d}A(w) \leq \frac{1}{\pi}\int_{1-|z|<|w|<3} \frac{1}{|w|^2}\,\mathrm{d}A(w) = 2\log\frac{3}{1-|z|}.$$

Putting all these estimates together, we obtain (6.2). □

Proof of Theorem 6.1. Let $f \in \mathcal{D}$. By Theorem 3.3 we can write $f = \mathcal{C}g$, where $g \in L^2(\mathbb{A})$. We then have $f(z) - f^*(\zeta) = \mathcal{C}g(z) - \mathcal{C}g(\zeta)$ for all $z \in \mathbb{D}$ and a.e. $\zeta \in \mathbb{T}$. For $\delta > 0$, let us set $g_\delta(w) := g(w)1_{\{1<|w|<1+\delta\}}$. Just as in the proof of Theorem 4.4, we have

$$|\mathcal{C}g(z) - \mathcal{C}g(\zeta)| \leq |\mathcal{C}(g-g_\delta)(z) - \mathcal{C}(g-g_\delta)(\zeta)| + |\mathcal{C}g_\delta(z)| + |\mathcal{C}g_\delta(\zeta)|,$$

where the first term tends to zero as $z \to \zeta$ unrestrictedly in $\mathbb{D}$, and the third term satisfies $|\mathcal{C}g_\delta(\zeta)| \leq \widetilde{\mathcal{C}}g(\zeta)$. As for the second term, if we write

$$\Omega_\kappa(\zeta) := \bigl\{z \in \mathbb{D} : |z-\zeta| < \kappa/\log\bigl(3/(1-|z|)\bigr)\bigr\},$$

then by Lemma 6.2

$$|\mathcal{C}g_\delta(z)| \leq 2\widetilde{\mathcal{C}}g_\delta(\zeta) + 2\bigl(\kappa(Mh_{g_\delta})(\zeta)\bigr)^{1/2} \qquad \bigl(z \in \Omega_\kappa(\zeta)\bigr).$$

Putting this all together, we deduce that

$$\limsup_{\substack{z\to\zeta\\ z\in\Omega_\kappa(\zeta)}} |f(z) - f^*(\zeta)| \leq 3\widetilde{\mathcal{C}}g_\delta(\zeta) + 2\bigl(\kappa(Mh_{g_\delta})(\zeta)\bigr)^{1/2}.$$

Now, by Theorems 4.5 and 2.15, we have

$$|\{\widetilde{\mathcal{C}}g_\delta > t\}| \leq Ae^{-Bt^2/\|g_\delta\|^2_{L^2(\mathbb{A})}},$$

and by (6.1)

$$|\{Mh_{g_\delta} > t\}| \leq A\|h_{g_\delta}\|_{L^1(\mathbb{T})}/t \leq A'\|g_\delta\|^2_{L^2(\mathbb{A})}/t,$$

where A, A', B are absolute constants. Also, clearly, $\|g_\delta\|_{L^2(\mathbb{A})} \to 0$ as $\delta \to 0$. Hence, given $\epsilon > 0$, we can choose $\delta > 0$ small enough to ensure that both $|\{\widetilde{\mathcal{C}}g_\delta > \epsilon\}| < \epsilon$ and $|\{Mh_{g_\delta} > \epsilon\}| < \epsilon$. It follows that

$$\limsup_{\substack{z\to\zeta \\ z\in\Omega_\kappa(\zeta)}} |f(z) - f^*(\zeta)| \le 3\epsilon + 2(\kappa\epsilon)^{1/2}$$

for all $\zeta \in \mathbb{T}$ outside a set of measure 2ϵ. Letting $\epsilon \to 0$, we deduce that

$$\lim_{\substack{z\to\zeta \\ z\in\Omega_\kappa(\zeta)}} f(z) = f^*(\zeta)$$

for all $\zeta \in \mathbb{T}$ outside a set of measure zero. □

7. Further results

In this section we discuss very briefly and, for the most part, without detailed proofs, some extensions of the principal results in this article, as well as some theorems concerning their sharpness.

The two main theorems on boundary limits, namely Beurling's theorem (Theorem 4.1) and the theorem of Nagel – Rudin – Shapiro (Theorem 6.1), are the simplest in a whole scale of results of this kind, with a playoff between the width of the approach region and the size of the exceptional set. The following result is taken from [36], where it is also shown to be optimal in a certain sense.

Theorem 7.1. *Let $f \in \mathcal{D}$.*

(i) *For all $\zeta \in \mathbb{T}$ outside a set of logarithmic capacity zero, $f(z) \to f^*(\zeta)$ as $z \to \zeta$ in each region*

$$\{z : |z - \zeta| < \kappa(1 - |z|)^\lambda\} \qquad (\kappa > 0, \lambda > 0).$$

(ii) *Given $0 < \gamma \le 1$, for all $\zeta \in \mathbb{T}$ outside a set of γ-dimensional Hausdorff content zero, $f(z) \to f^*(\zeta)$ as $z \to \zeta$ in each region*

$$\left\{z : |z - \zeta| < \kappa\left(\log\frac{1}{1 - |z|}\right)^{-1/\gamma}\right\} \qquad (\kappa > 0).$$

It is not possible to widen the approach region any further than in (ii). Indeed, given a function $\psi\colon (0,1) \to \mathbb{R}^+$ with $\lim_{t\to 0}\psi(t)\log(1/t) = \infty$, it is possible to construct $f \in \mathcal{D}$, $f \not\equiv 0$ with infinitely many zeros in $\Omega(\zeta) := \{z : |z-\zeta| < \psi(1-|z|)\}$ for each $\zeta \in \mathbb{T}$. Consequently, whenever $\lim_{z\to\zeta, z\in\Omega(\zeta)} f(z)$ exists, it must equal zero, and this can only happen for ζ in a set of measure zero. (This argument is taken from [28].)

The fact that logarithmic capacity is optimal in part (i) can be demonstrated using a construction of Carleson [10] (where it was actually used to prove something a little different). The construction shows that, given a compact subset E of $\mathbb{T}$ of logarithmic capacity zero, there exists $f \in \mathcal{D}$ such that $\lim_{r\to 1}|f(r\zeta)| = \infty$ for all $\zeta \in E$. By refining this construction, one can obtain the following more precise result, taken from [15].

We write d for arclength distance in $\mathbb{T}$ and $E_t := \{\zeta \in \mathbb{T} : d(\zeta, E) \le t\}$.

Theorem 7.2. *Let E be a closed subset of $\mathbb{T}$, and let $\eta\colon (0,\pi] \to \mathbb{R}^+$ be a decreasing continuous function. The following statements are equivalent:*

(i) *there exists* $f \in \mathcal{D}$ *such that*

$$\liminf_{z\to\zeta} |f(z)| \geq \eta\big(d(\zeta, E)\big) \quad (\zeta \in \mathbb{T});$$

(ii) *there exists* $f \in \mathcal{D}$ *such that* $|\operatorname{Im} f(z)| \leq 1$ *and*

$$\liminf_{z\to\zeta} \operatorname{Re} f(z) \geq \eta\big(d(\zeta, E)\big) \quad (\zeta \in \mathbb{T});$$

(iii) E *and* η *satisfy*

$$\int_0^\pi c(E_t)\,|\mathrm{d}\eta^2(t)| < \infty. \tag{7.1}$$

That (ii) implies (i) is obvious, and that (i) implies (iii) follows easily from the strong-type inequality, Theorem 5.4. The main thrust of the theorem is the implication that (iii) implies (ii), which can be considered as a partial converse to the strong-type inequality.

It is shown in [15] that (7.1) is implied by the condition

$$\int_0^1 |E_t|\eta'(t)^2\,\mathrm{d}t < \infty. \tag{7.2}$$

Thus (7.2) is sufficient for (i) and (ii) to hold. We conclude this section by giving a simple direct proof of this fact in the case when η is convex. It is a consequence of the following theorem (in which we write 'dist' for Euclidean distance).

Theorem 7.3. *Let* E *be a closed subset of* $\mathbb{T}$, *and let* $\eta\colon (0,2] \to \mathbb{R}^+$ *be a decreasing convex function. If* (7.2) *holds, then there exists* $f \in \mathcal{D}$ *such that*

$$\operatorname{Re} f(z) \geq \eta\big(\operatorname{dist}(z, E)\big) \quad (z \in \mathbb{D}). \tag{7.3}$$

Proof. Let $\mathbb{A} := \{w : 1 < |w| < 2\}$, and define $g\colon \mathbb{A} \to \mathbb{C}$ by

$$g(w) := -3w\eta'\big(\operatorname{dist}(w, E)/3)\big) \qquad (w \in \mathbb{A}).$$

We claim that $g \in L^2(\mathbb{A})$. Indeed, writing $w = re^{\mathrm{i}\theta}$ and denoting by d the arclength distance, we have

$$\operatorname{dist}(w, E) \geq \max\big\{(r-1),\ (2/\pi)d(e^{\mathrm{i}\theta}, E)\big\} \geq (r-1)/2 + d(e^{\mathrm{i}\theta}, E)/\pi.$$

Therefore, since $|\eta'(t)|$ is a decreasing function,

$$\begin{aligned}
\int_{\mathbb{A}} |g(w)|^2\,\mathrm{d}A(w) &\leq \int_{\theta=0}^{2\pi}\int_{r=1}^{2} 9r^2\left|\eta'\left(\frac{r-1}{6} + \frac{1}{3\pi}d(e^{\mathrm{i}\theta}, E)\right)\right|^2 r\,\mathrm{d}r\,\mathrm{d}\theta \\
&\leq 72\int_{\theta=0}^{2\pi}\int_{t=d(e^{\mathrm{i}\theta},E)/3\pi}^{1} |\eta'(t)|^2\,\mathrm{d}t\,\mathrm{d}\theta \\
&= 72\int_{t=0}^{1}\int_{d(e^{\mathrm{i}\theta},E)\leq 3\pi t} |\eta'(t)|^2\,\mathrm{d}\theta\,\mathrm{d}t \\
&= 72\int_{t=0}^{1} |E_{3\pi t}|\,|\eta'(t)|^2\,\mathrm{d}t \\
&\leq 216\pi\int_{t=0}^{1} |E_t|\,|\eta'(t)|^2\,\mathrm{d}t.
\end{aligned}$$

By (7.2), this last integral is finite, thereby justifying our claim. (In fact, a very similar calculation shows that (7.2) is also a necessary condition for $g \in L^2(\mathbb{A})$.)

Now set $f := \mathcal{C}g + \eta(1/3)$, where $\mathcal{C}$ denotes the Cauchy transform. Explicitly,

$$f(z) := \frac{1}{\pi}\int_{\mathbb{A}} \Big(\frac{3w}{w-z}\Big)\big|\eta'\big(\operatorname{dist}(w,E)/3\big)\big|\,\mathrm{d}A(w) + \eta(\tfrac{1}{3}) \qquad (z \in \mathbb{D}).$$

By Theorem 3.2, we have $f \in \mathcal{D}$. We shall show that f satisfies (7.3). Performing the change of variable $z \mapsto \mathrm{e}^{\mathrm{i}\theta}z$, and using the obvious rotation-invariance properties of f, it is enough to prove (7.3) in the case when $z = x \in [0,1)$. Furthermore, since $\operatorname{Re}\big(w/(w-x)\big)$ is always positive for $w \in \mathbb{A}$, it is clear that $\operatorname{Re} f(x) \geq \eta(\frac{1}{3})$, so (7.3) is certainly true if $\operatorname{dist}(x,E) \geq \frac{1}{3}$.

Suppose now that $z = x \in [0,1)$ and that $\operatorname{dist}(x,E) < \frac{1}{3}$. We make the change of variable $w = x + \rho\mathrm{e}^{\mathrm{i}\phi}$. Note that $w \in \mathbb{A}$ provided that $2\operatorname{dist}(x,E) < \rho < 1$ and $0 < \phi < \pi/3$. Indeed, under these conditions, we have

$$|x + \rho\mathrm{e}^{\mathrm{i}\phi}| \leq x + \rho < 2 \quad \text{and} \quad |x + \rho\mathrm{e}^{\mathrm{i}\phi}| \geq x + \rho\cos\phi \geq x + \operatorname{dist}(x,E) \geq 1.$$

Also $\operatorname{Re}\big(3w/(w-x)\big) = 3 + \operatorname{Re}(3x/\rho\mathrm{e}^{\mathrm{i}\phi}) \geq 1/\rho$ and $\operatorname{dist}(w,E) \leq \operatorname{dist}(x,E) + \rho$. Hence, after making the change of variable, we have

$$\begin{aligned}\operatorname{Re} f(x) &\geq \frac{1}{\pi}\int_0^{\pi/3}\int_{2\operatorname{dist}(x,E)}^1 \frac{1}{\rho}\big|\eta'\big((\operatorname{dist}(x,E)+\rho)/3\big)\big|\,\rho\,\mathrm{d}\rho\,\mathrm{d}\phi + \eta(1/3)\\ &= \frac{1}{3}\int_{2\operatorname{dist}(x,E)}^1 \big|\eta'\big((\operatorname{dist}(x,E)+\rho)/3\big)\big|\,\mathrm{d}\rho + \eta(\tfrac{1}{3}) \geq \eta\big(\operatorname{dist}(x,E)\big).\end{aligned}$$

This gives (7.3) and completes the proof. □

8. Bibliographical remarks

§2. Logarithmic capacity can be defined in many different ways: in terms of energy [2, 5, 19, 21, 24, 31, 33, 34], potentials [1, 3, 11, 22, 27], réduites [4, 13, 20], Green's functions [17, 23, 29], and transfinite diameter [30]. We have followed the energy definition. Though these definitions may give rise to different numbers for the logarithmic capacity of a set, they are all equivalent in the sense that any one of them can be made small if any other is sufficiently small. In particular, the sets of logarithmic capacity zero are the same, whichever definition is used.

§3. The representation theorem 3.3 is due to Dyn′kin [14]. The simple proof given here is taken from Borichev [8].

§4. Beurling's theorem (for radial limits) appeared in [7]. Our approach follows Borichev [8].

§5. The weak-type capacitary inequality is due to Beurling [7]. He had earlier obtained Corollary 5.2 in his thesis [6], showing that, if $f(0) = 0$ and $\mathcal{D}(f) \leq 1$, then $|\{|f^*| > t\}| \leq 2\pi\mathrm{e}^{1-t^2}$. This was later strengthened by Chang and Marshall [12] (see also [16, 25]), who proved that

$$\sup\Big\{\int_{\mathbb{T}} \exp(|f^*|^2)\,\mathrm{d}\theta : f(0) = 0, \mathcal{D}(f) \leq 1\Big\} < \infty.$$

The strong-type capacitary inequality is due to Hansson [18], extending earlier versions due to Maz′ya and Adams. Our proof is adapted from that given in [1], which also contains a detailed account of the history of this result..

§6. Theorem 6.1 is due to Nagel, Rudin and Shapiro [28]. Our approach is an adaptation of their ideas and those in [8].

§7. For more information about the playoff between the width of the approach region and the size of the exceptional set, we refer to [8, 9, 35, 36]. Theorem 7.2 is taken from [15], where it is stated in a slightly different form. Theorem 7.3 we believe to be new.

References

1. D. R. Adams and L. I. Hedberg, *Function spaces and potential theory*, Grundlehren Math. Wiss., vol. 314, Springer, Berlin, 1996.
2. L. V. Ahlfors, *Conformal invariants: topics in geometric function theory*, McGraw-Hill Series in Higher Mathematics, McGraw-Hill, New York, 1973.
3. H. Aikawa and M. Essén, *Potential theory — selected topics*, Lecture Notes in Math., vol. 1633, Springer, Berlin, 1996.
4. D. H. Armitage and S. J. Gardiner, *Classical potential theory*, Springer Monogr. Math., Springer, London, 2001.
5. C. A. Berenstein and R. Gay, *Complex variables: An introduction*, Grad. Texts in Math., vol. 125, Springer, New York, 1991.
6. A. Beurling, *Études sur un problème de majoration*, Dissertation, Uppsala University, Uppsala, 1933.
7. ———, *Ensembles exceptionnels*, Acta Math. **72** (1940), 1–13 (French).
8. A. Borichev, *A note on Dirichlet-type spaces*, Technical Report 11, Department of Mathematics, Uppsala University, 1993.
9. ———, *Boundary behavior in Dirichlet-type spaces*, Technical Report 3, Department of Mathematics, Uppsala University, 1994.
10. L. Carleson, *Sets of uniqueness for functions regular in the unit circle*, Acta Math. **87** (1952), 325–345.
11. ———, *Selected problems on exceptional sets*, Van Nostrand Mathematical Studies, vol. 13, Van Nostrand Co., Princeton, NJ, 1967.
12. S.-Y. A. Chang and D. E. Marshall, *On a sharp inequality concerning the Dirichlet integral*, Amer. J. Math. **107** (1985), no. 5, 1015–1033.
13. J. L. Doob, *Classical potential theory and its probabilistic counterpart*, Grundlehren Math. Wiss., vol. 262, Springer, New York, 1984.
14. E. M. Dyn′kin, *Continuations and integral representations of smooth functions of one complex variable*, Dissertation, Steklov Mathematical Institute, Leningrad, 1973 (Russian).
15. O. El-Fallah, K. Kellay, and T. Ransford, *Cyclicity in the Dirichlet space*, Ark. Mat. **44** (2006), no. 1, 61–86.
16. M. Essén, *Sharp estimates of uniform harmonic majorants in the plane*, Ark. Mat. **25** (1987), no. 1, 15–28.
17. J. B. Garnett and D. E. Marshall, *Harmonic measure*, New Math. Monogr., vol. 2, Cambridge Univ. Press, Cambridge, 2005.
18. K. Hansson, *Imbedding theorems of Sobolev type in potential theory*, Math. Scand. **45** (1979), no. 1, 77–102.
19. W. K. Hayman and P. B. Kennedy, *Subharmonic functions.* Vol. I, London Math. Soc. Monogr., vol. 9, Academic Press, London, 1976.
20. L. L. Helms, *Introduction to potential theory*, Pure Appl. Math., vol. 22, R. E. Krieger Pub. Co, Huntington, NY, 1975. Reprint of the 1969 edition.
21. E. Hille, *Analytic function theory.* Vol. II, Introductions to Higher Mathematics, Ginn and Co., Boston, MA, 1962.
22. J.-P. Kahane and R. Salem, *Ensembles parfaits et séries trigonométriques*, 2nd ed., with notes by J.-P. Kahane, T. W. Körner, R. Lyons, and S. W. Drury, Hermann, Paris, 1994.
23. P. Koosis, *The logarithmic integral.* Vol. II, Cambridge Stud. Adv. Math., vol. 21, Cambridge Univ. Press, Cambridge, 1992.
24. N. S. Landkof, *Foundations of modern potential theory*, translated by A. P. Doohovskoy, Grundlehren Math. Wiss., vol. 180, Springer, New York, 1972.

25. D. E. Marshall, *A new proof of a sharp inequality concerning the Dirichlet integral*, Ark. Mat. **27** (1989), no. 1, 131–137.
26. J. Mashreghi, *Representation theorems in Hardy spaces*, London Math. Soc. Stud. Texts, vol. 74, Cambridge Univ. Press, Cambridge, 2009.
27. N. G. Meyers, *A theory of capacities for potentials of functions in Lebesgue classes*, Math. Scand. **26** (1970), 255–292.
28. A. Nagel, W. Rudin, and J. H. Shapiro, *Tangential boundary behavior of functions in Dirichlet-type spaces*, Ann. of Math. (2) **116** (1982), no. 2, 331–360.
29. Z. Nehari, *Conformal mapping*, Dover, New York, 1975. Reprinting of the 1952 edition.
30. Ch. Pommerenke, *Boundary behaviour of conformal maps*, Grundlehren Math. Wiss., vol. 299, Springer, Berlin, 1992.
31. T. Ransford, *Potential theory in the complex plane*, London Math. Soc. Stud. Texts, vol. 28, Cambridge Univ. Press, Cambridge, 1995.
32. W. Rudin, *Real and complex analysis*, 3rd ed., McGraw-Hill, New York, 1987.
33. E. B. Saff and V. Totik, *Logarithmic potentials with external fields*, with Appendix B by T. Bloom, Grundlehren Math. Wiss., vol. 316, Springer, Berlin, 1997.
34. M. Tsuji, *Potential theory in modern function theory*, Chelsea, New York, 1975. Reprinting of the 1959 original.
35. J. B. Twomey, *Tangential limits for certain classes of analytic functions*, Mathematika **36** (1989), no. 1, 39–49.
36. ———, *Tangential boundary behaviour of harmonic and holomorphic functions*, J. London Math. Soc. (2) **65** (2002), no. 1, 68–84.

Département de Mathématiques, Université Mohamed V, B.P. 1014 Rabat, Morocco
E-mail address: elfallah@fsr.ac.ma

Institut de Mathématiques de Bordeaux, Université Bordeaux 1, 351 cours de la Libération, 33405 Talence cedex, France
E-mail address: karim.kellay@math.u-bordeaux1.fr

Département de mathématiques et de statistique, Université Laval, Québec, QC G1V 0A6, Canada
E-mail address, J. Mashreghi: javad.mashreghi@mat.ulaval.ca
E-mail address, T. Ransford: ransford@mat.ulaval.ca

Centre de Recherches Mathématiques
CRM Proceedings and Lecture Notes
Volume **55**, 2012

A Simple Numerical Approach to the Riemann Hypothesis

N. Tarkhanov

Abstract. The Riemann hypothesis is equivalent to the fact that the reciprocal function $1/\zeta(s)$ extends from the interval $(\frac{1}{2}, 1)$ to an analytic function in the quarter-strip $\frac{1}{2} < \Re s < 1$, $\Im s > 0$. Function theory allows one to rewrite the condition of analytic continuability in an elegant form amenable to numerical experiments.

Introduction

The Riemann hypothesis is that all zeros of the Riemann zeta function $\zeta(z)$ in the critical strip $0 < \Re z < 1$ belong to the critical line $\Re z = \frac{1}{2}$. This just amounts to saying that the function $1/\zeta(z)$ extends from the interval $(\frac{1}{2}, 1)$ to an analytic function in the quarter-strip $\frac{1}{2} < \Re z < 1$, $\Im z > 0$. Note that the restriction of $1/\zeta(s)$ to $(\frac{1}{2}, 1)$ is actually continuous on the closed interval. Hence the function theory allows one to rewrite the condition of analytic continuability in an elegant form which is amenable to numerical experiments. More precisely, one constructs an explicit sequence $\{c_n\}$ of complex numbers, such that the equality $\lim \sqrt[n]{|c_n|} = 1$ is fulfilled if and only if the Riemann hypothesis is true. The numbers c_n are integrals of $1/\zeta(s)$ over the interval $[\frac{1}{2}, 1]$ with explicit weight function depending on n. Computations with the newest versions of Mathematica, Maple and Matlab performed by my diploma students give certain evidence to the fact that the limit is 1 indeed. However, the standard computer programmes are not sufficient to evaluate the sequence $\sqrt[n]{|c_n|}$ with strict accuracy. The numerical data obtained in this way testify to $\lim \sqrt[n]{|c_n|} = 1$ not only for $1/\zeta(s)$ but also for other functions (e.g., $1/(s-3/4-\imath/2)$) which fail to be analytically extendable to the critical quarter-strip. Thus the paper gives rise to the problem of elaborating an efficient program which recognizes through the behavior of the sequence $\sqrt[n]{|c_n|}$ those continuous functions on $[\frac{1}{2}, 1]$ which extend to analytic functions in the quarter-strip.

2010 *Mathematics Subject Classification.* Primary 11M26; Secondary 11Mxx.
Key words and phrases. Zeta function, Carleman formulas, numerical methods.
The author wishes to express his thanks to S. Grudsky for several helpful comments concerning asymptotic formulas.
This is the final form of the paper.

1. The Riemann zeta function

In this section we gather necessary material about the Riemann zeta function. For complete proofs the reader is referred to [4, 6, 7].

For complex numbers $s = \Re s + \sqrt{-1}\Im s$ in the half-plane $\Re s > 1$ the Riemann function is defined by

$$\zeta(s) = \sum_{n=1}^{\infty} \frac{1}{n^s},$$

the series converging absolutely and uniformly in each half-plane $\Re s > s_0$ with $s_0 > 1$.

In 1737 Euler proved his product formula which displayed a deep connection of $\zeta(s)$ with the distribution of prime numbers.

Theorem 1.1. *If $\Re s > 1$ then*

$$\zeta(s) = \prod_{p} \Big(1 - \frac{1}{p^s}\Big)^{-1},$$

where the product runs over all prime numbers p ($p = 1$ is no prime number).

In order to extend $\zeta(s)$ to an analytic function on all of $\mathbb{C}$, one uses the analytic extension of the gamma function constructed by Weierstraß. More precisely,

$$\frac{1}{\Gamma(s)} = s e^{\gamma s} \prod_{n=1}^{\infty} \Big(1 + \frac{1}{n}\Big) e^{-s/n}$$

holds for $\Re s > 0$, where γ is the Euler–Mascheroni constant.

The right-hand side of this equality is an entire function of s vanishing at the points $s = 0, -1, -2, \ldots$.

Lemma 1.2. *If $\Re s > 1$ then*

$$\zeta(s) = \frac{\pi^{s/2}}{\Gamma(s/2)} \Big(\frac{1}{s(s-1)} + \int_1^{\infty} \big(x^{s/2-1} + x^{-s/2-1/2}\big) \sum_{n=1}^{\infty} e^{-\pi n^2 x}\, dx\Big).$$

The lemma shows that the Riemann zeta function extends to a meromorphic function in the whole complex plane with the only pole at $s = 1$ which is simple.

This function vanishes at $s = -2, -4, \ldots$, the other zeros of $\zeta(s)$ are known to lie in the critical strip $0 < \Re s < 1$.

B. Riemann conjectured (1869) that all zeros of $\zeta(s)$ in the critical strip belong to the line $\Re s = \frac{1}{2}$.

The restriction of $\zeta(s)$ to the critical strip is symmetric with respect to both the critical line $\Re s = \frac{1}{2}$ and the interval $(0, 1)$ of the real axis. Moreover, it is different from zero for all $s \in [0, 1]$.

Hence the Riemann hypothesis just amounts to saying that $\zeta(s)$ has no zeros in the quarter-strip $\frac{1}{2} < \Re s < 1$, $\Im s > 0$.

For real $x > 0$, let $\pi(x)$ denote the number of prime numbers p which satisfy $p \le x$.

B. Riemann showed a formula for the difference

$$\pi(x) - \int_0^x \frac{ds}{\log s}$$

in terms of x and zeros of $\zeta(s)$ lying in the critical strip.

If $\zeta(s)$ has no zeros with $\Re s > s_0$ for some $\frac{1}{2} \le s_0 < 1$, then the asymptotic formula

$$\pi(x) = \int_0^x \frac{\mathrm{d}s}{\log s} + O(x^{s_0} \log x)$$

holds. The Riemann hypothesis just amounts to this formula with $s_0 = \frac{1}{2}$.

Some textbooks in complex analysis include the so-called prime number theorem proved independently by J. Hadamard and Ch.-J. de la Vallée-Poussin (1896). It reads $\pi(x) \sim x/\log x$.

2. Analytic continuation in a lune

Denote by $\mathbb{D} := \{w \in \mathbb{C} : |w| < 1\}$ the open unit disk with center at the origin in the plane of complex variable w. Let S be a regular curve in $\mathbb{D}$, whose endpoints lie on the unit circle and which does not run through 0 (i.e., $0 \notin S$). The curve S divides the disk $\mathbb{D}$ into two domains and we write G for the subdomain of $\mathbb{D}$ that does not contain the origin 0. In this way we obtain a bounded domain with piecewise smooth boundary which is referred to as lune. The boundary of G consists of two parts, one of the two is the curve S and the other an arc of the circle $\partial\mathbb{D}$, see Figure 1.

In 1926 T. Carleman found a simple formula for analytic continuation in a corner. For this reason the following refined formula is named after him. This formula is well known, see for instance [1]. Since the proof is very simple we give it for completeness.

Theorem 2.1. *Suppose f is a holomorphic function in G continuous up to the boundary of G. Then*

$$f(w) = \lim_{n\to\infty} \int_S f(w') \left(\frac{w}{w'}\right)^n \frac{1}{2\pi\imath} \frac{\mathrm{d}w'}{w' - w} \tag{2.1}$$

for all $w \in G$.

PROOF. Fix an arbitrary $w \in G$. Since $0 \notin \overline{G}$ and f is holomorphic in G and continuous up to the boundary, the function

$$F(w') := f(w') \left(\frac{w}{w'}\right)^n$$

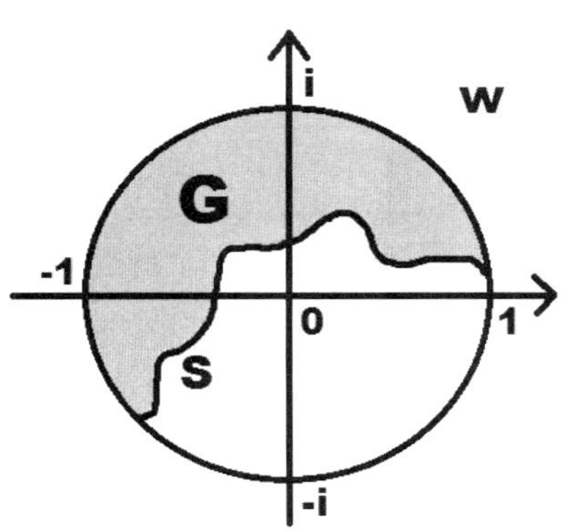

FIGURE 1. A basic domain.

is holomorphic in G and continuous on $\overline{G}$ for all $n = 1, 2, \ldots$. By the integral formula of Cauchy one gets

$$F(w) = \int_{\partial G} F(w') \frac{1}{2\pi\imath} \frac{dw'}{w' - w}$$

for all $w \in G$. Substituting F yields

$$f(w) = \int_{\partial G} f(w') \Big(\frac{w}{w'}\Big)^n \frac{1}{2\pi\imath} \frac{dw'}{w' - w}.$$

for each $n = 1, 2, \ldots$.

The integral on the right-hand side splits into two integrals, the first is over S and the second one over $\partial G \setminus S$. So

$$f(w) = \int_S f(w') \Big(\frac{w}{w'}\Big)^n \frac{1}{2\pi\imath} \frac{dw'}{w' - w} + \int_{\partial G \setminus S} f(w') \Big(\frac{w}{w'}\Big)^n \frac{1}{2\pi\imath} \frac{dw'}{w' - w}.$$

On letting $n \to \infty$ one obtains

$$f(w) = \lim_{n\to\infty} \int_S f(w') \Big(\frac{w}{w'}\Big)^n \frac{1}{2\pi\imath} \frac{dw'}{w' - w} + \lim_{n\to\infty} \int_{\partial G \setminus S} f(w') \Big(\frac{w}{w'}\Big)^n \frac{1}{2\pi\imath} \frac{dw'}{w' - w},$$

in the case when at least one of the limits exists.

We now show that the second limit exists and is precisely zero. Since $w \in G$ and $w' \in \partial G \setminus S$, we get

$$\Big|\frac{w}{w'}\Big| = \frac{|w|}{1} < 1.$$

It follows that the sequence $(w/w')^n$ converges to zero uniformly in $w' \in \partial G \setminus S$. So the limit of the second integral can be evaluated under the integral sign and one obtains

$$\lim_{n\to\infty} \int_{\partial G \setminus S} f(w') \Big(\frac{w}{w'}\Big)^n \frac{1}{2\pi\imath} \frac{dw'}{w' - w} = \int_{\partial G \setminus S} f(w') \Big(\lim_{n\to\infty} \Big(\frac{w}{w'}\Big)^n\Big) \frac{1}{2\pi\imath} \frac{dw'}{w' - w} = 0,$$

which establishes the desired formula. □

To our best knowledge, Theorem 2.1 gives the simplest explicit formula of analytic continuation in complex analysis. Based upon this formula, we show a criterion of analytic continuability into G for a function f_0 given on the part S of boundary ∂G. While polynomials of z are dense in the Banach space $C(\overline{S})$, those functions on $\overline{S}$ which extend analytically to G form a subspace of infinite codimension in $C(\overline{S})$. In particular, the continuous functions $f_0 \not\equiv 0$ of compact support in S fail to have analytic continuation to the domain G, which is a consequence of the uniqueness theorem.

The following result was published by Aizenberg in [2].

Theorem 2.2. *Let $f_0 \in C(\overline{S})$ satisfy $f_0 \not\equiv 0$. In order that there be a holomorphic function $f \in C(G \cup S)$ in G, such that $f(w) = f_0(w)$ for all $w \in S$, it is necessary and sufficient that*

$$\limsup_{n\to\infty} \sqrt[n]{\Big|\int_S \frac{f_0(w')}{(w')^{n+1}}\, dw'\Big|} = 1. \tag{2.2}$$

PROOF. *Necessity.* Given a nonzero function $f_0 \in C(\overline{S})$, we define the Cauchy-type integral of f_0 by

$$Cf_0(w) = \int_S f_0(w') \frac{1}{2\pi\imath} \frac{dw'}{w' - w}$$

for $w \notin \overline{S}$. This is a holomorphic function away from the closure of S, and we denote by $C^{\pm} f_0$ the restrictions of Cf_0 to G and $\mathbb{C} \setminus \overline{G}$, respectively. The Sokhotsky–Plemelj formula says that

$$\lim_{\varepsilon \to 0+} \Big(C^+ f_0\big(w' + \varepsilon\nu(w')\big) - C^- f_0\big(w' - \varepsilon\nu(w')\big) \Big) = f_0(w') \tag{2.3}$$

holds uniformly in w' on compact subsets of S, where $\nu(w')$ is the inward unit normal vector to S at a point $w' \in S$. In particular, if either of the functions $C^{\pm} f_0$ extends continuously to S then so does the other function. The limit in (2.3) is obviously zero, if $w' \in \partial G \setminus \overline{S}$.

Assume that there is a holomorphic function f in G which is continuous up to S and satisfies $f = f_0$ on S. A simple manipulation with the Cauchy integral formula for f shows that the difference $C^+ f_0 - f$ extends to a continuous (even C^∞) function on $G \cup S$. Since f is continuous on $G \cup S$, the integral $C^+ f_0$ extends to a continuous function on $G \cup S$. By the above, $C^- f_0$ extends continuously to $\mathbb{D} \setminus G$, too.

Consider the function

$$F(w) = \begin{cases} C^+ f_0(w) - f(w), & \text{if } w \in G \cup S, \\ C^- f_0(w), & \text{if } w \in \mathbb{D} \setminus \overline{G}, \end{cases}$$

in the disk $\mathbb{D}$. This function is holomorphic in $\mathbb{D} \setminus S$ and continuous on all of $\mathbb{D}$, for $C^+ f_0 - f = C^- f_0$ on S, which is due to Sokhotsky–Plemelj formula (2.3). From the Morera theorem we easily deduce that F is actually holomorphic in the unit disk $\mathbb{D}$. Hence, the Taylor series of this function around the origin converges in all of $\mathbb{D}$. The series looks like

$$F(w) = \sum_{n=0}^{\infty} c_n w^n \tag{2.4}$$

for $|w| < 1$, where

$$c_n = \frac{1}{2\pi\imath} \int_S \frac{f_0(w')}{(w')^{n+1}} dw',$$

for $F = C^- f_0$ nearby the origin. From the Cauchy–Hadamard formula for the convergence radius of power series we readily conclude that $\limsup \sqrt[n]{|c_n|} \le 1$. If this limit is less than 1, then the series (2.4) converges in a a disk about the origin of radius greater than 1. Hence, $C^- f_0$ extends to a holomorphic function in a neighborhood of the closure of S, and so does Cf_0. On applying the Sokhotsky–Plemelj formula once again we see that $f_0 \equiv 0$ on S, a contradiction. This establishes (2.2), as desired.

Sufficiency. To prove the converse theorem, let f_0 be a continuous function on the closure of S satisfying (2.2). By assumption, the integral $C^- f_0$ is holomorphic in a disk of sufficiently small radius $\varepsilon > 0$ around the origin (take $\varepsilon < \text{dist}(0, S)$). Hence, $C^- f_0$ expands in this small disk as a power series whose coefficients are given by

$$c_n = \frac{1}{2\pi\imath} \int_S \frac{f_0(w')}{(w')^{n+1}}\, dw',$$

cf. (2.4). Condition (2.2) forces the power series (2.4) to actually converge in the unit disk $\mathbb{D}$ to a holomorphic function F. By the uniqueness theorem, the integral $C^- f_0$ extends holomorphically to all of $\mathbb{D}$, and this analytic continuation is F. Hence it follows that the integral $C^+ f_0$ extends to a continuous function on $G \cup S$. We now set

$$f(w) := C^+ f_0(w) - F(w)$$

for $w \in G \cup S$, thus obtaining a holomorphic function in G which is continuous up to S and satisfies $f(w) = f_0(w)$ for all $w \in S$, as desired. □

3. A Carleman formula for a half-disk

The upper half-disk $\mathbb{D}' = \{z \in \mathbb{C} : |z| < 1, \Im(z) > 0\}$ is a canonical domain of the lune type corresponding to $S = (-1, 1)$. Since $0 \in (-1, 1)$, formula (2.1) is no longer applicable. To this end one needs a transformation $w = h(z)$ which maps $\mathbb{D}'$ conformally onto a lune like that in formula (2.1). We look for such a transformation in the group of fractional affine automorphisms of the unit disk $\mathbb{D}$. These have the form

$$h(z) = e^{\imath\varphi} \frac{z - a}{\bar{a}z - 1}$$

for $z \in \mathbb{D}$, with $\varphi \in [0, 2\pi)$ and $|a| < 1$. We pose two additional conditions on h, namely

(1) $h(0) = t\imath$, where $t \in (0, 1)$;
(2) $h(\imath) = \imath$.

The desired transformation is illustrated in Figure 2.

An easy computation shows that there is only one automorphism of the unit disk satisfying (1) and (2). This is

$$h(z) = \frac{z + t\imath}{1 - t\imath z}. \tag{3.1}$$

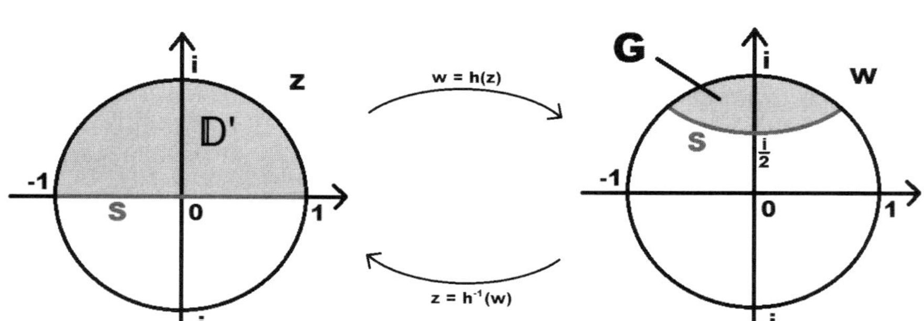

FIGURE 2. A conformal mapping of $\mathbb{D}'$ onto a lune for $t = \frac{1}{2}$.

Corollary 3.1. *Let f be a holomorphic function in the half-disk $\mathbb{D}'$ continuous up to the boundary of $\mathbb{D}'$. Then*

$$f(z) = \lim_{n\to\infty} \int_{-1}^{1} f(z') \left(\frac{h(z)}{h(z')}\right)^n \frac{1}{2\pi\imath} \frac{dh(z')}{h(z') - h(z)} \tag{3.2}$$

for all $z \in \mathbb{D}'$.

Proof. Set $F(w) := f\big(h^{-1}(w)\big)$ for all $w \in G$. Since $z = h^{-1}(w)$ maps the domain G conformally onto the half-disk $\mathbb{D}'$ and f is holomorphic in $\mathbb{D}'$, the function F is holomorphic in G. Furthermore, $z = h^{-1}(w)$ extends to a homeomorphism of the closure of G onto that of $\mathbb{D}'$. Hence, F is continuous up to the boundary of G. By formula (2.1),

$$F(w) = \lim_{n\to\infty} \int_S F(w') \left(\frac{w}{w'}\right)^n \frac{1}{2\pi\imath} \frac{dw'}{w' - w}$$

for all $w \in G$. On substituting $w = h(z)$ and $w' = h(z')$ and taking into account that S is the image of $(-1, 1)$ by h, we arrive at

$$F\big(h(z)\big) = \lim_{n\to\infty} \int_{-1}^{1} F\big(h(z')\big) \left(\frac{h(z)}{h(z')}\right)^n \frac{1}{2\pi\imath} \frac{dh(z')}{h(z') - h(z)}$$

for all $z \in \mathbb{D}'$, as desired. □

Replacing the function h in (3.2) by its expression (3.1) we write the formula for analytic continuation from the interval $(-1, 1)$ into the half-disk in explicit form. More precisely,

$$f(z) = \lim_{n\to\infty} \left(\frac{z + t\imath}{1 - t\imath z}\right)^n \int_{-1}^{1} f(z') \left(\frac{1 - t\imath z'}{z' + t\imath}\right)^n \frac{1 - t\imath z}{1 - t\imath z'} \frac{1}{2\pi\imath} \frac{dz'}{z' - z}$$

for all $z \in \mathbb{D}'$. Recall that t is any number in the interval $(0, 1)$. For $t = 0$ we recover formula (2.1), however, this value is prohibited, for the integrand becomes singular at $z' = 0$.

Under conformal map (3.1) Theorem 2.2 is also traced back to conditions of analytic continuability of functions from the interval $(-1, 1)$ to the upper half-disk $\mathbb{D}'$. We actually rewrite the same invariant object in other holomorphic coordinates. Let f_0 be a continuous function on $[-1, 1]$. Then $F_0 := f_0 \circ h^{-1}$ is a continuous function on S, the image of $(-1, 1)$ by $w = h(z)$. This is a regular curve in $\mathbb{D} \setminus \{0\}$ with endpoints

$$\pm\frac{1 - t^2}{1 + t^2} + \frac{2t}{1 + t^2}\,\imath$$

on the unit circle. Obviously, f_0 extends to a holomorphic function f in $\mathbb{D}'$, which is continuous up to $(-1, 1)$, if and only if F_0 extends to a holomorphic functions $F := f \circ h^{-1}$ in G continuous up to S. By Theorem 2.2 F_0 extends analytically to G if and only if

$$\lim_{n\to\infty} \sqrt[n]{|c_n|} = 1,$$

where

$$\begin{aligned} c_n &= \frac{1}{2\pi\imath}\int_S \frac{F_0(w)}{w^{n+1}}\,\mathrm{d}w \\ &= \frac{1}{2\pi\imath}\int_{-1}^{1} \frac{f_0(z)}{\big(h(z)\big)^{n+1}}\,\mathrm{d}h(z) \\ &= \frac{1}{2\pi\imath}\int_{-1}^{1} f_0(z)\Big(\frac{1-t\imath z}{z+t\imath}\Big)^n \frac{1-t^2}{(z+t\imath)(1-t\imath z)}\,\mathrm{d}z \end{aligned}$$

for $n = 0, 1, \ldots$, because $\mathrm{d}h(z) = (1-t^2)/(1-t\imath z)^2\,\mathrm{d}z$.

Corollary 3.2. *Let $f_0 \in C[-1,1]$ be a nonzero function. In order that there be a holomorphic function f in $\mathbb{D}'$ continuous up to $(-1,1)$ and satisfying $f = f_0$ on $(-1,1)$, it is necessary and sufficient that* $\limsup \sqrt[n]{|c_n|} = 1$, *where c_n are given by* (3.3).

Using the triangle inequality one finds readily an explicit estimate from above for the limit in question. Namely,

$$\begin{aligned} |c_n| &\le \Big(\sup_{z\in(-1,1)} \frac{1}{|h(z)|}\Big)^{n+1} \frac{1}{2\pi}\int_{-1}^{1} |f_0(z)||\mathrm{d}h(z)| \\ &\le \Big(\frac{1}{|h(0)|}\Big)^{n+1} \frac{1}{2\pi}\int_{-1}^{1} |f_0(z)|\frac{1-t^2}{1+t^2z^2}\,\mathrm{d}z \end{aligned}$$

for all $n = 0, 1, \ldots$, the last inequality being a consequence of the fact that the modulus of $h(z)$ takes on its global infimum in $(-1,1)$ at the point $z = 0$, as is easy to see from Figure 2. The right-hand side here is a constant multiple of $|h(0)|^{-n}$, and so

$$\limsup \sqrt[n]{|c_n|} \le \frac{1}{|h(0)|} = \frac{1}{t}. \tag{3.3}$$

Given any $\varepsilon > 0$, choose t close to 1 in the interval $(0,1)$, so that $1/t < 1+\varepsilon$. On applying Corollary 3.2 we see that $\limsup \sqrt[n]{|c_n|}$ belongs to the interval $[1, 1+\varepsilon)$ for each continuous function f_0 on $[-1,1]$ which is different from the identical zero. Hence, the condition $\limsup \sqrt[n]{|c_n|} = 1$ might be efficient numerically only if t is small enough.

4. Reduction of the Riemann hypothesis

Arguing as in Section 3, we look for a conformal mapping $z = k(v)$ of the critical quarter-strip $\mathbb{H}' := \{v \in \mathbb{C} : \Re v \in (\frac{1}{2}, 1), \Im v > 0\}$ onto the half-disk $\mathbb{D}'$. For this purpose we need several lemmata.

Lemma 4.1. *The function $z = \tan v$ maps the strip $-\pi/4 < \Re v < \pi/4$ conformally onto the unit disk $\mathbb{D}$.*

PROOF. As is well known, the function $z = \tan v$ maps the strip $-\pi/2 < \Re v < \pi/2$ conformally into the complex plane $\mathbb{C}$. It remains to specify the image of the strip $-\pi/4 < \Re v < \pi/4$ by this mapping. Given any v in the strip $-\pi/4 < \Re v < \pi/4$, we get

$$|\tan v| = \sqrt{\frac{\mathrm{e}^{2\Im v} - 2\cos(2\Re v) + \mathrm{e}^{-2\Im v}}{\mathrm{e}^{2\Im v} + 2\cos(2\Re v) + \mathrm{e}^{-2\Im v}}}.$$

By assumption, $-\pi/2 < 2\Re v < \pi/2$, whence $\cos(2\Re v) > 0$ and $\cos(2\Re v) = 0$ if and only if either $\Re v = -\pi/4$ or $\Re v = \pi/4$. So the quotient under the root sign is less than 1 in the open strip and it is equal to 1 if and only if $\Re v = -\pi/4$ or $\Re v = \pi/4$, as desired. □

From Lemma 4.1 we further deduce

Lemma 4.2. *The function* $z = \tan v$ *maps the half-strip*

$$\{v \in \mathbb{C} : -\pi/4 < \Re v < \pi/4, \Im v > 0\}$$

conformally onto the half-disk $\mathbb{D}'$. *The image of the interval* $(-\pi/4, \pi/4)$ *by this mapping is the interval* $(-1, 1)$.

PROOF. Using the Euler formula one easily obtains

$$\tan v = \frac{2\sin(2\Re v) + \imath(e^{2\Im v} - e^{-2\Im v})}{e^{2\Im v} + 2\cos(2\Re v) + e^{-2\Im v}}.$$

If $\Im v > 0$, then $e^{2\Im v} > e^{-2\Im v}$, and so $\Im \tan v > 0$, which shows the first part of the lemma. For the second part we assume $v \in (-\pi/4, \pi/4)$, so v is real. Then $z = \tan v$ is also real. Since the function $\tan v$ is strongly monotonic increasing on the interval $(-\pi/2, \pi/2)$ and $\tan(-\pi/4) = -1$ and $\tan(\pi/4) = 1$, the assertion is clear. □

In order to construct a conformal mapping of the half-disk $\mathbb{D}'$ onto the critical quarter-strip $\mathbb{H}'$, it suffices to take the composition of $z = \tan v$ with an affine transformation of the v-plane. That is

$$z := k(v) = \tan \pi\left(v - \tfrac{3}{4}\right) \tag{4.1}$$

with inverse $v = k^{-1}(z) := \frac{3}{4} + (1/\pi)\arctan z$, see Figure 3.

Lemma 4.3. *If* f *is a holomorphic function in the critical quarter-strip* $\mathbb{H}'$, *then the composition* $f \circ k^{-1}$ *is a holomorphic function in the upper half-disk* $\mathbb{D}'$. *If moreover the function* f *is continuous up to the interval* $(\frac{1}{2}, 1)$, *then* $f \circ k^{-1}$ *is continuous up to* $(-1, 1)$.

PROOF. The proof is obvious by the above. We formulate this lemma for convenience of references. □

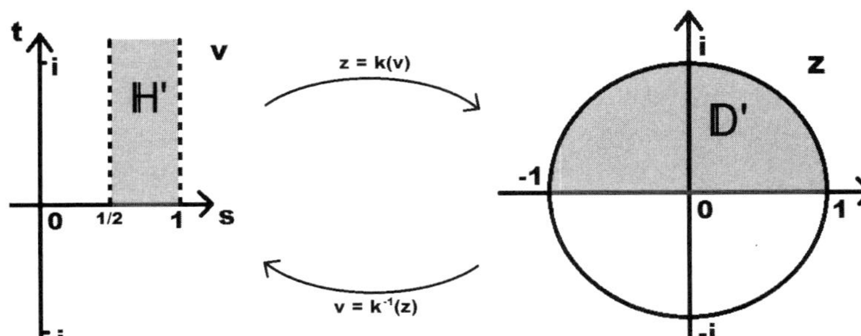

FIGURE 3. A conformal mapping of $\mathbb{H}'$ onto $\mathbb{D}'$.

The function $z = k(v)$ maps the segment $\Re v = 1$, $\Im v \geq 0$ homeomorphically onto the quarter-circle $|z| = 1$, $\arg z \in [0, \pi/2)$ and the segment $\Re v = \frac{1}{2}$, $\Im v \geq 0$ homeomorphically onto the quarter-circle $|z| = 1$, $\arg z \in (\pi/2, \pi]$. The inverse mapping $v = k^{-1}(z)$ extends continuously to the entire boundary of the upper half-disk $\mathbb{D}'$ except for the north pole $z = \imath$ where k^{-1} blows up. Thus, the mapping $v = k^{-1}(z)$ might be used to compactify the closure of $\mathbb{H}'$ by adding a "point at infinity" to it. A function f on the closure of $\mathbb{H}'$ is said to be continuous on such a one-point compactification of the closure of $\mathbb{H}'$ if the composition $f \circ k^{-1}$ is continuous on the closure of $\mathbb{D}'$. A Carleman-type formula with integration over the interval $[\frac{1}{2}, 1]$ still holds for functions f holomorphic in $\mathbb{H}'$ and continuous on the compactification of $\overline{\mathbb{H}}'$. However, the Riemann zeta function does not belong to this class, so we shall not discuss the Carleman formula for holomorphic functions in $\mathbb{H}'$. On the other hand, the criterion of analytic continuability does not require any continuity on the compactification of the closure of $\mathbb{H}'$. Therefore, it extends to holomorphic functions in the critical quarter-strip in much the same way as Corollary 3.2.

Theorem 4.4. *Let f_0 a continuous function on the interval $[\frac{1}{2}, 1]$. In order that there be a holomorphic function f in $\mathbb{H}'$, such that f is continuous up to the open interval $(\frac{1}{2}, 1)$ and coincides with f_0 on this interval, it is necessary and sufficient that*

$$\limsup \sqrt[n]{|c_n|} = 1,$$

where

$$c_n = \frac{1}{2\pi\imath} \int_{-1}^{1} f_0\left(\frac{3}{4} + \frac{1}{\pi} \arctan z\right) \left(\frac{1 - t\imath z}{z + t\imath}\right)^n \frac{1 - t^2}{(z + t\imath)(1 - t\imath z)}\, dz. \tag{4.2}$$

PROOF. By assumption,

$$F_0(z) := f_0\left(\frac{3}{4} + \frac{1}{\pi} \arctan z\right)$$

is a continuous function of $z \in [-1, 1]$. By Lemma 4.3, it extends to a holomorphic function in the half-disk $\mathbb{D}'$ continuous up to $(-1, 1)$ if and only if $f_0(v)$ extends to a holomorphic function in the quarter-strip $\mathbb{H}'$ continuous up to $(\frac{1}{2}, 1)$. The theorem now follows from Corollary 3.2. □

In Section 1 we have done an alternative formulation of the Riemann hypothesis. We now make it more precise.

Lemma 4.5. *The Riemann hypothesis is true if and only if there exists a holomorphic function in $\mathbb{H}'$ which is continuous up to $(\frac{1}{2}, 1)$ and equal to $1/\zeta(s)$ for all $s \in (\frac{1}{2}, 1)$.*

PROOF. Since the Riemann zeta function does not vanish on the interval $[\frac{1}{2}, 1]$ and has a simple pole at $s = 1$, the function $f_0(s) := 1/\zeta(s)$ is continuous on $[\frac{1}{2}, 1]$ and has a simple zero at $s = 1$. If $\zeta(s) \neq 0$ for all $s \in \mathbb{H}'$, then the function $1/\zeta(s)$ is actually holomorphic in the whole critical quarter-strip. Hence, the Riemann hypothesis just amounts to the fact that f_0 extends to a holomorphic function in $\mathbb{H}'$ continuous up to $(\frac{1}{2}, 1)$. □

Corollary 4.6. *The Riemann hypothesis holds if and only if* $\limsup \sqrt[n]{|c_n|} = 1$, *where*

$$c_n = \frac{1}{2\pi\imath} \int_{-1}^{1} \frac{1}{\zeta(3/4 + (1/\pi)\arctan z)} \left(\frac{1 - t\imath z}{z + t\imath} \right)^n \frac{1 - t^2}{(z + t\imath)(1 - t\imath z)}\, dz. \tag{4.3}$$

PROOF. This follows immediately from Lemma 4.5 and Theorem 4.4. □

The idea of this approach goes back at least as far as André Weil who first proved the analogue of the Riemann hypothesis for general curves over finite fields in 1942. As but one part of his proof was to show that the logarithmic derivative of the zeta function has no poles in the "critical strip." Weil proved crucial estimates for the radius of convergence of the power series in which the logarithmic derivative expands around the origin, see [5]. For the Riemann zeta function arguments of Weil were recovered by Aizenberg et al. in [3]. The main formula of [3] contains an unfortunate misprint. However, it remains still valid as L. Aizenberg explained in a private letter to the author after the preprint had been published.

5. Numerical experiments

Formula (4.2) can be rewritten in the form

$$c_n = \frac{1}{2\pi\imath} \int_S e^{n(-\log w)} f_0\Big((h \circ k)^{-1}(w)\Big) \frac{dw}{w},$$

where $S \subset \mathbb{D}$ is the image of $(-1, 1)$ by $w = h(z)$. Since the function $-\log w$ has no saddle points in the complex plane, the saddle-point method does not apply to construct asymptotics of c_n as $n \to \infty$. Hence, Theorem 4.4 seems to be of purely numerical interest.

My graduate students A. Bühmann (2009) and M. Albinus (2010) evaluated numerically several terms $\sqrt[n]{|c_n|}$ for the functions $1/\zeta(s)$ and $1/\big(s - (3/4 + \imath)\big)$ on the interval $[\frac{1}{2}, 1]$. Obviously, the latter function does not extend analytically to all of $\mathbf{H}'$. Hence, the limit $\limsup \sqrt[n]{|c_n|}$ for this function is larger than 1 (but $\leq 1/t$). Several numerical values of sequence $\sqrt[n]{|c_n|}$ corresponding to $t = \frac{1}{2}$ are given in Table 1.

Computations with the newest versions of Mathematica, Maple and Matlab give certain evidence to the fact that for the function $1/\zeta(s)$ the limit of $\sqrt[n]{|c_n|}$ is 1 indeed. However, on replacing $1/\zeta(z)$ by other functions (e.g., $1/\big(s - (3/4 + \imath)\big)$) which fail to be analytically extendable to the critical quarter-strip computer simulations still suggest that $\limsup \sqrt[n]{|c_n|} = 1$. So the standard computer programs don't allow one to specify by means of Theorem 4.4 those continuous functions on $[\frac{1}{2}, 1]$ which extend analytically to the critical quarter-strip, cf. [3]. A severe difficulty consists in rough evaluations of integrals depending on a parameter. For large values of the parameter the graph of the integrand function in (4.2) fills in a rectangle. The problem arising is to elaborate an effective program for numerical evaluation of the limit $\limsup \sqrt[n]{|c_n|}$, where c_n are given by formula (4.2). This features once again the transcendental character of the Riemann problem on zeros of zeta function.

Acknowledgements. The author gratefully acknowledges the help of Matthias Albinus with the excellent pictures used in this paper.

TABLE 1. A computer simulation of $\sqrt[n]{|c_n|}$ for $t = \frac{1}{2}$.

	$1/\zeta(s)$	$1/(s-(3/4+\imath))$
$n=1$	0,5361	1,5046
$n=2$	0,7322	1,1659
$n=3$	0,8124	1,0265
$n=4$	0,8557	0,9274
$n=5$	0,8827	0,8374
$n=6$	0,9013	0,7391
$n=7$	0,9148	0,6070
$n=8$	0,9250	0,5941
$n=9$	0,9330	0,7636
$n=10$	0,9395	0,8512
$n=11$	0,9449	0,8960
$n=12$	0,9493	0,9159
$n=13$	0,9531	0,9195
$n=14$	0,9564	0,9110
$n=15$	0,9592	0,8928
$n=16$	0,9617	0,8677

References

1. L. Aizenberg, *Carleman's formulas in complex analysis: Theory and applications*, Math. Appl., vol. 244, Kluwer, Dordrecht, 1993.
2. ———, *Carleman's formulas and conditions of analytic extendability*, Topics in Complex Analysis (Warsaw, 1992) (P. Jakóbczak and W. Pleśniak, eds.), Banach Center Publ., vol. 31, Polish Acad. Sci., Warsaw, 1995, pp. 27–34.
3. L. Aizenberg, V. Adamchik, and V. E. Levit, *One computational approach in support of the Riemann hypothesis*, Comput. Math. Appl. **37** (1999), no. 1, 87–94.
4. J. B. Conrey, *The Riemann hypothesis*, Notices Amer. Math. Soc. **50** (2003), no. 3, 341–353.
5. P. M. Gauthier and N. Tarkhanov, *On the instability of the Riemann hypothesis for curves over finite fields*, J. Approx. Theory **164** (2012), no. 4, 504–515.
6. A. A. Karatsuba and S. M. Voronin, *The Riemann zeta-function*, de Gruyter Exp. Math., vol. 5, de Gruyter, Berlin, 1992.
7. E. C. Titchmarsh, *The theory of the Riemann zeta-function*, Clarendon Press, Oxford, 1951.

INSTITUTE OF MATHEMATICS, UNIVERSITY OF POTSDAM, AM NEUEN PALAIS 10, 14469 POTSDAM, GERMANY

E-mail address: `tarkhanov@math.uni-potsdam.de`

Centre de Recherches Mathématiques
CRM Proceedings and Lecture Notes
Volume **55**, 2012

A Survey of Linear Extremal Problems in Analytic Function Spaces

Catherine Bénéteau and Dmitry Khavinson

ABSTRACT. The purpose of this survey paper is to recall the major benchmarks of the theory of linear extremal problems in Hardy spaces and to outline the current status and open problems remaining in Bergman spaces. We focus on the model extremal problem of maximizing the norm of the linear functional associated with integration against a polynomial of finite degree, and discuss known solutions of particular cases of that problem. We examine duality and its application in both Hardy and Bergman spaces. Finally, we discuss some recent progress on the finiteness of the Blaschke product of the extremal solution in Bergman spaces.

1. Introduction and historical remarks

Solving extremal problems has been one of the major stimuli for progress in complex analysis, starting with the Schwarz lemma in the late 19th century, followed by work on coefficients of bounded analytic functions by C. Carathéodory and L. Fejér, Landau, Szasz, and others. At the end of the First World War, F. Riesz considered a best approximation problem in the Hardy space H^1, and in 1926, Szasz associated this problem with a dual problem in H^∞. This duality was rediscovered by Geronimus and, in a more general framework, by Krein in 1938. Extremal problems in multiply connected domains were studied by Grunsky (1940), Heins (1940), Robinson (1943), Goluzin (1946), and Ahlfors (1947). Macintyre and Rogosinski (1950) gave a detailed survey of results related to extremal problems involving coefficients of functions in all Hardy classes. Systematic use of duality in linear extremal problems for analytic functions started with S. Ya. Khavinson (1949) and independently Rogosinski and Shapiro (1953). Further studies were undertaken by Bonsall, Royden, Read, Adamyan, Arov, Krein, Walsh, among others. For a full account of the history of the development of extremal problems and references, see [11, pp. 51–57].

2010 *Mathematics Subject Classification.* Primary 30H05; Secondary 30E05 .
Both authors would like thank the Institut Mittag-Leffler (Djursholm, Sweden) for support while working on this project.
The second author also gratefully acknowledges partial support from the National Science Foundation under the grant DMS-0855597.

Work on Bergman spaces began with Ryabych in the early 1960s, who started the investigation of the existence and regularity of solutions [21, 22]. In 1991, Osipenko and Stessin [19] solved an explicit optimization problem in Bergman spaces involving linear combinations of the value of a function and its derivative at a particular point in the disk. The theory of contractive divisors in Bergman spaces, initiated by Hedenmalm [7], followed by Duren, D. Khavinson, Shapiro and Sundberg [2, 3], prompted a burst of activity in Bergman spaces which gave insight into the structure of the z-invariant subspaces of Bergman spaces. These developments are recorded in two books on Bergman spaces [4, 8]. In 1997, D. Khavinson and Stessin made a deeper study of linear extremal problems in Bergman spaces [9]. Ferguson gave a simpler proof of Ryabych's regularity results and generalized them in 2009 and 2010 [5]. The paper [26] contains a nice discussion of results on extremal problems in Bergman spaces.

The purpose of this survey is to recall the major benchmarks of the theory in Hardy spaces and to outline the current status of developments in Bergman spaces as well as the obstacles that still remain there. The plan of the paper is as follows: we begin in Section 2 by defining Bergman spaces, state a model extremal problem, and investigate the existence and uniqueness of extremals. In Section 3, we give examples and known solutions of that extremal problem in special cases. Section 4 discusses the duality theorem, and in Section 5, we apply duality to see how to get the solutions in Hardy spaces. In Section 6, we tackle the Bergman space case, discuss the difficulties and examine the connection with partial differential equations. In Section 7, we give a proof of a new result that the Blaschke product of the extremal solution for Bergman spaces A^p for p close to 2 is finite.

2. A model extremal problem in Bergman spaces

Let us begin by examining a model extremal problem in the Bergman space.

Definition 2.1. For $0 < p < \infty$, define the Bergman space as

$$A^p = \left\{ f \text{ analytic in } \mathbb{D} : \left(\int_{\mathbb{D}} |f(z)|^p \, \mathrm{d}A(z) \right)^{1/p} =: \|f\|_{A^p} < \infty \right\},$$

where $\mathrm{d}A(z) = (1/\pi)\,\mathrm{d}x\,\mathrm{d}y$ denotes normalized area measure in the unit disk $\mathbb{D}$, $z = x + \mathrm{i}y$.

Consider the following **model extremal problem**: Fix $1 \leq p < \infty$. Given a nonzero polynomial

$$\omega(z) = \sum_{k=0}^{N} a_k z^k,$$

describe the extremal solutions of the problem:

$$\lambda_p := \sup\left\{ \mathrm{Re}\left(\int_{\mathbb{D}} f(z)\overline{\omega(z)}\, \mathrm{d}A(z) \right) : \|f\|_{A^p} \leq 1 \right\} \tag{2.1}$$

or, equivalently,

$$\sup\left\{ \mathrm{Re}\left(\sum_{0}^{N} c_k f^{(k)}(0) \right), c_k = \frac{a_k}{(k+1)!}, \|f\|_{A^p} \leq 1 \right\}. \tag{2.2}$$

Solving problem (2.1) is equivalent to solving the following problem:

$$\inf\left\{ \|F\|_{A^p} : \int_{\mathbb{D}} F\overline{\omega}\, \mathrm{d}A = 1 \right\}, \tag{2.3}$$

since it is easily checked that F^* is a solution to (2.3) if and only if f^* is a solution to (2.1), where $F^* = f^*/\lambda_p$. Let us examine questions of existence and uniqueness.

For $1 \leq p < \infty$, if $\{F_n\}$ is a sequence of A^p functions approaching the infimum in (2.3), then their A^p norms are bounded, and thus, thinking of these functions as linear functionals on A^q for $1/p + 1/q = 1$, by the weak* compactness of bounded sets in A^p, there exists a function $F^* \in A^p$ and a subsequence F_{n_k} of F_n such that F_{n_k} approaches F^* weak*. In particular,

$$1 = \int_{\mathbb{D}} F_{n_k}\overline{\omega}\, \mathrm{d}A \to \int_{\mathbb{D}} F^*\overline{\omega}\, \mathrm{d}A,$$

and therefore

$$\int_{\mathbb{D}} F^*\overline{\omega}\, \mathrm{d}A = 1,$$

and of course, $F_{n_k} \to F^*$ pointwise. Finally, by Fatou's theorem, $\|F^*\|_{A^p} \leq \liminf \|F_{n_k}\|_{A^p}$, and therefore

$$\|F^*\|_{A^p} = \inf\left\{ \|F\|_{A^p} : \int_{\mathbb{D}} F\overline{\omega}\, \mathrm{d}A = 1 \right\},$$

as desired.

If $p = 1$, the argument is similar but slightly more delicate, since to use weak* compactness, we must think of A^1 as a subset of the set of complex measures on $\overline{\mathbb{D}}$. In this case, for a sequence $\{F_n\}$ of A^1 functions approaching the infimum in (2.3), the measures $F_n\, \mathrm{d}A$ form a bounded sequence of measures on the disk, and therefore, by weak* compactness of bounded measures on $\overline{\mathbb{D}}$, there exists a measure $d\mu^*$ such that some subsequence $F_{n_k}\, \mathrm{d}A$ approaches $\mathrm{d}\mu^*$ weak*, that is

$$\int_{\mathbb{D}} F_{n_k} f\, \mathrm{d}A \to \int_{\mathbb{D}} f\, \mathrm{d}\mu,$$

for every f continuous in $\overline{\mathbb{D}}$. We now appeal to a version of the F&M Riesz theorem for A^1 proved by H. Shapiro [23, 24], which can be stated as follows.

Theorem 2.2. *Let Ω be any bounded open set with smooth boundary, and let $M(\overline{\Omega})$ be the Banach space of bounded complex measures on $\overline{\Omega}$, and suppose $f_n \in A^1(\Omega)$ is a sequence of functions such that $f_n\, \mathrm{d}A \to \mathrm{d}\mu$ weak*, for some $\mu \in M(\overline{\Omega})$. Then there exists $f \in A^1(\Omega)$ such that $\mathrm{d}\mu = f\, \mathrm{d}A$.*

Here, $A^1(\Omega)$ is naturally defined as the space of integrable analytic functions in the domain Ω. Note that in the original statement of this theorem, the domain Ω is allowed to have nonsmooth boundary points, and then the limit measure is of the form $f\, \mathrm{d}\mu + \mathrm{d}\nu$, where ν a singular measure supported on these non smooth boundary points. See [24, pp. 75 – 76] for details.

Now, getting back to the proof of existence, we see that by Theorem 2.2 applied to $\mathbb{D}$, the measure μ^* in question is absolutely continuous, and therefore, there exists a function F^* such that the measures $F_{n_k}\, \mathrm{d}A$ approach $F^*\, \mathrm{d}A$ weak*. The rest of the argument is the same as for $p > 1$, since ω is continuous.

Finally, the Bergman spaces A^p are strictly convex for all $1 \leq p < \infty$, (see, for example, [4, [pp. 28 – 29]), which implies that there can only be one element of minimal norm satisfying $\int_{\mathbb{D}} F\overline{\omega}\,\mathrm{d}A = 1$. Therefore the solution to problem 2.3 and therefore to problem 2.1 is unique. Note that for $1 < p < \infty$, the argument showing existence and uniqueness of an extremal for the model problem considered here immediately extends to $\omega \in A^q$, for $1/p + 1/q = 1$.

The main thrust of this work is to study the smoothness properties of the extremal functions. It is always expected that the solution of a "nice" extremal problem is much better than the generic function in the space. In particular, for ω a polynomial as in the above Model Problem (2.1), it turns out that the extremal solution f^* is a bounded analytic function, continuous in the closed disk. However, obtaining this regularity already involves significant technical difficulty and requires the use of deep results from nonlinear partial differential equations [5, 9, 22].

Let us now turn to some examples of known solutions to the model problem for particular cases.

3. Examples of known solutions

For some particular polynomials or for $p = 2$, the solutions to model problem (2.1) are known explicitly and are discussed briefly here.

Example 3.1. The simplest example is that of A^2. In that case, by the Cauchy-Schwarz inequality, the supremum

$$\sup\left\{\operatorname{Re}\left(\int_{\mathbb{D}} f(z)\overline{\omega(z)}\,\mathrm{d}A(z)\right) : \|f\|_{A^p} \leq 1\right\}$$

is attained by the function $f^* = \omega/\|\omega\|_{A^2}$, for any polynomial ω, and in fact, for any $\omega \in A^2$. Henceforth, we will assume that $p \neq 2$.

Example 3.2. If $\omega = z^N$, then Ryabych [21] showed that the extremal function for

$$\sup\{\operatorname{Re} f^{(N)}(0) : \|f\|_{A^p} \leq 1\}$$

is $f^*(z) = \{(Np+1)/2\}^{1/p} z^N$.

Example 3.3. If we would like to consider Example 3.2 in a more general setting by estimating values of a function at a point β instead of at the origin, we can allow ω to be a simple rational function, namely, $\omega(z) = (1-\overline{\beta}z)^{-2}$, for $|\beta| < 1$, the Bergman kernel at β. In that case, problem (2.1) becomes

$$\sup\{\operatorname{Re} f(\beta) : \|f\|_{A^p} \leq 1\}.$$

This problem was studied by Ryabych in [21] and also by Vukotić in [25], and the extremal solution has the form

$$f^*(z) = (1 - |\beta|^2)^{2/p}(1 - \overline{\beta}z)^{-4/p}.$$

Example 3.4. In 1991, Osipenko and Stessin [19] considered the case of a simple linear polynomial, $\omega(z) = a_0 + a_1 z$. Problem (2.1) then becomes

$$\sup\{\operatorname{Re}(c_0 f(0) + c_1 f'(0)) : \|f\|_{A^p} \leq 1\}.$$

They showed that

$$f^*(z) = C\frac{z-\beta}{1-\overline{\beta}z}(1 - \overline{\alpha}z)^{2/p}$$

where $|\beta| \leq 1$, $|\alpha| \leq 1$, and C is a constant. They also wrote down equations relating c_0 and c_1 to β and α. In addition, they considered the case when 0 is replaced by an arbitrary point $\zeta \in \mathbb{D}$, i.e., when ω is a linear combination of the

Bergman kernel and its derivative. Their results were already technically quite difficult, and there doesn't seem to be much hope of generalizing their approach, which was based on "guessing" the exact form of the extremal, to higher degree polynomials.

Example 3.5. Problems of the type considered in Examples 3.1 through 3.4 are connected to what are often called Carathéodory – Fejér type problems. An important example is the problem of finding, for given $|\beta_j| < 1$ for $j = 1, \dots, m$:

$$(3.1) \quad \inf\{\|f\|_{A^p} : f^{(N)}(0) = 1, \\ f(0) = \cdots = f^{(N-1)}(0) = f(\beta_1) = \cdots = f^{(k_m)}(\beta_m) = 0\}.$$

This problem is equivalent to the problem of finding, for given $|\beta_j| < 1$ for $j = 1, \dots, m$:

$$(3.2) \quad \sup\{\operatorname{Re} f^{(N)}(0) : \|f\|_{A^p} \le 1, \\ f(0) = \cdots = f^{(N-1)}(0) = f(\beta_1) = \cdots = f^{(k_m)}(\beta_m) = 0\},$$

whose extremal solution is by definition a "contractive divisor" in A^p. Contractive divisors in Bergman spaces were discovered and studied in the 1990s, first by Hedenmalm [7] for $p = 2$, and then by Duren, Khavinson, Shapiro, and Sundberg [2,3] for all p, and later by MacGregor and Stessin [14]. Contractive divisors turned out to be intimately connected to the theory of z invariant subspaces in Bergman spaces. For more on this subject, see [4,8].

The extremal functions are

$$f^*(z) = Cz^N \prod_1^m \left\{ \frac{z - \beta_j}{1 - \overline{\beta_j} z} \right\}^{k_j} R(z)^{2/p},$$

where R is a rational function with poles at ∞ and at $\{1/\bar{\beta}_j\}_{j=1}^m$ of degree less than or equal to $2N + \sum_{j=1}^m k_j$. The problem is equivalent to problem (2.1) with ω being a specific linear combination of the Bergman kernel $(1 - \overline{\zeta} z)^{-2}$ and its $\partial / \partial \overline{\zeta}$ derivatives at the β_j.

All of these particular examples hint at a more general theory and a simple form of the extremal solutions. Let us now examine this question and turn to the tools that helped establish the qualitative form of the solutions in spaces simpler than the Bergman spaces.

4. Duality

In the late 1940s and early 1950s, the systematic use of duality to solve extremal problems in complex function theory became prevalent. The simplest form of a duality statement is the following.

Theorem 4.1 (see, e.g., [11, p. 2]**).** *Let X be a normed linear space, and let X^* be the space of bounded linear functionals on X. Suppose $E \subset X$ is a subspace of X and l_0 is a given linear functional in X^*. Then*

$$\sup\{|l_0(f)| : f \in E, \|f\| \le 1\} = \inf\{\|l_0 - l\| : l \in E^\perp\} = \|l_0 - l^*\|$$

for some $l^* \in E^\perp$, where $E^\perp$ is the annihilator of E in X^*, defined to be the set $\{l \in X^* : l(f) = 0 \ \forall f \in E\}$.

Duality thus pairs a linear extremal problem with a problem of best approximation. This pairing allows one to gain a deeper understanding of each one in turn, as we shall see. The short proof follows immediately from the Hahn-Banach theorem and is included to emphasize how the best approximation in the dual problem appears.

PROOF. First note that for any $l \in E^{\perp}$ and for any $f \in E$ with $\|f\| \leq 1$, we have

$$|l_0(f)| = |(l_0 - l)(f)| \leq \|l_0 - l\|.$$

Therefore,

$$\|l_0\|_E = \sup\{|l_0(f)| : f \in E, \|f\| \leq 1\} \leq \inf\{\|l_0 - l\| : l \in E^{\perp}\}.$$

Now by the Hahn – Banach theorem, there exists a linear functional L on X such that $L|_E = l_0$ and $\|L\|_X = \|l_0\|_E$. Therefore $(l_0 - L) \in E^{\perp}$, and

$$\|l_0\|_E \leq \inf\{\|l_0 - l\| : l \in E^{\perp}\} \leq \|l_0 - (l_0 - L)\|_X = \|L\|_X,$$

and therefore the chain of inequalities is actually a chain of equalities. Putting $l^* = l_0 - L$ gives the desired result. □

One immediate interesting note is that the best approximation l^* therefore always exists, even in this general setting. Whether the supremum is actually achieved and gives rise to an extremal $f^* \in E$ is not always the case and depends on the spaces and linear functionals being considered. Finally, in order to be able to use the duality theorem, we need to have information about the annihilator spaces. To see how this works, let us, in the following section, apply duality to our model problem, but in Hardy spaces. (See, for example, [1, 10] for details.)

5. Duality applied to H^p-theory

Although Bergman spaces are defined in a natural way, the functions in those spaces are quite complicated, and their structure is not entirely understood. For example, Bergman space functions need not have any radial limits on the unit circle. On the other hand, the Hardy spaces are classical spaces whose structure is quite well understood, although their definition is in some sense less natural. We begin this section by defining the Hardy spaces.

Definition 5.1. For $0 < p < \infty$, define the Hardy space as

$$H^p := \left\{ f \text{ analytic in } \mathbb{D} : \sup_{0<r<1} \frac{1}{2\pi} \int_0^{2\pi} |f(re^{i\theta})|^p \, d\theta := \|f\|_{H^p}^p < \infty \right\}.$$

The Hardy space of bounded analytic functions is defined by

$$H^\infty = \left\{ f \text{ analytic in } \mathbb{D} : \|f\|_\infty := \sup_{z\in\mathbb{D}} |f(z)| < \infty \right\}.$$

It is well-known that Hardy space functions f have radial limits $\lim_{r\to 1^-} f(re^{i\theta}) =: f(e^{i\theta})$ for almost every $\theta \in \mathbb{T}$, and that $\|f\|_{H^p} = \|f(e^{i\theta})\|_{L^p}$, where L^p is the usual space of measurable and p integrable functions on the circle. For $\omega(z) = \sum_{k=0}^N a_k z^k$, let us now consider our model problem in the *Hardy space*, for $1 \leq p < \infty$, of finding

$$\lambda := \sup\left\{ \operatorname{Re} \int_0^{2\pi} f(e^{i\theta}) \, \overline{\omega(e^{i\theta})} \frac{d\theta}{2\pi}, \|f\|_{H^p} \leq 1 \right\}. \tag{5.1}$$

In order to apply the Duality Theorem 4.1 to this setting, we will consider the space $E = H^p$ as a subspace of the space $X = L^p(\mathbb{T})$. Then the dual space X^* is L^q, where $1/p + 1/q = 1$, and the annihilator space $E^\perp$ is the set of L^q functions g such that

$$\int_0^{2\pi} f(e^{i\theta})\overline{g(e^{i\theta})}\, d\theta = 0$$

for every $f \in H^p$. In particular,

$$\int_0^{2\pi} e^{in\theta}\overline{g(e^{i\theta})}\, d\theta = 0$$

for every integer $n \geq 0$, and therefore, by the classical F & M Riesz theorem, (see, for example, [1, p. 41]), $\overline{g} \in H^q$ and vanishes at the origin. (We will write the space of such functions as H^q_0.) Therefore, Problem (5.1) above can be written, in its dual form, as:

$$\lambda = \min\{\|\overline{\omega} - \phi^*\|_{L^q(\mathbb{T})}, \phi^* \in H^q_0\}.$$

By the general duality theorem, we know the best approximation ϕ^* exists, and one can also show that ϕ^* is unique whenever f^* exists. It turns out that the extremal f^* for the supremum problem exists and is unique for $p > 1$, but for $p = 1$ in general this is not the case: in fact, the existence of f^* here for $p = 1$ follows from the continuity of ω (as in the A^1 case discussed in Section 2). See [11],*Sections 4 and 6, for more details on questions of existence and uniqueness in the Hardy space case.

Now given these extremals f^* and ϕ^*, notice that we have the following chain of inequalities, which must actually be equalities:

$$\begin{aligned}\lambda = \operatorname{Re}\int_0^{2\pi} f^*(e^{i\theta})\,\overline{\omega(e^{i\theta})}\frac{d\theta}{2\pi} &= \operatorname{Re}\int_0^{2\pi} f^*(e^{i\theta})\big(\overline{\omega(e^{i\theta})} - \phi^*(e^{i\theta})\big)\frac{d\theta}{2\pi}\\ &\leq \int_0^{2\pi} |f^*(e^{i\theta})|\,|\overline{\omega(e^{i\theta})} - \phi^*(e^{i\theta})|\frac{d\theta}{2\pi}\\ &\leq \|f\|_{H^p}\|\overline{\omega} - \phi^*\|_{L^q}\\ &\leq \|\overline{\omega} - \phi^*\|_{L^q}\\ &= \lambda.\end{aligned}$$

Therefore, because we have equality in Hölder's inequality above, we must have that

$$\lambda^q|f^*(e^{i\theta})|^p = |\overline{\omega(e^{i\theta})} - \phi^*(e^{i\theta})|^q$$

almost everywhere on the circle. Taking qth roots gives that

$$\lambda|f^*(e^{i\theta})|^{p/q} = \lambda|f^*(e^{i\theta})|^{p-1} = |\overline{\omega(e^{i\theta})} - \phi^*(e^{i\theta})|$$

almost everywhere on $\mathbb{T}$. In addition, for equality above, we must have that $\|f\|_{H^p} = 1$ and

$$f^*(e^{i\theta})\big(\overline{\omega(e^{i\theta})} - \phi^*(e^{i\theta})\big) \geq 0$$

a.e. on the circle. Putting these together gives an equivalent characterization of extremality of f^*, ϕ^* as:

$$\lambda\frac{|f^*(e^{i\theta})|^p}{f^*(e^{i\theta})} = \overline{\omega(e^{i\theta})} - \phi^*(e^{i\theta}) \quad \text{a.e. on } \mathbb{T}. \tag{5.2}$$

Equivalently, the function $\lambda|f^*|^p/f^* - \overline{\omega}$ annihilates H^p.

Now notice that the function $f^*(z)\big(\overline{\omega(z)} - \phi^*(z)\big)$ is well-behaved inside the disk, in a neighborhood of the unit circle, since $f^* \in H^p$ and $\phi^* \in H^q$, and for $|z| = 1$, $\overline{\omega(z)} = \overline{a_0} + \sum_{k=1}^{N} \overline{a_k}/z^k$. Therefore the product $f^*(\overline{\omega} - \phi^*)$ can be thought of as a function that is in the Hardy class H^1 in an annulus inside the unit disk. Moreover, this function is positive (hence real) on the circle. Thus, one can apply the Schwarz reflection principle (see, for example, [18, pp. 183–185]) to extend the product to $\mathbb{C}\setminus\{0\}$. Moreover, at $z = 0$, the singularity is given by the behavior of $\overline{\omega}$, and hence is a pole of order N, and therefore by reflection a similar behavior occurs at ∞. Hence $f^*(\overline{\omega} - \phi^*)$ is a rational function on the Riemann sphere with two poles, each of order N at 0 and ∞. Moreover, since $f^*(\overline{\omega}-\phi^*)$ is positive on the unit circle, by the argument principle, the increment of the argument of the product is 0 on the circle, and hence the product has $2N$ zeros, which by the symmetry required by the reflection principle, must be divided equally and symmetrically inside and outside the disk. (Zeros on the circle come with even multiplicity.) Hence,

$$f^*(z)\big(\overline{\omega(z)} - \phi^*(z)\big) = Cz^{-N}\prod_{j=1}^{N}(z-\alpha_j)(1-\overline{\alpha_j}z), \quad |\alpha_j| \le 1$$

for some constant C. In particular, f^* has no singular part, and the zeros of f^* are among the α_j such that $|\alpha_j| < 1$, and therefore the Blaschke factor of f^* is

$$\prod_{j=1}^{k}\frac{z-\alpha_j}{1-\overline{\alpha_j}z},$$

where $k \le N$. Finally, the outer part of f^* is determined by the equation on the boundary that requires

$$\begin{aligned}\lambda|f^*(z)|^p = f^*(z)\big(\overline{\omega(z)} - \phi^*(z)\big) &= Cz^{-N}\prod_{1}^{N}(z-\alpha_j)(1-\overline{\alpha_j}z)\\ &= C\prod_{1}^{N}\Big[\frac{z-\alpha_j}{z}\Big](1-\overline{\alpha_j}z)\\ &= C\prod_{1}^{N}(1-\alpha_j\bar{z})(1-\overline{\alpha_j}z) = C\prod_{1}^{N}|1-\overline{\alpha_j}z|^2.\end{aligned}$$

Therefore the extremal f^* is of the form

$$f^*(z) = C\prod_{1}^{k}\frac{z-\alpha_j}{1-\overline{\alpha_j}z}\prod_{1}^{N}(1-\overline{\alpha_j}z)^{2/p}, \quad k \le N, \tag{5.3}$$

giving a qualitative solution to the extremal problem.

Let us now turn to a discussion of this problem in Bergman spaces.

6. Extremal problems in Bergman spaces

Recall the problem of finding and describing the extremal solutions of

$$\lambda_p := \sup\Big\{\mathrm{Re}\Big(\int_{\mathbb{D}} f(z)\overline{\omega(z)}\,\mathrm{d}A(z)\Big) : \|f\|_{A^p} \le 1\Big\}, \tag{6.1}$$

where ω is a polynomial of degree N. As seen in Section 3, for each $1 < p < \infty$, the solution f^* exists and is unique. For $p = 2$, the solution is $f^* = \omega/\|\omega\|$ by the Cauchy–Schwarz inequality. For $1 \le p < \infty$, and even for any $\omega \in H^q$,

$q : 1/p + 1/q = 1$, Ryabych proved ([22]) that the solution is in H^p, and, in particular, has boundary values at almost every point of the unit circle $\mathbb{T}$. T. Ferguson ([5]) recently refined that proof.

Moreover, deeper and more technical results in [9] imply that for $1 < p < \infty$, the solution f^* of (2.1), with ω a polynomial, is continuous in the closed disk, in fact is Lip $(\gamma, \overline{\mathbb{D}})$ for some γ that depends on p [9]. These results support the philosophy that extremal functions are much better than the generic functions of the space where the problem is set. The most that is known regarding the extremal solution to date is the following theorem of D. Khavinson and M. Stessin [9]:

Theorem 6.1. *Let ω be a polynomial of degree N and let $1 < p < \infty$. The extremal solution f^* to*

$$\lambda_p := \sup\left\{\mathrm{Re}\left(\int_{\mathbb{D}} f(z)\overline{\omega(z)}\,\mathrm{d}A(z)\right) : \|f\|_{A^p} \leq 1\right\}, \tag{6.2}$$

is in $\mathrm{Lip}(\gamma, \overline{\mathbb{D}})$ *and has the form*

$$f^*(z) = C\prod_{j=1}^{\infty}\frac{\beta_j - z}{1-\overline{\beta_j}z}\frac{|\beta_j|}{\beta_j}\prod_{j=1}^{N}(1-\overline{\alpha_j}z)^{2/p},$$

where C, $|\beta_j| < 1$, and $|\alpha_j| \leq 1$ are constants, and the β_j can only accumulate to those values of α_l that lie on the circle.

One of the missing pieces of the theory of extremal problems in Bergman spaces versus that in Hardy spaces is information about the Blaschke product. We thus state the following:

Conjecture 6.2. The Blaschke product of the extremal solution has at most N factors.

In the next section, we will show some progress in this direction, but the full conjecture is still open.

Let us now examine some key steps in the proof of Theorem 6.1 and see how the theory of PDEs comes into play. We will discuss here only the case $p > 1$, since for $p = 1$, these methods break down. The validity of Theorem 6.1 for $p = 1$ remains unknown.

Recall that we have already shown that there exists a unique solution f^* to problem (6.1). In order to apply duality, we consider A^p as a subspace of $L^p(\mathbb{D})$, and identify the annihilator of A^p by appealing to Khavin's lemma (see [6; 24, Lemma 4.2, p. 26]).

Lemma 6.3. *The annihilator of $A^p(\mathbb{D})$ inside $L^p(\mathbb{D})$ can be described by*

$$(A^p)^{\perp} \cong \left\{\frac{\partial v}{\partial \bar{z}} : v \in W_0^{1,q}(\mathbb{D})\right\},$$

where $W_0^{1,q}(\mathbb{D})$ is the Sobolev space of functions vanishing on $\mathbb{T}$ and with gradients in $L^q(\mathbb{D})$, with $1/p + 1/q = 1$ and $\partial v/\partial\bar{z} = \frac{1}{2}(\partial/\partial x + \mathrm{i}\partial/\partial y)$.

By following the same reasoning as in Section 5, and setting, for the sake of brevity, $\lambda := \lambda_p$, we then obtain that

$$\lambda\frac{|f^*|^p}{f^*} - \overline{\omega} = \frac{\partial u^*}{\partial \bar{z}} \quad \text{a.e. in } \mathbb{D}, \tag{6.3}$$

where $u^* \in W_0^{1,q}$. Now set $v(z) := u^*(z) + \overline{\Omega(z)}$, $\Omega(z) := \int_0^z \omega(\zeta)\, \mathrm{d}\zeta$, so that

$$\frac{\partial v}{\partial \bar{z}} = \lambda \frac{|f^*|^p}{f^*}.$$

Then v solves the *nonlinear* boundary value problem:

$$\begin{aligned} \frac{\partial}{\partial z}(|v_{\bar{z}}|^{q-2} v_{\bar{z}}) &= 0 \qquad \text{in } \mathbb{D}; \\ v &= \overline{\Omega} \qquad \text{on } \mathbb{T}. \end{aligned} \tag{6.4}$$

By results of Ch. Morrey, O. Ladyzhenskaya, and N. Ural′tseva (see the discussion in [9] and [12, 15–17]), the unique solution v of (6.4) belongs to $C^{1+\beta}(\overline{\mathbb{D}})$, $\beta = \beta(q)$. Since

$$f^* = \lambda^{q-1} \frac{|v_{\bar{z}}|^q}{v_{\bar{z}}},$$

we get that $f^* \in \mathrm{Lip}(\gamma, \overline{\mathbb{D}})$, $\gamma = \gamma(p)$. (See [9] for details.)

Remark 6.4. For values of p in any compact subset of $(1, \infty)$, the corresponding extremals f^* can all be taken to be $\mathrm{Lip}(\gamma, \overline{\mathbb{D}})$ for the same γ. This will turn out to be key later (see Section 7) when estimating the number of zeros of extremal functions.

Note that Conjecture 6.2 is equivalent to the following:

Conjecture 6.5. The solution of the BVP (6.4)

$$\begin{aligned} \frac{\partial}{\partial z}(|v_{\bar{z}}|^{q-2} v_{\bar{z}}) &= 0 \qquad \text{in } \mathbb{D}; \\ v &= \overline{\Omega} \qquad \text{on } \mathbb{T}. \end{aligned} \tag{6.5}$$

has at most N critical points in $\mathbb{D}$, that is, points where $\partial v / \partial \bar{z} = 0$.

J. Lewis [13] has proved a real-valued version of this result for q-Laplacians.

Now let us examine the outer part F of the extremal. We would like to show that $|F|^p = |f^*|^p$ on $\mathbb{T}$ is a positive trigonometric polynomial of degree less than or equal to N. First notice that the Hardy–Littlewood theorem ([1, Theorem 5.1]) implies that since $f^* \in \mathrm{Lip}(\gamma, \overline{\mathbb{D}})$, $(f^*)' \in A^{p_1}$, for some $p_1 > 1$. By duality, we get the so-called *orthogonality relationships*

$$\int_{\mathbb{D}} \left\{ \lambda \frac{|f^*|^p}{f^*} - \overline{\omega} \right\} g \, \mathrm{d}A = 0$$

for all $g \in A^1$. Note that, since ω is of degree N, $\int_{\mathbb{D}} \overline{\omega} z^{N+k} g \, \mathrm{d}A = 0$ for any integer $k \geq 1$ and any $g \in A^1$. Now, Green's formula gives for every integer $k \geq 1$:

$$\begin{aligned} \int_{\mathbb{T}} |f^*|^p z^{N+k} d\theta = \mathrm{i} \int_{\mathbb{T}} |f^*|^p z^{N+k+1} \, \mathrm{d}\bar{z} &= p \int_{\mathbb{D}} \frac{|f^*|^p}{f^*} (f^*)' z^{N+k+1} \, \mathrm{d}A \\ &\quad + 2 \int_{\mathbb{D}} \frac{|f^*|^p}{f^*} (N + k + 1) z^{N+k} f^* \, \mathrm{d}A. \end{aligned} \tag{6.6}$$

Each of these last two terms is 0, by the orthogonality relationships, and therefore, for each $k \geq 1$,

$$\int_{\mathbb{T}} |f^*|^p z^{N+k} \, \mathrm{d}\theta = 0,$$

showing that the outer part of f^* is the $(2/p)$th root of a polynomial of degree at most N. We therefore get the desired form of the extremal f^* as in Theorem 6.1.

The problem of showing that the Blaschke factor has at most N terms, or, even, indeed, is finite, still remains. We will discuss this more in the next section. Notice, though, that instead of problem (6.1), we can consider the problem with ω a rational function, to get point evaluations at points other than the origin. More specifically, given $\omega(z) := \sum_{k=1}^{N} a_k/(1-\overline{w_k}z)^2$, $|w_k|<1$, a linear combination of Bergman reproducing kernels, our problem becomes that of finding the extremal solutions to

$$\sup\left\{\operatorname{Re}\left(\sum_{1}^{N} a_k f(w_k)\right), |w_k|<1, \|f\|_{A^p}\le 1\right\}. \tag{6.7}$$

Then the results of Khavinson and Stessin [9], analogous to Theorem 6.1, imply that

$$f^*(z) = CB(z)\prod_{j=1}^{2N-2}(1-\overline{\alpha_j}z)^{2/p}\prod_{1}^{N}(1-\overline{w_j}z)^{-4/p},$$

where C is a constant, $|\alpha_j|\le 1$, $j=1,\dots,2N-2$, are constants and the zeros of the Blaschke product B may only accumulate to those α_j that lie on $\mathbb{T}$.

We now turn to a discussion of what more can be said about the Blaschke factor, at least for values of p close to 2.

7. A continuity approach

In an attempt to shed some light on Conjecture 6.2, we begin with the following lemma.

Lemma 7.1. *For $1\le p<\infty$, $\lambda(p):=\lambda_p$ is a decreasing function of p.*

PROOF. First note that one can easily use Jensen's inequality to show that the norms $\|f\|_p$ are increasing. Letting $1\le p<q<\infty$, let f_p^* be the extremal solution to problem (2.1) for p and f_q^* for q. Then $\|f_q^*\|_p\le\|f_q^*\|_q=1$, and therefore f_q^* is a competitor for the extremal problem (2.1) for p. Therefore

$$\operatorname{Re}\left(\int_{\mathbb{D}} f_q^*(z)\overline{\omega(z)}\,\mathrm{d}A(z)\right)\le\operatorname{Re}\left(\int_{\mathbb{D}} f_p^*(z)\overline{\omega(z)}\,\mathrm{d}A(z)\right)=\lambda(p),$$

or $\lambda(q)\le\lambda(p)$, as desired. □

Lemma 7.2. *Let $1<p<\infty$, and suppose that $p_n\to p$ and that $f_{p_n}^*(z)\to f(z)$ uniformly in $\overline{\mathbb{D}}$. Then $f=f_p^*$, that is, f is the extremal function for the problem (2.1) for p.*

PROOF. Since $f_{p_n}^*(z)\to f(z)$ uniformly in $\overline{\mathbb{D}}$, we have that

$$\int_{\mathbb{D}} f_{p_n}^*(z)\overline{\omega(z)}\,\mathrm{d}A(z)\to\int_{\mathbb{D}} f(z)\overline{\omega(z)}\,\mathrm{d}A(z).$$

Note that since $|f_{p_n}^*|\to|f|$ and since $p_n\to p$, $|f_{p_n}^*|^{p_n}\to|f|^p$, and therefore by Fatou's lemma and since $\|f_{p_n}^*\|_{p_n}=1$, $\|f\|_p\le 1$. Therefore, f is a competitor for the extremal problem (2.1) for p.

Now if f is not the solution to the extremal problem (2.1), then there exists $g = f_p^* \in A^p$ such that $\|g\|_p = 1$, g is continuous in $\overline{\mathbb{D}}$ (by Theorem 6.1), and, for some $\varepsilon > 0$,

$$\operatorname{Re}\Big(\int_{\mathbb{D}} g\overline{\omega}\,\mathrm{d}A\Big) > \operatorname{Re}\Big(\int_{\mathbb{D}} f\overline{\omega}\,\mathrm{d}A\Big) + \varepsilon.$$

Therefore, there exists N such that for $n \geq N$,

$$\operatorname{Re}\Big(\int_{\mathbb{D}} g\overline{\omega}\,\mathrm{d}A\Big) > \operatorname{Re}\Big(\int_{\mathbb{D}} f_n\overline{\omega}\,\mathrm{d}A\Big) + \frac{\varepsilon}{2}. \tag{7.1}$$

Now g is continuous in the closed unit disk, and therefore for $p_n \to p$, the functions $|g|^{p_n}$ are bounded above by some constant, and therefore (by the bounded convergence theorem),

$$\gamma_n^{p_n} := \int_{\mathbb{D}} |g|^{p_n}\,\mathrm{d}A \to \int_{\mathbb{D}} |g|^p\,\mathrm{d}A = 1.$$

Note that the functions g/γ_n have norm 1 in A^{p_n}.

Since $\gamma_n \to 1$ and by (7.1), there exists M such that for $m \geq M$ and for $n \geq N$,

$$\frac{1}{\gamma_m}\operatorname{Re}\Big(\int_{\mathbb{D}} g\overline{\omega}\,\mathrm{d}A\Big) > \operatorname{Re}\Big(\int_{\mathbb{D}} f_n^*\overline{\omega}\,\mathrm{d}A\Big) + \frac{\varepsilon}{4}.$$

Choosing a large enough $n = m$ satisfying this inequality leads to a contradiction of the extremality of f_n^*.

Therefore, we must indeed have that $g = f_p^*$, as desired. □

Note that the hypothesis in Lemma 7.2 that the functions $f_{p_n}^*$ converge uniformly to f in the closed disk is stronger than what is really necessary for the proof: what is required is that the measures $f_{p_n}^*\,\mathrm{d}A$ converge weakly to $f\,\mathrm{d}A$.

Corollary 7.3. *For $1 < p < \infty$, $\lambda(p)$ is a continuous function of p.*

Proof. If $p_n \to p$, then, by the remark after the statement of Lemma 6.3, the functions $f_{p_n}^*$ are all in $\operatorname{Lip}(\gamma, \overline{\mathbb{D}})$ for the some γ, and therefore form a uniformly bounded and equicontinuous family. Therefore, by the Arzelà–Ascoli theorem, there exists a subsequence $f_{p_{n_k}}^*$ that converges uniformly in $\overline{\mathbb{D}}$ to some function f. By Lemma 7.2, $f = f_p^*$. Therefore, by the bounded convergence theorem,

$$\int_{\mathbb{D}} f_{p_{n_k}}^*\overline{\omega}\,\mathrm{d}A(z) \to \int_{\mathbb{D}} f_p^*\overline{\omega}\,\mathrm{d}A(z).$$

Taking real parts, we get that $\lambda(p_{n_k}) \to \lambda(p)$. But since the function λ is monotone, $\lambda(p_n) \to \lambda(p)$. □

Theorem 7.4. *If ω has no zeros on the boundary of the disk, then there exists $\Delta > 0$ such that if $|p - 2| < \Delta$, the extremal function f_p^* has at most N zeros in $\overline{\mathbb{D}}$.*

Proof. If $p = 2$, then we know the solution is $f_2^* = \omega$, and ω has at most N zeros in the unit disk, because it is a polynomial of degree N.

First note that the extremal functions cannot have zeros that accumulate inside the unit disk (otherwise they would be identically zero) and therefore the zeros of the extremals can only accumulate to the boundary of the disk.

Let us first show that there exists Δ such that for p in a Δ neighborhood of 2, f_p^* has a finite number of zeros. Suppose not. Then there exists a sequence $p_n \to 2$ such that $f_{p_n}^*$ have infinitely many zeros in a compact neighborhood of the boundary of the disk. These zeros must have an accumulation point, and by the previous remark, this accumulation point must be on the boundary of the disk.

As in the proof of Corollary 7.3 and using Theorem 6.1, by passing to a subsequence if necessary, the $f^*_{p_n}$ converge uniformly in $\overline{\mathbb{D}}$ to ω by Lemma 7.2. But then ω must vanish at the accumulation point on the boundary of the disk, which is a contradiction. Therefore there exists some neighborhood of p for which f^*_p have finitely many zeros. Moreover by Hurwitz' theorem, there exists some neighborhood of 2 such that for all p in that neighborhood, all the f^*_p have the same number of zeros as ω inside $\mathbb{D}$, that is, at most N. □

Remark 7.5. This argument works for a more general ω, as long as we know ω has no zeros on the boundary and has at most N zeros in the disk. In particular, we have the following corollary.

Corollary 7.6. *For p sufficiently close to 2, the extremal solution to*

$$\sup\left\{\operatorname{Re}\left(\sum_{k=1}^{N} a_k f(w_k)\right), |w_k| < 1, \|f\|_{A^p} \leq 1\right\} \tag{7.2}$$

has at most $2N-2$ zeros in the unit disk, provided that $\sum_{k=1}^{N} a_k/(1-\bar{w}_k z)^2$ has no zeros on the unit circle.

The analogue to Conjecture 6.2 is thus that Corollary 7.6 holds for all $p > 1$.

The question remains how to deal with the case when ω has zeros on the circle. It is then natural to consider a sequence of functions ω_n without zeros on the circle that converge to ω. In addition, for a fixed p, one can then easily show that the extremals f^*_{p,ω_n} converge to $f_{p,\omega}$. However, the challenge remains to obtain uniform estimates on both p and n that would allow for a fixed neighborhood of $p = 2$ that does not depend on n.

References

1. P. Duren, *Theory of H^p spaces*, Pure Appl. Math., vol. 38, Academic Press, New York, 1970; 2nd ed., Dover, Mineola, NY, 2000.
2. P. Duren, D. Khavinson, H. S. Shapiro, and C. Sundberg, *Contractive zero-divisors in Bergman spaces*, Pacific J. Math. **157** (1993), no. 1, 37–56.
3. ———, *Invariant subspaces in Bergman spaces and the biharmonic equation*, Michigan Math. J. **41** (1994), no. 2, 247–259.
4. P. Duren and A. Schuster, *Bergman spaces*, Math. Surveys Monogr., vol. 100, Amer. Math. Soc., Providence, RI, 2004.
5. T. Ferguson, *Continuity of extremal elements in uniformly convex spaces*, Proc. Amer. Math. Soc. **137** (2009), no. 8, 2645–2653.
6. V. P. Havin, *Approximation by analytic functions in the mean*, Dokl. Akad. Nauk SSSR **178** (1968), 1025–1028 (Russian); English transl., Soviet Math. Dokl. **9** (1968), 245–258.
7. H. Hedenmalm, *A factorization theorem for square area-integrable analytic functions*, J. Reine Angew. Math. **422** (1991), 45–68.
8. H. Hedenmalm, B. Korenblum, and K. Zhu, *Theory of Bergman spaces*, Grad. Texts in Math., vol. 199, Springer, New York, 2000.
9. D. Khavinson and M. Stessin, *Certain linear extremal problems in Bergman spaces of analytic functions*, Indiana Univ. Math. J. **46** (1997), no. 3, 933–974.
10. S. Ya. Khavinson, *On an extremal problem of the theory of analytic functions*, Uspehi Matem. Nauk (N.S.) **4** (1949), no. 4(32), 158–159 (Russian).
11. ———, *Two papers on extremal problems in complex analysis*, Amer. Math. Soc. Transl. Ser. 2, vol. 129, Amer. Math. Soc., Providence, RI, 1986.

12. O. A. Ladyzhenskaya and N. N. Ural′tseva, *Linear and quasilinear elliptic equations*, Academic Press, New York, 1968.
13. J. L. Lewis, *On critical points of p harmonic functions in the plane*, Electron. J. Differential Equations **1994** (1994), no. 03, 1–4.
14. T. H. MacGregor and M. I. Stessin, *Weighted reproducing kernels in Bergman spaces*, Michigan Math. J. **41** (1994), no. 3, 523–533.
15. C. B. Morrey Jr., *On the solutions of quasi-linear elliptic partial differential equations*, Trans. Amer. Math. Soc. **43** (1938), no. 1, 126–166.
16. ———, *Second order elliptic equations in several variables and Hölder continuity*, Math. Z **72** (1959/1960), 146–164.
17. ———, *Multiple integrals in the calculus of variations*, Grundlehren Math. Wiss., vol. 130, Springer, New York, 1966.
18. Z. Nehari, *Conformal mapping*, McGraw-Hill, New York, Toronto, London, 1952.
19. K. Yu. Osipenko and M. I. Stesin, *Recovery problems in Hardy and Bergman spaces*, Mat. Zametki **49** (1991), no. 4, 95–104, 159 (Russian); English transl., Math. Notes **49** (1991), no. 3-4, 395–401.
20. W. W. Rogosinski and H. S. Shapiro, *On certain extremum problems for analytic functions*, Acta Math. **90** (1953), 287–318.
21. V. G. Ryabykh, *Certain extremal problems*, Nauchnye Soobscheniya R.G.U. (1965), 33–34 (Russian).
22. ———, *Extremal problems for summable analytic functions*, Sibirsk. Mat. Zh. **27** (1986), no. 3, 212–217, 226 (Russian).
23. H. S. Shapiro, *Some inequalities for analytic functions integrable over a plane domain*, Approximation and Function Spaces (Gdańsk, 1979), North-Holland, Amsterdam, 1981, pp. 645–666.
24. ———, *The Schwarz function and its generalization to higher dimensions*, Univ. Arkansas Lecture Notes Math. Sci., vol. 9, Wiley, New York, 1992.
25. D. Vukotić, *A sharp estimate for A^p_α functions in $\mathbf{C}^n$*, Proc. Amer. Math. Soc. **117** (1993), no. 3, 753–756.
26. ———, *Linear extremal problems for Bergman spaces*, Exposition. Math. **14** (1996), no. 4, 313–352.

Department of Mathematics and Statistics, University of South Florida, 4202 E. Fowler Ave., PHY114 , Tampa FL 33620, USA

E-mail address, C. Bénéteau: `cbenetea@usf.edu`

E-mail address, D. Khavinson: `dkhavins@usf.edu`

Centre de Recherches Mathématiques
CRM Proceedings and Lecture Notes
Volume **55**, 2012

A Unifying Construction for Measure-Valued Continuous and Discrete Branching Processes

Lucian Beznea, Oana Lupaşcu, and Andrei-George Oprina

ABSTRACT. We first emphasize conditions for a sub-Markovian resolvent of kernels to ensure that it is associated to a càdlàg Markov process. It turns out that this is a method of proving existence and path regularity for both continuous and discrete measure-valued branching Markov processes.

1. Introduction

Measure-valued branching processes are describing the time evolution of a system of particles, located in an Euclidean domain (or, more general, in a Lusin topological space) E: a spatial motion given by a Markov process with state space E is combined with a branching procedure.

Two types of semilinear partial differential equations are closely related to the branching processes: $\Delta u = u^\alpha$, with $1 < \alpha \leq 2$, and $\Delta u - u = -\sum_{k=1}^{\infty} q_k u^k$, where the coefficients q_k are positive and $\sum_{k=1}^{\infty} q_k \leq 1$. Instead of the Laplace operator we may consider the generator of a Markov process with state space E, the above mentioned spatial motion.

The associated branching processes have sets of measures as natural state spaces: the set $M(E)$ of all positive finite measures on E for the first equation (the continuous branching case, the processes are called *superprocesses*, in particular, when $\alpha = 2$, one obtains the super-Brownian motion) and the set S of all finite sums of Dirac measures on E (called the space of finite configurations of E) for the second equation (the discrete branching case). From the vast literature about the superprocesses we indicate the monographs [14, 15, 19, 20], the courses [12], and the articles [1, 16]. For the discrete branching processes we refer to the classical works of N. Ikeda, M. Nagasawa, S. Watanabe, and M. L. Silverstein (cf. [17, 18, 23]); a branching type process with state space S was constructed in [8].

2010 *Mathematics Subject Classification.* Primary 60J80, 60J45, 60J68; Secondary 60J35, 47D07.

Key words and phrases. Branching process, superprocess, excessive function, standard process, càdlàg trajectories, tight capacity.

The third named author acknowledges support from the Romanian Ministry of Education, Research, Youth and Sport (PN II Program, CNCSIS code 1186/2008).

This is the final form of the paper.

The first difficulty in the construction is the lack of the local compactness of the state spaces $M(E)$ and S of the forthcoming branching processes. The second one is the absence of a duality (reference) measure for the associated semigroup or resolvent family, in particular, the L^p and energy methods (e.g., the Dirichlet space theory) are not applicable.

The purpose of this paper is to show that it is possible to treat the construction and the proof of the path regularity for both continuous and discrete measure-valued branching Markov processes, using the same approach. Note that a nice link between the positive solutions of the two types of semilinear equations presented above was given in [13]. Our approach is based on new developments in constructing càdlàg strong Markov processes, given a resolvent of kernels; cf. [10]. The method uses analytic and probabilistic potential theory tools and it was tailored for finding martingale solutions for singular stochastic differential equations on Hilbert spaces and for proving the standardness property for infinite dimensional Lévy processes on Hilbert spaces (see [6]). It turns out that some of these techniques are applicable to the branching processes too, this observation was also followed in [9] for the continuous branching case. An example is given by the tightness of the capacity and the existence of the superharmonic functions having compact level sets (the so called compact Lyapunov functions), related to the path regularity of the process.

The organization of the paper is as follows. In Section 2 we present (Theorem 2.1) a version of a main result from [10] on the construction of standard processes associated to a resolvent of kernels, convenient for the announced applications. Section 3.1 is devoted to the branching kernels on $M(E)$ and S, and it is illustrated by typical examples. The application to the continuous branching case is contained in Section 3.2: Theorem 3.5 is a new approach for the existence and path regularity of the continuous branching processes studied in [1, 16]. The application to the discrete branching is stated in Section 3.3 (see [7] for more details). Connections between the discrete branching processes and the associated semilinear equations are exposed in Remark 3.7.

The first named author acknowledges enlightening discussions with P. J. Fitzsimmons, during the preparation of this paper.

2. Preliminaries on the resolvents of kernels and standard processes

Let E be a Lusin topological space (i.e., E is homeomorphic to a Borel subset of a compact metric space) with Borel σ-algebra $\mathcal{B}$. Let further $\mathcal{U} = (U_\alpha)_{\alpha>0}$ be a sub-Markovian resolvent of kernels on $(E, \mathcal{B})$.

We denote by $\mathcal{E}(\mathcal{U})$ the set of all $\mathcal{B}$-measurable $\mathcal{U}$-*excessive functions*: $u \in \mathcal{E}(\mathcal{U})$ if and only if u is a non-negative, numerical, $\mathcal{B}$-measurable function, $\alpha U_\alpha u \le u$ for all $\alpha > 0$ and $\lim_{\alpha\to\infty} \alpha U_\alpha u(x) = u(x)$ for all $x \in E$.

If $\beta > 0$, we denote by $\mathcal{U}_\beta$ the sub-Markovian resolvent of kernels $\mathcal{U}_\beta = (U_{\beta+\alpha})_{\alpha>0}$. If w is a $\mathcal{U}_\beta$-*supermedian function* (i.e., $\alpha U_{\beta+\alpha} w \le w$ for all $\alpha > 0$) then its $\mathcal{U}_\beta$-excessive regularization $\widehat{w}$ is given by $\widehat{w}(x) = \sup_\alpha \alpha U_{\beta+\alpha} w(x)$, $x \in E$.

We assume further in this section that for some $\beta > 0$ there exists a strictly positive constant k with $k \le U_\beta 1$ (in particular, this happens if the resolvent $\mathcal{U}$ is Markovian, i.e., $\alpha U_\alpha 1 = 1$).

Suppose that the resolvent $\mathcal{U}$ satisfies the following condition:

(h1) $\sigma(\mathcal{E}(\mathcal{U}_\beta)) = \mathcal{B}$ and all the points of E are non-branch points with respect to $\mathcal{U}_\beta$, that is $1 \in \mathcal{E}(\mathcal{U}_\beta)$ and if $u, v \in \mathcal{E}(\mathcal{U}_\beta)$ then for all $x \in E$ we have $\inf(u,v)(x) = \widehat{\inf(u,v)}(x)$.

Let $\mathrm{Exc}(\mathcal{U})$ be the set of all $\mathcal{U}$-*excessive measures* on E: $\xi \in \mathrm{Exc}(\mathcal{U})$ if and only if it is a σ-finite measure on $(E, \mathcal{B})$ such that $\xi \circ \alpha U_\alpha \leq \xi$ for all $\alpha > 0$. Recall that if $\xi \in \mathrm{Exc}(\mathcal{U})$ then actually $\xi \circ \alpha U_\alpha \nearrow \xi$ as $\alpha \to \infty$. We denote by $\mathrm{Pot}(\mathcal{U})$ the set of all *potential* $\mathcal{U}$-excessive measures: if $\xi \in \mathrm{Exc}(\mathcal{U})$ then $\xi \in \mathrm{Pot}(\mathcal{U})$ if $\xi = \mu \circ U$, where μ is a σ-finite measure on $(E, \mathcal{B})$.

If $\beta > 0$ then the *energy functional* $L_\beta\colon \mathrm{Exc}(\mathcal{U}_\beta) \times \mathcal{E}(\mathcal{U}_\beta) \to \overline{\mathbb{R}}_+$ is defined by

$$L_\beta(\xi, u) := \sup\{\nu(u) : \mathrm{Pot}(\mathcal{U}_\beta) \ni \nu \circ U_\beta \leq \xi\}.$$

For a family $\mathcal{G}$ of real valued functions on E we denote by $\sigma(\mathcal{G})$ (resp. by $\mathcal{T}(\mathcal{G})$) the σ-algebra (resp. the topology) generated by $\mathcal{G}$ and by $b\mathcal{G}$ (resp. $[\mathcal{G}]$) the subfamily of bounded functions from $\mathcal{G}$ (resp. the linear space spanned by $\mathcal{G}$).

We denote by $p\mathcal{B}$ the set of all positive $\mathcal{B}$-measurable functions on E.

Right and standard processes. A *transition function* on $(E, \mathcal{B})$ is a family $(p_t)_{t\geq 0}$ of kernels on $(E, \mathcal{B})$ which are sub-Markovian (i.e., $p_t 1 \leq 1$ for all $t \geq 0$), such that $p_0 f = f$ and $p_s(p_t f) = p_{s+t} f$ for all $s, t \geq 0$ and $f \in p\mathcal{B}$. We assume that for all $f \in p\mathcal{B}$ the function $(t, x) \mapsto p_t f(x)$ is $\mathcal{B}([0,\infty)) \otimes \mathcal{B}$-measurable. We denote by $\mathcal{U} = (U_\alpha)_{\alpha>0}$ the family of kernels on $(E, \mathcal{B})$ given by

$$U_\alpha f = \int_0^\infty \mathrm{e}^{-\alpha t}\, p_t f \,\mathrm{d}t.$$

Consequently, $\mathcal{U} = (U_\alpha)_{\alpha>0}$ is a sub-Markovian resolvent of kernels on $(E, \mathcal{B})$ and it is called *associated with* $(p_t)_{t\geq 0}$.

A *right process with state space* E (associated with the transition function $(p_t)_{t\geq 0}$) is a collection $X = (\Omega, \mathcal{G}, \mathcal{G}_t, X_t, \theta_t, P^x)$ where: $(\Omega, \mathcal{G})$ is a measurable space, $(\mathcal{G}_t)_{t\geq 0}$ is a family of sub σ-algebras of $\mathcal{G}$ such that $\mathcal{G}_s \subseteq \mathcal{G}_t$ if $s \leq t$; for all $t \geq 0$, $X_t\colon \Omega \to E_\Delta$ is a $\mathcal{G}_t/\mathcal{B}_\Delta$-measurable map such that $X_t(\omega) = \Delta$ for all $t > t_0$ if $X_{t_0}(\omega) = \Delta$, where Δ is a cemetery state adjoined to E as an isolated point of $E_\Delta := E \cup \{\Delta\}$ and $\mathcal{B}_\Delta$ is the Borel σ-algebra on E_Δ; we set $\zeta(\omega) := \inf\{t : X_t(\omega) = \Delta\}$; for each $t \geq 0$, the map $\theta_t\colon \Omega \to \Omega$ is such that $X_s \circ \theta_t = X_{s+t}$ for all $s > 0$; for all $x \in E_\Delta$, P^x is a probability measure on $(\Omega, \mathcal{G})$ such that $x \mapsto P^x(F)$ is universally $\mathcal{B}$-measurable for all $F \in \mathcal{G}$; $E^x(f \circ X_0) = f(x)$ and the following *Markov property* holds:

$$E^x(f(X_{s+t}) \cdot G) = E^x(p_t^\Delta f(X_s) \cdot G)$$

for all $x \in E_\Delta$, $s, t \geq 0$, $f \in p\mathcal{B}_\Delta$ and $G \in p\mathcal{G}_s$, where p_t^Δ is the Markovian kernel on $(E_\Delta, \mathcal{B}_\Delta)$ such that $p_t^\Delta 1 = 1$ and $p_t^\Delta|_E = p_t$; for all $\omega \in \Omega$ the function $t \mapsto X_t(\omega)$ is right continuous on $[0,\infty)$; the filtration $(\mathcal{G}_t)_{t\geq 0}$ is *right continuous* (i.e., $\mathcal{G}_t = \mathcal{G}_{t+} := \bigcap_{s>t} \mathcal{G}_s$) and *augmented* (i.e., $\mathcal{G}_t = \widetilde{\mathcal{G}}_t := \bigcap_\mu \mathcal{G}_t^\mu$, where for every probability measure μ on $(E, \mathcal{B})$, $\mathcal{G}^\mu$ is the completion of $\mathcal{G}$ with respect to the probability measure $P^\mu := \int P^x \mu(\mathrm{d}x)$ on $(\Omega, \mathcal{G})$ and $\mathcal{G}_t^\mu$ is the completion of $\mathcal{G}_t$ in $\mathcal{G}^\mu$ with respect to P^μ); we assume that for all $\alpha > 0$, every function u which is α-excessive with respect to the resolvent associated with $(p_t)_{t\geq 0}$ and each probability measure μ on $(E, \mathcal{B})$, the function $t \mapsto u \circ X_t$ is right continuous on $[0,\infty)$ P^μ-a.s.

We consider the *natural filtration* associated with X: $\mathcal{F} := \widetilde{\mathcal{F}}^0$, $\mathcal{F}_t := \widetilde{\mathcal{F}}_t^0$, where $\mathcal{F}^0 := \sigma(X_s \mid s < \infty)$, $\mathcal{F}_t^0 := \sigma(X_s \mid s \le t)$. It is known that always a right process may be considered with respect to its natural filtration:

$$X = (\Omega, \mathcal{F}, \mathcal{F}_t, X_t, \theta_t, P^x).$$

The sub-Markovian resolvent $\mathcal{U} = (U_\alpha)_{\alpha>0}$ associated with $(p_t)_{t\ge 0}$ is called the *resolvent of the process* X and for all $f \in p\mathcal{B}$, $\alpha > 0$ and $x \in E$ we have

$$U_\alpha f(x) = E^x \int_0^\zeta \mathrm{e}^{-\alpha t} f(X_t)\,\mathrm{d}t,$$

with the convention $f(\Delta) = 0$.

A *stopping time* is a map $T\colon \Omega \to \overline{\mathbb{R}}_+$ such that the set $[T \le t]$ belongs to $\mathcal{F}_t$ for all $t \ge 0$.

Recall that a right process $X = (\Omega, \mathcal{F}, \mathcal{F}_t, X_t, \theta_t, P^x)$ with state space E is called *standard* if for every finite measure μ on $(E, \mathcal{B})$ X has *càdlàg trajectories* under P^μ, i.e., it possesses left limits in E P^μ-a.e. on $[0, \zeta)$ and X is *quasi-left continuous up to* ζ P^μ-a.e., i.e., for every increasing sequence $(T_n)_n$ of stopping times with $T_n \nearrow T$ we have $X_{T_n} \to X_T$ P^μ-a.e. on $[T < \zeta]$; ζ being the life time of X.

For more details on right and standard processes see, e.g., [22].

The following result is a consequence of [10, Theorem 5.2, Corollary 5.3(ii), and Theorem 5.5(i)]. It turns out that the next theorem is convenient for the claimed applications to the measure-valued processes we shall present in the next section.

Theorem 2.1. *Suppose that* $\mathcal{U} = (U_\alpha)_{\alpha>0}$ *is a sub-Markovian resolvent of kernels on* $(E, \mathcal{B})$ *such that* (h1) *and the following two conditions are satisfied.*

(h2) *There exists a countable subset* $\mathcal{F}$ *of* $[b\mathcal{E}(\mathcal{U}_\beta)]$ *generating the topology of* E, $1 \in \mathcal{F}$, *and there exists* $u_o \in \mathcal{E}(\mathcal{U}_\beta)$, $u_o < \infty$, *such that if* ξ, η *are two finite* $\mathcal{U}_\beta$*-excessive measures with* $L_\beta(\xi + \eta, u_o) < \infty$ *and* $L_\beta(\xi, \varphi) = L_\beta(\eta, \varphi)$ *for all* $\varphi \in \mathcal{F}$, *then* $\xi = \eta$.

(h3) *For every* $x \in E$ *there exists* $v_x \in \mathcal{E}(\mathcal{U}_\beta)$ *such that* $v_x(x) < \infty$ *and the set* $[v_x \le n]$ *is relatively compact for all* n; *such a function* v_x *is called* compact Lyapunov function.

Then the following assertions hold.

(i) *There exists a càdlàg process* X *with state space* E *such that* $\mathcal{U}$ *is the resolvent of* X, *i.e., for all* $\alpha > 0$, $f \in bp\mathcal{B}$, *and* $x \in E$ *we have* $U_\alpha f(x) = E^x \int_0^\infty \mathrm{e}^{-\alpha t} f(X_t)\,\mathrm{d}t$.

(ii) *Assume in addition that* $\mathcal{U} = (U_\alpha)_{\alpha>0}$ *is the resolvent of kernels associated with a Markovian transition function* $(p_t)_{t\ge 0}$ *on* $(E, \mathcal{B})$ *and there exists a family* $\mathcal{K}$ *of bounded, continuous, real-valued functions on* E *which is multiplicative* (*i.e., if* $f, g \in \mathcal{K}$ *then* $fg \in \mathcal{K}$) *and generates* $\mathcal{B}$, *such that* $p_t f$ *is a continuous function for all* $f \in \mathcal{K}$. *Then the càdlàg process* X, *having the resolvent* $\mathcal{U}$, *is standard.*

Remark 2.2. (i) Conditions (h1) and (h3) are necessary in order to deduce that $\mathcal{U}$ is the resolvent family of a standard process. For the necessity of (h1) see [2, 5], while the necessity of (h3) is a particular case of [10, Remark (i) following Theorem 5.2]; for the reader's convenience we give some more details below.

(ii) Condition (h2) is a version of [10, (H7)] and it is an "uniqueness of charges type property." However, (h2) is slightly modified in the sense that in (H7) the function u_o has a particular form, $u_o = U_\beta g$ with $g \in p\mathcal{B}$, hence $L_\beta(\xi, u_o) = \int g\, d\xi$.

Before proving the theorem we present some necessary preliminaries on the tightness of the capacity induced by $\mathcal{U}$ and the enlargement of the space E.

Tight capacities. Assume that $\mathcal{U} = (U_\alpha)_{\alpha>0}$ is a sub-Markovian resolvent of kernels on $(E, \mathcal{B})$ satisfying condition (h1). If $M \in \mathcal{B}$ and $u \in \mathcal{E}(\mathcal{U}_\beta)$, then the *reduced function* (with respect to $\mathcal{U}_\beta$) *of u on M* is the function $R^M_\beta u$ defined by

$$R^M_\beta u := \inf\{v \in \mathcal{E}(\mathcal{U}_\beta) : v \geq u \text{ on } M\}.$$

The reduced function $R^M_\beta u$ is a universally $\mathcal{B}$-measurable $\mathcal{U}_\beta$-supermedian function.

Let further $u_0 := U_\beta f_0$, where f_0 is a bounded, strictly positive $\mathcal{B}$-measurable function, and fix a finite measure λ on $(E, \mathcal{B})$. Consider the functional $M \mapsto c_\lambda(M)$, $M \subset E$, defined as

$$c_\lambda(M) := \inf\{\lambda(R^G_\beta u_0) : G \text{ open}, M \subset G\}.$$

By [2] c_λ is a Choquet capacity on E.

Statement 2.3. *The following assertions are equivalent* (*see* [9, Prop. 2.1.1]):

(i) *The capacity c_λ is* tight, *i.e., there exists an increasing sequence $(K_n)_n$ of compact sets such that* $\inf_n c_\lambda(E \setminus K_n) = 0$.

(ii) *There exists a $\mathcal{U}_\beta$-excessive function v which is λ-integrable such that the set $[v \leq \alpha]$ is relatively compact for all $\alpha > 0$.* (*We say that v* has compact level sets.)

Proof of Remark 2.2, the necessity of (h3). Assume that $\mathcal{U}$ is the resolvent of a standard process. Because in particular the trajectories are càdlàg, it follows from [21] (see also [3]) that the capacity c_λ is tight. The fact that (h3) holds is now a consequence of Statement 2.3. □

Enlargement of the space. There exist a second Lusin measurable space $(E_1, \mathcal{B}_1)$ such that $E \subset E_1$, $E \in \mathcal{B}_1$, $\mathcal{B} = \mathcal{B}_1|_E$, and a resolvent of kernels $\mathcal{U}^1 = (U^1_\alpha)_{\alpha>0}$ on $(E_1, \mathcal{B}_1)$ satisfying (h1) on this larger space, $U^1_\beta(1_{E_1 \setminus E}) = 0$, $\mathcal{U}$ is the restriction of $\mathcal{U}^1$ to E (i.e., $U_\beta g = U^1_\beta(g^1)|_E$, where $g^1 \in p\mathcal{B}_1$ and $g^1|_E = g$). We clearly have $\mathrm{Exc}(\mathcal{U}_\beta) = \mathrm{Exc}(\mathcal{U}^1_\beta)$ and the following property holds (for one and therefore for all $\beta > 0$):

Statement 2.4. *Every $\xi \in \mathrm{Exc}(\mathcal{U}^1_\beta)$ with $L_\beta(\xi, 1) < \infty$ is a potential on E_1* (*with respect to $\mathcal{U}^1_\beta$*).

One can take for E_1 the set of all extreme points of the set $\{\xi \in \mathrm{Exc}(\mathcal{U}_\beta) : L_\beta(\xi, 1) = 1\}$, endowed with the σ-algebra $\mathcal{B}_1$ generated by the functionals $\tilde{u}$, $\tilde{u}(\xi) := L_\beta(\xi, u)$ for all $\xi \in E_1$ and $u \in \mathcal{E}(\mathcal{U}_\beta)$. Let $(E', \mathcal{B}')$ be a Lusin measurable space such that $E \subset E'$, $E \in \mathcal{B}'$, $\mathcal{B} = \mathcal{B}'|_E$, and there exists a proper sub-Markovian resolvent of kernels $\mathcal{U}' = (U'_\alpha)_{\alpha>0}$ on $(E', \mathcal{B}')$ stisfying (h1) on E', $U'_\beta(1_{E' \setminus E}) = 0$, Statement 2.4 holds with respect to $\mathcal{U}'$ on E', and $\mathcal{U}$ is the restriction of $\mathcal{U}'$ to E. Then the map $x \mapsto \varepsilon_x \circ U'_\beta$ is a measurable isomorphism between $(E', \mathcal{B}')$ and the measurable space $(E_1, \mathcal{B}_1)$.

Every excessive functions has an extension from E to E_1 as follows. If $\xi = \mu \circ U_\beta \in \mathrm{Pot}(\mathcal{U}_\beta)$ and $u \in \mathcal{E}(\mathcal{U}_\beta)$ then by [2, Theorem 1.4.5] we have

$$L_\beta(\xi, u) = \int u \, \mathrm{d}\mu. \tag{2.1}$$

For every $u \in \mathcal{E}(\mathcal{U}_\beta)$ we consider the function $\tilde{u}\colon E_1 \to \overline{\mathbb{R}}_+$ defined above,

$$\tilde{u}(\xi) := L_\beta(\xi, u), \quad \xi \in E_1.$$

Then by (2.1) we have $\tilde{u}(\varepsilon_x \circ U_\beta) = u(x)$ for all $x \in E$ and therefore, by the embedding of E in E_1,

$$\tilde{u}|_E = u.$$

In addition, $\tilde{u}$ is $\mathcal{U}^1_\beta$-excessive and it is the (unique) extension by fine continuity of u from E to E_1.

PROOF OF THEOREM 2.1. (i) Let us fix $x \in E$ and set $\lambda := \delta_x$. From (h3) and Statement 2.3 we deduce that there exists an increasing sequence $(K_n)_n$ of compact subsets of E such that $\inf_n c_\lambda(E \setminus K_n) = 0$.

We can argue now as in proof of [10, Corollary 5.3(ii)] to deduce that $\mathcal{U}$ is the resolvent of a right process with càdlàg trajectories. Recall that a main step in that proof is to show that the set $E_1 \setminus E$ is polar, by proving that it is λ-polar for every finite measure λ.

The property of $E_1 \setminus E$ to be a polar set still holds under our (weaker) hypothesis (h3) which ensures the tightness of the capacity c_λ only for $\lambda = \delta_x$ for all $x \in E$. Consequently the set $E_1 \setminus E$ is δ_x-polar for every $x \in E$, hence polar.

(ii) In order to deduce that the càdlàg process X is quasi-left continuous (hence standard), we can modify the proof of [10, Theorem 5.5(ii)]. More precisely, instead of using the fact that $U_\alpha f$ is a continuous function on E we can use the continuity of $p_t f$, $f \in \mathcal{K}$. Note also that $\lim_{\alpha\to\infty} \alpha U_\alpha f = f$ pointwise on E because f is continuous. □

3. Continuous and discrete branching processes

Let $M(E)$ be the set of all positive finite measures on E.

For a function $f \in p\mathcal{B}$ we shall consider the mappings $l_f\colon M(E) \to \overline{\mathbb{R}}_+$ and $e_f\colon M(E) \to [0,1]$, defined by

$$l_f(\mu) := \langle \mu, f \rangle := \int f \, \mathrm{d}\mu, \quad \mu \in M(E), e_f := \exp(-l_f).$$

$M(E)$ is endowed with the weak topology and note that the Borel σ-algebra $\mathcal{M}(E)$ on $M(E)$ is generated by $\{l_f : f \in bp\mathcal{B}\}$.

A second set of measures as state space for a forthcoming discrete branching process will be the set S of all positive measures μ on E which are finite sums of Dirac measures: $\mu = \sum_{k=1}^m \delta_{x_k}$, where $x_1, \dots, x_m \in E$, called the space of *finite configurations* of E (cf. [23]). The space S of all finite configurations of E is identified with the direct sum of all symmetric mth powers $E^{(m)}$ of E, hence

$$S = \bigoplus_{m\geq 1} E^{(m)};$$

$E^{(m)}$ is the factorization of the Cartesian product E^m by the equivalence relation induced by the permutation group σ^m; for details see, e.g., [8, 17]. The space $E^{(m)}$ is endowed with the quotient topology, where E^m is equipped with the product topology. The space S is equipped with the canonical topological structure and we denote by $\mathcal{B}(S)$ the Borel σ-algebra on S.

3.1. Branching kernels. Recall that if p_1, p_2 are two finite measures on $M(E)$, then their convolution $p_1 * p_2$ is the finite measure on $M(E)$ defined for every $F \in bp\mathcal{M}(E)$ by

$$\int p_1 * p_2(d\nu)F(\nu) := \int p_1(\mathrm{d}\nu_1) \int p_2(\mathrm{d}\nu_2)F(\nu_1 + \nu_2).$$

In particular, if $f \in p\mathcal{B}$ then

$$p_1 * p_2(e_f) = p_1(e_f)p_2(e_f). \tag{3.1}$$

Note that if p_1 and p_2 are concentrated on S then $p_1 * p_2$ has the same property.

According with [23], a kernel N on $(S, \mathcal{B}(S))$ (resp. on $(M(E), \mathcal{M}(E))$) which is sub-Markovian (i.e., $N1 \le 1$) is called *branching kernel* provided that for all $\mu, \nu \in S$ (resp. for all $\mu, \nu \in M(E)$) we have

$$N_{\mu+\nu} = N_\mu * N_\nu,$$

where N_μ denotes the measure on $(S, \mathcal{B}(S))$ (resp. on $(M(E), \mathcal{M}(E))$) such that $Ng(\mu) = \int g\, dN_\mu$ for all $g \in bp\mathcal{B}(S)$ (resp. $g \in bp\mathcal{M}(E)$).

A right (Markov) process with state space $M(E)$ or S is called *branching process* provided that its transition function is formed by branching kernels. For the probabilistic interpretation of this analytic branching property see, e.g., [16, p. 337].

Example of branching kernels on $M(E)$. We set $\mathcal{S} := bp\mathcal{B}$. Recall that a function $\varphi \colon \mathcal{S} \to \mathbb{R}$ is called *negative definite* provided that for all $n \ge 2$, $\{f_1, f_2, \dots, f_n\} \subset \mathcal{S}$, and $\{a_1, a_2, \dots, a_n\} \subset \mathbb{R}$ with $\sum_{i=1}^n a_i = 0$ we have

$$\sum_{i,j} a_i a_j \varphi(f_i + f_j) \le 0.$$

Statement 3.1. *Let $\varphi \colon \mathcal{S} \to \mathbb{R}_+$ be a negative definite function such that if $(f_n)_n$ is pointwise decreasing to f then $\big(\varphi(f_n)\big)_n$ is decreasing to $\varphi(f)$. Then there exists a unique finite measure $\overline{P}$ on $\big(M(E), \mathcal{M}(E)\big)$ such that*

$$\overline{P}(e_f) = \exp\big(-\varphi(f)\big) \quad \textit{for all } f \in bp\mathcal{B}$$

(see [16, (A.6)]*).*

Let now $V \colon bp\mathcal{B} \to bp\mathcal{B}$ be a map such that for every $x \in E$ the function $f \mapsto Vf(x)$ is negative definite. Applying Statement 3.1, there exists a unique kernel Q on $\big(M(E), \mathcal{M}(E)\big)$ such that

$$Q(e_f) = e_{Vf} \quad \text{for all } f \in bp\mathcal{B} \tag{3.2}$$

(see [1, Proposition 4.1]). We claim that Q is a branching kernel on $M(E)$. Indeed, we have to prove that for all $\mu, \nu \in M(E)$

$$Q_{\mu+\nu} = Q_\mu * Q_\nu.$$

By a monotone class argument it is sufficient to check the above equality for functions on $M(E)$ of the form e_f, $f \in bp\mathcal{B}$. Using (3.1) and (3.2) we then have

$$Q_\mu * Q_\nu(e_f) = Q_\mu(e_f) \cdot Q_\nu(e_f) = e_{Vf}(\mu) \cdot e_{Vf}(\nu) = Q_{\mu+\nu}(e_f).$$

Multiplicative functions and branching kernels on S. For every real-valued $\mathcal{B}$-measurable function φ consider the function $\widehat{\varphi}\colon \bigoplus_{m\geq 1} E^m \to \mathbb{R}$ defined as

$$\widehat{\varphi}(x) := \varphi(x_1) \cdot \dots \cdot \varphi(x_m) \quad \text{for } x = (x_1, \dots, x_m) \in E^m.$$

Note that the restriction of $\widehat{\varphi}$ to each E^m is invariant under the action of σ^m. Therefore $\widehat{\varphi}$ may be considered as a real-valued function defined on S. Such a function $\widehat{\varphi}\colon S \to \mathbb{R}$ is called *multiplicative* (cf. [23]; see also [8]).

With this notation a multiplicative function $\widehat{\varphi}$, $\varphi \in p\mathcal{B}$, $\varphi \leq 1$, is the restriction to S of an exponential function on $M(E)$,

$$\widehat{\varphi} = e_{-\ln\varphi}.$$

Statement 3.2. *For every sub-Markovian kernel $B\colon p\mathcal{B}(S) \to p\mathcal{B}$ there exists a branching kernel $\widehat{B}$ on $\big(S, \mathcal{B}(S)\big)$ such that for every $\mathcal{B}$-measurable function φ, $|\varphi| \leq 1$, we have*

$$\widehat{B}\widehat{\varphi} = \widehat{B\widehat{\varphi}}.$$

Conversely, if H is a branching kernel on $\big(S, \mathcal{B}(S)\big)$ then there exists a unique sub-Markovian kernel $B\colon p\mathcal{B}(S) \to p\mathcal{B}$ such that $H = \widehat{B}$ (see [8, Proposition 3.2]*).*

Example of branching kernels on S. Let $q_k \in p\mathcal{B}$, $k \geq 1$, satisfying $\sum_{k\geq 1} q_k \leq 1$. Consider the kernel $B\colon p\mathcal{B}(S) \to p\mathcal{B}$ defined as

$$Bg(x) := \sum_{k\geq 1} q_k(x) g_k(x, \dots, x),$$

$g \in bp\mathcal{B}(S)$, $x \in E$, with $g_k := g|_{E^{(k)}}$ for all $k \geq 1$. Using Statement 3.2 there exists a branching kernel $\widehat{B}$ on S such that for all $\varphi \in p\mathcal{B}$, $\varphi \leq 1$, we have

$$\widehat{B}\widehat{\varphi}|_E = \sum_{k\geq 1} q_k \varphi^k.$$

3.2. Continuous branching. Let $X = (\Omega, \mathcal{F}, \mathcal{F}_t, X_t, \theta_t, P^x)$ be a fixed Borel right Markov process with state space E and assume that the transition function $(T_t)_{t\geq 0}$ of X is Markovian (i.e., $T_t 1 = 1$).

We also fix a *branching mechanism*, that is, a function $\Phi\colon E \times [0, \infty) \to \mathbb{R}$ of the form

$$\Phi(x, \lambda) = -b(x)\lambda - c(x)\lambda^2 + \int_0^\infty (1 - \mathrm{e}^{-\lambda s} - \lambda s) N(x, \mathrm{d}s),$$

where $c \geq 0$ and b are bounded $\mathcal{B}$-measurable functions and $N\colon p\mathcal{B}\big((0,\infty)\big) \to p\mathcal{B}$ is a kernel such that $N(u \wedge u^2) \in bp\mathcal{B}$. Examples of branching mechanisms are $\Phi(\lambda) = -\lambda^\alpha$ for $1 < \alpha \leq 2$.

For each $f \in bp\mathcal{B}$ the equation

$$v_t(x) = T_t f(x) + \int_0^t T_s\big(x, \Phi(\cdot, v_{t-s})\big)\, \mathrm{d}s, \quad t \geq 0, x \in E,$$

has a unique solution $(t, x) \mapsto V_t f(x)$ jointly measurable in (t, x) such that for all $t > 0$ $\sup_{0\leq s\leq t} \|v_s\|_\infty < \infty$. The mapping $f \mapsto V_t f$, $t \geq 0$, form a nonlinear semigroup of operators on $bp\mathcal{B}$ and for each $x \in E$ the function $f \mapsto V_t f(x)$ is negative definite (see [1, Section 3, Proposition 3.2; 16]). Note that the above equation is formally equivalent with

$$\begin{cases} \frac{\mathrm{d}}{\mathrm{d}t} v_t(x) = \mathrm{L} v_t(x) + \Phi\big(x, v_t(x)\big) \\ v_0 = f, \end{cases}$$

where L is the infinitesimal generator of the semigroup $(T_t)_{t\geq 0}$.

The negative definiteness of the mapping $f \mapsto V_t f$ and (3.2) imply that for each $t \geq 0$ there exists a unique kernel Q_t on $\big(M(E), \mathcal{M}(E)\big)$ such that

$$Q_t(e_f) = e_{V_t f}, \quad f \in bp\mathcal{B}. \tag{3.3}$$

Since the family $(V_t)_{t\geq 0}$ is a (nonlinear) semigroup on $bp\mathcal{B}$, it follows that $(Q_t)_{t\geq 0}$ is a (linear) semigroup of kernels on $\big(M(E), \mathcal{M}(E)\big)$.

In the sequel, to simplify the notations, we assume that $b = 0$. This is possible by using the Feynman – Kac formula to perturb conveniently $(T_t)_{t\geq 0}$. At the level of the infinitesimal generator this transformation means to replace L by $\mathrm{L} - b'$, where $b' := b + \beta'$, with $\beta' > \|b^-\|_\infty$; see [1, p. 695].

Let $\overline{\mathcal{U}} = (\overline{U}_\alpha)_{\alpha>0}$ be the Markovian resolvent of kernels on $\big(M(E), \mathcal{M}(E)\big)$ generated by the semigroup $(Q_t)_{t\geq 0}$ given by (3.3) and denote by $\bar{L}_\beta$ the energy functional with respect to $\overline{\mathcal{U}}_\beta$.

There are two classes of $\overline{\mathcal{U}}_\beta$-excessive functions which are produced by the $\mathcal{U}_\beta$-excessive functions: the linear and the exponential ones; for details see [1, Section 4]. More precisely, the following assertion holds.

Statement 3.3. *If $u \in \mathcal{E}(\mathcal{U}_\beta)$ then l_u, $1 - e_u$, and $1 - e_{V_t u}$, $t \geq 0$, are $\overline{\mathcal{U}}_\beta$-excessive functions.*

Ray cones. If $\beta > 0$ then a *Ray cone* associated with $\mathcal{U}_\beta$ is a cone $\mathcal{R}$ of bounded $\mathcal{U}_\beta$-excessive functions such that: $U_\alpha(\mathcal{R}) \subset \mathcal{R}$ for all $\alpha > 0$, $U_\beta\big((\mathcal{R} - \mathcal{R})_+\big) \subset \mathcal{R}$, $\sigma(\mathcal{R}) = \mathcal{B}$, it is min-stable, separable in the supremum norm and $1 \in \mathcal{R}$. Such a Ray cone always exists. Below if we say Ray cone it is always meant to be associated with one fixed resolvent $\mathcal{U}_\beta$. If $\mathcal{U}$ is transient (i.e., there exists a strictly positive function $f_{\mathrm{o}} \in bp\mathcal{B}$ with $\sup_\alpha U_\alpha f_{\mathrm{o}} < \infty$, then one can take $\beta = 0$, that is, there is a Ray cone of $\mathcal{U}$-excessive functions). A *Ray topology* on E is a topology generated by a Ray cone. For more details see, e.g., [2, Chapter 1] and also [4] for the non-transient case.

We can state now the application of Theorem 2.1 to the continuous branching case. One obtains in this way a new approach for the existence and path regularity of the continuous branching processes studied in [1, Theorem 4.9(ii); 16]. As in [1] we need a regularity assumption:

Statement 3.4. *There exists a convex cone $\mathcal{C} \subset b\mathcal{E}(\mathcal{U}_\beta)$ which is separable in the supremum norm, inf-stable, separates the points of E, and there exists a sequence $(u_n)_n \subset \mathcal{C}$, $u_n \nearrow 1$, such that for all $t \geq 0$ $V_t(\mathcal{C})$ is a subset of the closure in the supremum norm of $[\mathcal{C}]$.*

By [1, Proposition 4.8] each of the following two conditions implies that Statement 3.4 holds:

- b, c and N do not depend on $x \in E$;
- $(T_t)_{t\geq 0}$ is a Feller semigroup (on the locally compact space E) and for all $t \geq 0$ $V_t\big(C_0(E)\big) \subset C_0(E)$.

Theorem 3.5. *If X is a standard process and Statement* 3.4 *holds, then there exists a branching standard process (called* (X, Φ)-superprocess*) with state space* $M(E)$ *endowed with the weak topology, having* $(Q_t)_{t\geq 0}$ *as transition function.*

PROOF. We have to verify conditions (h1) – (h3) for $\overline{\mathcal{U}}_\beta$. By [1, Proposition 4.5] (h1) holds for $\overline{\mathcal{U}}_\beta$.

We check now condition (h3). Theorem (47.10) in [22] implies that X has càdlàg trajectories in any Ray topology. Consider a Ray topology $\mathcal{T}(\mathcal{R})$ which is finer than the original topology and it is generated by a Ray cone $\mathcal{R}$, $\overline{\mathcal{R}} \supset \mathcal{C}$. We endow $M(E)$ with the weak Ray topology (cf. [16, p. 343]). Let $\lambda \in M(E)$. Remark (i), following [10, Theorem 5.2], implies the existence of a function $v \in \mathcal{E}(\mathcal{U}_\beta) \cap L^1(E, \lambda)$ having $\mathcal{T}(\mathcal{R})$-compact level sets. By Statement 3.3 $l_v \in \mathcal{E}(\overline{\mathcal{U}}_\beta)$ and it follows from [11] that l_v has compact level sets in the weak Ray topology. Clearly we have $l_v(\lambda) = \int v \,\mathrm{d}\lambda < \infty$ and we conclude that (h3) holds.

It remains to show that (h2) also holds for $\overline{\mathcal{U}}_\beta$. We take l_1 as the function u_o; by Statement 3.3 we have $l_1 \in \mathcal{E}(\overline{\mathcal{U}}_\beta)$ and clearly $l_1 < \infty$. Let $\mathcal{C}_o$ be a countable dense subset of $\mathcal{C}$. We may suppose that if $u, v \in \mathcal{C}_o$ then $u + v$ also belongs to $\mathcal{C}_o$. Let further $\mathcal{R}_o$ be a countable dense subset of $\mathcal{R}$ such that $\mathcal{C}_o \subset \mathcal{R}_o$ and $1 \in \mathcal{R}_o$. The countable set $\mathcal{F}$ from (h2) is defined as

$$\mathcal{F} := \{e_u : u \in \mathcal{R}_o\}$$

and observe that by Statement 3.3 we have $\mathcal{F} \subset [b\mathcal{E}(\overline{\mathcal{U}}_\beta)]$. Let further ξ, η be two finite $\overline{\mathcal{U}}_\beta$-excessive measures such that $\bar{L}_\beta(\xi, e_u) = \bar{L}_\beta(\eta, e_u)$ for all $u \in \mathcal{R}_o$ and

$$\bar{L}_\beta(\xi + \eta, l_1) < \infty. \tag{3.4}$$

To show that $\xi = \eta$ we can now proceed as in the proof of [1, Theorem 4.9, Step I, p. 699]. Because the σ-algebra $\mathcal{M}(E)$ is generated by $\mathcal{K} := \{e_u : u \in \mathcal{C}_o\}$, a monotone class argument implies that $\xi = \eta$ provided that

$$\xi(e_u) = \eta(e_u) \quad \text{for all } u \in \mathcal{C}_o.$$

By [1, Remark 3.6] the above equality holds if

$$\bar{L}_\beta\big(\xi, Q_t(e_u)\big) = \bar{L}_\beta\big(\eta, Q_t(e_u)\big) \quad \text{for all } u \in \mathcal{C}_o \text{ and } t \geq 0. \tag{3.5}$$

Let $M(E)_1$ be the enlargement of $M(E)$ as described in Section 2. Since $\bar{L}_\beta(\xi + \eta, 1) < \infty$, by Statement 2.4 there exist two measures μ and ν on $M(E)_1$ such that $\xi = \mu \circ \overline{U}_\beta$ and $\eta = \nu \circ \overline{U}_\beta$. Note that every $f \in [b\mathcal{E}(\overline{\mathcal{U}}_\beta)]$ has a finely continuous extension $\tilde{f}$ to $M(E)_1$. Let $\tilde{\mathcal{A}}$ be the linear space of bounded functions on $M(E)_1$ spanned by the set $\{\tilde{e}_u : u \in \mathcal{C}_o\}$. Because $\tilde{\mathcal{A}}$ is an algebra and $\mu(\tilde{e}_u) = \bar{L}_\beta(\xi, e_u) = \bar{L}_\beta(\eta, e_u) = \nu(\tilde{e}_u)$ for every $u \in \mathcal{C}_o$, by the monotone class theorem we have

$$\mu(F) = \nu(F) \quad \text{for all } F \in \sigma(\tilde{\mathcal{A}}). \tag{3.6}$$

If $u \in \mathcal{C}_o$ then by Statement 3.4 there exists a sequence $(f_n)_n \subset [\mathcal{C}_o]$ converging uniformly to $V_t u$. As a consequence, for every $\lambda \in M(E)$ we have $|e_{f_n}(\lambda) - e_{V_t u}(\lambda)| \leq \|f_n - V_t u\|_\infty \cdot l_1(\lambda)$, hence $|\tilde{e}_{f_n} - \tilde{e}_{V_t u}| \leq \|f_n - V_t u\|_\infty \cdot \tilde{l}_1$ on $M(E)_1$. It follows that $(\tilde{e}_{f_n})_n$ converges pointwise to $\tilde{e}_{V_t u}$ on the set $[\tilde{l}_1 < \infty] \in \sigma(\tilde{\mathcal{A}})$. From (3.4) we get $(\mu+\nu)(\tilde{l}_1) = \bar{L}_\beta(\xi+\eta, l_1) < \infty$, hence $\tilde{l}_1 < \infty$ $(\mu+\nu)$-a.e. Therefore, $1_{[\tilde{l}_1<\infty]} \cdot \tilde{e}_{V_t u}$ is $\sigma(\tilde{\mathcal{A}})$-measurable and by (3.6) we now deduce that $\mu(\tilde{e}_{V_t u}) = \nu(\tilde{e}_{V_t u})$ for all $u \in \mathcal{C}_o$. Using (3.3) we conclude that (3.5) holds, so $\xi = \eta$.

Because $\mathcal{C}_o \subset \mathcal{R}$, it follows that $\mathcal{K}$ is a family of bounded real-valued continuous (in the weak Ray topology) functions on $M(E)$ which is multiplicative and we already observed that $\mathcal{K}$ generates the σ-algebra $\mathcal{M}(E)$. In addition, if $u \in \mathcal{C}_o$ then by Statement 3.4 and since $Q_t(e_u) = e_{V_t u}$, we deduce that $Q_t(e_u)$ is also a weak Ray continuous function on $M(E)$ for all $t \geq 0$. Hence the hypothesis of assertion (ii) in Theorem 2.1 is also verified.

Applying Theorem 2.1 we deduce that $\overline{\mathcal{U}}_\beta$ is the resolvent of a standard process, the (X, Φ)-superprocess, with state space $M(E)$, endowed with the weak Ray topology. By [16, Corollary (3.6)] it follows that the (X, Φ)-superprocess is standard in the weak topology too. □

3.3. Discrete branching. As in Section 3.2 let $X = (\Omega, \mathcal{F}, \mathcal{F}_t, X_t, \theta_t, P^x)$ be a fixed Borel right process with state space E and suppose that the transition function of X is Markovian.

Let $B\colon bp\mathcal{B}(S) \to bp\mathcal{B}$ be a sub-Markovian kernel such that $\sup_{x\in E} Bl_1(x) < \infty$. We also fix a function $c \in bp\mathcal{B}$ and denote by $(T_t)_{t\geq 0}$ the transition function of the process obtained by killing X with the multiplicative functional

$$\left(\mathrm{e}^{-\int_0^t c(X_s)\,\mathrm{d}s}\right)_{t\geq 0}.$$

It is given by the Feynman – Kac formula:

$$T_t f(x) = E^x\left[\mathrm{e}^{-\int_0^t c(X_s)\,\mathrm{d}s} f(X_t)\right], \quad f \in bp\mathcal{B}, x \in E.$$

Note that if L is the infinitesimal generator of X, then the above killed process has the generator $\mathrm{L} - c$.

Denote by $\mathcal{B}_\mathrm{u}$ the set of all functions $\varphi \in p\mathcal{B}$ such that $\varphi \leq 1$.

Statement 3.6. *The following result was stated in* [8, Proposition 6.2]. *For every $\varphi \in \mathcal{B}_\mathrm{u}$ the equation*

$$w_t(x) = T_t\varphi(x) + \int_0^t T_s(cB\widehat{w}_{t-s})(x)\,\mathrm{d}s, \quad t \geq 0, x \in E$$

has a unique solution $(t, x) \mapsto H_t\varphi(x)$ jointly measurable in (t, x), such that $H_t\varphi \in \mathcal{B}_\mathrm{u}$ and the following assertions hold.

(i) *For each $t \geq 0$ there exists a sub-Markovian kernel $\mathbf{H}_t\colon bp\mathcal{B}(S) \to bp\mathcal{B}$ such that $H_t\varphi = \mathbf{H}_t\widehat{\varphi}$ for all $\varphi \in \mathcal{B}_\mathrm{u}$.*

(ii) *The family $(H_t)_{t\geq 0}$ is a nonlinear semigroup of operators on $\mathcal{B}_\mathrm{u}$.*

Using Statement 3.2, for every $t \geq 0$ there exists a branching kernel $\widehat{H}_t$ on $(S, \mathcal{B}(S))$ such that

$$H_t\varphi = \mathbf{H}_t\widehat{\varphi} = \widehat{H}_t\widehat{\varphi}|_E.$$

By assertion (ii) of Statement 3.6 one can prove that the family $(\widehat{H}_t)_{t\geq 0}$ is a sub-Markovian semigroup of branching kernels on $(S, \mathcal{B}(S))$.

Remark 3.7. (i) Assume that the killed process having $(T_t)_{t\geq 0}$ as transition function is transient and let L be the infinitesimal generator of the process X. If $v \in \mathcal{B}_\mathrm{u}$ is such that $\widehat{v}$ is an invariant function with respect to the branching semigroup $(\widehat{H}_t)_{t\geq 0}$ on S, then v belongs to the domain of L and

$$(\mathrm{L} - c)v + cB\widehat{v} = 0. \tag{3.7}$$

In particular, if B is as in the example of branching kernels on S at the end of Section 3.1, then v is the solution of the nonlinear equation

$$(\mathrm{L}-c)v + c\sum_{k\geq 1} q_k v^k = 0.$$

It turns out that $(\widehat{H}_t)_{t\geq 0}$ is the main tool in the treatment of the (nonlinear) Dirichlet problem associated with equation (3.7); see [8, Proposition 6.1 and Section 6.2].

(ii) For each $t \geq 0$ and $f \in bp\mathcal{B}$ define the function $V_t f \in p\mathcal{B}$ as $V_t f := -\ln H_t\big(\exp(-f)\big)$. Then $V_t f$ is a bounded function (see [7] for details), the family $(V_t)_{t\geq 0}$ is a (nonlinear) semigroup on $bp\mathcal{B}$, and the following equality similar to (3.3) holds

$$\widehat{H}_t(e_f) = e_{V_t f}, \quad f \in bp\mathcal{B}, t \geq 0. \tag{3.8}$$

Therefore the semigroup approach for the continuous branching presented in Section 3.2 (developed by E. B. Dynkin and P. J. Fitzsimmons) is analogue to the above construction of the branching semigroup (due to N. Ikeda, M. Nagasawa, S. Watanabe, and M .L. Silverstein) in the discrete branching case (cf. [8, Remark 6.3]).

We state now the second application of Theorem 2.1, namely, to the discrete branching case: under supplementary hypotheses similar to Statement 3.4, if we assume that the process X is standard, and X killed with c is transient, there exists a branching standard process with state space S, having $(\widehat{H}_t)_{t\geq 0}$ as transition function; we send to [7] for a detailed treatment.

As in the case of continuous branching, one shows first that all the points of S are non-branch points with respect to the resolvent of kernels on $\big(S, \mathcal{B}(S)\big)$ associated with the branching semigroup $(\widehat{H}_t)_{t\geq 0}$. It turns out that in this case too the convenient compact Lyapunov functions on S required by (h3) are linear functionals of compact Lyapunov functions on E of the process X. A main tool in proving (h2) is the exponential representation (3.8) of $\widehat{H}_t$.

Acknowledgement. This work was supported by a grant of the Romanian National Authority for Scientific Research, CNCS–UEFISCDI, project number PN-II-ID-PCE-2011-3-0045.

References

1. L. Beznea, *Potential-theoretical methods in the construction of measure-valued Markov branching processes*, J. Eur. Math. Soc. (JEMS) **13** (2011), no. 3, 685–707.
2. L. Beznea and N. Boboc, *Potential theory and right processes*, Math. Appl., vol. 572, Kluwer, Dordrecht, 2004.
3. ———, *On the tightness of capacities associated with sub-Markovian resolvents*, Bull. London Math. Soc. **37** (2005), no. 6, 899–907.
4. L. Beznea, N. Boboc, and M. Röckner, *Markov processes associated with L^p-resolvents and applications to stochastic differential equations on Hilbert space*, J. Evol. Equ. **6** (2006), no. 4, 745–772.
5. ———, *Quasi-regular Dirichlet forms and L^p-resolvents on measurable spaces*, Potential Anal. **25** (2006), no. 3, 269–282.
6. L. Beznea, A. Cornea, and M. Röckner, *Potential theory of infinite dimensional Lévy processes*, J. Funct. Anal. **261** (2011), no. 10, 2845–2876.
7. L. Beznea and O. Lupaşcu, *Measure-valued discrete branching Markov processes*, preprint.
8. L. Beznea and A.-G. Oprina, *Nonlinear PDEs and measure-valued branching type processes*, J. Math. Anal. Appl. **384** (2011), no. 1, 16–32.

9. L. Beznea and M. Röckner, *Applications of compact superharmonic functions: path regularity and tightness of capacities*, Complex Anal. Oper. Theory **5** (2011), no. 3, 731–741.
10. ———, *From resolvents to càdlàg processes through compact excessive functions and applications to singular SDE on Hilbert spaces*, Bull. Sci. Math. **135** (2011), no. 6-7, 844–870.
11. V. I. Bogachev, *Measure theory.* Vol. I, II, Springer, Berlin, 2007.
12. D. A. Dawson and E. Perkins, *Superprocesses at Saint-Flour*, Probability at Saint-Flour, Springer, 2012.
13. E. B. Dynkin, *Solutions of semilinear differential equations related to harmonic functions*, J. Funct. Anal. **170** (2000), no. 2, 464–474.
14. ———, *Diffusions, superdiffusions and partial differential equations*, Amer. Math. Soc. Colloq. Publ., vol. 50, Amer. Math. Soc., Providence, RI, 2002.
15. A. M. Etheridge, *An introduction to superprocesses*, Univ. Lecture Ser., vol. 20, Amer. Math. Soc., Providence, RI, 2000.
16. P. J. Fitzsimmons, *Construction and regularity of measure-valued Markov branching processes*, Israel J. Math. **64** (1988), no. 3, 337–361 (1989).
17. N. Ikeda, M. Nagasawa, and S. Watanabe, *Branching Markov processes.* I, J. Math. Kyoto Univ. **8** (1968), 233–278.
18. ———, *Branching Markov processes.* III, J. Math. Kyoto Univ. **9** (1969), 95–160.
19. J.-F. Le Gall, *Spatial branching processes, random snakes and partial differential equations*, Lectures Math. ETH Zürich, Birkhäuser, Basel, 1999.
20. Z. Li, *Measure-valued branching Markov processes*, Probab. Appl. (N. Y.), Springer, Heidelberg, 2011.
21. T. Lyons and M. Röckner, *A note on tightness of capacities associated with Dirichlet forms*, Bull. London Math. Soc. **24** (1992), no. 2, 181–184.
22. M. Sharpe, *General theory of Markov processes*, Pure Appl. Math., vol. 133, Academic Press, Boston, MA, 1988.
23. M. L. Silverstein, *Markov processes with creation of particles*, Z. Wahrscheinlichkeitstheorie und Verw. Gebiete **9** (1968), 235–257.

Simion Stoilow Institute of Mathematics of the Romanian Academy, P.O. Box 1-764, 014700 Bucharest, Romania
E-mail address: `lucian.beznea@imar.ro`

Laboratoire des Sciences des Procédés et des Matériaux, Université Paris 13, Campus de Villetaneuse, 99 av. Jean-Baptiste Clément, 93430 Villetaneuse, France; University Politehnica of Bucharest, Splaiul Independentei nr. 313, sector 6, Bucharest, 060042, Romania and Simion Stoilow Institute of Mathematics of the Romanian Academy, P.O. Box 1-764, 014700 Bucharest, Romania
E-mail address: `oana.lupascu@lspm.cnrs.fr`

Simion Stoilow Institute of Mathematics of the Romanian Academy, P.O. Box 1-764, 014700 Bucharest, Romania
E-mail address: `oandrei22@yahoo.com`

Centre de Recherches Mathématiques
CRM Proceedings and Lecture Notes
Volume **55**, 2012

Compactifications of the Plane and Extensions of the Disc Algebra

V. Nestoridis

Abstract. We investigate the uniform limits of the set of polynomials on the closed unit disc $\overline{D}$ with respect to the chordal metric χ. More generally we examine analogous questions replacing $\mathbb{C} \cup \{\infty\}$ by other metrizable compactifications of $\mathbb{C}$. We also extend Mergelyan's theorem with respect to any metrizable compactification of $\mathbb{C}$ in the case of any Jordan domain. The above result can be applied to universal Taylor series in the sense of Luh and Chui–Parnes.

1. Introduction

A classical approximation theorem in Complex Analysis is Mergelyan's theorem ([19]). It states that if $K \subset \mathbb{C}$ is a compact set with connected complement, $f\colon K \to \mathbb{C}$ is a function continuous on K and holomorphic in K^0 and $\varepsilon > 0$ is given, then there exists a polynomial P, such that $\sup_{z\in K}|f(z) - P(z)| < \varepsilon$. The proof of the above theorem is not easy; however, if $K = \overline{D} = \{z \in \mathbb{C} : |z| \leq 1\}$, then the proof is elementary (see for example [2]).

In the above consideration we have uniform approximation on $\overline{D}$ (or on K) by polynomials with respect to the Euclidean metric on $\mathbb{R}^2 = \mathbb{C}$. However, if we consider functions taking the value ∞, as it happens very often in Complex Analysis, then the Euclidean metric on $\mathbb{R}^2 = \mathbb{C}$, does not allow us to speak about uniform approximation. But, if we consider the one-point compactification $\mathbb{C}\cup\{\infty\}$ of $\mathbb{C}$ and the chordal distance χ on it ([1]), then we can speak about uniform approximation on a compact set of functions taking values in $\mathbb{C} \cup \{\infty\}$. Since $\mathbb{C} \cup \{\infty\}$ is compact any other metric giving the same topology is uniformly equivalent to the metric χ. Thus, most of the results obtained for χ remain valid for any other equivalent metric on $\mathbb{C} \cup \{\infty\}$.

In [4, 14, 15] the uniform limits of the polynomials on $\overline{D}$ (or on K) with respect to the metric χ are investigated and compared with those defined with respect to the Euclidean metric. One such limit is the constant function $f \equiv \infty$. The other limits

2010 *Mathematics Subject Classification.* Primary 30J99; Secondary 46A99,30E10, 30K05.

Key words and phrases. Disc algebra, Mergelyan's theorem, chordal metric, polynomial approximation, compactification, generic property.

I would like to thank E. Katsoprinakis for helpful discussions.

This is the final form of the paper.

are exactly all holomorphic functions $f\colon D \to \mathbb{C}$, such that, for every $\zeta \in \partial D = T$ the limit $\lim_{z\to\zeta, z\in D} f(z)$ exists in $\mathbb{C} \cup \{\infty\}$. The class of the above functions is an extension of the classical disc algebra $A(D)$ and is denoted by $\tilde{A}(D)$. The proof of the above result is fundamental for our paper and is included in Section 2. Properties of the elements of the class $\tilde{A}(D)$, as well as, topological properties of $\tilde{A}(D)$ endowed with its natural metric topology can be found in [14, 15]. One open question is whether every compact set $K \subset \partial D = T$ with zero Lebesgue measure is a compact set of interpolation for the class $\tilde{A}(D)$; that is, if $\varphi\colon K \to \mathbb{C} \cup \{\infty\}$ is any continuous function is it always true that there exists $f \in \tilde{A}(D)$ such that $f_{|K} = \varphi$, provided that $K \subset \partial D$ is a compact set with zero Lebesgue measure? Another open question relates to a possible extension of Mergelyan's theorem.

Question. Let $L \subset \mathbb{C}$ be a compact set with L^c connected, $f\colon L \to \mathbb{C} \cup \{\infty\}$ a continuous function such that, for every component V of L^0 it holds $f_{|V} \equiv \infty$ or $f(V) \subset \mathbb{C}$ and $f_{|V}$ holomorphic. Then, is it true that there exists a sequence of polynomials P_n so that $P_n \to f$ uniformly on L with respect to the metric χ?

We also mention that the previous study on $\overline{D}$ could be extended to the case of several complex variables in the polydisc or in the ball, but this will not be included in the present article.

After [15] was written N. Papadatos preferred to replace the one-point compactification $\mathbb{C} \cup \{\infty\}$ by another one more geometric in some sense. Now there does not exist a unique point at infinity. There are lots of points at infinity situated on a circle and the finite plane may be seen as the interior of that circle. Thus, this new compactification looks like the closed unit disc and we can replace the metric χ by a new metric induced by the Euclidean metric on the disc. An analogous study for this metric can be found in [16, 17] and is briefly described in Section 3. The new class of uniform limits of the polynomials on $\overline{D}$ is denoted by $\bar{A}(D)$; it contains functions of the finite type and functions of the infinite type. The functions of the finite type are exactly all holomorphic functions $f\colon D \to \mathbb{C}$, such that, for every $\zeta \in \partial D = T$ the limit $\lim_{z\to\zeta, z\in D} f(z)$ exists in this new compactification. The functions of the infinite type are continuous functions defined on $\overline{D}$ with values in the set of these new points at infinity, but not all of them. A characterization of these functions can be found in [16, 17]; see also Section 3.

The fact that a similar study for this new compactification of $\mathbb{C}$ is possible, may lead us to generalize this study to the case of an arbitrary compactification of $\mathbb{C}$ carrying a metric or a uniform structure. But are there any such compactifications apart from the two previously studied? I. Androulidakis gave the following example ([2]).

A torus in $\mathbb{R}^3$ endowed with the Euclidean metric from $\mathbb{R}^3$ induces a metrizable compactification of $\mathbb{C}$; because, if we cut the torus along two circles it becomes homeomorphic to an open rectangle which is homeomorphic to $\mathbb{C}$. More generally, any compact surface in $\mathbb{R}^3$ gives a metrizable compactification of $\mathbb{C}$, because it becomes homeomorphic to $\mathbb{C}$ if we cut it accordingly.

Once we saw the previous example we also thought of any bounded simply connected domain Ω in $\mathbb{C}$. By the Riemann mapping theorem Ω is homeomorphic to the D, hence to $\mathbb{C}$. Thus $\overline{\Omega}$ endowed with the Euclidean distance induces a metrizable compactification of $\mathbb{C}$. If Ω is unbounded we consider its closure in $\mathbb{C} \cup \{\infty\}$ endowed with the metric χ; this also induces a metrizable compactification of $\mathbb{C}$.

In the present paper we do a similar study for the arbitrary metrizable compactification of $\mathbb{C}$; we do not consider more general compactification of $\mathbb{C}$ carrying a uniform structure, although this is meaningful. In section 4 we identify the class $A(D,S)$ of uniform limits on $\overline{D}$ of all polynomials with respect to the metric ρ of the compactification S of $\mathbb{C}$. Clearly $A(D,S)$ does not depend on the specific choice of the metric ρ because S is compact. Then $A(D,S) = A^{\text{fin}}(D,S) \cup A^{\text{inf}}(D,S)$, where $A^{\text{fin}}(D,S)$ contains exactly all holomorphic functions $f\colon D \to \mathbb{C}$ such that $\lim_{z\to\zeta, z\in D} f(z)$ exists in S for all $\zeta \in \partial D = T$. The set $A^{\text{inf}}(D,S)$ contains continuous functions defined on $\overline{D}$ with values in the set of points at infinity in S, but not all of them in general. In each particular case of compactification S one should try to characterize the elements of $A^{\text{inf}}(D,S)$. In Section 4 we find a proof of the previous result for the arbitrary metrizable compactification S, as well as, some properties of the elements of $A(D,S)$, $A^{\text{fin}}(D,S)$, $A^{\text{inf}}(D,S)$.

In Section 5 we have a topological study of the classes $A(D,S)$, $A^{\text{fin}}(D,S)$, $A^{\text{inf}}(D,S)$ where they are endowed with their natural metric. In particular the classical disc algebra $A(D)$ is an open and dense subset of $A(D,S)$ and of $A^{\text{fin}}(D,S)$. The class $A(D,S)$ is a complete metric space. We generalize to $A^{\text{fin}}(D,S)$ a generic result about $A(D)$ of Stegenga and Stephenson [22]. We also show that for two metrizable compactifications S_1, S_2, the classes $A(D,S_1)$ and $A(D,S_2)$ are the "same" if and only if S_1 and S_2 are the "same". However, there are two distinct S_1 and S_2 such that $A^{\text{fin}}(D,S_1) = A^{\text{fin}}(D,S_2)$.

In Section 6 we state a possible extension of Mergelyan's theorem for the arbitrary metrizable compactification S of $\mathbb{C}$ and we prove that it holds for Jordan domains; See also [18] for two special cases.

In Section 7 we relate the previous study to universal Taylor series [3, 5, 9, 12]. A holomorphic function f on D is called universal (Taylor Series), if the partial sums of its Taylor expansion approximate uniformly every polynomial on lots of compact sets K with connected complement.

If this holds for all such K with $K \cap D = \varnothing$, then we write $f \in U$. If this holds for all such K with $K \cap \overline{D} = \varnothing$, then we write $f \in U_1$. We have $U \subset U_1 \subset H(D)$, $U \neq U_1$ and both classes U, U_1 are dense and G_δ in $H(D)$ endowed with the topology of uniform convergence on compacta ([3, 11–13]).

Firstly, we notice that the uniform approximation in the definition of universal Taylor series is with respect to the Euclidean metric in $\mathbb{R}^2 = \mathbb{C}$. If we change it to be with respect to χ (or more generally ρ) we find equivalent definitions and the classes U, U_1 do not change.

Secondly we investigate the intersections $U \cap \tilde{A}(D)$ and $U_1 \cap \tilde{A}(D)$. We have $U \cap \tilde{A}(D) = \varnothing$, because for every $f \in U$ and every $\zeta \in T = \partial(D)$ the limit $\lim_{z\to\zeta, z\in D} f(z)$ cannot exist in $\mathbb{C}$ ([10]), while for $f \in \tilde{A}(D)$, $f \not\equiv \infty$ the above limit exists in $\mathbb{C}$ for almost all $\zeta \in T$. Concerning $U_1 \cap \tilde{A}(D)$ this is non void, because $U_1 \cap A(D)$ is dense and G_δ in $A(D)$ ([3, 11]). We show that $U_1 \cap \tilde{A}(D)$ is dense and G_δ in $\tilde{A}(D)$ and in $\tilde{A}(D) \setminus \{f \equiv \infty\}$. This result can be extended if we replace $\mathbb{C} \cup \{\infty\}$, χ by the arbitrary metrizable compactification of $\mathbb{C}$ and if we replace D, $\overline{D}$ by any domain Ω in $\mathbb{C}$ and its closure $\overline{\Omega}$, provided $\{\infty\} \cup [\mathbb{C} \setminus \overline{\Omega}]$ is connected. An open question is whether there exists a universal Taylor series $f \in U_1$ belonging to $\tilde{A}(D) \setminus A(D)$. I believe that the answer is affirmative, but I do not have a proof. Finally we can show generic existence of universal Taylor series $f \in U_1 \cap \tilde{A}(D)$, so that, we have simultaneous approximation of f by the approximating partial sums of the Taylor expansion of f uniform on $\overline{D}$ with respect to the metric χ (or ρ more generally).

2. The one-point compactification

Consider the one-point compactification $\mathbb{C}\cup\{\infty\}$ of $\mathbb{C}$ endowed with the chordal metric χ on it ([1]). We have $\chi(z_1,z_2) = |z_1-z_2|/(\sqrt{1+|z_1|^2}\sqrt{1+|z_2|^2})$ for $z_1,z_2\in\mathbb{C}$ and $\chi(z,\infty)=1/\sqrt{1+|z|^2}$ for $z\in\mathbb{C}$ and $\chi(\infty,\infty)=0$.

Let D denote the open unit disc in $\mathbb{C}$ and $\overline{D}$ its closure.

Assume that a sequence of polynomials f_n, $n=1,2,\dots$, converges to a function $f\colon\overline{D}\to\mathbb{C}\cup\{\infty\}$ uniformly on $\overline{D}$ with respect to χ. Since uniform convergence preserves continuity, it follows that f is continuous when $\mathbb{C}\cup\{\infty\}$ is endowed with the metric χ. Suppose that $f(z_0)=\infty$ for some z_0, $|z_0|<1$. Then for some small $r>0$ we have $\chi\big(f(z_0),f(z)\big)<\frac{1}{3}$ for all $z\colon |z-z_0|<r$. Since $f_n\to f$ uniformly with respect to χ, there exists n_0 so that $\chi(f_n(z),\infty)<\frac{1}{2}$ for all $n\geq n_0$ and all $z\colon |z-z_0|<r$. Thus, $f_n(z)\in\mathbb{C}-\{0\}$ for all $n\geq n_0$ and $z\colon|z-z_0|<r$. Since f_n is holomorphic in $\mathbb{C}$, being a polynomial, it follows that $1/f_n$ is holomorphic in the disc $\{z:|z-z_0|<r\}$. As $\chi(1/z_1,1/z_2)=\chi(z_1,z_2)$ for all $z_1,z_2\in\mathbb{C}\cup\{\infty\}$, it follows that $1/f_n\to 1/f$ uniformly on $\{z:|z-z_0|<r\}$ with respect to the distance χ. Since $\chi\big(\infty,f(z)\big)<\frac{1}{3}$ it follows that $\chi\big(0,1/f(z)\big)<\frac{1}{3}$ for all $z\colon|z-z_0|<r$ and $\chi\big(0,1/f_n(z)\big)<\frac{1}{2}$ for all $z\colon|z-z_0|<r$ and $n\geq n_1$, for some $n_1\geq n_0$. Thus, there exists $M<+\infty$ so that $|1/f_n(z)|<M$ and $|1/f(z)|<M$ for all $n\geq n_1$ and $z\colon|z-z_0|<r$. It easily follows that

$$\sup_{z:|z-z_0|<r}\left|\frac{1}{f_n(z)}-\frac{1}{f(z)}\right|\leq\sup_{z:|z-z_0|<r}\chi\left(\frac{1}{f_n(z)},\frac{1}{f(z)}\right)\cdot C\quad\text{for all } n\geq n_1$$

with $C<+\infty$. Thus, the sequence of holomorphic mappings $1/f_n$, $n\geq 1$, converges uniformly on $\{z:|z-z_0|<r\}$ to $1/f$ with respect to the usual Euclidean distance on $\mathbb{C}=\mathbb{R}^2$. But $1/f_n(z)\neq 0$ for all $z\colon|z-z_0|<r$, $n\geq n_1$ and $1/f(z_0)=0$. Hurwitz theorem implies $f\equiv\infty$ on $\{z:|z-z_0|<r\}$. Thus the set $f^{-1}(\infty)\cap D$ is open. The same set is relatively closed by continuity. Since D is connected, it follows $f\equiv\infty$. Conversely, for $P_n(z)\equiv n$ the limit is ∞. So the constant function infinity is one possible limit. If f is not identically equal to ∞ and f is a possible limit, then $f(D)\subset\mathbb{C}$ and the value ∞ is taken only on the boundary $\partial D=T$ where $T=\{z\in\mathbb{C}:|z|=1\}$ is the unit circle. It is also easily seen, by the previous discussion, that $f_n\to f$ uniformly on compact subsets of D with respect to the usual Euclidean metric in $\mathbb{C}=\mathbb{R}^2$. Thus, f is holomorphic in D.

Theorem 2.1. *Let $f\colon\overline{D}\to\mathbb{C}\cup\{\infty\}$ be a continuous function such that $f(D)\subset\mathbb{C}$ and $f_{|D}\colon D\to C$ is a holomorphic function. Then, there exists a sequence of polynomials $f_n(z)=\sum_{j=0}^{N_n}a_{j,n}z^j$, $a_{j,n}\in\mathbb{C}$, $N_n\in\{0,1,2,\dots\}$ such that $f_n\to f$ uniformly on $\overline{D}$ with respect to the metric χ.*

PROOF. Let $\varepsilon>0$. Continuity of f on the compact set $\overline{D}$ implies uniform continuity with respect to the metric χ. Thus, there exists r, $0<r<1$, so that $\chi\big(f(z),f(rz)\big)<\varepsilon/2$ for all $z\in\overline{D}$. We consider the Taylor expansion of f in D with center zero $f(z)=\sum_{j=0}^{\infty}a_jz^j$. The convergence is uniform on $\{z:|z|\leq r\}$ with respect to the usual Euclidean metric in $\mathbb{C}=\mathbb{R}^2$. Thus, there exists N such that

$$\left| f(\tau) - \sum_{j=0}^{N} a_j \tau^j \right| < \frac{\varepsilon}{2}, \quad \text{for all } \tau \text{ with } |\tau| \leq r.$$

But $\chi(z_1, z_2) \leq |z_1 - z_2|$ and $|rz| \leq r$ for all $z \in \overline{D}$. It follows that

$$\chi\left(f(rz), \sum_{j=0}^{N} a_j r^j z^j \right) < \frac{\varepsilon}{2} \quad \text{for all } z \in \overline{D}.$$

The triangle inequality implies $\chi(f(z), P(z)) < \varepsilon$ for all $z \in \overline{D}$ where P is the polynomial $P(z) = \sum_{j=0}^{N} a_j r^j z^j$. This completes the proof. □

Definition 2.2. $\tilde{A}(D)$ denotes the set of all functions $f : \overline{D} \to \mathbb{C} \cup \{\infty\}$ such that f is identically equal to ∞ or f is continuous with respect to the metric χ, $f(D) \subset \mathbb{C}$ and $f_{|D} \colon D \to \mathbb{C}$ is holomorphic in D.

We have proven the following; See also [4, 14, 15].

Theorem 2.3. *The class $\tilde{A}(D)$ coincides with the set of all uniform limits with respect to the chordal metric χ of polynomials on the closed unit disc $\overline{D}$.*

The function $1/(1-z)$ belongs to $\tilde{A}(D)$ but not to the classical $A(D)$. In [14, 15], some properties of the elements of $\tilde{A}(D)$ are investigated. Privalov's theorem implies that either the set $(\partial D) \cap f^{-1}(\infty)$ is a compact set with zero length or $f \equiv \infty$. Conversely, every compact set $K \subset \partial D$ with zero length coincides with $f^{-1}(\infty)$ for an element $f \in \tilde{A}(D)$, $f \not\equiv \infty$,

A natural question is to search for a characterization of all compact sets $K \subset \partial D$ which are compact sets of interpolation for $\tilde{A}(D)$, that is, for all continuous functions $\varphi \colon K \to \mathbb{C} \cup \{\infty\}$, is it true that there exists $f \in \tilde{A}(D)$ such that $f_{|K} = \varphi$. If $K \subset \partial D$ has positive length, then we know that K is not a compact set of interpolation for $\tilde{A}(D)$. We do not know if every compact set $K \subset \partial D$ with zero length is a compact set of interpolation of $\tilde{A}(D)$. More generally every result for the classical disc algebra $A(D)$ may be examined if it holds for the class $\tilde{A}(D)$.

It is also true that for $f, g \in \tilde{A}(D)$ the quantity $\sup_{z \in \overline{D}} \chi(f(z), g(z))$ is not controlled by $\sup_{\zeta \in \partial D} \chi(f(\zeta), g(\zeta))$. Thus, the maximum principle fails. However, if $f(\zeta) = g(\zeta)$ for all $\zeta \in \partial(D)$ (or for a compact set of positive length in ∂D) for some $f, g \in \tilde{A}(D)$ then $f \equiv g$.

If f and g belong to $\tilde{A}(D)$, then the natural distance between f and g is $\chi(f, g) = \sup_{z \in \overline{D}} \chi(f(z), g(z))$. With this metric $\tilde{A}(D)$ is a complete metric space. The classical disc algebra $A(D)$ is an open and dense subset of $\tilde{A}(D)$. Furthermore, the relative topology of $A(D)$ from $\tilde{A}(D)$ coincides with the usual topology of $A(D)$. For other topological properties of $\tilde{A}(D)$ see Section 5 below.

Replacing $\overline{D}$ by other compact sets $L \subset \mathbb{C}$ with L^c connected we arrive to the following.

Question. Let $L \subset \mathbb{C}$ be a compact set with L^c connected, $f \colon L \to \mathbb{C} \cup \{\infty\}$ a continuous function such that, for every component V of L^0 it holds $f_{|V} \equiv \infty$ or $f(V) \subset \mathbb{C}$ and $f_{|V}$ holomorphic. Then is it true that there exists a sequence of polynomials P_n so that $P_n \to f$ uniformly on L with respect to the metric χ?

We are able to answer in the affirmative the previous question in some particular cases. The general case remains open.

3. Another particular compactification

In [16, 17] the plane $\mathbb{C}$ is identified with D by the homeomorphism $\mathbb{C} \ni z \to z/(1+|z|) \in D$. This way $\overline{D}$ becomes an obvious compactification of D endowed with the usual Euclidean metric. This induces a compactification of $\mathbb{C}$, where there are several points at ∞ situated on a circle: $\overline{\mathbb{C}} = \mathbb{C} \cup \mathbb{C}^\infty$, $\mathbb{C}^\infty = \{\infty e^{i\vartheta} : \vartheta \in [0, 2\pi)\}$. The metric d on $\overline{\mathbb{C}}$ is $d(z, w) = |z/(1+|z|) - w/(1+|w|)|$, for $z, w \in \mathbb{C}$, $d(z, \infty e^{i\vartheta}) = |z/(1+|z|) - e^{i\vartheta}|$ for $z \in \mathbb{C}$, $\vartheta \in \mathbb{R}$ and $d(\infty e^{i\varphi}, \infty e^{i\vartheta}) = |e^{i\varphi} - e^{i\vartheta}|$, $\varphi, \vartheta \in \mathbb{R}$.

The set of uniform limits $\tilde{A}(D)$ of the polynomials on $\overline{D}$ with respect to d consists of the finite type and the infinite type. The finite type contains all holomorphic functions $f: D \to \mathbb{C}$ such that, for every $\zeta \in \partial D$ the limit $\lim_{z\to\zeta, z\in D} f(z)$ exists in $\overline{\mathbb{C}} = \mathbb{C} \cup \mathbb{C}^\infty$.

The infinite type contains all continuous functions $f: \overline{D} \to \mathbb{C}^\infty$, $f(z) = \infty e^{i\vartheta(z)}$, where $\vartheta: \overline{D} \to \mathbb{R}$ is continuous on $\overline{D}$ and harmonic in D. To give a reason for this let us consider a sequence of polynomials P_n, $n = 1, 2, \ldots$ converging to $f(z) = \infty e^{i\vartheta(z)}$, $\vartheta(z) \in \mathbb{R}$. Then $\inf_{|z|\leq 1}|P_n(z)| \to +\infty$ and $\operatorname{Arg} P_n(z) \to \vartheta(z)$ uniformly on $\overline{D}$, where the branches of $\operatorname{Arg} P_n$ are adjusted and ϑ is any continuous version of the argument of f. Since $\operatorname{Arg} P_n = \Im \log P_n$ is harmonic it follows that θ is also harmonic on D and continuous on $\overline{D}$. This is a rough sketch of the proof. For details we refer to [16, 17].

Furthermore, in [16, 17] one finds an investigation of properties of the elements of $\tilde{A}(D)$, topological properties of $\tilde{A}(D)$ endowed with its natural metric topology, as well as possible extensions of Mergelyan's theorem.

4. Arbitrary metrizable compactifications

We consider an arbitrary metrizable compactification $S = \mathbb{C} \cup \mathbb{C}^\infty$ of $\mathbb{C}$; then, the complex plane $\mathbb{C}$ is an open dense subset of S, because $\mathbb{C}$ is locally compact ([6]). We also mention that $\mathbb{C}^\infty$ is disjoint from $\mathbb{C}$, it contains all points at infinity and it is a closed subset of S. We denote by ρ the metric on S; however, because S is compact, any other equivalent metric ρ' on S is in fact uniformly equivalent with ρ; thus, the uniform limits of any set of functions with values in S do not depend on the metric ρ but they depend only on the compactification S.

As we mentioned in the Introduction there are plenty of examples of metrizable compactifications of $\mathbb{C}$. In the previous sections we examined the cases $S = \mathbb{C} \cup \{\infty\}$ as well as $\mathbb{C} \cup \mathbb{C}^\infty$, $C \approx D \equiv \{z \in \mathbb{C} : |z| < 1\}$ and $\mathbb{C}^\infty \approx T$ with $T = \{z \in \mathbb{C} : |z| = 1\}$. Another useful example is the real projective space of dimension 2. One model of this is $\mathbb{C} \approx D$ and $\mathbb{C}^\infty \approx T_{|\sim}$ where the equivalence relation $\sim$ on T identifies w and $-w$, for all $w \in T$. This compactification is metrizable ([21]). More generally if we consider any continuous surjection $\pi: \overline{D} \to D \cup \mathbb{C}^\infty$ where $\pi_{|D} = id_D$ and $\pi(T) \subset \mathbb{C}^\infty$, then $D \cup \mathbb{C}^\infty$ is a metrizable compactification with $\mathbb{C} \approx D$. In particular we may have $\mathbb{C} \approx D$ and $\mathbb{C}^\infty \approx T_{|\approx}$ where the equivalence relation $\approx$ on T identifies all points of the closed arc $I = \{e^{i\vartheta} : 0 \leq \vartheta \leq \pi/4\}$. Finally we mention that the compactifications $S_1 = \mathbb{C} \cup \mathbb{C}^\infty$ with $\mathbb{C} \approx D$ and $\mathbb{C}^\infty \approx T$ and $S_2 = \mathbb{C} \cup \mathbb{C}'^\infty$ with $\mathbb{C} \approx D - [0, 1)$ and $\mathbb{C}'^\infty = T \cup [0, 1)$ are different from each other, although both are represented by $\overline{D}$; the reason is that the points at infinity are different.

We fix a metrizable compactification $S = \mathbb{C} \cup \mathbb{C}^\infty$ of $\mathbb{C}$ endowed with a metric ρ and we denote by $A(D, S)$ the set of uniform limits with respect to ρ of the polynomials on $\overline{D}$. Let $\phi\colon S = \mathbb{C} \cup \mathbb{C}^\infty \to \mathbb{C} \cup \{\infty\}$ be the function $\phi(w) = w$ for $w \in \mathbb{C}$ and $\phi(w) = \infty$ for $w \in \mathbb{C}^\infty$. It is easily seen that ϕ is continuous. Since S is compact, it follows that ϕ is uniformly continuous. Thus, if P_n are polynomials and $P_n \to f$ uniformly on $\overline{D}$ with respect to ρ, it follows that $P_n = \phi(P_n) \to \phi(f)$ uniformly on $\overline{D}$ with respect to the chordal metric χ. Combining this with the results of Section 2 we see that $A(D, S)$ consists of the finite type $A^{\text{fin}}(D, S)$ and the infinite type $A^{\text{inf}}(D, S) : A(D, S) = A^{\text{fin}}(D, S) \cup A^{\text{inf}}(D, S)$, with $A^{\text{fin}}(D, S) \cap A^{\text{inf}}(D, s) = \varnothing$. A function $f : \overline{D} \to S$ belongs to $A^{\text{fin}}(D, S)$ if it is continuous, $f(D) \subset \mathbb{C}$ and $f_{|D}$ is holomorphic. We notice that such a function f is always ρ-approximable by polynomials. Indeed since $\overline{D}$ is compact it follows that f is uniformly continuous; thus, if $\varepsilon > 0$ is given there exists r, $0 < r < 1$, so that $\rho\big(f(z), f(rz)\big) < \varepsilon/2$ for all $z : |z| \le 1$. Since $f(\{w \in \mathbb{C} : |w| \le r\})$ is a compact subset of $\mathbb{C}$, the partial sums $S_N(f, 0)(w) = \sum_{n=0}^{N} f^{(n)}(0)w^n/n!$, $N = 0, 1, 2, \ldots$ converge to f uniformly on $\{w \in \mathbb{C} : |w| \le r\}$ with respect to the Euclidean distance and since the Euclidean distance is uniformly equivalent to ρ on each compact subset of $\mathbb{C}$, it follows that there exists $N \in \mathbb{N}$ such that $\rho\big(f(rz), S_N(f, 0)(rz)\big) < \varepsilon/2$, for every $z\colon |z| \le 1$. The triangle inequality implies $\sup_{|z|\le 1} \rho\big(f(z), P(z)\big) < \varepsilon$, where P is the polynomial $P(z) = \sum_{n=0}^{N} \big[f^{(n)}(0)r^n/n!\big] z^n$. Concerning $A^{\text{inf}}(D, S)$ it contains functions $f\colon D \to \mathbb{C}^\infty$ which are continuous, but not all of them in general, as it can be seen by the results of Section 3. The only fact that we know is that $A^{\text{inf}}(D, S) = \{f \in A(D, S) : f(z) \in \mathbb{C}^\infty \text{ for all } z \in \overline{D}\}$ is non void. This holds because $A(D, S)$ contains all constant functions in S: If $b \in \mathbb{C}$, then $A(D, S)$ contains the constant polynomial b. If $b \in \mathbb{C}^\infty$, let $b_n \in \mathbb{C}$ with $b_n \to b$, as $n \to +\infty$. If f_{b_n} denotes the constant polynomial equal to b_n, then $f_{b_n} \to f_b$ uniformly on $\overline{D}$, where $f_b(z) = b$ for all $z \in \overline{D}$. Thus $A^{\text{inf}}(D, S)$ contains f_b and therefore $A^{\text{inf}}(D, S) \ne \varnothing$. We also have $A(D, S) \supset A^{\text{fin}}(D, S) \supset A(D)$, where $A(D)$ denotes the classical disc algebra. We also note that $A^{\text{fin}}(D, \mathbb{C} \cup \{\infty\})$ contains $A^{\text{fin}}(D, S)$, for all metrizable compactifications S, where we have identified every $f \in A(D, S)$ with its restriction $f_{|D}$.

Theorem 4.1. *If $f \in A^{\text{fin}}(D, S)$, then $f^{-1}(\mathbb{C}^\infty)$ is a subset of T with zero Lebesgue measure.*

PROOF. This follows from the results of Section 2 combined with the fact that $\phi(f) \in \tilde{A}(D)$. □

Theorem 4.2. *If $f, g \in A^{\text{fin}}(D, S)$ satisfy $f(\zeta) = g(\zeta)$ for a compact subset of $T = \partial D$ of positive Lebesgue measure, then $f \equiv g$. In particular this holds if $f, g \in A^{\text{fin}}(D, S)$ and $f_{|T} = g_{|T}$.*

PROOF. The functions $f_{|D}$ and $g_{|D}$ take complex values and the function $f - g$ is holomorphic in D with zero limits on a compact subset of $T = \partial D$ with positive Lebesgue measure. Privalov's theorem ([8]) implies $f \equiv g$. □

Counterexample 4.3. For some metrizable compactification S of $\mathbb{C}$ we can have $f, g \in A^{\text{inf}}(D, S)$ $f_{|T} \equiv g_{|T}$ but $f \not\equiv g$.

PROOF. We consider the compactification $S = \mathbb{C} \cup \mathbb{C}^\infty$ with $\mathbb{C} \approx D$ and $\mathbb{C}^\infty \approx T/\approx$, where the equivalence relation $\approx$ on T identifies all points in $I =$

$\{e^{i\vartheta} : 0 \le \vartheta \le \pi/4\}$; let B denotes this new point. Let $h : \partial D \to T/\approx$ be defined as $h(e^{i\vartheta}) = e^{i\theta}$ for all $e^{i\vartheta} \in T \setminus I$ and $h(e^{i\vartheta}) = B$ for all $e^{i\vartheta} \in I$. Let $\Theta_1 : T \to \mathbb{R}$ be defined by $\Theta_1(e^{i\vartheta}) = \vartheta$ for $0 \le \vartheta \le \pi$ and $\Theta_1(e^{i\vartheta}) = 2\pi - \vartheta$ for $\pi \le \vartheta \le 2\pi$. The harmonic extension $\widehat{\Theta}_1$ of Θ_1 satisfies $\widehat{\Theta}_1(0) = \pi/2$ and defines an element $\infty\widehat{\Theta}_1$ belonging to $A^{\inf}(D, \widetilde{S})$ where $\widetilde{S}$ the compactification of Section 3.

Let $J : \mathbb{R} \to \mathbb{R}$ continuous such that $J([0, \pi/4]) \subset [0, \pi/4]$, $J(0) = 0$ $J(\pi/4) = \pi/4$ and $J(x) = x$ for all $x \in \mathbb{R} \setminus [0, \pi/4]$.

We set $\Theta_2 = J \circ \Theta_1 : T \to \mathbb{R}$. We can easily modify the map J so that the harmonic extension $\widehat{\Theta}_2$ of Θ_2 satisfies $\widehat{\Theta}_2(0) \ne \pi/2$. We set $f = \infty h(\widehat{\Theta}_1)$ and $g = \infty h(\widehat{\Theta}_2)$. Then $f, g \in A^{\inf}(D, S)$, $f_{|I} \equiv g_{|I} \equiv \infty B$, $|I| > 0$ but $f \not\equiv g$, because $\infty h(\widehat{\Theta}_1(0)) = \infty h(\pi/2) = \infty \cdot e^{i\pi/2} \ne \infty B$ and $\infty h\big(\widehat{\Theta}_2(0)\big) \ne \infty \cdot e^{i\pi/2}$. □

Definition 4.4. Let $S = \mathbb{C} \cup \mathbb{C}^\infty, \rho$ be a metrizable compactification of $\mathbb{C}$. A compact set $K \subset \partial D = T$ is called a compact set of interpolation for $A(D, S)$, if for every continuous function $h : K \to S$ there exists $f \in A(D, S)$ with $f_{|K} = h$.

Theorems 4.1 and 4.2 easily imply that if the compact set $K \subset \partial D = T$ has positive Lebesgue measure, then K is not a compact set of interpolation for $A(D, S)$. It is an open question to characterize the compact sets of interpolation for $A(D, S)$. One could also consider compact sets of interpolation for each of the classes $A^{\text{fin}}(D, S)$ and $A^{\inf}(D, S)$.

5. Some topological results

Let $S = \mathbb{C} \cup \mathbb{C}^\infty, \rho$ be a metrizable compactification of $\mathbb{C}$ with its metric ρ and $A(D, S)$, $A^{\text{fin}}(D, S)$, $A^{\inf}(D, S)$ the classes of functions defined in Section 4. For $f, g \in A(D, S)$ the natural metric is $\rho(f, g) = \sup_{z \in \overline{D}} \rho\big(f(z), g(z)\big)$.

Theorem 5.1. *Under the above assumptions and notation we have the following.*

(1) *$(A(D, S), \rho)$ is a complete metric space.*
(2) *$A^{\text{fin}}(D, S)$ is an open and dense subset of $A(D, S)$.*
(3) *$A^{\inf}(D, S)$ is closed nowhere dense in $A(D, S)$.*
(4) *The classical disc algebra $A(D)$ is an open and dense subset of $A(D, S)$.*
(5) *The relative topology of $A(D)$ from $A(D, S)$ coincides with the usual topology of $A(D)$ as a Banach algebra.*

Proof. Because (S, ρ) is complete, it follows that the space $W = \{f : \overline{D} \to S\}$ endowed with the metric $\rho(f, g) = \sup_{z \in \overline{D}} \rho\big(f(z), g(z)\big)$, $f, g \in W$ is complete. Now $A(D, S)$ is closed in W, because it is the set of limits in W of a subset of W (precisely of the set of polynomials). Thus, $A(D, S)$ is complete. This gives (1). Since $A^{\text{fin}}(D, S)$ contains the set of polynomials it follows that $A^{\text{fin}}(D, S)$ is dense in $A(D, S)$. To show that $A^{\text{fin}}(D, S)$ is open in $A(D, S)$, let $f \in A^{\text{fin}}(D, S)$; then $f(0) \in \mathbb{C}$. Since $\mathbb{C}$ is open in $S = \mathbb{C} \cup \mathbb{C}^\infty$, there exists $\delta > 0$ so that $\rho(f(0), w) > \delta$ for all $w \in \mathbb{C}^\infty$. Let $g \in A(D, S)$ with $\rho(f, g) < \delta$. Then $\rho(f(0), g(0)) < \delta$ which implies $g(0) \in \mathbb{C}$. Thus $g \in A^{\text{fin}}(D, S)$. This proves (2). Part (3) is equivalent to (2). Since $A(D)$ contains the set of polynomials it is dense in $A(D, S)$. It is also open and the proof is similar to the proof that $A^{\text{fin}}(D, S)$ is open with the difference that we prove $g(z) \in \mathbb{C}$ for every $z \in \overline{D}$ and not only for $z = 0$. It suffices to set $\delta > 0$ to be the distance of the compact set $f(\overline{D}) \subset \mathbb{C}$ from the compact set $\mathbb{C}^\infty$ $\big(f \in A(D)\big)$. This gives (4).

It remains to prove (5). Let $f_n \in A(D)$, $n = 1, 2, \ldots,$

$$f \in A(D) \qquad \text{and} \qquad \sup_{z\in\overline{D}}|f_n(z) - f(z)| \to 0, \quad \text{as } n \to +\infty.$$

One easily concludes that there exist a compact $K \subset \mathbb{C}$ and a natural number n_0 so that $f_n(z) \in K$ for all $n \geq n_0$ and $f(z) \in K$ for all $z \in \overline{D}$. Since on K the Euclidean distance is uniformly equivalent to ρ, it follows $\rho(f_n, f) \to 0$, as $n \to +\infty$.

Reversely, let $f_n \in A(D)$, $n = 1, 2, \ldots,$ $f \in A(D)$ and $\rho(f_n, f) \to 0$ as $n \to +\infty$. Let $\delta = \min\{\rho(f(z), w) : z \in \overline{D}, w \in \mathbb{C}^\infty\}$.

Then $\delta > 0$. The set $B = \{x \in S : \min\{\rho(x, f(z)) : z \in \overline{D}\} \leq \delta/2\}$ is a compact subset of $\mathbb{C}$. There exists a natural number n_0, so that $f_n(\overline{D}) \subset B$ for all $n \geq n_0$.

Obviously $f(\overline{D}) \subset B$. On B the Euclidean metric is uniformly equivalent to ρ; thus, $\sup_{z\in\overline{D}}|f_n(z) - f(z)| \to 0$, as $n \to +\infty$. This completes the proof. □

In [22] the authors consider an arbitrary Hausdorff measure function h. It is proven that the set of $f \in A(D)$ such that $\Lambda_h(E_f) = 0$ is dense and G_δ in $A(D)$, where $E_f = \{\zeta \in T : f(\zeta) \notin f(D)\}$ and Λ_h is defined in [22]. According to Theorem 5.1(4) and (5) the disc algebra $A(D)$ with its usual topology is an open and dense subset of $A^{\text{fin}}(D, S)$. It follows that the set $\{f \in A(D) : \Lambda_h(E_f) = 0\}$ is G_δ and dense in $A^{\text{fin}}(D, S)$. Therefore the bigger set $\{f \in A^{\text{fin}} D, S) : \Lambda_h(E_f) = 0\}$ is residual in $A^{\text{fin}}(D, S)$, where $E_f = \{\zeta \in T : f(\zeta) \notin f(D)\}$. We will show that this set is also G_δ in $A^{\text{fin}}(D, S)$. According to the first part of the proof of [22, Theorem 3.1], it suffices to show that the sets $\{f \in A^{\text{fin}}(D, S) : M_h(E_f) < 1/N\}$ are open (for the definition of M_h see [22]).

Fix $N \in \mathbb{N}$ and $f \in S_N$; there exists a countable collection of discs D_m so that $E_f \subset \bigcup_m D_m$ and $\Sigma h(r_m) < 1/N$, where r_m is the radius of D_m. Since $E_f \supset f^{-1}(\mathbb{C}^\infty)$, by the continuity of f and the compactness of $T \setminus \bigcup_m D_m$, there exists $M < +\infty$ so that $|f(\zeta)| < M$ for all $\zeta \in T \setminus \cup D_m$.

We consider a finite set of disjoint open intervals (α_i, β_i), $i = 1, \ldots, n$, $\alpha_1 < \beta_1 < \alpha_2 < \beta_2 < \cdots < \alpha_n < \beta_n < \alpha_1 + 2\pi$ and an r, $0 < r < 1$ so that $f^{-1}(\mathbb{C}^\infty) \subset \bigcup_{i=1}^n \{e^{i\vartheta} : \alpha_i < \vartheta < \beta_i\} \subset \bigcup_{i=1}^n \{e^{i\vartheta} : \alpha_i \leq \vartheta \leq \beta_i\} \subset \bigcup_m D_m$ and for every $z \in S = \left\{te^{it} : r \leq t \leq 1, \vartheta \in \bigcup_{i=1}^n [\alpha_i, \beta_i]\right\}$ we have $|f(z)| > \widetilde{M}$ for some $\widetilde{M}$, $M < \widetilde{M} < +\infty$ (where we agree that if $f(z) \in \mathbb{C}^\infty$ then $|f(z)| = +\infty$).

Obviously for $\zeta \in T \setminus \bigcup_m D_m$ we have $f(\zeta) \notin f(S)$ and $\zeta \notin E_f$.

It follows that $f(\zeta) \in f(D \setminus S)$.

Following [22] we use the definition and properties of $\delta_f(w, G)$ where $w \in \mathbb{C}$, $G \subset \mathbb{C}$ open and f holomorphic in G. It is clear that $\delta_f(w, G)$ is well defined in [22] if $w \in f(G)$; however, for $w \in \mathbb{C} \setminus f(G)$ it is meant in [22] that $\delta_f(w, G) = 0$. It is proven that if $\delta_f(w_0, G) = +\infty$ for some $w_0 \in \mathbb{C}$, then $\delta_f(w, G) \equiv +\infty$ for all $w \in \mathbb{C}$. Otherwise $\delta_f(w, G) \in [0, +\infty)$ for all $w \in \mathbb{C}$ and $|\delta_f(w_1, G) - \delta_f(w_2, G)| \leq |w_1 - w_2|$ for all $w_1, w_2 \in \mathbb{C}$. It is also proven in [22] that $\delta_f(w, G) > 0 \Leftrightarrow w \in f(G)$. Finally if f and g are bounded and holomorphic in G, then for every $z \in G$ we have $\left|\delta_f\big(f(z)\big) - \delta_g\big(g(z)\big)\right| \leq 2\sup_{x\in G}|f(x) - g(x)|$.

We set $G = D \setminus S$ which is an open set.

Since for $\zeta \in T \setminus \bigcup_m D_m$ we have $f(\zeta) \in f(D \setminus S)$, it follows $\delta_f(f(\zeta), D \setminus S) > 0$. By the continuity of δ_f and the compactness of $T \setminus \bigcup_m D_m$ we have $\eta = \inf\{\delta_f(f(\zeta); D \setminus S); \zeta \in T \setminus \bigcup_m D_m\} > 0$. If a function $g \in A^{\text{fin}}(D, S)$ satisfies

$\sup_{D\setminus S}|f(z) - g(z)| < \eta/4$, it follows $\delta_g(g(\zeta), D \setminus S) > 0$ for all $\zeta \in T \setminus \bigcup_m D_m$. Thus, $g(\zeta) \in g(S \setminus S) \subset g(D)$. It follows $E_g \subset \bigcup_m D_m$ which implies $g \in S_N$.

The set $L = f(\overline{D - S})$ is a compact subset of $\mathbb{C}$. Thus, $\inf \rho(L, \mathbb{C}^\infty) = \tau > 0$. The set $L' = \{x \in S : \rho(x, L) \leq \tau/2\}$ is a compact subset of $\mathbb{C}$ containing L. On L' the Euclidean metric is uniformly equivalent to ρ. Thus, there exists $\varepsilon > 0$, $\varepsilon < \tau/2$ such that ($\alpha \in L'$, $\beta \in L'$, $\rho(\alpha, \beta) < \varepsilon$ implies $|\alpha - \beta| < \eta/8$).

Let $g \in A^{\text{fin}}(D, S)$ satisfy $\rho(f, g) < \varepsilon$. We will show $g \in S_N$. This will complete the proof. Indeed, for $z \in D \setminus S$ we have $f(z) \in L \subset L'$, $g(z) \in L'$ and $\rho\big(f(z), g(z)\big) < \varepsilon$. This implies $|f(z) - g(z)| < n/8$. It follows $\sup_{z\in\overline{D}}|f(z) - g(z)| < n/4$ which implies that g is in S_N and the proof is completed. Thus, we have proved the following.

Theorem 5.2. *Let h be any Hausdorff measure function and (S, ρ) any metrizable compactification of $\mathbb{C}$ with metric ρ. The set of all $f \in A^{\text{fin}}(D, S)$ such that $\Lambda_h(E_f) = 0$ is dense and G_δ in $A^{\text{fin}}(D, S)$ and in $A(D, S)$, where $E_f = \{\zeta \in T : f(\zeta) \notin f(D)\}$ and Λ_h is defined in* [22].

Next we consider the sets $Y(S) = \{f \in A(D, S) : f(D) \subset f(T)\}$ and $W(S) = \{f \in A(D, S) : f(T) = S = \mathbb{C} \cup \mathbb{C}^\infty\}$. Obviously $W(S) \subset Y(S)$.

Theorem 5.3. *The sets $Y(S)$ and $W(S)$ are closed nowhere dense subsets of $A(D, S)$ and $Y(S) \neq \varnothing$.*

Proof. Let $f_n \in Y(S)$, $f \in A(D, S)$ and $\rho(f_n, f) \to 0$ as $n \to +\infty$. Let $z \in D$. We will show $f(z) \in f(T)$. This will show that $Y(S)$ is closed.

Let $\zeta_n \in T$ such that $f_n(\zeta_n) = f_n(z)$. By compactness of T there exists a subsequence ζ_{k_n} converging to some $\zeta \in T$. By the uniform convergence we have $f_{k_n}(\zeta_{k_n}) \to f(\zeta)$. But $f_{k_n}(\zeta_{k_n}) = f_{k_n}(z) \to f(z)$. Thus, $f(z) = f(\zeta) \in f(T)$ and $Y(S)$ is closed. To prove that $W(S)$ is closed, it suffices to prove that $A(D, S) \setminus W(S)$ is open. Let $f \in A(D, S) \setminus W(S)$; thus, there exist $w_0 \in S$ with $w_0 \notin f(T)$. Since $f(T)$ is compact in S, it follows that $\inf_{\zeta\in T} \rho(w_0, f(\zeta)) = \varepsilon > 0$. If $g \in A(D, S)$ satisfies $\rho(f, g) < \varepsilon$, then one easily sees that $w_0 \notin g(T)$; thus, $g \in A(D, S) \setminus W(S)$. Therefore $A(D, S) \setminus W(S)$ is open and $W(S)$ is closed.

The interiors of $Y(S)$ and $W(S)$ are empty, because $Y(S)$, $W(S)$ do not contain any polynomial and the set of polynomials is dense in $A(D, S)$. Therefore, the sets $Y(S)$ and $W(S)$ are closed nowhere dense in $A(D)$. To show that $Y(S) \neq \varnothing$ it suffices to construct $f \in A(D) \subset A(D, S)$, such that $f(T) = \overline{D}$ and $|f(z)| \leq 1$ for all $z \in \overline{D}$. Let $K \subset T$ be a Cantor set with zero Lebesgue measure. Then K is a compact set of interpolation for $A(D)$ ([7]). It is well known that there exists a continuous mapping $\phi\colon K \to [0, 1]$ with $\phi(K) = [0, 1]$. Consider a continuous Peano curve $\Gamma\colon [0, 1] \to \overline{D}$ with $\Gamma([0, 1]) = \overline{D}$.

Define $h = \Gamma \circ \phi\colon K \to \overline{D}$ which is continuous and $h(K) = \overline{D}$. By the Rudin–Carleson theorem ([7, pp. 81–82], see also [20]), there exists $f \in A(D)$ with $f_{|K} = h$ and $|f(z)| \leq 1$ for all $z \in \overline{D}$. This shows $Y(S) \neq \varnothing$ and the proof is complete. $\square$

Remark 5.4. For $S = \mathbb{C} \cup \{\infty\}$ it has been proven in [4] that $W(\mathbb{C} \cup \{\infty\})$ is non void. For the general metrizable compactification S, ρ of $\mathbb{C}$ we do not know whether $W(S) \neq \varnothing$ or not. If we assume that a Cantor set K as in the previous proof is a compact of interpolation for $A(D, S)$ and if we assume that there exists a continuous surjection $\pi\colon \overline{D} \to S$, then it follows $W(S) \neq \varnothing$. Because, as in the previous proof, the function $h = \Gamma \circ \phi\colon K \to \overline{D}$ is continuous and $h(K) = \overline{D}$.

Then $\pi \circ h\colon K \to S$ is a continuous surjection. By our assumption that K is a compact set of interpolation for $A(D, S)$, it follows that there exists $f \in A(D, S)$ with $f_{|K} = \pi \circ h$.

Therefore $f(T) \supset f(K) = \pi \circ h(K) = S$ and $f \in W(S)$. We mention that it happens often that there exists a continuous surjection $\pi\colon \overline{D} \to S$; this is the case when S is the real projective space of dimension 2; however, it is not always true. A conterexample can be constructed with $S \subset \mathbb{C}$ provided that $\partial(S)$ contains $E = \{(x, \sin(1/x)) : 0 < x \leq 1\}$, because $\pi(T)$ should be a curve, while $E \cup \{(0,t) : -1 \leq t \leq 1\}$ is not.

Theorem 5.5. *Let $(S = \mathbb{C} \cup \mathbb{C}^\infty, \rho)$ and $(S' = \mathbb{C} \cup \mathbb{C}'^\infty, \rho')$ be two metrizable compactifications of $\mathbb{C}$, where ρ and ρ' denote the corresponding metrics. Then the following are equivalent.*

(1) *There exists a homeomorphism $T\colon S \to S'$ such that $T(w) = w$ for all $w \in \mathbb{C}$.*
(2) *There exists a homeomorphism $F\colon A(D,S) \to A(D,S')$ such that $F(P) = P$ for every polynomial P.*

Proof. (1) $\Rightarrow$ (2). T and T^{-1} are uniformly continuous. Let P_n, $n = 1, 2, \ldots$ be a sequence of polynomials and $f \in A(D,S)$ and assume $\rho(P_n, f) \to 0$, as $n \to +\infty$. Then $P_n = T \circ P_n$ converges in the ρ' metric to $T \circ f$. Thus, $T \circ f \in A(D, S')$. It suffices to set $F(f) = T \circ f$ for all $f \in A(D,S)$. One easily verifies that $F^{-1}(g) = T^{-1} \circ g$ for $g \in A(D,S')$ and that F and F^{-1} are continuous.

Since $T(w) = w$ for all $w \in \mathbb{C}$, it follows that for any polynomial P we have $F(P) = T \circ P = P$.

This proves the direction (1) $\Rightarrow$ (2).

(2) $\Rightarrow$ (1). Let $w \in S$. Because $\mathbb{C}$ is dense in S, there exists a sequence $w_n \in \mathbb{C}$ converging to w with respect to the metric ρ. We consider the constant polynomials $P_n(z) \equiv w_n$ for all $z \in \mathbb{C}$. We have $P_n \to f_w$, uniformly in ρ metric, where $f_w(z) \equiv w$ for all $z \in \overline{D}$.

Thus $f_w \in A(D,S)$. Applying F we find $P_n = F(P_n) \to F(f_w) \in A(D,S')$. Since P_n is constant it follows that $F(f_w)$ is also constant. Thus, there exist $w' \in S'$ such that $F(f_w)(z) = w'$ for all $z \in \overline{D}$. We set $T(w) = w'$. One easily verifies that for $w \in \mathbb{C}$ we have $T(w) = w' = w$. We observe that $T(w) = w' = [F(f_w)](0)$. This easily yields the continuity of T.

If the sequence $w_n \in \mathbb{C}$ is ρ-Cauchy, then there exists $w \in S$ with $w_n \xrightarrow{\rho} w$. By the continuity of T we have $w_n = T(w_n) \xrightarrow{\rho'} T(w)$. Thus, we conclude that every ρ-Cauchy sequence in $\mathbb{C}$ is also a ρ'-Cauchy sequence. Repeating the same arguments but using F^{-1} instead of F we conclude that every ρ'-Cauchy sequence in $\mathbb{C}$ is also a ρ-Cauchy sequence. Thus, S and S' being two completions of $(\mathbb{C}, \rho)$ and (C, ρ') respectively, they are homeomorphic by a homeomorphism $\Lambda\colon S \to S'$ satisfying $\Lambda(w) = w$ for all $w \in \mathbb{C}$. By continuity $\Lambda = T$. □

One wonders if the condition $A^{\text{fin}}(D, A) = A^{\text{fin}}(D, S')$ implies i) of the previous theorem. The answer is negative. Let S be the compactification $S = \mathbb{C} \cup \mathbb{C}^\infty$ with $\mathbb{C} \approx D$ and $\mathbb{C}^\infty \approx \partial D = T$ (the example of Section 3). Let S' be the example of the real projective space of dimension 2; that is $S' = \mathbb{C} \cup \mathbb{C}'^\infty$ with $\mathbb{C} \approx D$ and $\mathbb{C}'^\infty$ is the quotient of ∂D where all points w and $-w$ in ∂D are identified. Let $F : D \to \mathbb{C}$ be holomorphic and $\zeta \in \partial D$. Suppose $\lim_{z \to \zeta, z \in D} f(z)$ exists in $\mathbb{C}'^\infty$ and is $w \approx -w$,

$|w| = 1$. Let $V_\delta = \{z \in D : |\zeta - z| < \delta\}$ which is connected. If $\delta > 0$ is small enough, then $f(V_\delta)$ is contained in $E_1 \cup E_2$, where $E_1 = \{z \in D : |w - z| < \frac{1}{2}\}$ and $E_2 = \{z \in D : |(-w) - z| < \frac{1}{2}\}$. By the continuity of f the set $f(V_\delta)$ must be connected. Therefore, it is contained in one of the sets E_1 and E_2. It follows easily that $\lim_{z\to\zeta, z\in D} f(z)$ exists in $\mathbb{C}^\infty$. Thus, $A^{\text{fin}}(D, S) = A^{\text{fin}}(D, S')$ but $S \neq S'$ in the sense of Theorem 5.5.

An open question is whether there exists a metrizable compactification S of $\mathbb{C}$ such that $A^{\text{fin}}(D, S) = A(D)$.

6. Mergelyan's theorem

So far we have investigated the classes $A(D, S)$ of ρ-uniform limits of polynomials on the closed unit disc $\overline{D}$. It is interesting to replace $\overline{D}$ by other compact subsets L of $\mathbb{C}$, especially when L^c is connected. That would generalize the classical Mergelyan's theorem. A related question is the following.

Question. Let S, ρ be a metrizable compactification of $\mathbb{C}$ with its metric ρ. Let $L \subset \mathbb{C}$ be a compact set with L^c connected and a function $f\colon L \to S$ continuous, such that $f(L^0) \subset \mathbb{C}$ and $f_{|L^0}$ holomorphic. Does there exist a sequence of polynomials P_n, such that $P_n \to f$ uniformly on L with respect to the metric ρ?

In the special case where L is the closure of a Jordan domain Ω in $\mathbb{C}$, the answer to the above question is affirmative. Let $\varphi\colon D \to \Omega$ be a Riemann map. According to a theorem of Carathéodory ([8]), φ extends to a homeomorphism $\varphi\colon \overline{D} \to \overline{\Omega} \equiv L$.

Theorem 6.1. *Under the above assumptions and notation a function $f\colon L = \overline{\Omega} \to S$ is ρ-uniform limit of a sequence of polynomials on $\overline{\Omega}$ if and only if $f \circ \varphi \in A(D, S)$.*

Proof. Let P_n, $n = 1, 2, \dots$ be a sequence of polynomials such that

$$\sup_{w\in L} \rho(P_n(w), f(w)) \to 0 \quad \text{as } n \to +\infty.$$

Then $\sup_{z\in\overline{D}} \rho(P_n \circ \varphi(z), f \circ \varphi(z)) \to 0$, as $n \to +\infty$. We have $P_n \circ \varphi \in A(D) \subset A(D, S)$. Thus, there exists a polynomial Q_n such that $\sup_{z\in\overline{D}} \rho(Q_n(z), P_n\circ\varphi(z)) < 1/n$.

It follows $\sup_{z\in\overline{D}} \rho(Q_n(z), f \circ \varphi(z)) \to 0$, as $n \to +\infty$.

Therefore $f \circ \varphi \in A(D, S)$.

For the reverse direction we assume that $g = f \circ \varphi \in A(D, S)$. Thus, there exists a sequence of polynomials P_n such that $\sup_{z\in\overline{D}} \rho(P_n(z), f \circ \varphi(z)) \to 0$, as $n \to +\infty$. Setting $\varphi(z) = w \in L$ we find $\sup_{w\in L} \rho(P_n \circ \varphi^{-1}(w), f(w)) \to 0$, as $n \to +\infty$. Fix $n \in N$. Then $P_n \circ \varphi^{-1} \in A(L)$ and $P_n \circ \varphi^{-1}(L)$ is a compact subset of $\mathbb{C}$. The set $\{x \in \mathbb{C} : \inf_{w\in L} |x - P_n \circ \varphi^{-1}(w)| \leq 1\} = B$ is also a compact subset of $\mathbb{C}$ containing $P_n \circ \varphi^{-1}(L)$.

Therefore on B the Euclidean metric is uniformly equivalent to ρ. Thus, there exist $\delta > 0$, $\delta < 1$ such that ($\alpha, \beta \in B$ with $|\alpha - \beta| < \delta$ implies $\rho(\alpha, \beta) < 1/n$).

By the classical Mergelyan's theorem, there exists a polynomial Q_n, such that $\sup_{w\in L} |Q_n(w) - P_n \circ \varphi^{-1}(w)| < \delta$. Since $Q_n(w)$, $P_n \circ \varphi^{-1}(w) \in B$ for all $w \in L$, it follows $\sup_{w\in L} \rho(Q_n(w), P_n \circ \varphi^{-1}(w)) \leq 1/n$. The triangle inequality implies $\sup_{w\in L} \rho\big(Q_n(w), f(w)\big) \to 0$, as $n \to +\infty$.

Thus f is a ρ-uniform limit of polynomials on L. This completes the proof. $\square$

Taking into account the description of $A(D,S)$, Theorem 6.1 implies the following.

Theorem 6.2. *Let S, ρ be a metrizable compactification of $\mathbb{C}$ with metric ρ. Let Ω be a Jordan domain in $\mathbb{C}$ and $f\colon \overline{\Omega} \to S$ be a continuous function, such that $f(\Omega) \subset \mathbb{C}$ and $f_{|\Omega}$ holomorphic. Let $\varepsilon > 0$. Then there exists a polynomial P such that* $\sup_{z\in\overline{\Omega}} \rho(P(z), f(z)) < \varepsilon$.

For special cases of the above result see [18].

7. Relation to universal Taylor series

We denote by $H(D)$ the set of holomorphic functions in the open unit disc D. For $f \in H(D)$ we denote $S_N(f,0)(z) = \sum_{n=0}^{n} f^{(n)}(0)z^n/n!$, $N = 0,1,2,\dots$ the sequence of partial sums of the Taylor expansion of f with respect to 0. The class U_1, of universal Taylor series in D in the sense of Luh and Chui – Parnes is defined as follows

Definition 7.1. A holomorphic function $f \in H(D)$ belongs to the class U_1, if for every compact set $K \subset \mathbb{C}$, $K \cap \overline{D} = \varnothing$ with K^{c} connected and every function $h\colon K \to \mathbb{C}$ continuous on K and holomorphic in K^0, there exists a sequence $\lambda_n \in \{0,1,2,\dots\}$, $n = 1,2,\dots$ such that

$$\sup_{z\in K} |S_{\lambda_n}(f,0)(z) - h(z)| \to 0, \quad \text{as } n \to +\infty.$$

It is known that $U_1 = \bigcap_{m,j,s=1}^{\infty} \bigcup_n E(m,j,s,n)$ ([11]); see also [12]. Let us explain the notation.

The sequence f_j, $j = 1,2,\dots$ is an enumeration of the polynomials with coefficients in $Q + iQ$. The sequence K_m, $m = 1,2,\dots$ is given by the following.

Lemma 7.2 ([9]). *There exists a sequence K_m, $m = 1,2,\dots$ of compact sets $K_m \subset \mathbb{C}$, $K_m \cap \overline{D} = \varnothing$ with K_m^{c} connected with the following property: For every compact set $K \subset \mathbb{C}$, $K \cap \overline{D} = \varnothing$ with K^{c} connected, there exists $m \in \{1,2,\dots\}$ such that $K \subset K_m$.*

The parameter s takes values in the set $\{1,2,\dots,\}$ and the parameter n takes values in the set $\{0,1,2,\dots\}$.

Definition 7.3. $E(m,j,s,n) = \{f \in H(D) : \sup_{z\in K_m} |S_{\lambda_n}(f,0)(z) - f_j(z)| < 1/s\}$.

Theorem 7.4 ([11]). $U_1 = \bigcap_{m,j,s=1}^{\infty} \bigcup_{n=0}^{\infty} E(m,j,s,n)$.

Theorem 7.5 ([11]). *The class $U_1 \cap A(D)$ is dense and G_δ in $A(D)$ endowed with its usual topology.*

Combining Theorem 7.5 with the fact that $A(D)$ is open and dense in $\tilde{A}(D)$ and in $\tilde{A}(D) - \{f \equiv \infty\} = A^{\mathrm{fin}}(D, \mathbb{C} \cup \{\infty\})$ (Theorem 5.1) we easily conclude that $U_1 \cap A(D)$ is dense and G_δ in $\tilde{A}(D)$ and in $A^{\mathrm{fin}}(D, \mathbb{C} \cup \{\infty\})$. Thus the set $U_1 \cap \tilde{A}(D)$ is residual in $\tilde{A}(D)$ and in $A^{\mathrm{fin}}(D, \mathbb{C} \cup \{\infty\})$.

Theorem 7.6. *$U_1 \cap \tilde{A}(D)$ is dense and G_δ in $\tilde{A}(D)$ and in $A^{\mathrm{fin}}(D, \mathbb{C} \cup \{\infty\})$.*

PROOF. We have already seen that the set $U_1 \cap \tilde{A}(D)$ is residual. It remains to see that it is G_δ. Taking into account Theorem 7.4 it suffices to show that the set $E(m,j,s,n) \cap \tilde{A}(D)$ is open in $A^{\text{fin}}(D, \mathbb{C} \cup \{\infty\})$.

Let $f \in E(m,j,s,n) \cap \tilde{A}(D)$. Cauchy estimates give the existence of a positive number δ, such that $\big(g \in H(D), \sup_{|z|\le 1/2}|f(z)-g(z)| < \delta$ imply $g \in E(m,j,s,n)\big)$; see [12]. We fix such a $\delta > 0$. We are looking for an $\varepsilon > 0$ such that $(g \in \tilde{A}(D), \chi(f,g) < \varepsilon$ imply $\sup_{|z|\le 1/2}|f(z) - g(z)| < \delta)$.

The set $B = \big\{f(z) : |z| \le \frac{1}{2}\big\}$ is compact in $\mathbb{C}$. We set $\tau = \inf_{x\in B} \chi(x,\infty) > 0$. The $\varepsilon > 0$ that we will chose will satisfy $\varepsilon < \tau/2$. If $g \in \tilde{A}(D)$ satisfies $\chi(f,g) < \varepsilon < \tau/2$, then for all $z\colon |z| \le \frac{1}{2}$ we have $g(z) \in \Gamma = \{w \in \mathbb{C} : \chi(w,B) \le \tau/2\}$ and $B \subset \Gamma$. Since $\Gamma \subset \mathbb{C}$ is compact the chordal metric χ is uniformly equivalent to the Euclidean distance on Γ. Thus, there exists $\varepsilon > 0$, $\varepsilon < \tau/2$ such that $(w_1, w_2 \in \Gamma$ with $\chi(w_1,w_2) < \varepsilon$ imply $|w_1 - w_2| < \delta)$.

Let $g \in \tilde{A}(D)$ satisfying $\chi(f,g) < \varepsilon$. Then one easily sees that $g \not\equiv \infty$ and $\sup_{|z|\le 1/2}|f(z) - g(z)| < \delta$. This implies $g \in E(m,j,s,n) \cap \tilde{A}(D)$.

Therefore $E(m,j,s,n) \cap \tilde{A}(D)$ is open in $A^{\text{fin}}(D, \mathbb{C} \cup \{\infty\})$. The proof is complete. □

Question. Is there any universal Taylor series $f \in U_1$ belonging in $\tilde{A}(D) \setminus A(D)$?

I believe that the answer to the previous question is affirmative but I do not have a proof.

Remark 7.7. Theorem 7.6 can be generalized if $\mathbb{C} \cup \{\infty\}$, χ is replaced by any arbitrary metrizable compactification S, ρ of $\mathbb{C}$ and if $\overline{D}$ and D are replaced by $\overline{\Omega}$, Ω where $\Omega \subset \mathbb{C}$ is any domain (bounded or unbounded) satisfying that $\{\infty\} \cup [\mathbb{C} \setminus \overline{\Omega}]$ is connected (here the closure $\overline{\Omega}$ is taken in $\mathbb{C}$).

Remark 7.8. Theorem 7.6 can also be strengthen in the following way. We consider the set $X = \{f \in A^{\text{fin}}(D, \mathbb{C} \cup \{\infty\})$: For every $K \subset \mathbb{C}$ compact with K^c connected and every polynomial h there exists a sequence $\lambda_n \in \{0,1,2,\dots\}$, $n = 1, 2, \dots$ such that $\sup_{z\in K}|S_{\lambda_n}(f,0)(z) - h(z)| \to 0$, as $n \to +\infty$ and $\chi\big(f, S_{\lambda_n}(f,0)(z)\big) \to 0$, as $n \to +\infty\}$.

Then one can show that X is dense and G_δ in $A^{\text{fin}}(D, \mathbb{C} \cup \{\infty\})$. This also extends if we replace $\mathbb{C} \cup \{\infty\}, \chi$ by the arbitrary metrizable compactification S, ρ of $\mathbb{C}$.

References

1. L. V. Ahlfors, *Complex analysis: An introduction of the theory of analytic functions of one complex variable*, 2nd ed., McGraw-Hill, New York, 1966.
2. I. Androulidakis and V. Nestoridis, *Extensions of the disc algebra and of Mergelyan's theorem*, C. R. Math. Acad. Sci. Paris **349** (2011), no. 13-14, 745–748.
3. F. Bayart, K.-G. Grosse-Erdmann, V. Nestoridis, and C. Papadimitropoulos, *Abstract theory of universal series and applications*, Proc. Lond. Math. Soc. (3) **96** (2008), no. 2, 417–463.
4. L. Brown, W. Hengartner, and P. M. Gauthier, *Continuous boundary behaviour for functions defined in the open unit disc*, Nagoya Math. J. **57** (1975), 49–58.
5. C. K. Chui and M. N. Parnes, *Approximation by overconvergence of a power series*, J. Math. Anal. Appl. **36** (1971), 693–696.
6. J. Dugundji, *Topology*, 4th ed., Allyn and Bacon, Boston, MA, 1968.
7. K. Hoffman, *Banach spaces of analytic functions*, Prentice-Hall Series in Modern Analysis, Prentice-Hall, Englewood Cliffs, NJ, 1962.

8. P. Koosis, *Introduction to H_p spaces*, London Math. Soc. Lecture Note Ser., vol. 40, Cambridge Univ. Press, Cambridge, 1980.
9. W. Luh, *Approximation analytischer Funktionen durch überkonvergente Potenzreihen und deren Matrix-Transformierten*, Mitt. Math. Sem. Giessen Heft **88** (1970).
10. A. Melas and V. Nestoridis, *Universality of Taylor series as a generic property of holomorphic functions*, Adv. Math. **157** (2001), no. 2, 138–176.
11. A. Melas, V. Nestoridis, and I. Papadoperakis, *Growth of coefficients of universal Taylor series and comparison of two classes of functions*, J. Anal. Math. **73** (1997), 187–202.
12. V. Nestoridis, *Universal Taylor series*, Ann. Inst. Fourier (Grenoble) **46** (1996), no. 5, 1293–1306 (English, with English and French summaries).
13. ———, *An extension of the notion of universal Taylor series*, Computational Methods and Function Theory (Nicosia, 1997) (N. Papamichael, St. Ruscheweyh, and E. B. Saff, eds.), Ser. Approx. Decompos., vol. 11, World Sci. Publ., River Edge, NJ, 1999, pp. 421–430.
14. ———, *An extension of the disc algebra.* I, Bull. London Math. Soc., to appear.
15. ———, *An extension of the disc algebra*, available at `arXiv:1009.5364`.
16. V. Nestoridis and N. Papadatos, *An extension of the disc algebra.* II, submitted.
17. ———, *Another extension of the disc algebra*, available at `arXiv:1012.3674`.
18. V. Nestoridis and I. Papadoperakis, *A remark on two extensions of the disc algebra and Mergelyan's theorem*, available at `arXiv:1104.0833`.
19. W. Rudin, *Real and complex analysis*, 3rd ed., McGraw-Hill Book Co., New York, 1987.
20. ———, *Boundary values of continuous analytic functions*, Proc. Amer. Math. Soc. **7** (1956), 808–811.
21. M. Spivak, *A comprehensive introduction to differential geometry.* Vol. I, M. Spivak, Brandeis Univ., Waltham, MA, 1970.
22. D. A. Stegenga and K. Stephenson, *Generic covering properties for spaces of analytic functions*, Pacific J. Math. **119** (1985), no. 1, 227–243.

Department of Mathematics, University of Athens, 157 84 Athens, Greece
E-mail address: `vnestor@math.uoa.gr`

Centre de Recherches Mathématiques
CRM Proceedings and Lecture Notes
Volume **55**, 2012

Examples of Quantitative Universal Approximation

Thomas Kalmes, Markus Nieß, and Thomas Ransford

Abstract. Let $\mathcal{L} := (L_j)$ be a sequence of continuous maps from a complete metric space $(\mathcal{X}, d_{\mathcal{X}})$ to a separable metric space $(\mathcal{Y}, d_{\mathcal{Y}})$. An element $x \in \mathcal{X}$ is called *$\mathcal{L}$-universal* for a subset $\mathcal{M}$ of $\mathcal{Y}$ if $F(x, \mathcal{M}, \varepsilon) < \infty$ for all $\varepsilon > 0$, where

$$F(x, \mathcal{M}, \varepsilon) := \sup_{y \in \mathcal{M}} \inf\{j \in \mathbb{N} : d_{\mathcal{Y}}(y, L_j x) < \varepsilon\}.$$

In this article we obtain quantitative estimates for $F(x, \mathcal{M}, \varepsilon)$ in a variety of examples arising in the theory of universal approximation.

1. Introduction

Let $(\mathcal{X}, d_{\mathcal{X}})$ be a complete metric space, let $(\mathcal{Y}, d_{\mathcal{Y}})$ a separable metric space, and let $\mathcal{L} := (L_j)_{j \in \mathbb{N}}$ be a sequence of continuous mappings $L_j \colon \mathcal{X} \to \mathcal{Y}$. An element $x \in \mathcal{X}$ is called *$\mathcal{L}$-universal* if

$$\forall n \in \mathbb{N}\ \forall y \in \mathcal{Y}\ \exists N \in \mathbb{N} : \quad d_{\mathcal{Y}}(y, L_N x) < \frac{1}{n}.$$

We denote the set of all $\mathcal{L}$-universal elements by $\mathcal{U}(\mathcal{L})$. It is a G_δ-set, due to the separability of $\mathcal{Y}$. The sequence $\mathcal{L}$ is called *universal* if $\mathcal{U}(\mathcal{L}) \neq \varnothing$.

Given a subset $\mathcal{M} \subset \mathcal{Y}$, one might ask how fast the elements of $\mathcal{M}$ can be approximated by some $\mathcal{L}$-universal element x, that is, how many elements of the sequence $(L_j x)_{j \in \mathbb{N}}$ are needed to cover $\mathcal{M}$ by $B(L_j x, \varepsilon)$, $j = 1, \dots, N$, the ε-balls around $L_j x$? Evidently, the answer will be expressed in terms of the numbers:

$$F(x, \mathcal{M}, \varepsilon) = F(x, \varepsilon) := \sup_{y \in \mathcal{M}} \inf\{j \in \mathbb{N} \colon d_{\mathcal{Y}}(y, L_j x) < \varepsilon\}.$$

Note that $F(x, \mathcal{M}, \varepsilon)$ also depends on the metric $d_{\mathcal{Y}}$. Obviously, if $F(x, \mathcal{M}, \varepsilon)$ is finite for every $\varepsilon > 0$, then $\mathcal{M}$ must be *totally bounded* (that is, $\mathcal{M}$ can be covered by a finite number of ε-balls for every $\varepsilon > 0$).

When $\mathcal{Y}$ is a Fréchet space, a natural metric to consider is

$$d_{\mathcal{Y}}(y, z) := \sup_{n \in \mathbb{N}} \left(\min\left\{ p_n(y - z), \frac{1}{n} \right\} \right), \tag{1.1}$$

2010 *Mathematics Subject Classification.* Primary 30K05; Secondary 30D45, 30E10.
Second author supported by grants from DAAD.
Third author supported by grants from NSERC and the Canada Research Chairs program.
This is the final form of the paper.

where $(p_n)_{n\in\mathbb{N}}$ is an increasing sequence of seminorms defining the topology on $\mathcal{Y}$. In this case $d_{\mathcal{Y}}(y,z) < 1/n$ if and only if $p_n(y-z) < 1/n$. If $\mathcal{Y}$ is a Fréchet space, then the totally bounded subsets $\mathcal{M}$ of $\mathcal{Y}$ are precisely the relatively compact ones.

The above question was first studied in [10] for sequences of *composition* and *differentiation operators* on spaces $H(\Omega)$ of holomorphic functions on a simply connected domain Ω equipped with the compact-open topology. This is the Fréchet-space topology defined by the seminorms

$$p_n(f) := \|f-g\|_{K_n} := \max_{z\in K_n} |f(z)-g(z)|, \tag{1.2}$$

where $\mathcal{K} := (K_n)_{n\in\mathbb{N}}$ is a compact exhaustion of Ω, (i.e., $K_n \subseteq \Omega$ compact, K_n is contained in the interior of K_{n+1} for each $n\in\mathbb{N}$, and $\bigcup_{n\in\mathbb{N}} K_n = \Omega$). Recall that, in this situation, the totally bounded subsets of $H(\Omega)$ are exactly the *normal families.*

Consider the sequence $\mathcal{C} := (C_n)_{n\in\mathbb{N}}$ of composition operators, defined by

$$C_n\colon H(\Omega_2) \to H(\Omega_1),\ f \mapsto f\circ\varphi_n,$$

where $(\varphi_n)_{n\in\mathbb{N}}$ is a sequence of injective holomorphic mappings $\varphi_n\colon \Omega_1 \to \Omega_2$ between open subsets Ω_1,Ω_2 of $\mathbb{C}$. Recall that (φ_n) is called *runaway* if, for every pair of compact sets $K\subseteq\Omega_1$, $L\subseteq\Omega_2$, there exists an $N\in\mathbb{N}$ with $\varphi_N(K)\cap L = \varnothing$. This property characterizes the existence of $\mathcal{C}$-universal elements when $\Omega_1 = \Omega_2$ and Ω_1 is not conformally equivalent to $\mathbb{C}\setminus\{0\}$, cf. [3].

Now consider the sequence of differentiation operators $\mathcal{D} := (D^n)_{n\in\mathbb{N}}$, where

$$D\colon H(\Omega)\to H(\Omega),\ f\mapsto f'.$$

In this case, the existence of $\mathcal{D}$-universal elements is equivalent to Ω being simply connected, cf. [15].

In order to summarize the main results from [10] we introduce the following notation which will be used throughout this article. For a totally bounded subset $\mathcal{M}$ of an arbitrary metric space $\mathcal{Y}$ we define the *nth covering number*

$$\lambda_n := \lambda_n(\mathcal{M}) := \min\left\{ l\in\mathbb{N}\colon \exists\, y_1,\dots,y_l\in\mathcal{Y}\colon \mathcal{M}\subseteq \bigcup_{j=1}^{l} B(y_j,1/n)\right\}.$$

Obviously, the sequence $\big(\lambda_n(\mathcal{M})\big)_{n\in\mathbb{N}}$ measures the size of $\mathcal{M}$ in a metrical sense.

For totally bounded subsets $\mathcal{M}$ of $\mathcal{Y} = H(\Omega)$, i.e., for normal families over Ω, we need two more sequences. The first one, $(\gamma_n)_{n\in\mathbb{N}} = \big(\gamma_n(\mathcal{M})\big)_{n\in\mathbb{N}}$ measures the approximative behavior of the Taylor/Faber expansions and is defined as the smallest integers with

$$\|T_{\gamma_n}f - f\|_{K_n} < \frac{1}{n} \quad \forall f\in\mathcal{M},$$

where $T_k f$ denotes the kth Taylor/Faber polynomial on the compact set K_n. The second sequence $(\sigma_n)_{n\in\mathbb{N}} = \big(\sigma_n(\mathcal{M})\big)_{n\in\mathbb{N}}$ measures the speed of convergence of the anti-derivatives to 0 and is defined as the smallest integers with

$$\|(T_m f)^{(-j)}\|_{K_n} < \frac{1}{n^2} \quad \forall f\in\mathcal{M}, m\in\mathbb{N}\cup\{0\}, j\geq\sigma_n.$$

Using this notation, the main results in [10] are summarized in the following theorem.

Theorem 1. (i) *In case of* $\mathcal{C}$ *(composition operators): For any normal family* $\mathcal{M}$, *there exists a* $\mathcal{C}$*-universal function* f *with*

$$F(f,\mathcal{M},2/n) \leq n(\lambda_n+1) \quad (n\in\mathbb{N}).$$

The set of all $\mathcal{C}$*-universal functions satisfying the above estimate contains a* G_δ*-set, but is never dense. The set of* $\mathcal{C}$*-universal functions* f *satisfying*

$$F(f,\mathcal{M},2/n) = O(n\lambda_n) \quad (n\to\infty)$$

is dense.

(ii) *In case of* $\mathcal{D}$ *(differentiation operators): Let* Ω *be bounded. For any normal family* $\mathcal{M}$, *there exists a* $\mathcal{D}$*-universal function* f *with*

$$F(f,\mathcal{M},3/n) \leq n(\lambda_n+1)(\gamma_n+\sigma_{n(\lambda_n+1)}) \quad (n\in\mathbb{N}).$$

For $\Omega=\mathbb{D}$, *the unit disk,* $\gamma_n = O\big(n\log(nM_{2n+1})\big)$ *and* $\sigma_n = O\big(\log(n^2M_{2n+1})\big)$ *as* $n\to\infty$, *where* $M_n := \sup_{f\in\mathcal{M}}\|f\|_{K_n}$. *Hence, in this case,*

$$F(f,\mathcal{M},1/n) = O\big(n^2\lambda_{3n}\log(n\lambda_{3n}\max\{1,M_{12n\lambda_{3n}+1}\})\big) \quad (n\to\infty).$$

We introduce a special kind of fast approximating universal behavior.

Definition 2. A family of operators $\mathcal{L}$ is called *m-polynomial universal for* $\mathcal{M}$ if there is a $\mathcal{L}$-universal element x such that

$$F(x,\mathcal{M},1/n) = O(n^m) \quad (n\to\infty).$$

For a totally bounded set $\mathcal{M}\subseteq\mathcal{Y}$ with covering numbers λ_n, i.e., λ_n functions $f_1^{(n)},\dots,f_{\lambda_n}^{(n)}\in\mathcal{Y}$ cover $\mathcal{M}$ with their $(1/n)$-neighborhoods, the set of all m-polynomial universal functions is given by

$$\bigcup_{c\in\mathbb{N}}\left(\bigcap_{n\in\mathbb{N}}\bigcap_{j=1}^{\lambda_n}\bigcup_{N=1}^{c\cdot n^m} L_N^{-1}(B(f_j^{(n)},1/n))\cap\mathcal{U}(\mathcal{L})\right).$$

This is a $G_{\delta\sigma}$-set. It is unknown if it is also a G_δ-set.

In Section 2 we consider the above question for sequences of composition operators on kernels of differential operators and obtain exactly the same estimates as in the holomorphic case (compare Theorem 1 and Theorem 5). Section 3 contains an investigation of similar questions for universal Taylor series and comparisons of the results with those from [10] for differentiation operators. Finally, in Section 4, we consider some classic examples of normal families, like the set of normalized univalent functions S, and their covering numbers.

2. Composition operators on kernels of differential operators

In this section, let $\Omega\subset\mathbb{R}^d$ be open and let $P\in\mathbb{C}[X_1,\dots,X_d]$ be a non-zero polynomial. As usual, we equip $C^\infty(\Omega)$ with the Fréchet-space topology induced by the family of semi-norms

$$q_{K_n,n}(f) := \max_{\substack{x\in K_n\\ |\alpha|\leq n}} |\partial^\alpha f(x)|,$$

where $(K_n)_{n\in\mathbb{N}}$ is a compact exhaustion of Ω. We denote this Fréchet space by $\mathscr{E}(\Omega)$ and the metric defined in (1.1) by d. As the differential operator $P(D)$ is continuous on $\mathscr{E}(\Omega)$, it follows that the kernel of $P(D)$ in $\mathscr{E}(\Omega)$, namely

$$\mathscr{N}_P(\Omega) := \{f\in\mathscr{E}(\Omega) : P(D)f = 0\},$$

is a closed subspace of $\mathscr{E}(\Omega)$, and hence is itself a Fréchet space in a natural way. As is well known, $\mathscr{E}(\Omega)$ is separable, so the same is true for $\mathscr{N}_P(\Omega)$.

In case of P being hypoelliptic, the above mentioned Fréchet-space topology of $\mathscr{N}_P(\Omega)$ is induced by the family of semi-norms $(q_{K_n,0})_{n\in\mathbb{N}}$, see for example [8, Theorem 4.4.2]. We denote the corresponding metric defined in (1.1) by d_0. In particular, when dealing with the Cauchy – Riemann operator or the Laplace operator, we consider the spaces of holomorphic functions and harmonic functions respectively, equipped with the compact-open topology. As is well known, $\mathscr{N}_P(\Omega)$ is a Montel space if P is hypoelliptic (this follows for example from [8, Theorem 4.4.2]), so in this case $\mathcal{M} \subset \mathscr{N}_P(\Omega)$ is relatively compact if and only if $\mathcal{M}$ is bounded, i.e., if and only if for every compact $K \subset \Omega$ we have

$$\sup_{f\in\mathcal{M}} q_{K,0}(f) < \infty.$$

Definition 3. (i) Let $\varphi\colon \Omega \to \Omega$ be a C^∞-diffeomorphism. Then P is called *φ-invariant* if, for any $f \in C^\infty(\Omega)$, we have $f \circ \varphi \in \mathscr{N}_P(\Omega)$ whenever $f \in \mathscr{N}_P(\Omega)$. If P is φ-invariant and φ^{-1}-invariant, then we call P *completely φ-invariant.*

(ii) An open subset $U \subset \Omega$ is called *P-approximable in Ω* if $\{f|_U : f \in \mathscr{N}_P(\Omega)\}$ is dense in $\mathscr{N}_P(U)$.

Remark 4. (i) If P is φ-invariant, then the mapping

$$C_\varphi\colon \mathscr{N}_P(\Omega) \to \mathscr{N}_P(\Omega),\ f \mapsto f\circ\varphi$$

is well-defined and linear. Moreover, for compact $K \subset \mathbb{R}^d$ and $n \in \mathbb{N}_0$, we obviously have $q_{K,n}(C_\varphi f) \leq M q_{\varphi(K),n}(f)$ for all $f \in \mathscr{E}(\Omega)$, where $M > 0$ is a suitable constant depending on K and n. Thus C_φ is a continuous, linear operator on $\mathscr{N}_P(\Omega)$.

(ii) If, for the C^∞-diffeomorphism $\varphi\colon \Omega \to \Omega$, there is $g \in \mathscr{E}(\Omega)$ such that the set $\{x \in \Omega : g(x) = 0\}$ is nowhere dense in Ω and $P(D)\big(C_\varphi(f)\big) = g\, C_\varphi(P(D)f)$ for every $f \in \mathscr{E}(\Omega)$, then it follows immediately that P is completely φ-invariant. In case of $P(D)$ being the Cauchy – Riemann, Laplace or heat operator, it is shown in [9, Proposition 3.6] that this condition on φ is already necessary for P to be φ-invariant. Moreover, the same is true in case of $P(D)$ being the wave operator, under the mild additional assumption that φ does not mingle the time variable with the space variables and *vice versa.* It should be noted that in [9] the term "φ-invariance" is used for what we call complete φ-invariance here. Nevertheless, the proof of [9, Proposition 3.6] uses only that $f\circ\varphi \in \mathscr{N}_P(\Omega)$ for every $f \in \mathscr{N}_P(\Omega)$.

Let $(\varphi_n)_{n\in\mathbb{N}}$ be a sequence of C^∞-diffeomorphisms of Ω such that P is completely φ_n-invariant for every $n \in \mathbb{N}$. There are several articles dealing with the existence of universal functions for $(C_{\varphi_n})_{n\in\mathbb{N}}$ for special $P(D)$, in particular for the Cauchy – Riemann or the Laplace operator, see, e.g., [3, 4, 6]. For arbitrary P, a characterization is given in [9] for the case that Ω has convex components.

Our first result in this section is the following theorem.

Theorem 5. *Let $(\varphi_m)_{m\in\mathbb{N}}$ be a sequence of C^∞-diffeomorphisms on Ω such that P is completely φ_m-invariant for every m in $\mathbb{N}$. Assume that, for every compact subset K of Ω, there are a bounded open neighborhood $U \subset \Omega$ of K with $\overline{U} \subset \Omega$ and $m \in \mathbb{N}$ such that $\varphi_m(U) \cup U$ is P-approximable and $\varphi_m(U) \cap U = \varnothing$. Then there is a strictly increasing sequence of natural numbers $(m_n)_{n\in\mathbb{N}}$ such that, for* *any $\mathcal{M} \subset \mathscr{N}_P(\Omega)$ relatively compact, there is a universal function u for $(C_{\varphi_{m_n}})_{n\in\mathbb{N}}$ such that*

$$F(u, \mathcal{M}, 2/n) \leq n(\lambda_n + 1) \quad \forall n \in \mathbb{N}.$$

In order to make the proof of the above theorem more transparent, we first prove the following lemma.

Lemma 6. *Under the hypotheses of Theorem 5, for any compact exhaustion $(K_n)_{n\in\mathbb{N}}$ of Ω, there is a strictly increasing sequence $(m_n)_{n\in\mathbb{N}}$ of natural numbers such that, for any sequence $(f_n)_{n\in\mathbb{N}}$ in $\mathscr{N}_P(\Omega)$, there is $v \in \mathscr{N}_P(\Omega)$ with*

$$q_{K_n,n}\big(f_n - C_{\varphi_{m_n}}(v)\big) < \frac{1}{n} \quad \forall n \in \mathbb{N}.$$

Proof. Fix a compact exhaustion $(K_n)_{n\in\mathbb{N}}$ of Ω and a sequence $(f_n)_{n\in\mathbb{N}}$ in $\mathscr{N}_P(\Omega)$. We simply write C_n in place of C_{φ_n}.

We start by constructing a sequence of bounded, open subsets $(U_n)_{n\in\mathbb{N}}$ of Ω, sequences of natural numbers $(m_n)_{n\in\mathbb{N}}$ and $(r_n)_{n\in\mathbb{N}}$, and a sequence $(M_n)_{n\in\mathbb{N}}$ in $(1,\infty)$, such that:

(i) $\forall n \in \mathbb{N}$: $K_n \subset U_n \subset \overline{U_n} \subset \Omega$,
(ii) $\forall n \in \mathbb{N}$: $\varphi_{m_n}(U_n) \cap U_n = \varnothing$ and $\varphi_{m_n}(U_n) \cup U_n$ is P-approximable in Ω,
(iii) $(m_n)_{n\in\mathbb{N}}$ and $(r_n)_{n\in\mathbb{N}}$ are strictly increasing, with $r_n \geq n+1$ for each $n \in \mathbb{N}$,
(iv) $(M_n)_{n\in\mathbb{N}}$ is nondecreasing,
(v) $\forall n \in \mathbb{N}$, $f \in \mathscr{N}_P(\Omega)$: $q_{K_n,n}\big(C_{m_n}(f)\big) \leq M_n q_{K_{r_n},n}(f)$,
(vi) $\forall n \in \mathbb{N}$: $K_{r_n} \subset U_{n+1}$.

By hypothesis, there exists a bounded open neighborhood $U_1 \subset \Omega$ of K_1 with $\overline{U_1} \subset \Omega$, and there exists $m_1 \in \mathbb{N}$ with $\varphi_{m_1}(U_1) \cap U_1 = \varnothing$ and $\varphi_{m_1}(U_1) \cup U_1$ being P-approximable in Ω. Moreover, by the continuity of C_{m_1}, there are $r_1 \in \mathbb{N}, r_1 \geq 2$ and $M_1 > 1$ with $q_{K_1,1}\big(C_{m_1}(f)\big) \leq M_1\, q_{K_{r_1},1}(f)$.

Assume that $U_1, \ldots, U_n, m_1, \ldots, m_n, r_1, \ldots, r_n$ and $M_1, \ldots, M_n$ have already been constructed. For the compact set

$$K := \overline{U_n} \cup K_{r_n+1} \cup \bigcup_{j=1}^{m_n} \varphi_j(\overline{U_n}),$$

there exist, by hypothesis, a bounded open neighborhood $U_{n+1} \subset \overline{U_{n+1}} \subset \Omega$ and $m_{n+1} \in \mathbb{N}$ with $U_{n+1} \cap \varphi_{m_{n+1}}(U_{n+1}) = \varnothing$ and $U_{n+1} \cup \varphi_{m_{n+1}}(U_{n+1})$ being P-approximable in Ω. From $U_{n+1} \cap \varphi_{m_{n+1}}(U_{n+1}) = \varnothing$ and the definition of K, it follows that $m_{n+1} > m_n$. By the continuity of $C_{m_{n+1}}$, there are M_{n+1} and r_{n+1} with

$$q_{K_{n+1},n+1}\big(C_{m_{n+1}}(f)\big) \leq M_{n+1} q_{K_{r_{n+1}},n+1}(f)$$

for any $f \in \mathscr{N}_P(\Omega)$, where, without loss of generality, we may assume that $M_{n+1} \geq M_n$ and $r_{n+1} > \max\{r_n, n+2\}$.

We observe that, by (iii) and (vi), we have $K_{n+1} \subseteq K_{r_n} \subseteq U_{n+1}$ for every $n \in \mathbb{N}$.

Next, we recursively construct a sequence $(v_n)_{n\in\mathbb{N}}$ in $\mathscr{N}_P(\Omega)$ such that:

(a) $\forall n \in \mathbb{N}$: $q_{K_n,n}\big(f_n - C_{m_n}(v_n)\big) < 1/(2n)$,
(b) $\forall n \in \mathbb{N}$: $q_{K_{r_n},n}(v_{n+1} - v_n) < 1/(2^{n+1}M_{n+1})$.

Indeed, for $n = 1$, consider

$$w_1\colon U_1 \cup \varphi_{m_1}(U_1) \to \mathbb{C}, \quad w_1(x) := \begin{cases} 0, & \text{if } x \in U_1, \\ f_1\big(\varphi_{m_1}^{-1}(x)\big), & \text{if } x \in \varphi_{m_1}(U_1). \end{cases}$$

Since $U_1 \cap \varphi_{m_1}(U_1) = \varnothing$, the map w_1 is well-defined, and $w_1 \in \mathscr{N}_P\big(U_1 \cup \varphi_{m_1}(U_1)\big)$ follows from the complete φ_{m_1}-invariance of P. Fix $\psi_1 \in \mathscr{D}(U_1)$ such that $\psi_1 = 1$ in a neighborhood of K_1. Obviously, $\psi_1 \circ \varphi_{m_1}^{-1} \in \mathscr{D}\big(\varphi_{m_1}(U_1)\big)$, so that, for any $f \in \mathscr{N}_P\big(U_1 \cup \varphi_{m_1}(U_1)\big)$, we have $(\psi_1 \circ \varphi_{m_1}^{-1})f \in C^\infty(\Omega)$ in a natural way. Therefore,

$$p_1(f) := q_{K_{r_1},1}\big((\psi_1 \circ \varphi_{m_1}^{-1})f\big)$$

defines a continuous semi-norm on $\mathscr{N}_P\big(U_1 \cup \varphi_{m_1}(U_1)\big)$. The P-approximability of $U_1 \cup \varphi_{m_1}(U_1)$ in Ω and the continuity of the seminorm p_1 imply the existence of $v_1 \in \mathscr{N}_P(\Omega)$ with

$$p_1(v_1 - w_1) < \frac{1}{4M_1}.$$

Since, from the definition of w_1, we have $(\psi_1 \circ \varphi_{m_1}^{-1})w_1 = (\psi_1 \circ \varphi_{m_1}^{-1})(f_1 \circ \varphi_{m_1}^{-1})$, and because $\psi_1 = 1$ in a neighborhood of K_1, this implies

$$\begin{aligned} q_{K_1,1}\big(f_1 - C_{m_1}(v_1)\big) &= q_{K_1,1}\Big(C_{m_1}\big((\psi_1 \circ \varphi_{m_1}^{-1})(f_1 \circ \varphi_{m_1}^{-1} - v_1)\big)\Big) \\ &\le M_1 q_{K_{r_1},1}\big((\psi_1 \circ \varphi_{m_1}^{-1})(f_1 \circ \varphi_{m_1}^{-1} - v_1)\big) \\ &= M_1 p_1\big((w_1 - v_1)\big) < \tfrac{1}{4}, \end{aligned}$$

where we used (v) in the second step.

Assuming that $v_1, \ldots, v_n$ have already been constructed, we consider

$$w_{n+1}\colon U_{n+1} \cup \varphi_{m_{n+1}}(U_{n+1}) \to \mathbb{C},$$

$$w_{n+1}(x) := \begin{cases} v_n(x), & \text{if } x \in U_{n+1}, \\ f_{n+1}\big(\varphi_{m_{n+1}}^{-1}(x)\big), & \text{if } x \in \varphi_{m_{n+1}}(U_{n+1}). \end{cases}$$

Then, as for w_1, we have $w_{n+1} \in \mathscr{N}_P\big(U_{n+1} \cup \varphi_{m_{n+1}}(U_{n+1})\big)$. Fix $\psi_{n+1} \in \mathscr{D}(U_{n+1})$ such that $\psi_{n+1} = 1$ in a neighborhood of $K_{r_n} \supseteq K_{n+1}$. As above,

$$p_{n+1}(f) := q_{K_{r_{n+1}},n+1}\big((\psi_{n+1} \circ \varphi_{m_{n+1}}^{-1})f\big) + q_{K_{r_n},n}(f)$$

defines a continuous semi-norm on $\mathscr{N}_P\big(U_{n+1} \cup \varphi_{m_{n+1}}(U_{n+1})\big)$, so that the P-approximability of $U_{n+1} \cup \varphi_{m_{n+1}}(U_{n+1})$ in Ω yields $v_{n+1} \in \mathscr{N}_P(\Omega)$ with

$$p_{n+1}(v_{n+1} - w_{n+1}) < \frac{1}{2^{n+1}M_{n+1}}.$$

Again, since $(\psi_{n+1} \circ \varphi_{m_{n+1}}^{-1})w_{n+1} = (\psi_{n+1} \circ \varphi_{m_{n+1}}^{-1})(f_{n+1} \circ \varphi_{m_{n+1}}^{-1})$, and as $\psi_{m_{n+1}} = 1$ in a neighborhood of $K_{r_n} \supseteq K_{n+1}$, this implies

$$\begin{aligned} &q_{K_{n+1},n+1}\big(f_{n+1} - C_{m_{n+1}}(v_{n+1})\big) \\ &\quad = q_{K_{n+1},n+1}\Big(C_{m_{n+1}}\big((\psi_1 \circ \varphi_{m_{n+1}}^{-1})(f_{n+1} \circ \varphi_{m_{n+1}}^{-1} - v_{n+1})\big)\Big) \\ &\quad \le M_{n+1} q_{K_{r_{n+1}},n+1}\big((\psi_{n+1} \circ \varphi_{m_{n+1}}^{-1})(f_{n+1} \circ \varphi_{m_{n+1}}^{-1} - v_{n+1})\big) \\ &\quad = M_{n+1} q_{K_{r_{n+1}},n+1}\big((\psi_{n+1} \circ \varphi_{m_{n+1}}^{-1})(w_{n+1} - v_{n+1})\big) \\ &\quad \le M_{n+1} p_{n+1}(v_{n+1} - w_{n+1}) < \frac{1}{2^{n+1}} < \frac{1}{2(n+1)}, \end{aligned}$$

where we used (v) in the second step. Moreover, since $K_{r_n} \subset U_{n+1}$, and since, by definition, $v_n|_{U_{n+1}} = w_{n+1}|_{U_{n+1}}$, we obtain

$$q_{K_{r_n},n}(v_{n+1} - v_n) = q_{K_{r_n},n}(v_{n+1} - w_{n+1}) \leq p_{n+1}(v_{n+1} - w_{n+1}) < \frac{1}{2^{n+1}M_{n+1}},$$

thereby finishing the construction of $(v_n)_{n\in\mathbb{N}}$.

Because of the inclusion $K_{r_n} \supseteq K_n$, the fact that $M_n \geq 1$ and (b), we have

$$\forall n \in \mathbb{N}: \quad q_{K_n,n}(v_{n+1} - v_n) < \frac{1}{2^{n+1}},$$

so that $(v_n)_{n\in\mathbb{N}}$ is a Cauchy sequence in $\mathscr{N}_P(\Omega)$, and hence convergent. We set $v := \lim_{n\to\infty} v_n$, and observe that $v = v_n + \sum_{j=n}^{\infty}(v_{j+1} - v_j)$ for every $n \in \mathbb{N}$. From the continuity of C_{m_n}, and using (a), (v), (b), and (iv), we finally get that, for $n \in \mathbb{N}$,

$$\begin{aligned} q_{K_n,n}\big(f_n - C_{m_n}(v)\big) &= q_{K_n,n}\left(f_n - C_{m_n}(v_n) - \sum_{j=n}^{\infty} C_{m_n}(v_{j+1} - v_j)\right) \\ &\leq \frac{1}{2n} + \sum_{j=n}^{\infty} q_{K_n,n}\big(C_{m_n}(v_{j+1} - v_j)\big) \\ &\leq \frac{1}{2n} + \sum_{j=n}^{\infty} M_n q_{K_{r_n},n}(v_{j+1} - v_j) \\ &\leq \frac{1}{2n} + \sum_{j=n}^{\infty} M_n q_{K_{r_j},j}(v_{j+1} - v_j) \\ &\leq \frac{1}{2n} + \sum_{j=n}^{\infty} \frac{M_n}{2^{j+1}M_{j+1}} < \frac{1}{n}. \end{aligned}$$

This completes the proof of the lemma. □

PROOF OF THEOREM 5. Let $(K_n)_{n\in\mathbb{N}}$ be the compact exhaustion of Ω defining the metric d on $\mathscr{N}_P(\Omega)$. For $n \in \mathbb{N}$, let $f_1^{(n)}, \ldots, f_{\lambda_n}^{(n)} \in \mathscr{N}_P(\Omega)$ be such that

$$\mathcal{M} \subseteq \bigcup_{j=1}^{\lambda_n} B(f_j^{(n)}, 1/n),$$

and let $(g_n)_{n\in\mathbb{N}}$ be a dense sequence in $\mathscr{N}_P(\Omega)$. We define $(f_n)_{n\in\mathbb{N}}$ to be the sequence

$$f_1^{(1)}, \ldots, f_{\lambda_1}^{(1)}, g_1, f_1^{(2)}, \ldots, f_{\lambda_2}^{(2)}, g_2, f_1^{(3)}, \ldots, f_{\lambda_3}^{(3)}, g_3, \ldots$$

Applying Lemma 6 gives an increasing sequence of natural numbers $(m_n)_{n\in\mathbb{N}}$ and $u \in \mathscr{N}_P(\Omega)$ such that

$$q_{K_n,n}\big(f_n - C_{\varphi_{m_n}}(u)\big) < \frac{1}{n}.$$

Since $\{g_n : n \in \mathbb{N}\}$ is dense in $\mathscr{N}_P(\Omega)$, it follows that u is universal for $(C_{\varphi_{m_n}})_{n\in\mathbb{N}}$. Now fix $f \in \mathcal{M}$ and $n \in \mathbb{N}$. Then $d(f, f_j^{(n)}) < 1/n$ for some $1 \leq j \leq \lambda_n$. Because $f_j^{(n)} = f_N$ for some $n \leq N \leq \sum_{j=1}^{n}(\lambda_j + 1) \leq n(\lambda_n + 1)$, and because

$$q_{K_N,N}\big(f_N - C_{\varphi_{M_N}}(u)\big) < \frac{1}{N},$$

that is

$$d(f_j^{(n)}, C_{\varphi_{m_N}}(u)) < \frac{1}{N},$$

the result follows. □

In order to verify the hypothesis of Theorem 5 in some concrete situations we recall the following results about approximation of zero solutions of differential equations. Part (i) of the next theorem is the Malgrange–Lax Theorem, cf. [8, Theorem 4.4.5], while part (ii) is due to Hörmander, see, e.g., [8, Theorem 10.5.2].

Theorem 7. *Let $U \subseteq \Omega$ be open.*

(i) *Assume that P is elliptic. If $\Omega \backslash U$ is not the disjoint union $F \cup K$, where K is compact and nonempty and F is closed in Ω, then U is P-approximable in Ω.*

(ii) *Suppose that every $\mu \in \mathscr{E}'(\overline{\Omega})$ with $\operatorname{supp} P(-D)\mu \subset U$ already belongs to $\mathscr{E}'(U)$. Then U is P-approximable in Ω.*

Remark 8. (i) Let $\hat{\Omega}$ denote the one-point compactification of Ω. It is easily seen that the condition in (i) of Theorem 7 is equivalent to $\hat{\Omega} \setminus U$ being connected while part (ii) immediately implies the P-approximability in Ω of every $U \subset \Omega$ with convex components.

(ii) It is shown in [9, Proof of Corollary 4.6] that, if φ satisfies the condition under (ii) of Remark 4, and if $K \subset \Omega$ is compact, has only convex components and satisfies $\varphi(K) \cap K = \varnothing$, then $\varphi(\mathring{K}) \cup \mathring{K}$ is P-approximable in Ω, where $\mathring{K}$ denotes the interior of K.

(iii) Assume that Ω has only convex components and that every element of the sequence $(\varphi_m)_{m\in\mathbb{N}}$ of C^∞-diffeomorphisms satisfies the condition (ii) of Remark 4. Then it follows from (ii) above that the assumption of Theorem 5 is fulfilled if and only if, for every compact subset K of Ω, there is $m \in \mathbb{N}$ with $\varphi_m(K) \cap K = \varnothing$.

Corollary 9. *Let $(\varphi_m)_{m\in\mathbb{N}}$ be a sequence of C^∞-diffeomorphisms of Ω such that P is completely φ_m-invariant for every $m \in \mathbb{N}$. Assume, further, that for any compact subset $K \subset \Omega$, there is $m \in \mathbb{N}$ with $\varphi_m(K) \cap K = \varnothing$. Then there is an increasing sequence of natural numbers $(m_n)_{n\in\mathbb{N}}$ for which the following hold:*

(i) *Assume that Ω is contractible and that Ω has the complementation property, i.e., given any compact subset $K \subset \Omega$, there is at most one component of $\Omega \setminus K$ whose closure in Ω is not compact. If P is elliptic, then, for any relatively compact subset $\mathcal{M}$ of $\mathscr{N}_P(\Omega)$, there is a universal function $u \in \mathscr{N}_P(\Omega)$ for $(C_{\varphi_{m_n}})_{n\in\mathbb{N}}$ such that $F(u, \mathcal{M}, 2/n) \leq n(\lambda_n + 1)$ for each $n \in \mathbb{N}$.*

(ii) *If P is arbitrary, each φ_m satisfies the condition* (ii) *from Remark 4, and Ω has only convex components, then, for any relatively compact subset $\mathcal{M}$ of $\mathscr{N}_P(\Omega)$, there is a universal function $u \in \mathscr{N}_P(\Omega)$ for $(C_{\varphi_{m_n}})_{n\in\mathbb{N}}$ such that $F(u, \mathcal{M}, 2/n) \leq n(\lambda_n + 1)$ for each $n \in \mathbb{N}$.*

PROOF. Part (ii) follows from the hypothesis, Remark 8, and Theorem 5.

In order to show (i), it is straightforward to verify that, with

$$U_n := \{x \in \Omega : |x| < n \text{ and } \operatorname{dist}(x, \Omega^c) > 1/n\},$$

the set $\Omega \setminus U_n$ is not the disjoint union of a nonempty, compact set K and a set F closed in Ω. As φ_m is a homeomorphism, the same holds for $\varphi_m(U_n)$ for arbitrary m. By hypothesis, there is m_0 such that $U_n \cap \varphi_{m_0}(U_n) = \varnothing$. The contractibility of Ω

easily gives that every continuous mapping $g\colon \Omega \to S^1$ is homotopic to a constant. Together with the complementation property of Ω, this implies the unicoherence of $\hat{\Omega}$ (see, e.g., [5, Theorem 4.12]), so that for the two connected and closed sets $\hat{\Omega} \setminus U_n$ and $\hat{\Omega} \setminus \varphi_{m_0}(U_n)$ covering $\hat{\Omega}$, their intersection $\hat{\Omega} \setminus \big(U_n \cup \varphi_{m_0}(U_n)\big)$ is also connected. Therefore, $U_n \cup \varphi_{m_0}(U_n)$ is P-approximable in Ω, by Theorem 7(i). Part (i) now follows from this and from Theorem 5. □

3. Universal Taylor series

Let $\Omega \subseteq \mathbb{C}$ be a simply connected domain. For $L \subset \mathbb{C} \setminus \Omega$ compact with connected complement and $\zeta \in \Omega$, we consider the sequence $\mathcal{T}_L^\zeta = (T_{L,n}^\zeta)_{n\in\mathbb{N}}$ of linear operators

$$T_{L,n}^{(\zeta)}\colon H(\Omega) \to A(L), \quad f \mapsto T_{L,n}^{(\zeta)} f(z) := T_n^\zeta f(z) := \sum_{\nu=0}^{n} a_\nu^{(\zeta)} (z-\zeta)^\nu,$$

where $a_\nu^{(\zeta)}$ denotes the νth Taylor coefficient of f expanded about ζ, and $A(L)$ denotes the space of all continuous functions on L that are holomorphic in the interior of L. Endowing $A(L)$ with the sup-norm $\|f\|_L$, it follows from Mergelyan's theorem that $\{f|_L : f \in A(\tilde{L})\}$ is dense in $A(L)$ for any compact superset $\tilde{L}$ of L.

As shown in [13, Lemma 2.1], there exists a sequence $(L_k)_{k\in\mathbb{N}}$ of compact sets $L_k \subset \mathbb{C} \setminus \Omega$ with connected complement such that, for every compact subset $L \subset \mathbb{C} \setminus \Omega$ with connected complement, there is $k_0 \in \mathbb{N}$ with $L \subset L_{k_0}$. The set of all universal Taylor series in the sense of [13] is then given by

$$\mathcal{U}(\zeta) := \bigcap_{k\in\mathbb{N}} \mathcal{U}(\mathcal{T}_{L_k}^{(\zeta)}),$$

and it is shown in [12, Theorem 2] that

$$\mathcal{U}(\zeta_1) = \mathcal{U}(\zeta_2)$$

for any $\zeta_1, \zeta_2 \in \Omega$. Abusing our former notation we simply write $\mathcal{U}(\mathcal{T})$ for these equal sets, that is, $f \in \mathcal{U}(\mathcal{T})$ if and only if the set of the Taylor polynomials of f expanded about an arbitrary $\zeta \in \Omega$ is dense in any $A(L)$, where $L \subset \mathbb{C} \setminus \Omega$ is compact and has connected complement.

Our first aim is to compare how fast a normal family $\mathcal{M}$ may be approximated by the partial sums of a universal Taylor series $f \in \mathcal{U}(\mathcal{T})$ with the speed of approximation by the derivatives of a function $g \in \mathcal{U}(\mathcal{D})$. In a second step we then estimate the possible speed of approximation for $f \in \mathcal{U}(\mathcal{T})$. To help us in pursuit of these goals, we introduce the following notion:

Definition 10. Let $\Omega \subseteq \mathbb{C}$ be open, let $L_k \subset \mathbb{C} \setminus \Omega$ be compact, and let $f_k \in A(L_k)$. We say $f \in H(\Omega)$ has a *uniformly universal power series in* $\zeta_1 \in \Omega$ *for* $(f_k, L_k)_{k\in\mathbb{N}}$ if there is a sequence of natural numbers $(N_k)_{k\in\mathbb{N}}$ such that

$$\forall\, 1 \le j \le k\ \exists\, 1 \le n \le N_k : \quad \|f_j - T_n^{(\zeta_1)} f\|_{L_j} < \frac{1}{j^2}.$$

Let $\mathcal{P}_\mathbb{Q}$ be the set of all polynomials with coefficients in $\mathbb{Q}+\mathrm{i}\mathbb{Q}$, and let $\mathcal{K}_{\mathbb{C}\setminus\Omega} := (L_k)$ be a sequence of compact sets in $\mathbb{C}\setminus\Omega$ as above. If Ω is simply connected and if (f_k, L_k) contains each element $(p, L) \in \mathcal{P}_\mathbb{Q} \times \mathcal{K}_{\mathbb{C}\setminus\Omega}$ infinitely often, then a uniformly universal power series f for (f_k, L_k) is a universal Taylor series, i.e., $f \in \mathcal{U}(\mathcal{T})$.

Remark 11. (i) Let $\Omega = \mathbb{D}$, let f be a uniformly universal power series in $\zeta_1 = 0$ for (f_k, L_k), $k > 2$, with

$$f_k \equiv 0, \quad L_k := \left\{z\colon \left|z - \tfrac{3}{2}k\right| \leq k\right\},$$

and let (N_k) be a sequence of numbers as in Definition 10. Assume only that $\|T_{N_k}^{(0)} f\|_{L_k} \leq 1$ for each $k > 2$. Then the Taylor coefficients satisfy

$$|a_\nu|^{1/\nu} \leq k^{\log \frac{5}{2} - 1} \quad \text{for all } \nu \text{ with } \widetilde{N}_k := \left[\frac{N_k}{\log k}\right] + 1 \leq \nu \leq N_k,$$

cf. [7, p. 84]. Thus approximation by partial sums occurs with rather large blocks of small coefficients. Assume, further, that $f \in \mathcal{U}(\mathcal{T})$, so in particular the radius of convergence of f is 1. Since

$$\limsup_{\substack{\nu \to \infty \\ \nu \in I}} |a_\nu|^{1/\nu} = 0, \quad I := \mathbb{N} \cap \bigcup_{k \in \mathbb{N}} [\widetilde{N}_k, N_k],$$

for every $\varepsilon > 0$ the power series of f must also have infinitely many Taylor coefficients a_ν with $|a_\nu|^{1/\nu} \geq 1 - \varepsilon$, $\nu \in \mathbb{N} \setminus I$. More precisely, the set of indices

$$\kappa := \{k \in \mathbb{N} : \exists\, \nu \in (N_{k-1}, \widetilde{N}_k) \text{ with } |a_\nu|^{1/\nu} > k^{\log(5/2)-1}\} \subset \{k \in \mathbb{N} : N_{k-1} < \widetilde{N}_k\}$$

is infinite. Thus, on the infinite set κ we have

$$\frac{N_k}{N_{k-1}} \geq \log k, \quad k \in \kappa.$$

The same holds if f has finite radius of convergence, without necessarily belonging to $\mathcal{U}(\mathcal{T})$.

(ii) We compare the above quotient N_k/N_{k-1} with a similar one for a function $g \in \mathcal{U}(\mathcal{D})$. In [10, Theorem 8], a function $g \in \mathcal{U}(\mathcal{D}) \cap H(\Omega)$ (where $H(\Omega)$ is endowed with the natural metric as in (1.1) and seminorms as in (1.2)) is constructed, which is fast approximating for a normal family $\mathcal{M}$. For appropriate functions f_j, $j = 1, \dots, k$, define (N_k) to be a sequence of natural numbers with

$$\forall\, 1 \leq j \leq k \;\exists\, 1 \leq n \leq N_k : \quad \|f_j - g^{(n)}\|_{K_j} < \frac{1}{j}.$$

For the constructed function $g \in \mathcal{U}(\mathcal{D})$, we obtain from [10, Proof of Theorem 8] that $N_k \leq N_{k-1} + \sigma_k + \gamma_k$, where σ_k and γ_k are defined as in the paragraph preceding Theorem 1. Considering $\mathcal{M} = \{0\}$, i.e., $f_j \equiv 0$, as in (i), $\sigma_k = \gamma_k := k$ is a possible choice, and so is $N_k := k(k+1)$. Hence

$$\frac{N_k}{N_{k-1}} = \frac{k+1}{k-1},$$

which is bounded, and not strictly increasing to ∞ on a subsequence κ, as is the case for $f \in \mathcal{U}(\mathcal{T})$.

This simple example already illustrates the tremendous difference between the speeds of approximation by $f \in \mathcal{U}(\mathcal{T})$ and $g \in \mathcal{U}(\mathcal{D})$. To elucidate this difference, we remark that successive derivatives of a function may change rather quickly, while in universal approximation successive partial sums change rather slowly, which is expressed by large blocks of rather small coefficients, namely so-called Ostrowski gaps, cf. [7]. Even the boundedness of the partial sums on a nonpolar set $E \subset \mathbb{C} \setminus \overline{\mathbb{D}}$ causes small coefficients, in this case so-called Hadamard–Ostrowski gaps, as recently shown in [2].

Nevertheless, we also want to give results in the other direction by showing which speeds of approximation are possible, this time by estimating possible upper bounds, not for $F(f,\mathcal{M},1/n)$, but for the numbers N_k as defined in Definition 10. In order to construct a universal function f with small $F(f,\mathcal{M},1/n)$, we first find a sequence (f_k) containing appropriately chosen centers, whose balls $B(f_k,1/n)$ cover $\mathcal{M}$. Their number is $\lambda_n(\mathcal{M})$, the n-covering number of $\mathcal{M}$. Then these centers f_k will be approximated by the first N_k Taylor polynomials of f, i.e., by $T_j^{(\zeta)}f$, $j \in \{1,\dots,N_k\}$. Finally, $F(f,\mathcal{M},1/n)$ and N_k are connected, since $F(f,\mathcal{M},1/n) \le N_k$ for some k which may depend on $\lambda_n(\mathcal{M})$.

With regard to estimate N_k, we start by recalling some results on best polynomial approximation, cf. [1]. For a continuous complex-valued function f on a compact set K in the plane, let

$$d_n := d_n(f,K) := \inf\{\|f-p\|_K : p \in \mathcal{P}_n\},$$

where $\mathcal{P}_n$ is the vector space of complex polynomials of degree at most n. Recall that a Green's function g_K for $\mathbb{C}\setminus K$ is a continuous function $g_K\colon \mathbb{C}\to[0,+\infty)$ which is identically equal to zero on K, harmonic on $\mathbb{C}\setminus K$, and has a logarithmic singularity at infinity, in the sense that $g_K(z)-\log|z|$ is harmonic at infinity.

Theorem 12 (Walsh). *Let K be a compact subset of the plane such that $\mathbb{C}\setminus K$ is connected and has a Green's function g_K. For $R>1$, let $D_R := \{z\in\mathbb{C} : g_K < \log R\}$. Let f be continuous on K. Then $\limsup_{n\to\infty} d_n(f,K)^{1/n} \le 1/R$ if and only if f is the restriction to K of a function holomorphic in D_R.*

The proof of the "if" part of this theorem for the case $K=[-1,1]$, given in [1, Section 2] by the use of duality theory, in fact provides the following result which will be crucial for our considerations. We include its proof here for the reader's convenience.

Lemma 13. *Let Ω be an open subset of $\mathbb{C}$, and let K be a compact subset of Ω such that $\mathbb{C}\setminus K$ is connected and has a Green's function g_K. Let $R>1$ be such that $\overline{D_R}\subset\Omega$. Then, for every $f\in H(\Omega)$, we have*

$$\forall\, 1<r<\rho<R:\quad d_n(f,K) \le \|f\|_{\partial D_R}\left(\frac{r}{\rho}\right)^n \frac{8\lambda(D_R\setminus D_\rho)}{\pi\operatorname{dist}(\partial D_R, D_\rho)\operatorname{dist}(\partial D_r, K)},$$

where λ denotes Lebesgue measure on $\mathbb{C}$.

Proof. Let $1<r<\rho<R$. Choose $\phi\in\mathscr{D}(\Omega)$ with $\operatorname{supp}\phi\subseteq D_R$ and $\phi=1$ in a neighborhood of D_ρ, and set $F:=\phi f\in\mathscr{D}(\Omega)\subset\mathscr{D}(\mathbb{R}^2)$. Then it follows, as in [1, Section 2], that

$$d_n = d_n(f,K) = \int_{D_R\setminus D_\rho} \tilde\mu(z)\frac{\partial}{\partial\bar z}F(z)\,\mathrm{d}\lambda(z), \tag{3.1}$$

where λ denotes Lebesgue measure on $\mathbb{C}$, and $\tilde\mu\in H(\mathbb{C}\setminus K)$ satisfies

$$\forall\, z\in\mathbb{C}\setminus D_r:\quad |\tilde\mu(z)| \le \frac{1}{\pi\operatorname{dist}(\partial D_r,K)}\Big(\exp\big(\log r - g_K(z)\big)\Big)^n.$$

In particular, for all $z\in\mathbb{C}\setminus D_\rho$, we have

$$|\tilde\mu(z)| \le \frac{1}{\pi\operatorname{dist}(\partial D_r,K)}\big(\exp(\log r-\log\rho)\big)^n = \frac{1}{\pi\operatorname{dist}(\partial D_r,K)}\left(\frac{r}{\rho}\right)^n,$$

so that, by (3.1), by the identity $(\partial/\partial\bar{z})F(z) = f(z)(\partial/\partial\bar{z})\phi(z)$ and by the maximum principle applied to f, we have

$$\text{(3.2)} \qquad \begin{aligned} d_n &\leq \frac{1}{\pi\,\mathrm{dist}(\partial D_r, K)}\left(\frac{r}{\rho}\right)^n \int_{D_R\setminus D_\rho}\left|f(z)\frac{\partial}{\partial\bar{z}}\phi(z)\right|\mathrm{d}\lambda(z) \\ &\leq \|f\|_{\partial D_R}\left(\frac{r}{\rho}\right)^n \frac{1}{\pi\,\mathrm{dist}(\partial D_r, K)}\sup_{z\in D_R}\left|\frac{\partial}{\partial\bar{z}}\phi(z)\right|\lambda(D_R\setminus D_\rho). \end{aligned}$$

Let $\delta := \mathrm{dist}(\partial D_R, D_\rho)$ be the distance from D_ρ to the complement of D_R. According to [8, Proof of Theorem 1.4.2], we can choose ϕ with

$$\forall\,\alpha\in\mathbb{N}^2, |\alpha| = k,\ x\in\mathbb{R}^2: \quad |\partial^\alpha\phi(x)| \leq 8^k/(\delta_1\cdots\delta_k),$$

where $(\delta_j)_{j\in\mathbb{N}}$ is any decreasing sequence of positive numbers with $\sum_{j=1}^\infty \delta_j < \delta$. In particular, we can choose ϕ such that

$$\forall\,z\in\mathbb{C}: \quad \left|\frac{\partial}{\partial\bar{z}}\phi(z)\right| \leq \frac{8}{\delta_1},$$

with $0 < \delta_1 < \delta$ arbitrary. Combining this with (3.2) gives

$$\forall\,0<\delta_1<\delta: \quad d_n \leq \|f\|_{\partial D_R}\left(\frac{r}{\rho}\right)^n \frac{8}{\pi\,\mathrm{dist}(\partial D_r, K)\delta_1}\lambda(D_R\setminus D_\rho),$$

and, letting δ_1 tend to δ, we have

$$\forall\,1<r<\rho<R: \quad d_n \leq \|f\|_{\partial D_R}\left(\frac{r}{\rho}\right)^n \frac{8\lambda(D_R\setminus D_\rho)}{\pi\,\mathrm{dist}(\partial D_R, D_\rho)\,\mathrm{dist}(\partial D_r, K)}.$$

This completes the proof. □

To formulate our next result conveniently, we introduce the following notion. Let K, L be two non-empty, disjoint, compact subsets of $\mathbb{C}$ such that $\mathbb{C}\setminus(K\cup L)$ has a Green's function g. We call $R > 1$ *separating for* K *and* L if no component of $D_R := \{z\in\mathbb{C} : g(z) < \log R\}$ contains elements of both K and L. That is, if U_R is the union of the components of D_R intersecting K, and if $V_R := D_R\setminus U_R$, then U_R, V_R are open, disjoint neighborhoods of K, L, respectively with $U_R\cup V_R = D_R$.

Proposition 14. *Let $\Omega\subseteq\mathbb{C}$ be open and simply connected, let $\zeta\in\Omega$, and let $(K_k)_{k\in\mathbb{N}}$ be a compact exhaustion of Ω such that $\zeta\in K_1$ and $\mathbb{C}\setminus K_k$ is connected for every $k\in\mathbb{N}$. Also, for $k\in\mathbb{N}$, let $\Omega_k\subset\mathbb{C}$ be open, let $L_k\subset\Omega_k$ be compact, and let $f_k\in H(\Omega_k)$. Assume that $\mathbb{C}\setminus L_k$ is connected, that $K_k\cap L_k = \varnothing$, and that $\mathbb{C}\setminus(K_k\cup L_k)$ has a Green's function g_k for every $k\in\mathbb{N}$. Let $R_k > 1$ be separating for K_k, L_k, and suppose further that $D_k := D_{R_k} = \{z\in\mathbb{C} : g_k(z) < \log R_k\}\subset\Omega\cup\Omega_k$.*

Then, for every choice of $1 < r_k < \rho_k < R_k$ ($k\in\mathbb{N}$), there exists $f\in H(\Omega)$ with uniformly universal power series in ζ for $(f_k, L_k)_{k\in\mathbb{N}}$ such that

$$\forall\,k\geq 2: \quad N_k < N_{k-1} + \frac{\log^+\left(k^2\|f_k - T^{(\zeta)}_{N_{k-1}}f\|_{\overline{V_k}}\,q_k^{N_{k-1}}C_k\right)}{\log(\rho_k/r_k)} + 1,$$

where

$$V_k := V_{R_k}, \quad q_k := \frac{\mathrm{diam}(K_k\cup L_k)}{\mathrm{dist}(K_k, \overline{V_k})},$$

and

$$C_k := C(r_k,\rho_k,R_k) := \frac{8\lambda(D_{R_k}\setminus D_{\rho_k})}{\pi\,\mathrm{dist}(\partial D_{R_k}, D_{\rho_k})\,\mathrm{dist}(\partial D_{r_k}, K_k\cup L_k)}.$$

PROOF. Like in [11, Proof of Theorem 2] we begin by constructing a sequence of polynomials $(P_k)_{k\in\mathbb{N}_0}$ and a strictly increasing sequence of integers $(N_k)_{k\in\mathbb{N}_0}$ with the following properties: the degree of P_k satisfies $\deg P_k = N_k$, the point ζ is a zero of P_k of multiplicity at least N_{k-1}, and

$$\forall\, k \geq 1: \quad \|P_k\|_{K_k} < \frac{1}{k^2}, \tag{3.3}$$

as well as

$$\forall\, k \geq 1: \quad \left\| \sum_{\nu=0}^{k} P_\nu - f_k \right\|_{L_k} < \frac{1}{k^2}. \tag{3.4}$$

We set $P_0(z) \equiv 1$ and $N_0 = 0$. Suppose that, for some $k \in \mathbb{N}$, the polynomials $P_0, \dots, P_{k-1}$ and the integers $N_0, \dots, N_{k-1}$ have already been determined. Because R_k is separating for K_k and L_k, we have, with $U_k := U_{R_k}$ and $V_k := V_{R_k}$, disjoint open neighborhoods of K_k and L_k with $U_k \cup V_k = D_k$. Consider the function

$$h_k\colon U_k \cup V_k \to \mathbb{C}, \qquad z \mapsto \begin{cases} 0, & \text{if } z \in U_k, \\ \dfrac{f_k(z) - \sum_{\nu=0}^{k-1} P_\nu(z)}{(z-\zeta)^{N_{k-1}}}, & \text{if } z \in V_k, \end{cases}$$

which is well-defined and holomorphic. From Lemma 13, we obtain that

$$d_n(h_k, K_k \cup L_k) \leq \|h_k\|_{\partial D_k} \left(\frac{r_k}{\rho_k}\right)^n C_k \leq \|h_k\|_{\overline{V_k}} \left(\frac{r_k}{\rho_k}\right)^n C_k,$$

where, in the last step, we used the maximum principle and the fact that $h_k|_{U_k} = 0$. Hence, in order to have

$$d_n(h_k, K_k \cup L_k) < \frac{1}{k^2 \max_{K_k \cup L_k} |z-\zeta|^{N_{k-1}}}, \tag{3.5}$$

it suffices that

$$k^2 \max_{K_k \cup L_k} |z-\zeta|^{N_{k-1}} \|h_k\|_{\overline{V_k}} C_k < \left(\frac{\rho_k}{r_k}\right)^n.$$

The latter is obviously the case if

$$k^2 \left\| f_k - \sum_{\nu=0}^{k-1} P_\nu \right\|_{\overline{V_k}} \left(\frac{\max_{K_k \cup L_k} |z-\zeta|}{\min_{\overline{V_k}} |z-\zeta|} \right)^{N_{k-1}} C_k < \left(\frac{\rho_k}{r_k}\right)^n.$$

Moreover, $\min_{\overline{V_k}} |z-\zeta| \geq \operatorname{dist}(K_k, \overline{V_k})$ and $\max_{K_k \cup L_k} |z-\zeta| \leq \operatorname{diam}(K_k \cup L_k)$, the diameter of $K_k \cup L_k$, so that (3.5) is satisfied if

$$n \geq \frac{\log^+(k^2 \|f_k - \sum_{\nu=0}^{k-1} P_\nu\|_{\overline{V_k}} q_k^{N_{k-1}} C_k)}{\log(\rho_k/r_k)} =: c(k). \tag{3.6}$$

By the above, if we fix $n \in \mathbb{N} \cap [c(k), c(k)+1]$, then there is $\Pi_n \in \mathcal{P}_n$ satisfying

$$\|\Pi_n\|_{K_k} < \frac{1}{k^2 \cdot \max_{K_k \cup L_k} |z-\zeta|^{N_{k-1}}}$$

and

$$\left\| \Pi_n - \frac{f_k - \sum_{\nu=0}^{k-1} P_\nu}{(z-\zeta)^{N_{k-1}}} \right\|_{L_k} < \frac{1}{k^2 \cdot \max_{K_k \cup L_k} |z-\zeta|^{N_{k-1}}}.$$

By adding a sufficiently small multiple of the identity to Π_n, we can assume without loss of generality that $\deg \Pi_n \geq 1$. Setting $P_k(z) := (z-\zeta)^{N_{k-1}} \Pi_n(z)$, we thus obtain that ζ is a zero of P_k of multiplicity at least N_{k-1}, that $N_k := \deg P_k \leq N_{k-1} + n$ and $\deg P_k > N_{k-1}$, and that P_k fulfils (3.3) and (3.4).

With the P_k constructed, we now define $f\colon \Omega \to \mathbb{C}$, $z \mapsto \sum_{k=0}^{\infty} P_k(z)$. Because of (3.3), the function f is well-defined and holomorphic in Ω. Since $\deg P_k = N_k$ and $P_k(z) = (z-\zeta)^{N_{k-1}}\Pi_k(z)$ for some polynomial Π_k of strictly positive degree, it follows that $T_{N_k}^{(\zeta)} f = \sum_{\nu=0}^{k} P_\nu$ for every $k \in \mathbb{N}$. On the one hand, by (3.4), this implies that

$$\forall\, k \geq 1: \quad \|f_k - T_{N_k}^{(\zeta)} f\|_{L_k} < \frac{1}{k^2}, \tag{3.7}$$

and on the other hand, by (3.6) and the maximum principle, we have

$$\begin{aligned} N_k &= \deg P_k \leq N_{k-1} + n \\ &\leq N_{k-1} + \frac{\log^+(k^2 \|f_k - T_{N_{k-1}}^{(\zeta)} f\|_{\overline{V_k}}\, q_k^{N_{k-1}} C_k)}{\log(\rho_k / r_k)} + 1. \end{aligned}$$

Thus f has all the required properties. □

Obviously, the result stated in Proposition 14 contains too many unknown quantities in order to allow an explicit (nonrecursive) estimate for the growth of N_k. But nevertheless, in the general context, we already see that the N_k grow slower if L_k is farther away from Ω (respectively K_k), since q_k is smaller then.

Let us say that $f \in H(\mathbb{D})$ has a *universal Taylor series in* 0 *for* $H(\Omega)$ if the Taylor polynomials of f about 0 are dense in $H(\Omega)$, where $\Omega \subset \mathbb{C} \setminus \mathbb{D}$ is open. Instead of constructing a holomorphic function f with a universal Taylor series about the origin in the sense of [13], we construct $f \in H(\mathbb{D})$ having a universal Taylor series in 0 for $H(c + \mathbb{D})$ for some $c \in \mathbb{C}$, and we investigate how fast the elements of a given normal family $\mathcal{M}$ in $H(c + \mathbb{D})$ can be approximated by the Taylor polynomials of f.

Also in this situation the N_k grow slower, as we will see later, since the sets L_k and the functions f_k to approximate on L_k can be chosen closer to their predecessors L_{k-1} and f_{k-1}. Indeed, by (3.7), $T_{N_{k-1}}^{(\zeta)} f$ is close to f_{k-1} on L_{k-1}. If additionally L_k is close to L_{k-1} and f_k is close to f_{k-1}, then $\|f_k - T_{N_{k-1}}^{(\zeta)} f\|_{\overline{V_k}}$ remains rather small.

We consider the standard compact exhaustions of $\mathbb{D}$ and $c + \mathbb{D}$, respectively, that is $\mathcal{K} = (K_n)_{n\in\mathbb{N}}$ and $\mathcal{K}_c = (K_{c,n})_{n\in\mathbb{N}}$, where

$$K_n := \frac{n}{n+1}\overline{\mathbb{D}}, \quad K_{c,n} := c + K_n, \qquad n \in \mathbb{N}. \tag{3.8}$$

Since we are now dealing with disks, we have the following approximation result at our disposal, which will replace the use of Lemma 13.

Lemma 15. *Let* $L = L_1 \cup L_2 := \overline{D}(a_1, R_1) \cup \overline{D}(a_2, R_2)$ *be the union of two disjoint closed disks. Let* $K = K_1 \cup K_2 := \overline{D}(a_1, r_1) \cup \overline{D}(a_2, r_2)$, *where* $0 < r_j < R_j$ $(j = 1, 2)$. *Given* f *holomorphic on a neighborhood of* L *and* $n \geq 1$, *there exists a polynomial* p *such that* $\deg p < 2n$ *and*

$$\|f - p\|_K \leq \|f\|_L \frac{2\alpha^n}{(1-\alpha)} \left(\frac{\operatorname{diam}(K)}{\operatorname{dist}(L_1, L_2)}\right)^n,$$

where $\alpha := \max\{r_1, r_2\} / \min\{R_1, R_2\}$.

PROOF. Set $q(z) := (z-a_1)^n(z-a_2)^n$. We consider the special kind of Hermite interpolation polynomial

$$p(w) := \frac{1}{2\pi i}\int_{\partial L}\frac{f(z)}{q(z)}\frac{q(z)-q(w)}{z-w}\,dz,$$

and we shall show that this works.

Since $\big(q(z)-q(w)\big)/(z-w)$ is a polynomial in z, w of degree at most $2n-1$ in each variable, it follows that $p(w)$ is a polynomial of degree at most $2n-1$.

Also, by Cauchy's integral formula, if $w \in K$, then

$$f(w) = \frac{1}{2\pi i}\int_{\partial L}\frac{f(z)}{z-w}\,dz,$$

and so

$$f(w) - p(w) = \frac{1}{2\pi i}\int_{\partial L}\frac{f(z)}{z-w}\frac{q(w)}{q(z)}\,dz.$$

It follows that

$$\|f-p\|_K \le \frac{\|f\|_L\|q\|_K}{2\pi\,\mathrm{dist}(\partial L, K)}\int_{\partial L}\frac{|dz|}{|q(z)|}.$$

Now, if $w \in K_1$, then $|q(w)| \le r_1^n(\operatorname{diam} K)^n$. An analogous estimate holds for $w \in K_2$. Hence

$$\|q\|_K \le \max\{r_1, r_2\}^n(\operatorname{diam} K)^n.$$

Also, we clearly have

$$\mathrm{dist}(\partial L, K) \ge \min\{R_1 - r_1, R_2 - r_2\}.$$

Further, if $z \in \partial L_1$, then $|q(z)| \ge R_1^n\,\mathrm{dist}(L_1, L_2)^n$. Hence

$$\int_{\partial L_1}\frac{|dz|}{|q(z)|} \le \frac{2\pi R_1}{R_1^n\,\mathrm{dist}(L_1, L_2)^n}.$$

An analogous estimate holds for the integral over ∂L_2. Putting together these estimates, we get

$$\|f-p\|_K \le \|f\|_L\frac{\max\{r_1, r_2\}^n(\operatorname{diam} K)^n}{\min\{R_1 - r_1, R_2 - r_2\}\,\mathrm{dist}(L_1, L_2)^n}\left(\frac{1}{R_1^{n-1}} + \frac{1}{R_2^{n-1}}\right).$$

If we set $r := \max\{r_1, r_2\}$ and $R := \min\{R_1, R_2\}$, then we obtain

$$\|f-p\|_K \le \|f\|_L\frac{r^n(\operatorname{diam} K)^n}{(R-r)\,\mathrm{dist}(L_1, L_2)^n}\left(\frac{2}{R^{n-1}}\right).$$

The result follows from this. □

In the situation of the above lemma, let us choose $K = K_k \cup K_{c,k}$, that is $a_1 = 0$, $a_2 = c \in \mathbb{C}$, $r_1 = r_2 = k/(k+1) < 1$, and let $R_1 = R_2 =: R > 1$. The disks L_1, L_2 are disjoint as long as $\mathrm{dist}(L_1, L_2) = |c| - 2R > 0$. Further, we obtain

$$\alpha\frac{\mathrm{diam}(K)}{\mathrm{dist}(L_1, L_2)} < \frac{1}{R}\frac{|c|+2}{|c|-2R} =: q(R, c),$$

and the inequality in Lemma 15 reads

$$\|f-p\|_K \le \|f\|_L\frac{2R}{R-1}q(R, c)^n. \tag{3.9}$$

Lemma 16. *Let $R > 1$ and $c \in \mathbb{C}$ with $|c| > 2R$ and $q(R, c) < 1$. For any sequence of polynomials $(f_k)_{k\in\mathbb{N}}$, there is $f \in H(\mathbb{D})$ having a uniformly universal Taylor series in the origin for $(f_k, K_{c,k})_{k\in\mathbb{N}}$ such that the corresponding sequence $(N_k)_{k\in\mathbb{N}}$ is strictly increasing, $T_{N_k}^{(0)} f$ and N_k depend only on $f_1, \dots, f_{k-1}$ and, for $k \geq 7$, we have*

$$N_k \leq N_{k-1} + 1 + \frac{2\log^+(A)}{\log(q(R,c)^{-1})}, \tag{3.10}$$

where

$$A := \frac{2R}{R-1}(2R)^{\max\{\deg f_k, N_{k-1}\}}\left(\frac{2|c|+2}{|c|-R}\right)^{N_{k-1}}(\|f_k - f_{k-1}\|_{K_{c,k-1}} + \frac{1}{(k-1)^2}).$$

PROOF. The proof is very similar to that of Proposition 14, only we use Lemma 15 instead of Lemma 13.

As in the proof of Proposition 14, we construct recursively a sequence of polynomials $(P_k)_{k\in\mathbb{N}_0}$ and a strictly increasing sequence of integers $(N_k)_{k\in\mathbb{N}_0}$ such that the degree of P_k satisfies $\deg P_k = N_k$, the origin is a zero of P_k of multiplicity at least N_{k-1}, and

$$\forall\, k \geq 1: \qquad |P_k|_{K_k} < \frac{1}{k^2} \quad \text{and} \quad \left\|\sum_{\nu=0}^{k} P_\nu - f_k\right\|_{K_{c,k}} < \frac{1}{k^2}. \tag{3.11}$$

Let $P_0(z) \equiv 1$, and $N_0 := 0$. If, for some $k \in \mathbb{N}$, the polynomials $P_0, \dots, P_{k-1}$ and the integers $N_0, \dots, N_{k-1}$ have already been constructed, then we consider

$$h_k\colon \overline{B(0,R)} \cup \overline{B(c,R)} \to \mathbb{C}, \quad z \mapsto \begin{cases} 0, & \text{if } z \in \overline{B(0,R)}, \\ \dfrac{f_k(z) - \sum_{\nu=0}^{k-1} P_\nu(z)}{z^{N_{k-1}}}, & \text{if } z \in \overline{B(c,R)}, \end{cases}$$

which is well-defined and holomorphic in a neighborhood of $\overline{B(0,R)} \cup \overline{B(c,R)}$, since $|c| > 2R$. Lemma 15 and inequality (3.9) yield, for any $n \in \mathbb{N}$, the existence of a polynomial Π_n of degree not exceeding $2n-1$, with

$$\|h_k - \Pi_n\|_{K_k \cup K_{c,k}} \leq \|h_k\|_{\overline{B(0,R)} \cup \overline{B(c,R)}} \frac{2R}{R-1} q(R,c)^n = \|h_k\|_{\overline{B(c,R)}} \frac{2R}{R-1} q(R,c)^n,$$

where we used $h_k|_{\overline{B(0,R)}} = 0$. As $\max_{K_k \cup K_{c,k}} |z| = \max_{K_{c,k}} |z| = |c| + k/(k+1) < |c| + 1$, as well as $\min_{\overline{B(c,R)}} |z| = |c| - R$, as in the proof of Proposition 14 we obtain that, in order to have

$$\|h_k - \Pi_n\|_{K_k \cup K_{c,k}} < \frac{1}{k^2 \max_{K_k \cup K_{c,k}} |z|^{N_{k-1}}},$$

it is sufficient to have

$$k^2 \left\| f_k - \sum_{\nu=0}^{k-1} P_\nu \right\|_{\overline{B(c,R)}} \left(\frac{|c|+1}{|c|-R}\right)^{N_{k-1}} \frac{2R}{R-1} < q(R,c)^{-n}. \tag{3.12}$$

An application of Bernstein's lemma, cf. [14, Theorem 5.5.7], and (3.11) yields

$$\begin{aligned}\Big\| f_k - \sum_{\nu=0}^{k-1} P_\nu \Big\|_{\overline{B(c,R)}} &\le \Big\| f_k - \sum_{\nu=0}^{k-1} P_\nu \Big\|_{K_{c,k-1}} \left(\frac{Rk}{k-1}\right)^{\max\{\deg f_k, N_{k-1}\}} \\ &\le \left(\|f_k - f_{k-1}\|_{K_{c,k-1}} + \frac{1}{(k-1)^2} \right) (2R)^{\max\{\deg f_k, N_{k-1}\}}.\end{aligned}$$

Therefore, $n \in \mathbb{N}$ satisfies (3.12) if $n > \alpha(k)$, where $\alpha(k)$ is given by

$$\frac{\begin{array}{r}\log^+\big(2Rk^2/(R-1)(2R)^{\max\{\deg f_k, N_{k-1}\}}\big((|c|+1)/(|c|-R)\big)^{N_{k-1}} \\ \times \left(\|f_k - f_{k-1}\|_{K_{c,k-1}} + 1/(k-1)^2\right)\big)\end{array}}{\log(q(R,c)^{-1})}. \tag{3.13}$$

Fixing $n \in \mathbb{N} \cap [\alpha(k), 1 + \alpha(k)]$, we continue as in the proof of Proposition 14, to construct P_k and $N_k := \deg P_k \le N_{k-1} + 2n - 1$.

As in the proof of Proposition 14, it follows that $f\colon \mathbb{D} \to \mathbb{C}$, $z \mapsto \sum_{k=0}^{\infty} P_k(z)$ is holomorphic and has a uniformly universal Taylor series in 0 for $(f_k, K_{c,k})_{k\in\mathbb{N}}$. For the corresponding sequence $(N_k)_{k\in\mathbb{N}}$, we have

$$N_k = \deg P_k \le N_{k-1} + 2n - 1 \le N_{k-1} + 1 + 2\alpha(k).$$

If $k \ge 7$, then, because $k^2 \le 2^{k-1} (\le 2^{N_{k-1}})$, we obtain from (3.13) that $\alpha(k)$ is majorized by

$$\frac{\begin{array}{r}\log^+\big(2R/(R-1)(2R)^{\max\{\deg f_k, N_{k-1}\}}\big((2|c|+2)/(|c|-R)\big)^{N_{k-1}} \\ \times \left(\|f_k - f_{k-1}\|_{K_{c,k-1}} + 1/(k-1)^2\right)\big)\end{array}}{\log(q(R,c)^{-1})}.$$

This completes the proof. □

Remark 17. Proposition 14 and Lemma 16 show how small the values of the sequence $(N_k)_{k\in\mathbb{N}}$ can be chosen. Nevertheless, inspection of their proofs gives that, at each step, N_k can be chosen arbitrarily large.

Lemma 18. *Suppose that $\mathcal{M}$ be a normal family in $H(c+\mathbb{D})$. Let $M_n := 1 + \sup_{f\in\mathcal{M}} \|f\|_{K_{c,n}}$ and let $(\lambda_n)_{n\in\mathbb{N}}$ be the covering numbers of $\mathcal{M}$, i.e., there are functions $f_1^{(n)}, \dots, f_{\lambda_n}^{(n)} \in H(c+\mathbb{D})$ whose $(1/n)$-neighborhoods cover $\mathcal{M}$. If*

$$m \ge \tau_{n+1}(\mathcal{M}) := (n+1)^2 \log\big((n+1)^3 M_{n+1}\big),$$

then

$$\|g - T_m^{(c)} g\|_{K_{c,n}} < \frac{1}{n} \quad \forall\, g \in \mathcal{M} \cup \{f_1^{(n+1)}, \dots, f_{\lambda_{n+1}}^{(n+1)}\}.$$

PROOF. Let $g = f_j^{(n+1)}$ for some $1 \le j \le \lambda_{n+1}$. Then there is $f \in \mathcal{M}$ such that $\|g - f\|_{K_{c,n+1}} \le 1/(n+1)$, which implies

$$\|g\|_{K_{c,n+1}} \le \|f\|_{K_{c,n+1}} + \frac{1}{n+1} \le M_{n+1}.$$

This last estimate obviously also holds for $g \in \mathcal{M}$. By Cauchy's estimate, we get

$$\|g - T_m^{(c)} g\|_{K_{c,n}} \le M_{n+1} \sum_{\nu=m+1}^{\infty} \left(\frac{n(n+2)}{(n+1)^2}\right)^{\nu} = M_{n+1}\, n(n+2) \left(\frac{n(n+2)}{(n+1)^2}\right)^{m}.$$

The last term is less than $1/n$ provided that

$$m \geq (n+1)^2 \log\big((n+1)^3 M_{n+1}\big) \geq \frac{\log\big(n^2(n+2)\,M_{n+1}\big)}{\log\Big((n+1)^2/\big(n(n+2)\big)\Big)}.$$

This completes the proof. □

By choosing an appropriate sequence of polynomials (f_k) in Lemma 16, we can construct a function $f \in H(\mathbb{D})$ which has a universal Taylor series in 0 for $H(c+\mathbb{D})$ for some center $c \in \mathbb{C}$, which is fast approximating for a normal family $\mathcal{M} \subset H(c+\mathbb{D})$, and such that the corresponding sequence (N_k) has bounded quotients N_k/N_{k-1} (compare with Remark 11).

Theorem 19. *Let $c \in \mathbb{C}$ with $|c| > 10$. Let $\mathcal{M}$ be a normal family in $H(c+\mathbb{D})$ with covering numbers (λ_n), and suppose that $M_n = O\big(\exp(n^l)\big)$ for some $l \in \mathbb{N}\cup\{0\}$. Then there exists a function $f \in H(\mathbb{D})$ which has a universal Taylor series for $H(c+\mathbb{D})$, which is fast approximating for $\mathcal{M}$ in the sense that*

$$F(f, \mathcal{M}, 2/n) \leq N_{(n+1)(\lambda_{n+1}+1)},$$

and which has bounded quotients N_k/N_{k-1}.

Proof. For $R = 2$, we have $|c| > 10 > 2R$ and $q(2,c) < 1$. Let $f_1^{(n)}, \ldots, f_{\lambda_n}^{(n)} \in H(c+\mathbb{D})$ be functions whose $(1/n)$-neighborhoods cover $\mathcal{M}$. Let (q_n) be a sequence of polynomials which is dense in $H(c+\mathbb{D})$, and consider the sequence (g_k) given by

$$f_1^{(1)}, \ldots, f_{\lambda_1}^{(1)}, q_1, f_1^{(2)}, \ldots, f_{\lambda_2}^{(2)}, q_2, \ldots, f_1^{(n)}, \ldots, f_{\lambda_n}^{(n)}, q_n, \ldots$$

If $g_k := f_j^{(n)}$ for some j, n, then n is uniquely determined by k, and we have $n \leq k$. In this case, we set $f_k := T^{(c)}_{\tau_n(\mathcal{M})} g_k$. Using Lemma 18, we deduce

$$\deg f_k = n^2 \log(n^3 M_n) \quad \text{and} \quad \|g_k - f_k\|_{K_{c,n-1}} < \frac{1}{n-1}. \tag{3.14}$$

If, on the other hand, g_k is one of the (q_n), then we define $f_k := g_k$. Without loss of generality $\deg f_k \leq k$ and $\|f_k\|_{K_{c,k}} \leq k$.

Now let $f \in H(\mathbb{D})$ be the function constructed in Lemma 16, that is, f has a uniformly universal Taylor series in the origin for $(f_k, K_{c,k})$ with the corresponding sequence (N_k) satisfying (3.10). In view of (3.10), we can estimate the degree of f_k in case $g_k = f_j^{(n)}$:

$$\deg f_k = n^2 \log(n^3 M_n) \leq k^2 \log(k^3 M_k) = O(k^{l+3}),$$

since $n \leq k$ and $M_k = O\big(\exp(k^l)\big)$. Hence, by Bernstein's lemma, cf. [14, Theorem 5.5.7],

$$\|f_k\|_{K_{c,k}} \leq \|f_k\|_{K_{c,n}} \left(\frac{k}{k+1}\,\frac{n+1}{n}\right)^{\deg f_k} \leq M_n\, \mathrm{e}^{n^2 \log(n^3 M_n)}.$$

Thus, independently of whether $g_k = f_j^{(n)}$ or $g_k = q_n$, we obtain

$$\log\left(\|f_k - f_{k-1}\|_{K_{c,k-1}} + \frac{1}{(k-1)^2}\right) = O(k^{l+3}).$$

Hence, using the fact that $\max\{\deg f_k, N_{k-1}\} \leq \deg f_k + N_{k-1}$, inequality (3.10) reads

$$N_k \leq \alpha_1 N_{k-1} + \alpha_2 k^{l+3},$$

where α_1, α_2 are constants. As mentioned in Remark 17, it is possible to increase N_k at each step. So, let us choose

$$N_k := \max\{\alpha_1 N_{k-1} + \alpha_2 k^{l+3}, (k+1)^{l+3}\}, \tag{3.15}$$

which guarantees $N_{k-1} \geq k^{l+3}$ for every $k \in \mathbb{N}$. Depending on where the maximum in (3.15) is attained, N_k/N_{k-1} is either bounded by the constant $\alpha_1 + \alpha_2$ or by $\big((k+1)/k\big)^{l+3}$. Either way, N_k/N_{k-1} remains bounded as $k \to \infty$.

To each $g \in \mathcal{M}$ is associated $g_k = f_j^{(n+1)}$ with $\|g_k - g\|_{K_{c,n+1}} < 1/(n+1)$. Using (3.14), we have $\|f_k - g_k\|_{K_{c,n}} < 1/n$, which implies $\|f_k - g\|_{K_{c,n}} \leq 2/n$. By the choice of f and the sequences $(f_k), (g_k)$, we obtain $k \leq (n+1)(\lambda_{n+1}+1)$ and hence $F(f, \mathcal{M}, 2/n) \leq N_k$.

Since (q_n) is dense in $H(c+\mathbb{D})$, by construction so are the Taylor polynomials of f about 0, and hence f has a universal Taylor series for $H(c+\mathbb{D})$. □

In the next section we shall encounter several examples of normal families $\mathcal{M}$ for which $M_n = O(1)$, and so Theorem 19 is applicable with $l = 0$.

4. Normal families and covering numbers

In order to get some impression of the speed of approximation, we conclude with some examples of normal families in $H(\mathbb{D})$. Throughout this section, we suppose ε to be an arbitrarily small positive number, and consider the standard compact exhaustion (3.8) of $\mathbb{D}$ to define the seminorms in (1.2) and hence the natural metric on $H(\mathbb{D})$ in (1.1).

Let $E := \{f_1, \ldots, f_k\}$ be a finite subset of $H(\mathbb{D})$, let $B^\infty := \{f \in H(\mathbb{D}) : \sup_{\mathbb{D}}|f| \leq 1\}$, and let

$$S := \{f \in H(\mathbb{D}) : f \text{ one-to-one}, f(0) = 0, f'(0) = 1\}.$$

For each of these three normal families, we obtain the existence of $\mathcal{C}$- or $\mathcal{D}$-universal functions f with rates of approximation as follows:

$\mathcal{M}$	$F(f,\mathcal{M},1/n)$ (for $\mathcal{C}$)	$F(f,\mathcal{M},1/n)$ (for $\mathcal{D}$)
E	$O(n)$	$O\big(n^2\log(n\max\{1, M_{12kn+1}\})\big)$
B^∞	$O(n\lambda_{2n})$	$O\big(n(n\lambda_{3n})^{1+\varepsilon}\big)$
S	$O(n\lambda_{2n})$	$O\big(n^2\lambda_{3n}\log(n\lambda_{3n})\big)$

where $M_n := \sup_{f\in\mathcal{M}}\|f\|_{K_n}$ and λ_n denotes the nth covering number of $\mathcal{M}$.

Furthermore, $\mathcal{C}$ is 8- and $\mathcal{D}$ is $(9+\varepsilon)$-polynomial universal for the automorphism group

$$\mathrm{Aut}(\mathbb{D}) = \left\{f_{\gamma,a}(z) := \mathrm{e}^{\mathrm{i}\gamma}\frac{z-a}{1-\bar{a}z} : \gamma \in [0,2\pi), a \in \mathbb{D}\right\},$$

and $\mathcal{C}$ is $(2+\varepsilon)$-and $\mathcal{D}$ $(4+\varepsilon)$-polynomial universal for the set of all Koebe extremal functions

$$K := \{f_\alpha = \mathrm{e}^{-\mathrm{i}\alpha} f_0(\mathrm{e}^{\mathrm{i}\alpha}z) : \alpha \in [0,2\pi)\} \subseteq S, \quad f_0(z) = \frac{z}{(1-z)^2}.$$

For all this and further details, see [10].

In this context, the question arises, interesting in its own right, to estimate the nth covering number λ_n for S, or, going back one step, to estimate the minimal number $N(\delta)$ of balls of radius δ required to cover S. The following theorem provides upper and lower bounds for $N(\delta)$.

Theorem 20. *There exist constants $c, C > 0$ such that*

$$e^{c/\sqrt{\delta}} \le N(\delta) \le e^{(C/\delta)\log^2(1/\delta)}.$$

In particular, $N(\delta)$ grows faster than any power of $1/\delta$ as $\delta \to 0$. The proof of the upper bound is given in [10]. We give here the proof of the lower bound. It is based on an elementary lemma.

Lemma 21. *Let $f(z) = z + \sum_{k=2}^{\infty} a_k z^k$, where $\sum_{k=2}^{\infty} k|a_k| < 1$. Then $f \in S$.*

PROOF. Let $z, w \in \mathbb{D}$. Then $|z^k - w^k| \le k|z - w|$ for all k, so

$$|f(z) - f(w)| \ge |z - w| - \sum_{k=2}^{\infty} |a_k|\,|z^k - w^k| \ge |z - w|\left(1 - \sum_{k=2}^{\infty} k|a_k|\right).$$

It follows that f is injective. Thus $f \in S$. □

PROOF OF THE LOWER BOUND. Let $f, g \in H(\mathbb{D})$, say $f(z) = \sum_{k=0}^{\infty} a_k z^k$, and $g(z) = \sum_{k=0}^{\infty} b_k z^k$. By the maximum principle and the standard Cauchy estimates, for each $k \in \mathbb{N}$ we have

$$\|f - g\|_{K_k} = \max_{|z| = k/(k+1)} |f(z) - g(z)| \ge |a_k - b_k| \left(\frac{k}{k+1}\right)^k \ge \frac{|a_k - b_k|}{4},$$

and consequently

$$d(f, g) \ge \sup_{k \in \mathbb{N}} \min\left(\frac{|a_k - b_k|}{4}, \frac{1}{k}\right). \tag{4.1}$$

Now let $n \ge 2$, and consider the family $\mathcal{F}_n$ of polynomials f of the form

$$f(z) := z + \frac{1}{n^2} \sum_{k=2}^{n} \varepsilon_k z^k, \quad (\varepsilon_k \in \{-1, 1\},\ k = 2, \ldots, n).$$

By Lemma 21, we clearly have $\mathcal{F}_n \subset S$. Also, by (4.1), the distance between distinct polynomials $f, g \in \mathcal{F}_n$ is at least $1/(2n^2)$. Thus, in any covering by balls of radius $1/(4n^2)$, each of the polynomials of $\mathcal{F}_n$ must belong to a different ball. There are 2^{n-1} elements in the family $\mathcal{F}_n$. Therefore

$$N(1/4n^2) \ge 2^{n-1}.$$

As this holds for each $n \ge 2$, the lower bound follows. □

Acknowledgment. We want to thank the referee for a careful reading of the article and for his comments and suggestions which helped to improve the presentation of the results.

References

1. T. Bagby and N. Levenberg, *Bernstein theorems*, New Zealand J. Math. **22** (1993), no. 1, 1–20.
2. P. Beise, T. Meyrath, and J. Müller, *Universality properties of Taylor series inside the domain of holomorphy*, J. Math. Anal. Appl. **383** (2011), no. 1, 234–238.
3. L. Bernal González and A. Montes-Rodríguez, *Universal functions for composition operators*, Complex Variables Theory Appl. **27** (1995), no. 1, 47–56.
4. M. C. Calderón-Moreno and J. Müller, *Universal holomorphic and harmonic functions with additional properties*, Acta Math. Hungar. **105** (2004), no. 1-2, 1–15.
5. A. García-Máynez and A. Illanes, *A survey on unicoherence and related properties*, An. Inst. Mat. Univ. Nac. Autónoma México **29** (1989), 17–67 (1990).

6. P. M. Gauthier and M. R. Pouryayevali, *Universal (pluri)subharmonic functions*, Analysis (Munich) **27** (2007), no. 2-3, 273–284.
7. W. Gehlen, W. Luh, and J. Müller, *On the existence of O-universal functions*, Complex Variables Theory Appl. **41** (2000), no. 1, 81–90.
8. L. Hörmander, *The analysis of linear partial differential operators.* Vol. I: *Distribution theory and Fourier analysis*, Grundlehren Math. Wiss., vol. 256, Springer, Berlin, 1983; Vol. II: *Differential operators with constant coefficients*, Grundlehren Math. Wiss., vol. 257, 1983.
9. T. Kalmes and M. Nieß, *Universal zero solutions of linear partial differential operators*, Studia Math. **198** (2010), no. 1, 33–51.
10. ———, *Composition and differentiation operators and fast approximation*, J. Approx. Theory **164** (2012), no. 1, 57–76.
11. W. Luh, *Universal approximation properties of overconvergent power series on open sets*, Analysis **6** (1986), no. 2-3, 191–207.
12. J. Müller, V. Vlachou, and A. Yavrian, *Universal overconvergence and Ostrowski-gaps*, Bull. London Math. Soc. **38** (2006), no. 4, 597–606.
13. V. Nestoridis, *Universal Taylor series*, Ann. Inst. Fourier (Grenoble) **46** (1996), no. 5, 1293–1306.
14. T. Ransford, *Potential theory in the complex plane*, London Math. Soc. Stud. Texts, vol. 28, Cambridge Univ. Press, Cambridge, 1995.
15. J. H. Shapiro, *Simple connectivity and linear chaos*, Rend. Circ. Mat. Palermo (2) Suppl. **56** (1998), 27–48.

Fachbereich IV — Mathematik, Universität Trier, 54286 Trier, Germany
E-mail address: kalmesth@uni-trier.de

Mathematisch-Geographische Fakultät, Katholische Universität Eichstätt-Ingolstadt, 85071 Eichstätt, Germany
E-mail address: markus.niess@ku-eichstaett.de

Département de mathématiques et de statistique, Université Laval, Québec, QC G1V 0A6, Canada
E-mail address: ransford@mat.ulaval.ca

Centre de Recherches Mathématiques
CRM Proceedings and Lecture Notes
Volume 55, 2012

Harmonic Mappings with Quadrilateral Image

Jane McDougall

ABSTRACT. A harmonic mapping is a univalent harmonic function. Poisson extensions of piecewise constant complex functions on the unit circle provide well known examples of harmonic mappings onto a polygonal domain when the boundary function maps to the vertices of a convex polygon. It has been conjectured however that there are nonconvex polygons for which Poisson extensions of piecewise constant functions are never univalent. Here we show that there exists a piecewise constant map from the unit circle to the vertices of an arbitrary quadrilateral that has a univalent Poisson extension, thus solving Sheil-Small's mapping problem for quadrilaterals.

1. Introduction.

A *harmonic mapping* is a univalent complex valued function of a single complex variable that is harmonic. Such functions were of interest during the 20th century to differential geometers through their relationship with minimal surfaces. The 1984 paper of Clunie & Sheil-Small [4] led to an interest in establishing a general theory of univalent harmonic functions analogous to that of univalent analytic functions.

We prove an existence theorem for a particular class of harmonic mappings onto any simple closed quadrilateral. In view of the Riemann mapping theorem, there is no restriction in considering the unit disc as our domain. We narrow our attention further to the harmonic functions that are the Poisson extension to the unit disc of a piecewise constant boundary function. In his 1989 article [10], Sheil-Small considered functions of this form and introduced the mapping problem: For any given simple closed polygon Π, does there exist a piecewise constant function on the unit disc whose Poisson extension is a harmonic mapping onto Π? For convex polygons this question is answered completely by the Radó – Kneser – Choquet theorem. For nonconvex polygons however the situation remains unresolved. It is conjectured in [10] that there exist simple closed polygons for which the mapping problem cannot be solved; we show here that we can find a solution for any quadrilateral.

We now introduce some notation and state our objectives more formally. Denote by $\mathbb{D}$ the unit disc, and by $\mathbb{T} = \partial\mathbb{D}$ the boundary of the disc. Consider the

2010 *Mathematics Subject Classification.* Primary 30C55, 14E20; Secondary 30C85, 53A10.
Key words and phrases. Differential geometry.
This is the final form of the paper.

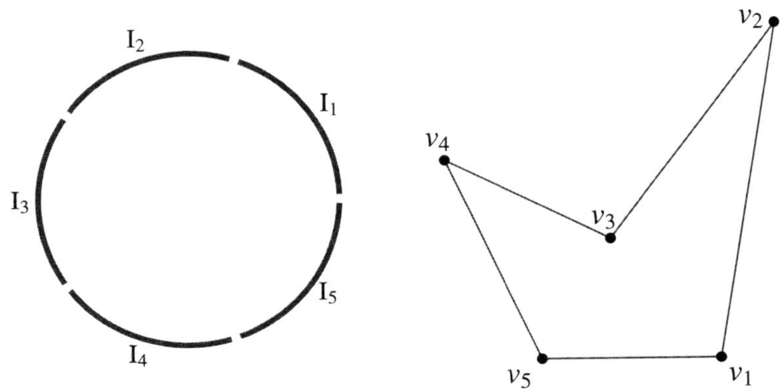

FIGURE 1. Boundary correspondence $\hat{f}$ onto the nonconvex pentagon Π_5.

N-gon Π with vertices $\{v_1, v_2, \dots, v_N\}$. We say that Π is *positively oriented* if the vertices lie on a positively oriented simple closed curve, with v_{i-1} preceding v_i, $1 < i \le N$. Let E be a set of N arc endpoints on the circle, thereby defining open arcs $\{I_1, I_2, \dots, I_N\}$ that partition $\mathbb{T} \setminus E$. Assume the arcs are indexed so as to be ordered by increasing argument. We will sometimes need to refer to the endpoints of a particular arc and so write $I_k = (I_k^-, I_k^+)$, $1 \le k \le N$. Let $\hat{f}$ be defined on $\mathbb{T} \setminus E$ by

$$\hat{f}(I_k) = v_k, \quad 1 \le k \le N. \tag{1.1}$$

For a positively oriented polygon Π with vertices $\{v_1, v_2, \dots, v_N\}$, the *mapping problem* is stated as follows: Find a set of open arcs $\{I_1, I_2, \dots, I_N\}$ with endpoints E, such that the set of arcs are indexed in order of increasing argument and partition $\mathbb{T} \setminus E$, and such that the Poisson extension f of $\hat{f}$ in (1.1) to the unit disc $\mathbb{D}$ is a harmonic mapping onto Π. The mapping problem was first stated in [10, p. 471].

Example 1.1. Let Π_5 be the nonconvex pentagon with vertices

$$v_1 = 1 - \mathrm{i}/2, \quad v_2 = 1.4 + 2\mathrm{i}, \quad v_3 = 0.2 + 0.4\mathrm{i}, \quad v_4 = -1 + \mathrm{i}, \quad v_5 = -0.3 - \mathrm{i}/2,$$

as shown in Figure 1. Suppose we choose arcs of equal length $2\pi/5$ so that the set of endpoints is $E = \{e^{2k\pi\mathrm{i}/5}, 1 \le k \le 5\}$, and that the arcs are mapped by $\hat{f}$ to successive vertices as in (1.1). Then the Poisson extension f of $\hat{f}$ solves the mapping problem for Π_5 only if the Poisson extension of $\hat{f}$ to f is univalent — we shall see in Section 3 however that it is not.

The main result is Theorem 4.1 in Section 4: *The mapping problem can be solved for any positively oriented quadrilateral.* Section 2 gives a brief overview of the theory of complex valued harmonic functions. Section 3 contains necessary details about harmonic functions that are Poisson extensions of piecewise constant functions on the unit circle.

2. Harmonic functions

We note some basic properties about harmonic functions. For a more detailed exposition see [5]. A harmonic function defined on a domain in $\mathbb{C}$ can be written in terms of analytic functions h and g

$$f = h + \bar{g},$$

known as the *canonical decomposition.*

The functions h and $\bar{g}$ are known as the *analytic part* and *anti-analytic part* of f respectively and the decomposition is unique if, for example, we normalize so that $g(0) = 0$. The decomposition is easily computed through the observation that $h' = f_z$ and $g' = f_{\bar{z}}$. The Jacobian is found to be $|h'|^2 - |g'|^2$. By Lewy's theorem, the function f is locally univalent when the Jacobian is nonzero. The function f is said to be *sense preserving* if the Jacobian is positive, or equivalently if $|h'| > |g'|$.

A harmonic map differs from a conformal map in that its real and imaginary parts are not necessarily connected by the Cauchy Riemann equations. In Example 2.1 we combine the real part of the analytic function z^3 with the imaginary part of the analytic function z^5.

For a sense-preserving locally univalent map, define the *second complex dilatation*

$$\omega_f(z) = \frac{\overline{f_{\bar{z}}}}{f_z} = \frac{g'(z)}{h'(z)}.$$

When the dilatation is identically zero, f is a conformal map. The dilatation ω_f serves as a measure of how far f is from being conformal. The second complex dilatation is distinct from the dilatation $\mu_f = f_{\bar{z}}/f_z$ used in theory of quasiconformal mappings; the function f is a k-quasiconformal map if $|\omega_f| \leq k < 1$ for some positive constant k. Since here we will only use the first mentioned sense of the word, we will refer to ω_f simply as the dilatation.

Example 2.1. The harmonic function

$$f(z) = \Re(z^3) + \mathrm{i}\Im(z^5)$$

can be rewritten

$$f(z) = \frac{z^3 + \bar{z}^3}{2} + \mathrm{i}\frac{z^5 - \bar{z}^5}{2i}$$

so $h = \frac{1}{2}z^3 + \frac{1}{2}z^5$, $g = \frac{1}{2}z^3 - \frac{1}{2}z^5$ and $\omega_f = g'/h' = (3 - 5z^2)/(3 + 5z^2)$ for $z \neq 0$. Then f is sense preserving if $|g'| < |h'|$. A computation shows that this inequality is equivalent to $\Re(z)^2 - \Im(z)^2 > 0$. Therefore the mapping is sense preserving in the first quadrant of the unit disc where $\Im(z) < \Re(z)$, or where $\arg z < \pi/4$, and sense reversing if $\arg z > \pi/4$. In these two disjoint regions f is locally univalent, and in fact univalent (see Figure 2).

3. Poisson extensions of piecewise constant maps

For a complex valued boundary function $\hat{f}(e^{it})$ defined almost everywhere on the unit circle we state the Poisson integral formula for the harmonic extension f to the unit disc:

$$f(z) = \mathcal{P}[\hat{f}](z) = \frac{1}{2\pi}\int_0^{2\pi} \frac{1 - |z|^2}{\left|\mathrm{e}^{\mathrm{i}t} - z\right|^2}\hat{f}(\mathrm{e}^{\mathrm{i}t})\,\mathrm{d}t, \quad z \in \mathbb{D}$$

or in polar form,

$$f(z) = \frac{1}{2\pi}\int_0^{2\pi} \frac{1 - r^2}{1 - 2r\cos(\theta - t) + r^2}\hat{f}(\mathrm{e}^{\mathrm{i}t})\,\mathrm{d}t, \quad z = r\mathrm{e}^{\mathrm{i}\theta} \in \mathbb{D}.$$

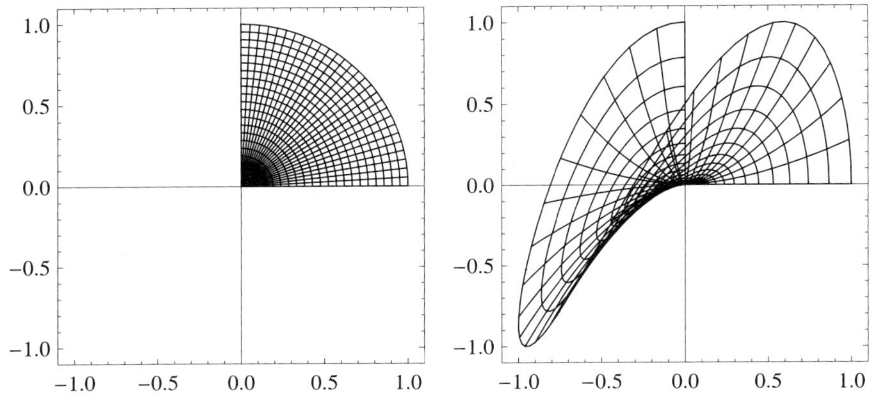

FIGURE 2. Image of a regular polar grid in the first quadrant of $\mathbb{D}$ under $f(z) = \Re(z^3) + i\Im(z^5)$

3.1. The Radó – Kneser – Choquet theorem. When the image of $\hat{f}$ bounds a convex region, univalence of the extended function is guaranteed by the Radó – Kneser – Choquet theorem. The theorem arises from a problem originally posed by Radó [9], solved by Kneser [7] and generalized independently by Choquet [3].

Theorem (Radó – Kneser – Choquet — basic form [7]**).** *Let Ω be convex and $\hat{f} \colon \partial\mathbb{D} \to \partial\Omega$ be a homeomorphism onto the Jordan curve $\partial\Omega$. Then the Poisson extension to the disc $\mathbb{D}$ is a univalent harmonic map onto Ω.*

An extended result requires only that the boundary correspondence $\hat{f} \colon \partial\mathbb{D} \to \partial\Omega$ be continuous, allowing for arcs of constancy (see [3]). For piecewise constant functions, we must rely on a further extension of the theorem: the conclusion remains true even if there are discontinuities in $\hat{f}$, provided that $\hat{f}$ is monotonic and $\hat{f}(\mathbb{D})$ does not lie on a line (see [5, Section 3.3]).

Our attention for the rest of this article will be restricted to boundary functions that are piecewise constant, and to their extensions. The generalized form of the Radó – Kneser – Choquet theorem implies that for a convex polygon, any choice of arcs (ordered by increasing argument) mapping to consecutive vertices of a polygon will lead to a univalent Poisson extension. However the Radó – Kneser – Choquet theorem says nothing if the target polygon is nonconvex. In Figure 3 we see how arcs of equal size, when mapped to the vertices of a specific nonconvex pentagon, produce a boundary function $\hat{f}$ that fails to have a univalent Poisson extension. The right side of Figure 3 represents the image of a regular polar grid in the unit disc under the Poisson extension f.

3.2. Coefficients in terms of angles. We now examine the Poisson integral formula for a piecewise constant boundary function in more detail. Let I be an arc of $\mathbb{T}$, and χ_I its characteristic function. By computing the antiderivative in the Poisson integral formula we obtain the well known

$$P[\chi_I](z) = \frac{1}{\pi}\left(\arg\frac{z - I^+}{z - I^-} - \frac{|I|}{2}\right),$$

where $|I|$ is the arglength of I, namely $\arg(I^+) - \arg(I^-)$.

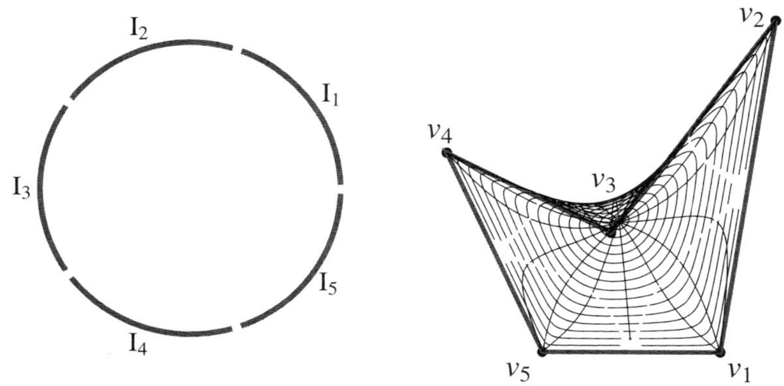

FIGURE 3. Non-univalent map onto a nonconvex pentagon

In terms of the *Euclidean angle* $E_I(z)$ subtended by I at z this becomes

$$P[\chi_I](z) = \frac{1}{\pi}\left(E_I(z) - \frac{|I|}{2}\right).$$

Recall well-known facts about angles subtended by an arc I at the circle center and boundary, as shown in Figure 4. We have the inequalities

$$|I|/2 \le E_I(z) \le \pi + |I|/2.$$

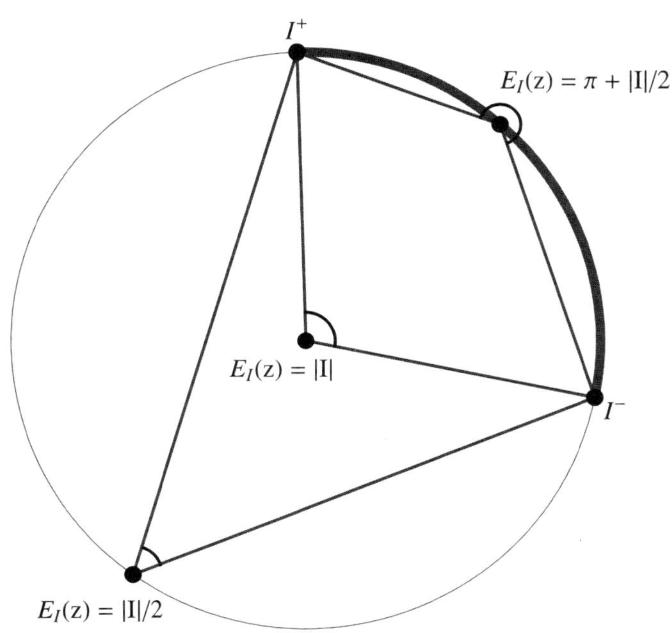

FIGURE 4. Euclidean angle subtended by an arc $I = (I^-, I^+)$.

The extremes of $E_I(z)$ occur when z is on $\mathbb{T}$ in the interior of either I or $\mathbb{T} \setminus I$. Thus

$$0 \leq P[\chi_I](z) \leq 1$$

If, using the Poincare disc model, we consider hyperbolic rather than Euclidean angles, we find a more natural interpretation for $P[\chi_I](z)$. Hyperbolic angles can be conveniently measured using a Möbius function that is a self-map of the unit disc.

Definition 3.1. Let $\alpha \in \mathbb{C}$, and define the Möbius transformation ϕ_α by

$$\phi_\alpha(z) = \frac{z - \alpha}{1 - \bar{\alpha} z}. \tag{3.1}$$

It is well know that for $\alpha \in \mathbb{D}$, ϕ_α is a disc automorphism that preserves the disc boundary, hyperbolic angles between curves, and hyperbolic distances between pairs of points in the unit disc. Note that for $|\alpha| < 1$, ϕ_α is also a known as a *Blaschke factor*. To measure the *hyperbolic angle* $H_I(z)$ subtended by an arc I at $z \in \mathbb{D}$, observe that ϕ_z maps the point z to the origin, circular arcs through z to radial lines through the origin, and the arc endpoints I^- and I^+ to the points $\phi_z(I^-)$ and $\phi_z(I^+)$ on the unit circle respectively. The transformation ϕ_α preserves angle, so for any $\alpha \in \mathbb{D}$ we have $H_I(z) = H_{\phi_\alpha(I)}(\phi_\alpha(z))$ as illustrated in Figure 5. Again, because the disc automorphism ϕ_z preserves hyperbolic angle for a fixed $z \in \mathbb{D}$ we have

$$H_I(z) = \arg \frac{\phi_z(I^+)}{\phi_z(I^-)}. \tag{3.2}$$

Presumably the following result is well known. It is included here because it is an important step in solving the the mapping problem for arbitrarily chosen quadrilaterals (Section 4).

Lemma 3.2. *Let $I = (I^-, I^+)$ be an open arc of the unit circle. The Euclidean and hyperbolic angles subtended by I at z are related by*

$$H_I(z) = 2E_I(z) - |I|.$$

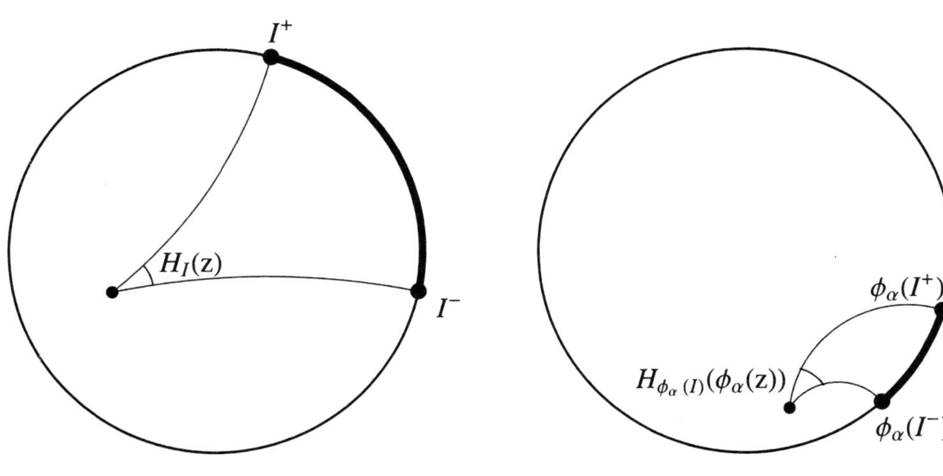

FIGURE 5. Hyperbolic angle subtended by I at z preserved under a disc automorphism ϕ_α

Moreover the Poisson integral of the characteristic function for I is the proportion of the circle subtended hyperbolically by the arc I at z

$$P[\chi_I](z) = \frac{1}{2\pi} H_I(z).$$

PROOF. Let $z \in \mathbb{D}$. A calculation shows that

$$\arg \frac{\phi_z(I^+)}{\phi_z(I^-)} = \arg(\mathrm{e}^{-\mathrm{i}|I|}) + 2\arg(\overline{I^-} - \bar{z}) + 2\arg(I^+ - z)$$

This can be rewritten

$$H_I(z) = 2\arg \frac{I^+ - z}{I^- - z} - |I| = 2E_I(z) - |I|. \qquad \square$$

We now show that the hyperbolic angles subtended by the arcs at a point z become weights in a convex combination of the vertices that produces $f(z)$.

Lemma 3.3. *Consider a set of open arcs $\{I_k : 1 \le k \le N\}$ with endpoints E, ordered by argument and partitioning $\mathbb{T} \setminus E$. Let Π be a positively oriented polygon with vertices $\{v_k : 1 \le k \le N\}$, and let $\hat{f} = \sum_{k=1}^N v_k \chi_{I_k}$ define a piecewise constant boundary function from the arcs onto Π. Then for each $z \in \mathbb{D}$, the Poisson extension f of $\hat{f}$ has the form*

$$f(z) = \sum_{k=1}^N \lambda_k(z) v_k, \quad \lambda_k(z) = \frac{H_{I_k}(z)}{2\pi}.$$

Thus $f(z)$ is a convex combination of the vertices of Π.

PROOF. Let $z \in \mathbb{D}$. The harmonic extension f is then

$$f(z) = P\left[\sum_{k=1}^N v_k \chi_{I_k}\right](z) = \sum_{k=1}^N v_k P[\chi_{I_k}](z) = \frac{1}{2\pi}\sum_{k=1}^N v_k H_{I_k}(z) = \sum_{k=1}^N \lambda_k v_k$$

with λ_k as in the lemma's statement. The total angle subtended by the arcs $\{I_k : 1 \le k \le N\}$ at z is 2π, so

$$\sum_{k=1}^N \lambda_k = \sum_{k=1}^N \frac{H_{I_k}(z)}{2\pi} = 1. \qquad \square$$

An immediate consequence of $f(z)$ being a convex combination of the vertices of a target polygon Π, is that $f(z)$ lies in the convex hull of the vertices of Π (a fact proved elsewhere using other methods — see [10] for example). Moreover, by Theorem 3 of [10], the image of f includes the interior of Π. The harmonic function in Figure 3 fails to be univalent in that points near the arc I_3 map to the exterior of Π_5 near v_3. In proving univalence we must therefore show that the image of f does not include any points in the exterior of Π.

We look at an example in order to to gain some insight as to how to modify the arc sizes and obtain a univalent map. In Figure 6 we see a point z_0 that maps to the exterior of the pentagon Π_5 of Example 1.1, where the arcs in the mapping are of equal length. The vertices v_2 and v_4 are weighted relatively lightly, as seen by the hyperbolic angles subtended by the arcs I_2 and I_4 at z_0. By increasing $|I_3|$ and reducing $|I_2|$ and $|I_4|$ we increase the corresponding weight of the vertex v_3 and reduce those of v_2 and v_4. For the modified arcs shown in Figure 7, $f(z_0)$ lies in

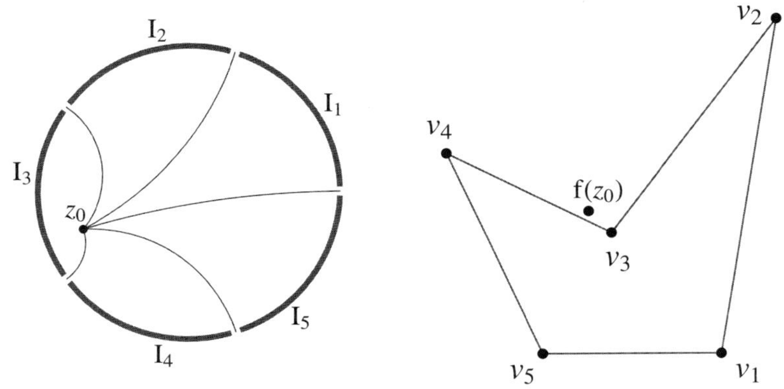

FIGURE 6. The point $z_0 = -0.7 - 0.25\mathrm{i}$ maps to the exterior of Π_5. The image $f(z_0)$ is a convex combination of v_1, v_2, v_3, v_4 and v_5 with weights proportional to the approximate angles 11°, 23°, 237°, 74° and 14° subtended by the arcs at z_0 respectively.

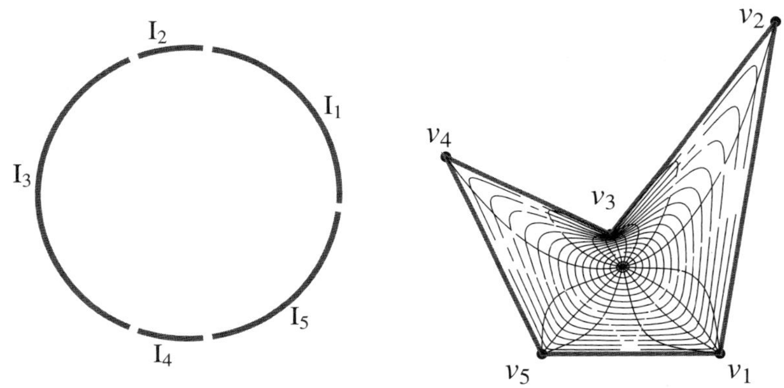

FIGURE 7. Modified arcs that solve the mapping problem for the pentagon Π_5 of Example 1.1.

the interior of Π_5, and in fact these arcs provide a solution to the mapping problem for Π_5.

Remark 3.4. In solving the mapping problem we see that uniqueness is only possible up to precomposition by a disc automorphism of the form $e^{i\lambda}\phi_\alpha$ where $\alpha \in \mathbb{D}$ and λ is real. Suppose for a given polygon Π, that the set of arcs $\{I_1, I_2, \ldots, I_n\}$ provide a boundary correspondence $\hat{f}$ to the vertices of Π with a univalent extension, and that g' has zero set W. Then the set of arcs $\{\phi(I_1), \phi(I_2), \ldots, \phi(I_n)\}$ provide a boundary correspondence $\tilde{f} = \hat{f} \circ \phi^{-1}$ with dilatation zeros $\phi(W)$. In Figure 8 we see the function from Figure 7 precomposed with the automorphism ϕ_α, where $\alpha = -0.1 - 0.3\mathrm{i}$. This provides a second set of arcs solving the mapping problem for the pentagon Π_5.

3.3. Role of the dilatation. For the Poisson extensions we are considering, the dilatation is, up to a multiple of magnitude 1, a product of finitely many

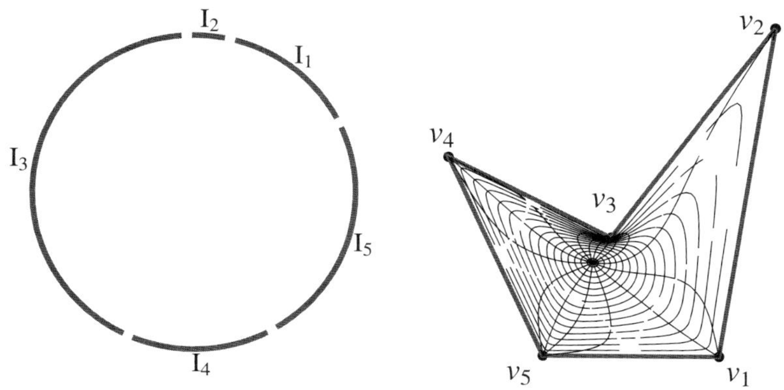

FIGURE 8. Univalent map from Figure 7 precomposed with a disc automorphism.

Blaschke factors or their reciprocals ([10, Theorem 1]). The dilatation provides a useful criterion to determine whether the Poisson extension is univalent.

Theorem 5 (Sheil-Small [10]). *Let f be a harmonic map obtained as the Poisson extension $P[\hat{f}]$ of a piecewise constant boundary function $\hat{f}$. Then f is univalent in the disc if and only if $\omega_f(z)$ is a product of finitely many Blaschke factors, that is if and only if all the zeros of ω_f lie inside the unit disc.*

As an illustration of this theorem, consider the nonconvex pentagon Π_5 introduced in Figure 6. The zeros of the dilatation for the nonunivalent function in Figure 3 are shown on the left of Figure 9, and those for the univalent function in Figure 7 appear on the right of Figure 9.

In order to apply this criterion for univalence to our Poisson extension

$$f(z) = \sum_{k=1}^{N} \frac{v_k}{\pi}\left(\arg \frac{z - I_k^+}{z - I_k^-} - \frac{|I_k|}{2}\right),$$

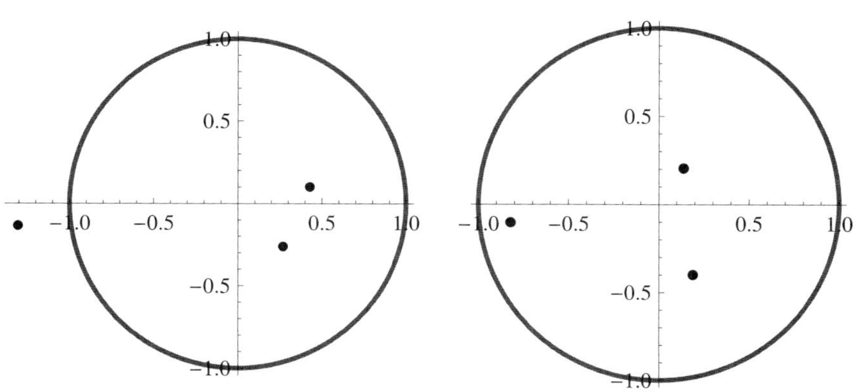

FIGURE 9. Dilatation zeros for the nonunivalent function in Figure 3 (left), and for the univalent function in Figure 7 (right).

we must compute the dilatation $\omega_f = g'/h'$ of $f = h + \bar{g}$ and then locate the dilatation zeros. We easily obtain

$$h'(z) = \sum_{k=1}^{N} \frac{v_k}{2\pi \mathrm{i}} \left(\frac{1}{z - I_k^+} - \frac{1}{z - I_k^-} \right) \quad \text{and} \quad g'(z) = \sum_{k=1}^{N} \frac{\overline{v_k}}{2\pi \mathrm{i}} \left(\frac{1}{z - I_k^+} - \frac{1}{z - I_k^-} \right).$$

The zeros of g' include the zeros of ω_f; g' may have additional zeros on $\mathbb{T}$ not present in the zero set of ω_f. We can rewrite

$$g'(z) = \frac{P(z)}{\prod\{(z - \zeta) : \zeta \in E\}}$$

where $P(z)$ is a polynomial, the degree of which is proved in [10, Section 1] to be less than or equal to $N - 2$.

3.4. Angles. A further connection between the dilatation of a mapping onto a polygon and the target polygon itself is that half the sum of the hyperbolic angles subtended by a given arc I_k at the dilatation zeros is equal to the corresponding interior angle θ_k of Π.

Lemma 3.5. *Consider a set of open arcs $\{I_k : 1 \le k \le N\}$ with endpoints E, indexed in order of increasing argument and partitioning $\mathbb{T}\setminus E$. Let Π be a positively oriented polygon with vertices $\{v_k : 1 \le k \le N\}$ and f be the univalent harmonic map obtained as the Poisson extension $P[\hat{f}]$ of the step function $\hat{f}$ that maps the arc I_k to v_k. Suppose further that ω_f has $N - 2$ zeros inside the unit disc, where the zeros are counted according to their multiplicities. Writing θ_k as the interior angle of Π at v_k, we have*

$$\theta_k = \frac{1}{2} \sum \{H_{I_k}(w) : w \text{ is a zero of } \omega_f\}, \tag{3.3}$$

where summands are repeated according to the multiplicities of the zeros.

Proof. Define $\Delta_I \arg \sqrt{\omega_f}$ as the total change in argument $\arg(\sqrt{\omega_f(I^+)}) - \arg(\sqrt{\omega_f(I^-)})$ over the arc I, where $\omega_f(I^+)$ and $\omega_f(I^-)$ are, respectively, the limits of $\omega_f(\mathrm{e}^{\mathrm{i}s})$ as s approaches the endpoints of I from the right or from the left respectively. Corollary 2.2a of [6] states that for each t on the unit circle

$$\Im\left\{\sqrt{\omega_f(\mathrm{e}^{\mathrm{i}t})}\left(\hat{f}(\mathrm{e}^{\mathrm{i}t+}) - \hat{f}(\mathrm{e}^{\mathrm{i}t-})\right)\right\} = 0$$

where $\hat{f}(\mathrm{e}^{\mathrm{i}t+})$ and $\hat{f}(\mathrm{e}^{\mathrm{i}t-})$ are, respectively, the limits of $f(\mathrm{e}^{\mathrm{i}s})$ as s approaches the endpoints of I from the right or from the left of t. This leads to the corner condition

$$\theta_i \equiv \Delta_I \arg \sqrt{\omega_f} \pmod{\pi}$$

(see [1] or for a more detailed explanation [5, Section 7.4] or the description in [8, Section 2.4]). Theorem 2.13 of [1] shows that if the harmonic extension f is univalent, then $\Delta_I \arg \sqrt{\omega_f}$ cannot exceed the interior angle θ_i. At a convex vertex then, there is only one possibility for $\Delta_I \arg \sqrt{\omega_f}$ mod π, namely

$$\theta_k = \Delta_{I_k} \arg \sqrt{\omega_f}. \tag{3.4}$$

However, in the case that v_k is not a convex vertex, i.e., if $\pi \le \theta_k \le 2\pi$, then there are two possibilities: if (3.4) holds we follow [1] calling v_k a *full resting point*, but it is also possible that

$$\theta_k = \pi + \Delta_I \arg \sqrt{\omega_f} \tag{3.5}$$

Since $\omega_f(z) = \mathrm{e}^{\mathrm{i}\lambda}\Pi\{\phi_w(z) : w \text{ is a zero of } \omega_f\}$, where λ is real and factors in the product are repeated corresponding to multiplicities of the zeros of ω_f, we have

$$\sum_{k=1}^{N} \Delta_{I_k} \arg \sqrt{\omega_f} = \frac{1}{2}(N-2)2\pi = (N-2)\pi.$$

A polygon with N sides has interior angle sum $(N-2)\pi$, so $\sum_{k=1}^{N} \theta_k = (N-2)\pi$. Since over the full circle $\Delta \arg \sqrt{\omega_f} = (N-2)\pi$, equation (3.4) rather than (3.5) must hold at each vertex of Π. Thus (3.4) becomes

$$\begin{aligned}
\theta_k &= \Delta_{I_k} \arg \sqrt{\Pi\{\phi_w(z) : w \text{ is a zero of } \omega_f\}} \\
&= \frac{1}{2}\sum\{\Delta_{I_k} \arg \phi_w(z) : w \text{ is a zero of } \omega_f\} \\
&= \frac{1}{2}\sum\{\arg(\phi_w(I^+)/\phi_w(I^-)) : w \text{ is a zero of } \omega_f\} \\
&= \frac{1}{2}\sum(H_{I_i}(w) : w \text{ is a zero of } \omega_f).
\end{aligned}$$

□

Remark 3.6. Arguments involving the interior angle sum to conclude whether vertices are full resting points have been made in the context of minimal surfaces in [2,8]. When f is univalent and ω has fewer than $N-2$ zeros inside the unit disc then some of the vertices necessarily fail to be full resting points.

Although from the Radó – Kneser – Choquet theorem the arcs in the boundary mapping can be arbitrarily chosen, we see that there is a constraint on the location of the dilatation zeros through consideration of the interior angles of Π alone: If a single arc endpoint is fixed and if the dilatation zeros are also fixed, then a necessary condition to obtain interior angles $\theta_1, \theta_2, \ldots, \theta_n$ in the target polygon is that the boundary arcs be chosen uniquely so that equation (3.3) is satisfied.

Example 3.7. Consider a boundary mapping $\hat{f}$ of the arcs $\{I_1, I_2, I_3\}$ to the vertices $\{1, \mathrm{e}^{\mathrm{i}\pi/4}, -\mathrm{i}\}$ under the constraints that $\omega_f = z$, and $1 \in I_1$, $\mathrm{i} \in I_2$, $-\mathrm{i} \in I_3$ (see Figure 10. Theorem 4.1 of [6] shows that the Poisson extension of such a boundary correspondence cannot be univalent. The same conclusion can be reached using Lemma 3.5. We see that the interior angle $\theta_1 > \pi/2$, but $|I_1| < \pi$. Therefore $\theta_1 = \frac{1}{2}H_{I_1}(0) = \frac{1}{2}E_{I_1}(0) = |I_1|/2$ cannot hold. Moreover, since the only dilatation zero is 0, Lemma 3.5 dictates that each arclength must be twice the corresponding interior angle: $|I_1| = 2\theta_1 \cong 3.92$, $|I_2| = 2\theta_2 \cong 1.57$, and $|I_3| = 2\theta_3 \cong 0.79$.

Remark 3.8. The context of Example 3.7, as it appeared in [6], was used to show that the two normalizations for harmonic mappings onto a triangle (i) $\omega_f = z$ and (ii) prescribing boundary values at exactly three points on the boundary are incompatible.

Lemma 3.5 allows us to characterize the harmonic maps under consideration onto a triangle.

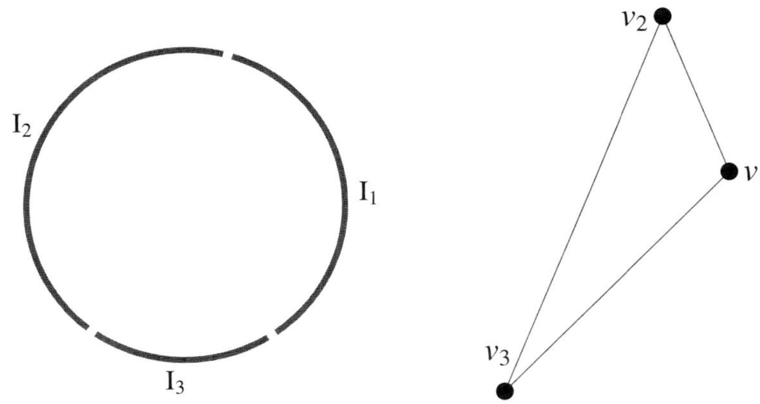

FIGURE 10. Example from Theorem 4.1 of [6].

Theorem 3.9. *Let Π be a positively oriented triangle with vertices v_1, v_2, and v_3 such that the interior angle at v_k is θ_k, $1 \leq k \leq 3$. Then if I_1, I_2, and I_3 are consecutive arcs on the unit circle such that I_k has length $2\theta_k$, the boundary correspondence that maps I_k to v_k, $1 \leq k \leq 3$ has a Poisson extension that is univalent with dilatation $\omega_f = z$. Moreover, these are the only solutions to the mapping problem for Π, up to precompositon with a disc automorphism.*

3.5. Side lengths of the target polygon. In solving the mapping problem, it would seem that we ought to begin with a target polygon Π and choose arcs so that the Poisson extension function has all of its dilatation zeros inside the unit disc. A related problem is to prescribe arcs and dilatation zeros, and determine the resulting polygon. In order to prove our main result in Section 4, it is this related problem that we will consider. In [10], Sheil-Small provides a constructive solution to this latter problem that applies to quite general choices of arcs and dilatation. It is stated in part below.

Theorem 1 (Sheil-Small [10]). *Suppose we prescribe a set of N open arcs $\{I_k : 1 \leq k \leq N\}$ with endpoints E, such that the set of arcs are indexed in order of increasing argument and partition $\mathbb{T}\backslash E$. Prescribe also a set $W = \{w_1, w_2, \ldots, w_m\}$ of up to $N-2$ zeros, possibly with repetitions. Then there is a polygon Π with vertices $\{v_k : 1 \leq k \leq N\}$ for which the Poisson extension f of the boundary function $\hat{f}$ that maps arcs to corresponding vertices has dilatation*

$$\omega_f = \mathrm{e}^{\mathrm{i}\lambda} \prod \{\phi_{w_k} : 1 \leq k \leq m\},$$

for some real constant λ. Moreover, the polygon Π is uniquely determined up to a rigid motion.

In order to solve the mapping problem for quadrilaterals we will make use of this construction. In particular, we need to obtain an expression for the ratio of successive side lengths of the resulting polygon. This expression is easily deduced by following the proof of [10]. Set

$$S(z) = \prod \{(z - \zeta) : \zeta \in E\} \quad \text{and} \quad Q(z) = \prod \{(z - w) : w \in W\}.$$

By formula (1.15) and the formula preceding (1.13) in [10],

$$|\overline{v_{k-1}v_k}| = \frac{1}{2\pi}\frac{Q(I_{k-1}^+)}{S'(I_{k-1}^+)} = \frac{1}{2\pi}\frac{Q(I_k^-)}{S'(I_k^-)}.$$

Thus the ratio of successive sides is given by

$$\frac{|\overline{v_k v_{k+1}}|}{|\overline{v_{k-1}v_k}|} = \frac{Q(I_k^+)}{S'(I_k^+)}\frac{S'(I_k^-)}{Q(I_k^-)},$$

where $1 \le k \le N$. Note that we have identified symbols indexed by 0 and 1 with those indexed by N and $N+1$ respectively. We can rewrite this side length ratio in terms of the set E of arc endpoints and their distances from the dilatation zeros. For $1 \le k \le N$ we can compute

$$\frac{|\overline{v_k v_{k+1}}|}{|\overline{v_{k-1}v_k}|} = \prod_{t\in E\setminus\{I_1^-, I_1^+\}} \frac{|t - I_k^-|}{|t - I_k^+|} \prod\left\{\frac{|w - I_k^+|}{|w - I_k^-|} : w \text{ is a zero of } \omega_f\right\}. \tag{3.6}$$

The second product in (3.6) is clearly the ratio $Q(I_k^+)/Q(I_k^-)$. A calculation shows that the first product is the ratio of the derivatives $S'(I_k^-)/S'(I_k^+)$.

Remark 3.10. The zeros of ω_f are equal to W and repeated in the same multiplicity. The map f will be of course be univalent if the dilatation zeros are chosen to lie inside the unit disc at the outset.

Remark 3.11. Theorem 1 is stated and applied here in less generality than in [10] where it is possible to prescribe the zeros of g' for the resulting harmonic extension map $f = h + \bar{g}$. Zeros of g' may be selected to fall on the disc boundary $\mathbb{T}$, and will not be present in ω_f.

4. Mapping problem for quadrilaterals

Observe that a quadrilateral is uniquely determined (up to a rigid motion and scale) by its interior angles and the ratio of lengths of any two adjacent sides. We now solve the mapping problem for an arbitrary quadrilateral Π. The proof outline is as follows. We solve the mapping problem for a polygon Π' with interior angles identical to those of Π, and with prescribed dilatation zeros. Then by varying one of the dilatation zeros in the construction, we will see that the ratio of two adjacent sides of Π' takes on every positive real number, while the interior angles remain fixed and the mapping onto Π' remains univalent. We thus show that we can obtain every quadrilateral with a given set of interior angles.

Theorem 4.1. *The mapping problem can be solved for any positively oriented quadrilateral.*

PROOF. Suppose we begin with an arbitrary quadrilateral Π with vertices $\{v_1, v_2, v_3, v_4\}$ such that the interior angle at the vertex v_k is θ_k, $1 \le k \le 4$. If Π is convex then the Radó – Kneser – Choquet theorem provides a solution, so assume Π is nonconvex. Then Π has a single interior angle larger than π, say $\theta_1 > \pi$.

We will apply the construction from Theorem 1 of [10] with $N = 4$ to obtain a map onto a polygon Π' with the correct interior angles at its vertices. We obtain a significant simplification in choosing our arcs if we stipulate our dilatation zeros to be $\{0, w_1\}$, where $w_1 \in \mathbb{D}$. A dilatation zero at 0 can always be accomplished by precomposition with a disc automorphism. We will see that in order to obtain a polygon Π' with interior angles θ_k, $1 \le k \le 4$, we must simply choose I_k so that $E_{w_1}(I_k) = \theta_k$. This remarkable simplification holds only for quadrilaterals.

Since $\theta_1 > \pi$, there exists a $\delta > 0$ such that

$$2\pi - 2\max\{\theta_2 + \theta_3, \theta_3 + \theta_4\} - \delta > 2(\theta_1 - \pi).$$

This follows because $\theta_1 - \pi = \pi - (\theta_2 + \theta_3 + \theta_4)$, so

$$\theta_1 - \pi < \pi - (\theta_2 + \theta_3) \quad \text{and} \quad \pi - (\theta_3 + \theta_4). \tag{4.1}$$

Now choose an arc I_1 of $\mathbb{T}$ so that $|I_1|$ lies in the interval

$$(2(\theta_1 - \pi), 2\pi - 2\max\{\theta_2 + \theta_3, \theta_3 + \theta_4\} - \delta). \tag{4.2}$$

The arc $|I_1|$ is therefore large enough so that $2\theta_1 - |I_1|$ lies in the range $(0, 2\pi)$. Therefore we can choose $w_1 \in \mathbb{D}$ so that $H_{w_1}(I_1) = 2\theta_1 - |I_1|$. The upper bound on $|I_1|$ allows for enough room remaining on $\mathbb{T}$ for the arcs I_2, I_3 and I_4 so that the side ratios for Π' can take on all possible positive values at the end of the proof. By the relationship between Euclidean and hyperbolic angle in Lemma 3.2

$$H_{w_1}(I_1) = 2E_{w_1}(I_1) - |I_1|,$$

we have

$$E_{w_1}(I_1) = \theta_1.$$

We now define arcs I_2, I_3, and I_4 so that $E_{w_1}(I_k) = \theta_k$: extend four rays from w_1, separated by the angles $\{\theta_k : 1 \le k \le 4\}$ in counterclockwise order, such that the two rays separated by θ_1 pass through $\mathbb{T}$ at I_1^- and I_1^+ (see Figure 11). Define the arcs I_2, I_3, and I_4, ordered by increasing argument, with endpoints obtained from the points of intersection of the four rays with the unit circle. The arc endpoints are then $E = \{I_1^+, I_2^+, I_3^+, I_4^+\}$ (or equivalently $\{I_1^-, I_2^-, I_3^-, I_4^-\}$). Proceeding with the construction from Theorem 1 of [10] using the arcs $\{I_1, I_2, I_3, I_4\}$ and $W = \{0, w_1\}$, we set $Q(z) = z(z - w_1)$ and $S(z) = \prod\{(z - \zeta) : \zeta \in E\}$.

We obtain a harmonic function f onto a polygon Π' with vertices $\{v_i' : 1 \le k \le 4\}$. Since by construction the dilatation zeros $\{0, w_1\}$ lie in the unit disc, f is also univalent. By Lemma 3.5, we can compute the interior angle at v_k' to be

$$\tfrac{1}{2}\big(H_0(I_k) + H_{w_1}(I_k)\big) = \tfrac{1}{2}\big(|I_k| + H_{w_1}(I_k)\big) = E_{w_1}(I_k) = \theta_k, \quad 1 \le k \le 4.$$

Thus we have a univalent map onto a polygon Π' with the same interior angles as Π, but not necessarily the correct side lengths.

We now vary our construction. Define by $C = C_{\theta_1 I_1}$ the circle determined by the points I_1^+, I_1^- and w_1. We will move w_1 on the arc $\Lambda = C_{\theta_1 I_1} \cap \mathbb{D}$, noting that the angle $H_{w_1}(I_1)$ is unchanged (again, see Figure 11). The arc I_1 also remains fixed but in order to maintain a polygon with the prescribed interior angles, the arcs I_2, I_3 and I_4 all change as w_1 moves along Λ.

By equation (3.6) with $k = 1$,

$$\frac{|\overline{v_1' v_2'}|}{|\overline{v_4' v_1'}|} = \prod_{t \in E \setminus \{I_1^-, I_1^+\}} \frac{|t - I_1^-|}{|t - I_1^+|} \prod \left\{ \frac{|w - I_1^+|}{|w - I_1^-|} : w \text{ is a zero of } \omega_f \right\}.$$

This first product has two factors and the second just one (the factor corresponding to the zero 0 of the dilatation ω_f reduces to 1). For an arc I of the the unit circle

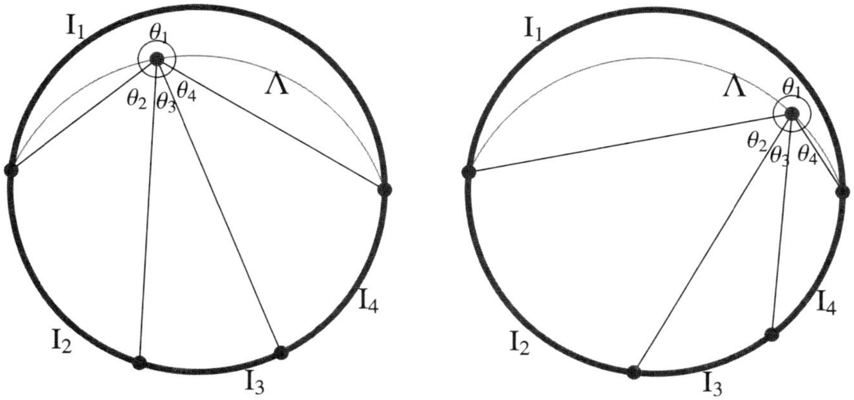

FIGURE 11. Arcs I_2, I_3, and I_4 determined by the Euclidean interior angles θ_i, $1 \leq i \leq 4$ of Π for two different values of the dilatation zero w_1 on the circular path Λ.

with length $|I|$, simple geometry shows that the length of the corresponding chord is $\frac{1}{2}\sin(|I|/2)$. Therefore the expression simplifies to

$$
\begin{aligned}
\frac{|\overline{v_1'v_2'}|}{|\overline{v_4'v_1'}|} &= \left(\prod_{t\in\{I_2^+, I_3^+\}} \frac{|t - I_1^-|}{|t - I_1^+|}\right)\frac{|w_1 - I_1^+|}{|w_1 - I_1^-|} \qquad (4.3)\\
&= \left(\frac{\sin(|I_1 + I_2|/2)}{\sin(|I_2|/2)}\frac{\sin(|I_4|/2)}{\sin(|I_1 + I_4|/2)}\right)\frac{|w_1 - I_1^+|}{|w_1 - I_1^-|}
\end{aligned}
$$

We show that all positive real numbers are taken on by the sidelength ratio $|\overline{v_1'v_2'}|/|\overline{v_4'v_1'}|$ at v_1'. Figure 12 depicts the ratio of sidelengths in diagram form: lengths of the three solid lines form the three numerator terms in (4.3), and the

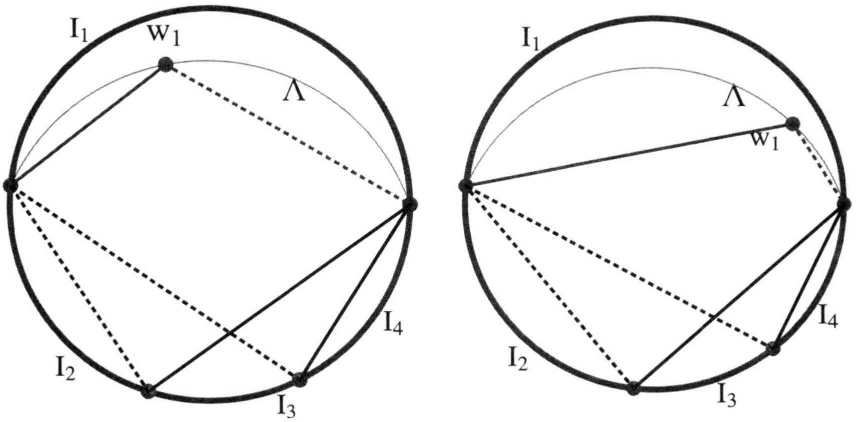

FIGURE 12. Sidelength ratios at vertex v_1' of Π' for two different values of w_1 on Λ. The lengths of solid lines represent factors in the numerator of equation (4.3) and the lengths of dashed lines the factors in the denominator.

lengths of the three dashed lines represent the denominator terms. Clearly we have

$$\lim_{\substack{w_1 \to I_1^+ \\ w_1 \text{ on } \Lambda}} \frac{|w_1 - I_1^+|}{|w_1 - I_1^-|} = 0 \quad \text{and} \quad \lim_{\substack{w_1 \to I_1^- \\ w_1 \text{ on } \Lambda}} \frac{|w_1 - I_1^+|}{|w_1 - I_1^-|} = \infty$$

when w_1 is restricted to lie on Λ as it approaches endpoints of I_1. We now show that the quantity in square brackets in (4.3) remains bounded from ∞ and from 0—that is, neither $|I_2|$ nor $|I_4|$ shrinks to zero. Note that the arc I_k subtends an angle of at most $2\theta_k$ ($k \in \{2, 3, 4\}$), which we see simply by considering the possible range of values of $E_{I_k}(w_1)$ for w_1 in the disc.

We show that $|I_2|$ is bounded from zero as $w_1 \to I_1^+$ by showing that $|I_3+I_4+I_1|$ is bounded above by a constant less than 2π, where $I_3 + I_4 + I_1$ is the concatenated arc (I_3^-, I_1^+).

$$\begin{aligned}|I_1 + I_3 + I_4| &= |I_1| + |I_3| + |I_4| = |I_1| + 2\theta_3 + 2\theta_4 - H_{w_1}(I_3) - H_{w_1}(I_4)\\ &\le |I_1| + 2(\theta_3 + \theta_4) \le 2\pi - 2(\theta_3 + \theta_4) - \delta + 2(\theta_3 + \theta_4) = 2\pi - \delta.\end{aligned}$$

The last inequality holds because of the upper bound on our initial choice of $|I_1|$ in (4.2).

Similarly as $w_1 \to I_1^-$ along Λ, we show that $|I_4|$ is bounded from zero.

$$\begin{aligned}|I_1 + I_2 + I_3| &= |I_1| + |I_2| + |I_3| = |I_1| + 2\theta_2 + 2\theta_3 - H_{w_1}(I_2) - H_{w_1}(I_3)\\ &\le |I_1| + 2(\theta_2 + \theta_3) \le 2\pi - 2(\theta_2 + \theta_3) - \delta + 2(\theta_2 + \theta_3) = 2\pi - \delta.\end{aligned}$$

Thus $|\overline{v_1'v_2'}|/|\overline{v_4'v_1'}|$ takes on all positive values as w_1 traverses Λ; one of these values must be $|\overline{v_1v_2}|/|\overline{v_4v_1}|$ so that $\Pi' = \Pi$, up to a rigid motion and scaling. ☐

Remark 4.2. Rather than a single solution for a given quadrilateral, the proof of the main theorem produces an infinite family of distinct maps that solve the mapping problem, indexed by some allowable range of $|I_1|$ that satisfies (4.2). If $|I_1|$ is chosen too large, then only a restricted range of side length ratios for the quadrilateral Π' can be obtained. If $|I_1|$ is chosen too small, then the dilatation zero w_1 falls outside the unit disc in order to obtain a large enough interior angle θ_1 for Π', and the corresponding mapping fails to be univalent.

Remark 4.3. The relationship $\theta_k = E_{w_1}(I_k)$ of interior angles of Π to Euclidean angles subtended by the arcs at w_1 is unique to the case $N = 4$, and choosing one factor in the dilatation to be z. However, a similar simplification arises in constructing a minimal surface, in which case the dilatation must have an analytic square root, over a hexagon. We can therefore consider the dilatation to be, up to a multiple of magnitude 1, $\omega_f(z) = z^2\phi_{w_1}^2(z)$, where w_1 lies in the closed unit disc. The relationship $2\theta_k = E_{w_1}(I_k)$ then holds, and and we can then describe all harmonic mappings with fourth order dilatation that lift to minimal surfaces of Jenkins – Serrin type (JS surfaces) over a hexagon.

Open Question 1. What are the restrictions on the location of dilatation zeros for a harmonic mapping that is the Poisson extension of a piecewise constant boundary function on the unit circle, for a given quadrilateral?

Open Question 2. The mapping problem remains unsolved. Can the mapping problem be solved for polygons with more than 4 sides, or is there a polygon Π for which the mapping problem cannot be solved (as conjectured in [10])?

References

1. D. Bshouty and W. Hengartner, *Boundary values versus dilatations of harmonic mappings*, J. Anal. Math. **72** (1997), 141–164.
2. D. Bshouty and A. Weitsman, *On the Gauss map of minimal graphs*, Complex Var. Theory Appl. **48** (2003), no. 4, 339–346.
3. G. Choquet, *Sur un type de transformation analytique généralisant la représentation conforme et définie au moyen de fonctions harmoniques*, Bull. Sci. Math. (2) **69** (1945), 156–165.
4. J. Clunie and T. Sheil-Small, *Harmonic univalent functions*, Ann. Acad. Sci. Fenn. Ser. A I Math. **9** (1984), 3–25.
5. P. Duren, *Harmonic mappings in the plane*, Cambridge Tracts in Math., vol. 156, Cambridge Univ. Press, Cambridge, 2004.
6. W. Hengartner and G. Schober, *On the boundary behavior of orientation-preserving harmonic mappings*, Complex Variables Theory Appl. **5** (1986), no. 2-4, 197–208.
7. H. Kneser, *Lösung der Aufgabe* 41, Jber. Deutsch. Math.-Verein. **39** (1926), no. 1, 123–124.
8. J. McDougall and L. Schaubroeck, *Minimal surfaces over stars*, J. Math. Anal. Appl. **340** (2008), no. 1, 721–738.
9. T. Radó, *Aufgabe* 41, Jber. Deutsch. Math.-Verein. **35** (1926), no. 1, 49.
10. T. Sheil-Small, *On the Fourier series of a step function*, Michigan Math. J. **36** (1989), no. 3, 459–475.

Department of Mathematics and Computer Science, Colorado College, 14 East Cache La Poudre St., Colorado Springs, CO 80903, USA

E-mail address: JMcDougall@ColoradoCollege.edu

Centre de Recherches Mathématiques
CRM Proceedings and Lecture Notes
Volume **55**, 2012

Hartogs Phenomenon on Unbounded Domains — Conjectures and Examples

Al Boggess, Roman J. Dwilewicz, and Zbigniew Słodkowski

ABSTRACT. In this paper we consider the Hartogs type extension problem for unbounded domains Ω in $\mathbb{C}^2$. The conjecture is that if the closure $\overline{\Omega}$ does not contain any closed pseudoconcave subset K of $\mathbb{C}^2$, then any function that satisfies the tangential Cauchy–Riemann equations on the boundary $b\Omega$ can be holomorphically extended to Ω. The conjecture is proved for Reinhardt tube-like domains and several related examples are given.

1. Introduction

In this paper we give a review of results on holomorphic extension of Cauchy–Riemann (CR) or holomorphic functions, starting with a classical theorem of Hartogs up to very recent results and examples of the authors. We mainly concentrate on the case of unbounded domains in the complex vector space $\mathbb{C}^2$, however most of the results can be generalized to n-dimensions, $n \geq 2$.

FIGURE 1. Friedrich Hartogs (1874–1943). Courtesy of the Archives of the Mathematisches Forschungsinstitut Oberwolfach

2010 *Mathematics Subject Classification.* 32V25.
Key words and phrases. Hartogs extension, CR functions, holomorphic functions.
The second author was supported in part by the Missouri Research Board grant 2011-2013.
This is the final form of the paper.

The following classical holomorphic extension result of Hartogs [14] in 1906 is considered by most researchers as the formal beginning of Complex Analysis of several variables.

Theorem (Hartogs [14], 1906). *Let Ω be a bounded domain in $\mathbb{C}^n$, $n \geq 2$, with connected boundary* $\mathrm{b}\Omega$. *Then any holomorphic function in a neighborhood U of the boundary* $\mathrm{b}\Omega$ *can be holomorphically extended to Ω (see illustration in Figure* 2).

Interestingly, it was not until 2007 that the above theorem was proved in its full generality by Joël Merker and Egmont Porten [22] using the same analytic disc methods used by Hartogs in 1906. A very beautiful recollection of results and historical corrections are given in the paper by Michael Range [24]. We refer the reader to these papers ([22], [24]) for very detailed explanation of historical context of the Hartogs theorem. Here we only mention the main contributors to the proof of the theorem, however using different ideas than those of Hartogs: R. Fueter [13], E. Martinelli [20], [21], S. Bochner [3], G. Fichera [12], L. Ehrenpreis [11].

The Hartogs extension theorem has been generalized by many authors (see below) to the case of CR functions which are defined only on the boundary $b\Omega$ and satisfy the tangential Cauchy-Riemann equations (instead of being analytically defined on a neighborhood of the boundary). The precise definition of CR (or Cauchy-Riemann) functions is given in the next section. Of course in this case some regularity assumption, usually of class C^1, on the boundary should be imposed. The extensions in the spirit of the Hartogs theorem we will call *Hartogs type extension.* Holomorphic extensions of CR functions from the boundary, we will call *Hartogs tangential type extension.*

Historically, the above mentioned generalization for CR functions was proved by Francesco Severi [27] in 1931 (real analytic case), Hellmuth Kneser [17] in 1936 (strictly pseudoconvex), Gaetano Fichera [12] in 1957 (class $C^{1+\varepsilon}$), and finally by Enzo Martinelli [21] in 1961 (C^1); see the paper by Michael Range [24] for more detailed account.

Since the time when the above mentioned results appeared, some variations were proved, just to list a few: Kohn – Rossi [18], Ivashkovich [16], Laurent-Thiébaut [19], Harvey – Lawson [15], Dolbeault – Henkin [7], Sarkis [25, 26], Andreotti – Hill [1], Merker – Porten [23], Colţoiu – Ruppenthal [6], Dwilewicz [8, 9], Dwilewicz – Merker [10] and many others. In all of the above mentioned results, the domain is assumed to be bounded. There seems to be no general results in the case of unbounded domains, which is the main subject of this note. The problem we consider can be formulated as follows:

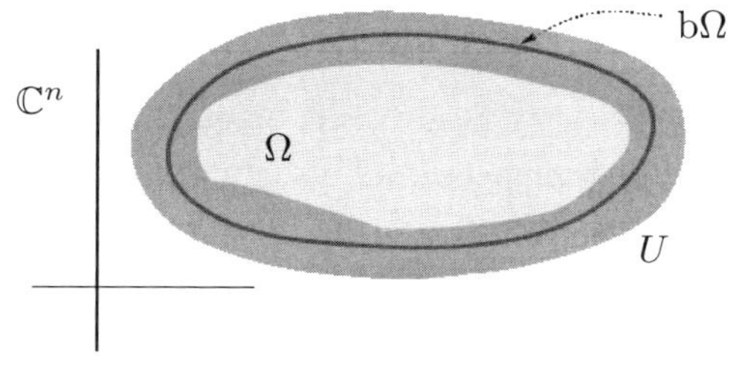

FIGURE 2. Illustration of the Hartogs extension theorem

Global Extension Problem in $\mathbb{C}^2$. *Let $\Omega \subset \mathbb{C}^2$ be an unbounded domain (open, connected) with connected C^1 boundary* $\mathrm{b}\Omega$. *Under what conditions on Ω is it true that any Cauchy–Riemann function defined on* $\mathrm{b}\Omega$ *can be holomorphically extended to Ω?*

The above problem can be refined. Looking at the classical Hartogs extension phenomenon we see that the main reason that the Hartogs type extension holds in a bounded domain in $\mathbb{C}^n$ is that there is no complex variety (of positive dimension) contained with its closure in the domain. A similar observation can be obtained from the examples given in this paper and also from Theorem 6.1. Because of that, the following conjecture can be formulated:

Conjecture for domains in $\mathbb{C}^2$ not containing closed pseudoconcave subsets. *Let Ω be a domain (open, connected) in $\mathbb{C}^2$ with connected boundary $M = b\Omega$ of class C^1. Assume that $\overline{\Omega}$ does not contain any closed pseudoconcave subset K of $\mathbb{C}^2$, i.e., the complement $\mathbb{C}^2 \setminus K$ is pseudoconvex. Then the tangential Hartogs extension holds for Ω.*

For tube-like domains (i.e., the projection of the domain along a 2-dimensional real plane is bounded in a transversal real plane), the above conjecture can be formulated in more clear terms:

Conjecture for tube-like domains in $\mathbb{C}^2$. *Let $\Omega \subset \mathbb{C}^2$ be a tube-like domain with connected boundary (resp., C^1 boundary). The Hartogs (resp., Hartogs tangential) type extension holds for Ω if and only if there is no complex line contained in Ω (resp., in $\overline{\Omega}$).*

The organization of the paper is as follows. After giving basic definitions in Section 2, we describe briefly the foliation method for extensions of holomorphic and CR functions in Section 3 and give its application to tube-like domains in Section 4. In Section 5 we apply a modified foliation method to tube-like domains along a complex line. Finally, in Section 6 we prove the Conjecture for Reinhardt tube-like domains and in the last section we show that the assumption in the conjecture cannot be relaxed to get the Hartogs tangential type extension.

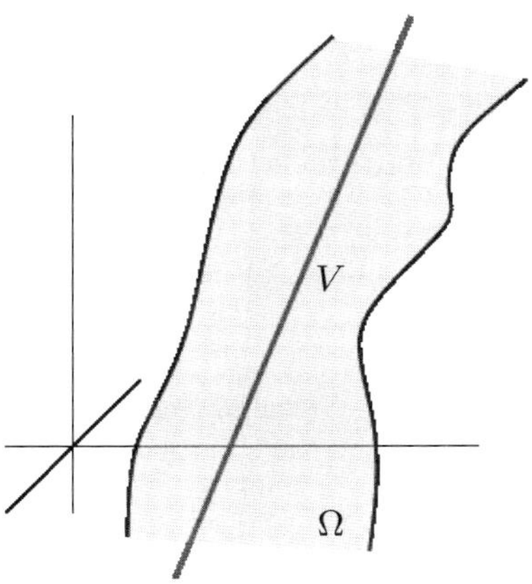

FIGURE 3. Illustration of the Conjecture for tube-like domains. If a complex line V is contained in Ω, then the Hartogs type extension does not hold.

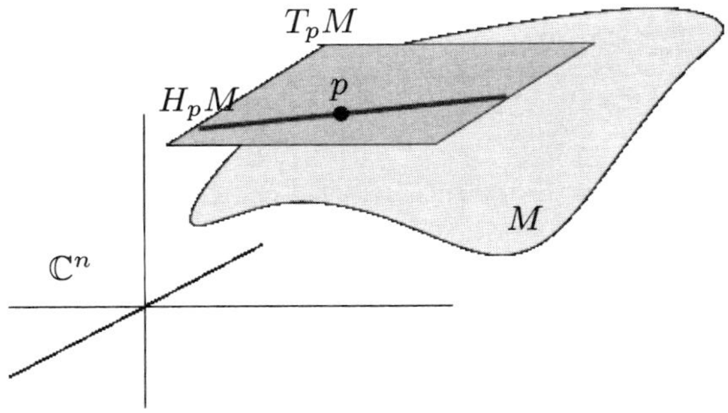

FIGURE 4. Tangent spaces to a real submanifold in $\mathbb{C}^n$.

2. Basic definitions and notation

Let M be a real embedded submanifold of class C^1 in $\mathbb{C}^n$. Let T_pM be the real tangent space to M at p. We can consider T_pM as a real vector subspace of the complex space $T_p\mathbb{C}^n$. We say that M is a *Cauchy–Riemann* (CR) *manifold* if

$$\dim_{\mathbb{C}}(H_pM) = \dim_{\mathbb{C}}(T_pM \cap \sqrt{-1}T_pM) \equiv \text{constant on } M.$$

This constant is called *CR dimension* of M and denoted by $\dim_{\mathrm{CR}} M$. If $\dim_{\mathrm{CR}} M = 0$, then M is called a *totally real manifold.*

By CR functions we mean functions $f\colon M \to \mathbb{C}$ of class C^1 that satisfy the tangential Cauchy–Riemann equations,

$$\bar{L}f = 0 \quad \text{for any } \bar{L} = a_1\frac{\partial}{\partial \bar{z}_1} + \cdots + a_n\frac{\partial}{\partial \bar{z}_n} \text{ tangent to } M.$$

Obviously, the restriction $f = F|_M$ of a holomorphic function $F\colon U \to \mathbb{C}$, $M \subset U$, to M is CR, but the opposite is not true in general.

By "smooth" (manifold or function) we shall mean of class C^1, unless otherwise specified. Usually the regularity conditions can be weakened, for instance, to consider CR functions in the distributional sense (e.g., continuous CR functions), however, it is not our intention here to formulate and prove results under the weakest hypotheses.

3. The foliation method

Let $\Omega = \Omega^+$ be a domain in $\mathbb{C}^n$, $n \geq 2$, with a smooth and connected boundary $M = \mathrm{b}\Omega$. Also we denote and assume that $\Omega^- = \mathbb{C}^n \setminus \overline{\Omega}$ is connected. Suppose that there is a holomorphic foliation of $\mathbb{C}^n$ by complex submanifolds X_s, each biholomorphic to $\mathbb{C}^k$, $1 \leq k \leq n-1$, $s \in \mathbb{C}^{n-k}$, such that the intersection $X_s \cap \overline{\Omega}$ is either compact or empty. Also assume that the foliation can be *straightened*, i.e., there exists a biholomorphic mapping of $\mathbb{C}^n$ onto itself that maps each leaf X_s onto $\mathbb{C}^k \times \{s\}$.

Example 3.1. Not all foliations can be straightened. Let $h(z_1, z_2) = P(z_1) + \mathrm{e}^{z_2}$, where $P = P(z_1)$ is a polynomial of degree at least two. It is possible to show that the foliation $L_s\colon P(z_1) + \mathrm{e}^{z_2} = s$ cannot be straightened.

We formulate a proposition which follows immediately from [9].

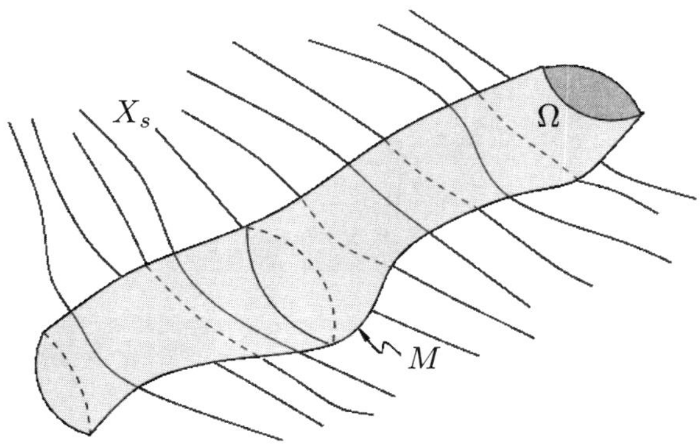

FIGURE 5. Unbounded domain Ω and foliation.

Proposition 3.2. *Under the above assumptions on the foliation, the following holds*:

(1) *If the dimension of the leaves of the foliation is $k \geq 2$, then any CR function f defined on M can be holomorphically extended to $F\colon \Omega \to \mathbb{C}$.*
(2) *If the dimension of the leaves of the foliation is $k = 1$ and that there exists $s_0 \in \mathbb{C}^{n-1}$ such that for s from a neighborhood of s_0 we have $X_s \cap \overline{\Omega} = \varnothing$, then any CR function defined on M can be holomorphically extended to $F\colon \Omega \to \mathbb{C}$.*

4. Application of the foliation method to tube-like domains

We apply the foliation method to tube-like domains. To avoid technicalities, we work in $\mathbb{C}^2$, but the ideas can be generalized to $\mathbb{C}^n$.

Let $(z_1, z_2) = (x_1 + iy_1, x_2 + iy_2)$ be coordinates in $\mathbb{C}^2$. Any tube in $\mathbb{C}^2$ with a two-dimensional (real) fiber is of the form

$$\rho(a_1 z_1 + \bar{a}_1 \bar{z}_1 + a_2 z_2 + \bar{a}_2 \bar{z}_2, a_3 z_1 + \bar{a}_3 \bar{z}_1 + a_4 z_2 + \bar{a}_4 \bar{z}_2) = 0, \tag{4.1}$$

where $\rho = \rho(c_1, c_2)$ is a real-valued smooth function in $\mathbb{R}^2$ with nonzero gradient.

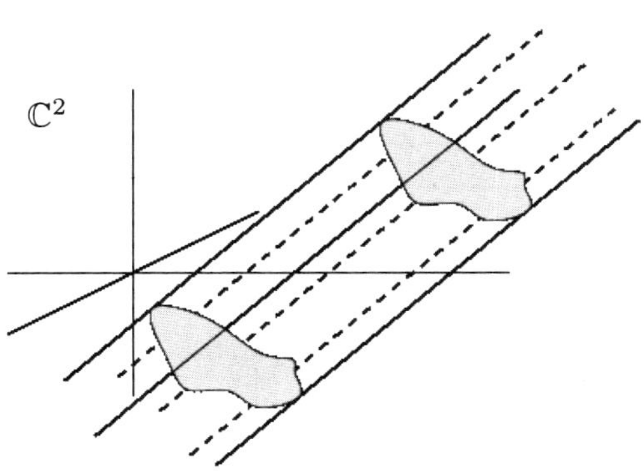

FIGURE 6. Tube in $\mathbb{C}^2$.

The fiber is given by the equations

$$\begin{cases} a_1 z_1 + \bar{a}_1 \bar{z}_1 + a_2 z_2 + \bar{a}_2 \bar{z}_2 = c_1 \\ a_3 z_1 + \bar{a}_3 \bar{z}_1 + a_4 z_2 + \bar{a}_4 \bar{z}_2 = c_2 \end{cases}$$

where c_1 and c_2 are real constants and the rank of the matrix

$$\operatorname{rank} \begin{bmatrix} a_1 & \bar{a}_1 & a_2 & \bar{a}_2 \\ a_3 & \bar{a}_3 & a_4 & \bar{a}_4 \end{bmatrix} = 2.$$

There are two cases: the fiber is totally real or is complex. If

$$\operatorname{rank} \begin{bmatrix} a_1 & a_2 \\ a_3 & a_4 \end{bmatrix} = 2,$$

then the fiber is totally real; otherwise is complex.

4.1. Tube-like domains along a totally real fiber. If the fiber is totally real, then there is a $\mathbb{C}$-linear isomorphism of $\mathbb{C}^2$ that maps the fiber onto $\mathrm{i}\mathbb{R}^2$ and the tube is given by

$$\rho(x_1, x_2) = 0. \tag{4.2}$$

In this case, it is known that any CR function from such tube can be holomorphically extended to the tube over the convex hull of its base (see [5]), and actually it is a generalization of the classical Bochner tube theorem [2].

However, using Proposition 3.2 we can get not only another proof (of extension to the inside of the tube) for tubes over closed Jordan curves, Figure 7 but also a generalization for *tube like* domains, see Figure 8.

For standard tubes as in (4.2) take the foliation of $\mathbb{C}^2$ given by

$$z_2 = z_1^2 + c, \quad \text{where } c = c_1 + \mathrm{i}c_2 \in \mathbb{C} \text{ is a constant.}$$

The intersection with the tube is

$$\begin{cases} \rho(x_1, x_2) = 0 \\ z_2 = z_1^2 + c \end{cases}$$

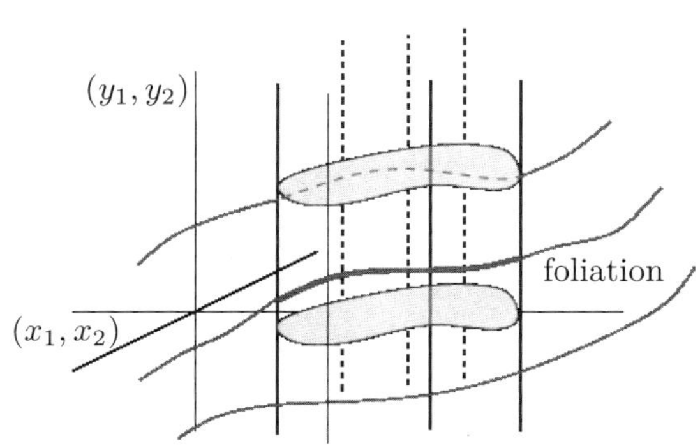

FIGURE 7. Foliation and a standard tube given by $\rho(x_1, x_2) = 0$.

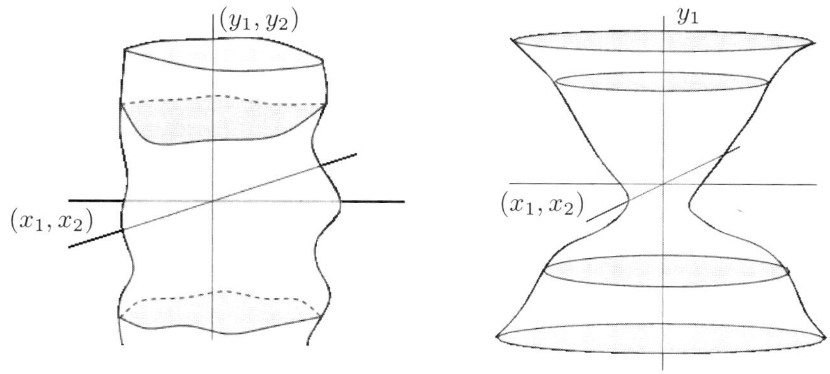

FIGURE 8. "Wavy" and "widening" tubes.

or

$$\begin{cases} \rho(x_1, x_2) = 0 \\ x_2 = x_1^2 - y_1^2 + c_1 \\ y_2 = 2x_1y_1 + c_2 \end{cases} \quad \text{or equivalently} \quad \begin{cases} \rho(x_1, x_2) = 0 \\ y_1^2 = x_1^2 - x_2 + c_1 \\ y_2 = 2x_1y_1 + c_2 \end{cases}$$

Since (x_1, x_2) is restricted to a compact set, then taking $c_1 < 0$ with $|c_1|$ large we get that the second equation does not have a solution. For each c, the solution set is compact or empty since (x_1, x_2) is restricted to a compact set and the above equations force (y_1, y_2) to be restricted to a compact set as well. Because the foliation $z_2 = z_1^2 + c$ can be straightened, Proposition 3.2 implies that any CR function on the tube can be holomorphically extended to the inside of the tube.

In a similar way, the extension theorem can be proved for tube-like domains, see Figure 8. For instance

$$x_1^2 + x_2^2 \leq g(y_1, y_2),$$

where g is a bounded smooth function, or

$$x_1^2 + x_2^2 \leq h(y_1^2),$$

where $h = h(t)$ is any smooth function that satisfies

$$\lim_{t \to \infty} \frac{h(t)}{t} = 0.$$

In the latter case, the domain is not tube-like since the projection along the y_1y_2-plane is not bounded, but still the Hartogs-tangential extension holds.

4.2. Tube-like domains along a complex fiber. The method described in the previous subsection cannot be applied in the same way for tube-like domains along complex fibers. First of all, the holomorphic extension doest not hold, in general, for such tubular domains, which can be seen in the following simple example:

Any tube with complex line fiber, is biholomorphically equivalent to the tube with fiber $\{0\} \times \mathbb{C}$ and can be given by the equation

$$\rho(x_1, y_1) = 0.$$

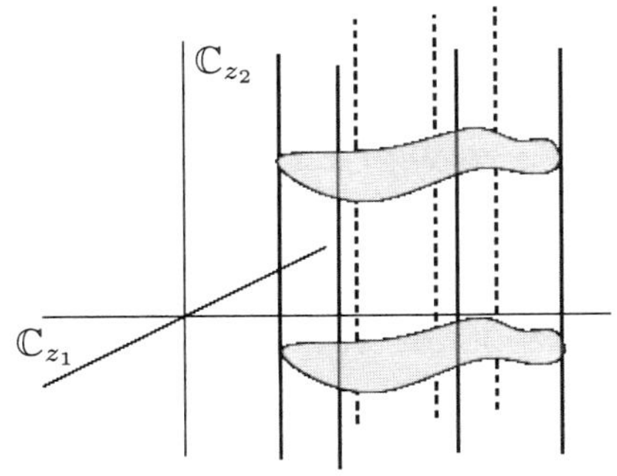

FIGURE 9. Tube along a complex fiber.

In particular,

$$\rho(x_1, y_1) = x_1^2 + y_1^2 = 1 \quad \text{or} \quad |z_1| = 1, \tag{4.3}$$

the function $f(z_1, z_2) = 1/z_1$ cannot be holomorphically extended to the tube over the solid disc.

To get holomorphic extension of CR functions defined on the boundary of the tube, we have to exclude the situation that a complex line lies inside the tube. Even if this additional assumption is satisfied, in general, it is not easy to construct a foliation that satisfies the conditions from Proposition 3.2. As an example, consider the tube (presented in Figure 10)

$$M_\varphi := \{(z_1, z_2); |z_1 - \varphi(z_2)| \leq \delta\} \quad (\delta < 1), \tag{4.4}$$

where

$$\varphi(z_2) = \begin{cases} 0 & \text{for } |z_2| \geq 1 \\ 1 - |z_2|^2 & \text{for } |z_2| < 1. \end{cases}$$

We have

Lemma 4.1. *For the tube* (4.4), *any holomorphic nonsingular curve given either by* $z_1 = h(z_2)$ *or* $z_2 = h(z_1)$, *where* h *is an entire function, intersects the tube.*

5. Tubes along a complex line: with a "bump" and with "dents"

The above Lemma 4.1 shows that there is a difficulty to apply the foliation method to the extension problem for tube-like domains along a complex line. In this section we will show how to overcome this difficulty.

5.1. Tube with a "bump". Consider $\mathbb{C}^2$ with coordinates z_1, z_2 and let C be the two real dimensional curve given by the union of the z_2 axis with $|z_2| \geq 1$ and the curve

$$\{z_1 = 1 - |z_2|^2; |z_2| \leq 1\}.$$

Let Ω be a δ-neighborhood of C in $\mathbb{C}^2$ with C^k boundary M, k high enough. The goal is to show the following extension theorem for M. In the following theorem and the remaining part of this section, "smooth" will mean of class C^k for high enough k.

Theorem 5.1. *Every smooth CR function on M extends to a unique holomorphic function on Ω.*

PROOF. Define the following Cauchy Kernels (on a line)

$$C_1(\zeta, z) = \frac{d(\zeta_1 - z_1)}{(\zeta_1 - z_1)} d(\zeta_1 - z_1) \wedge [\zeta_2 = z_2]$$

$$C_2(\zeta, z) = \frac{d(\zeta_2 - z_2)}{(\zeta_2 - z_2)} d(\zeta_2 - z_2) \wedge [\zeta_1 = z_1]$$

Here, $[\zeta_1 = z_1]$ is the current in $\mathbb{C}\times\mathbb{C}$ (with coordinates (ζ_1, z_1)) given by integration over the diagonal on the set $\{\zeta_1 = z_1\}$. For a fixed $z = (z_1, z_2) \in \mathbb{C}^2$, the ζ-support of $C_1(\zeta, z)$ is the complex line $\{(\zeta_1, z_2); \zeta_1 \in \mathbb{C}\}$. Similar considerations hold for C_2. Both C_1 and C_2 are fundamental solutions to $\bar{\partial}$ in $\mathbb{C}^2$. See the book [4] for explanation of the notions of current and fundamental solution.

STEP 1. Let $A = \{(z_1, z_2); |z_2| < 1 - 2\delta\}$, where $\delta > 0$ small enough. Let f be the given CR function on M. Our first goal is to extend f to a holomorphic function F_2 on $A \cap \Omega$. To do this, first extend f to a smooth function $\tilde{f}$ on $\overline{\Omega}$ with $\bar{\partial}\tilde{f} = 0$ on M to high order. Let v be the form defined by $\bar{\partial}\tilde{f}$ on Ω and 0 on the complement of Ω. Let $u_2 = C_2(v)$. The function u_2 is defined on the set $B = \{(z_1, z_2); \operatorname{Re} z_1 > \delta\}$ since the ζ-support of C_2 intersects $\overline{\Omega}$ in a compact set if $\operatorname{Re} z_1 > \delta$. Since C_2 is a fundamental solution to $\bar{\partial}$, we have $\bar{\partial}u_2 = v$ on B; u_2 is also

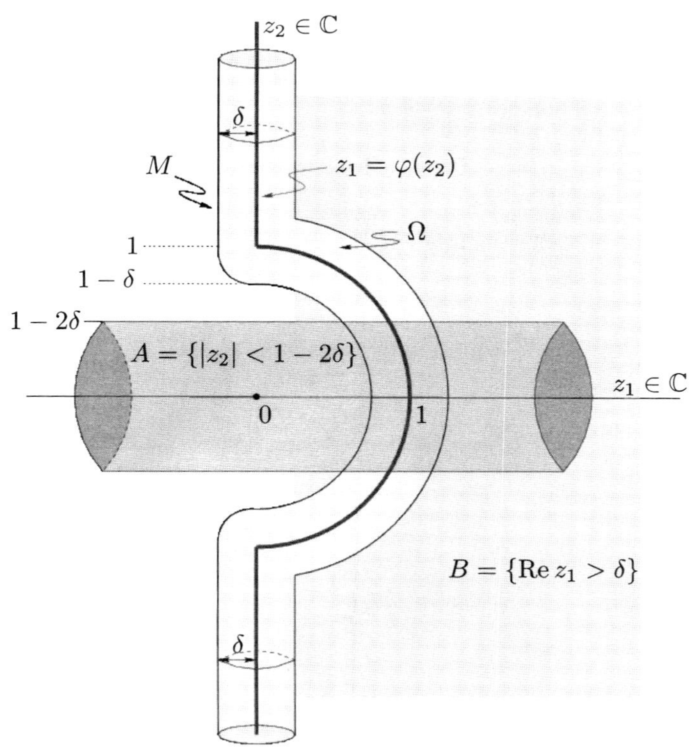

FIGURE 10. Tube along a complex line with "bump."

smooth (since v is smooth). Since v is zero for $\operatorname{Re} z_1 > 1+\delta$, clearly u_2 is also zero for $\operatorname{Re} z_1 > 1+\delta$ in view of the support property of C_2. Also u_2 is holomorphic on A intersect the complement of Ω; so u_2 is zero on all of A intersect the complement of Ω by the identity theorem for analytic functions. Let $F_2 = \tilde{f} - u_2$ on $A \cap \Omega$. F_2 is holomorphic on $A \cap \Omega$. Since $\tilde{f} = f$ and $u_2 = 0$ on M, clearly $F_2 = f$ on M and thus F_2 is the desired holomorphic extension of f to $A \cap \Omega$.

STEP 2. Continue F_2 as a smooth function on all of $\overline{\Omega}$ with $F_2 = f$ on M, $\bar{\partial} F_2 = 0$ on M to high order, and $\bar{\partial} F_2 = 0$ on $\Omega \cap A$.

STEP 3. Let v be the one-form defined by $\bar{\partial} F_2$ on Ω and 0 on the complement of Ω. Note that v is smooth. Let $u = C_1(v)$. Note that u is well defined and smooth on all of $\mathbb{C}^2$ since the support of C_1 always intersects $\overline{\Omega}$ in a compact set. As in Step 1, $\bar{\partial} u = v$ and furthermore, $u = 0$ on A since $\bar{\partial} F_2 = 0$ on $A \cap \Omega$ and the support property of C_1. Since u is holomorphic on the complement of Ω, and vanishes on A, clearly $u = 0$ on all of the complement of Ω by the identity theorem. Now let $F = F_2 - u$. Note that F is holomorphic on Ω. Since $u = 0$ on M and $F_2 = f$ on M, we have $F = f$ on M. So, F is the desired holomorphic extension of f to Ω. □

5.2. Tube with "dents". Here we only mention another example of a tube along a complex line with "dents." We do not give a proof of the tangential Hartogs phenomenon since some elements of the proof are similar to the case of the tube with a "bump" and other elements can be taken from the proof of Theorem 6.1.

Let a tube Ω with two dents be given as in Figure 11 and let f be a smooth CR function defined on the boundary $M = \partial\Omega$ of the tube. The vertex of A occurs at $x = -\delta$ and the vertex of B occurs at $x = \delta$ for some $0 < \delta < 1$ and also the edges should be smoothed off. To be more clear, the wedge A which is removed, is of the form

$$\{(z, w) \in \mathbb{C}^2; x > 4(\delta+1)(|w| - \tfrac{3}{2})^2 - \delta\},$$

where $z = x + iy$. Similarly for the wedge B. For such a tube, any smooth CR function f can be holomorphically extended to Ω with wedges filled out, i.e., to the tube $\{(z, w); |z| < 1\}$.

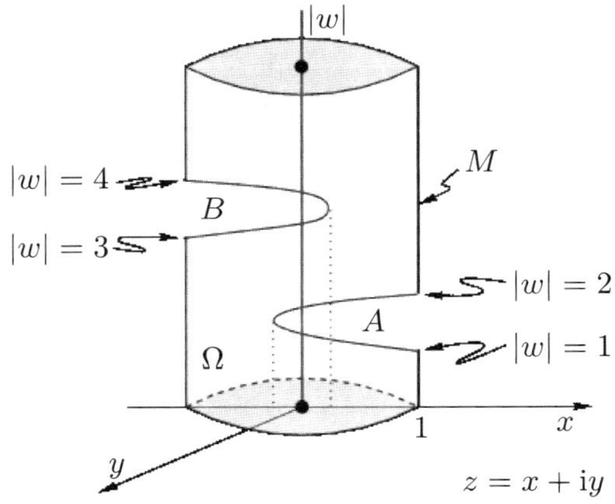

FIGURE 11. Tube with two wedges removed.

6. Conjecture

Again in this section, let *smooth* mean of class C^k for enough high k.

Conjecture. *The conjecture is in* $\mathbb{C}^2$ *(see Figure 12), where the coordinates are* (z, w), $z = x + \mathrm{i}y$, $w = u + \mathrm{i}v$. *Let* Ω *be a tube-like domain along the* w*-axis with the following properties*:

(1) Ω *is open and connected and with smooth connected boundary* $M = \mathrm{b}\Omega$.
(2) *The complement of* $\overline{\Omega}$ *is connected.*
(3) *Let* $\pi\colon \mathbb{C}^2 \to \mathbb{C}$, $\pi(z, w) = z$. *The projection* $\pi(\Omega)$ *is a bounded domain in* $\mathbb{C}_z$.
(4) $\pi(\overline{\Omega}) \subset \pi(\mathbb{C}^2 \setminus \overline{\Omega})$.

With the above assumptions, any smooth CR function f *on* M *can be holomorphically extended to* Ω.

It is elementary to prove that the condition (4) is equivalent to the property that there is no 1-dimensional complex variety in $\mathbb{C}^2$ contained entirely in $\overline{\Omega}$.

Before we formulate a theorem we give the definition of Reinhardt tube-like domain with respect to w:

Definition. Let Ω be a tube-like domain as in the formulation of the Conjecture. We say that Ω is Reihardt with respect to w if together with a point $(z, w) \in \Omega$ also the points $(z, \mathrm{e}^{\mathrm{i}\theta} w) \in \Omega$, $\theta \in \mathbb{R}$, are in the domain.

Tube-like Reihardt domains can be drawn in an $\mathbb{R}^3$ coordinate system $(x, y, |w|)$, as in Figure 13.

Theorem 6.1. *The conjecture holds in the case of Reinhardt tube-like domains with respect to the* w *variable.*

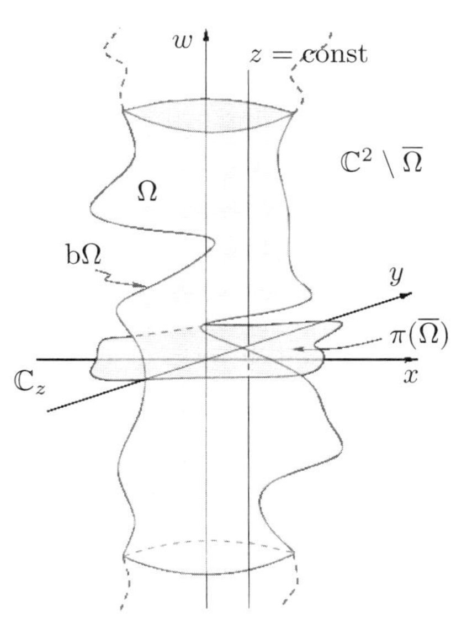

FIGURE 12. Illustration of the conjecture.

PROOF. First an elementary observation

$$\pi(\overline{\Omega}) \subset \overline{\pi(\Omega)},$$

and also we note that $\pi(\Omega)$ and $\pi(\mathbb{C}^2 \setminus \overline{\Omega})$ are open sets. However, $\pi(\overline{\Omega})$ is bounded but not necessarily compact. If we cut-off the tube by $|w| \leq R$, then its image is compact. Since Ω is a Reinhardt tube-like domain with respect to w, therefore the complement $\mathbb{C}^2 \setminus \overline{\Omega}$ is also a Reihardt domain with respect to w.

Pick up any point A which is "far away" from the tube, in particular, $\pi(A)$ is outside of $\overline{\pi(\Omega)}$. Next, take any point $z_0 \in \pi(\overline{\Omega})$. By the assumption of the Conjecture, namely $\pi(\overline{\Omega}) \subset \pi(\mathbb{C}^2 \setminus \overline{\Omega})$, we can find a point $P \in \mathbb{C}^2 \setminus \overline{\Omega}$ such that $\pi(P) = z_0$. By another assumption, the open set $\mathbb{C}^2 \setminus \overline{\Omega}$ is connected, consequently we can find a smooth curve γ joining point A with P. This curve can be approximated by continuous broken line whose segments are parallel to one of the axes: x-axis, y-axis, or $|w|$-axis. Along each segment of the broken line, we can construct a tube (in $\mathbb{R}^3$), as in Figure 13.

Let f be a smooth CR function defined on the boundary $M = \mathrm{b}\Omega$. We know that it can be represented as the difference of two holomorphic functions

$$f = f^+ - f^-, \quad f^+ \in \mathcal{H}(\Omega) \cap C^k(\overline{\Omega}), f^- \in \mathcal{H}(\mathbb{C}^2 \setminus \overline{\Omega}) \cap C^k(\mathbb{C}^2 \setminus \Omega),$$

where k is as at the beginning of this section.

Take the function f^- and expand it into the Laurent series with respect to w,

$$f^-(z, w) = \sum_{j=-\infty}^{\infty} a_j(z) w^j \tag{6.1}$$

in the domain where it is possible. We note that the coefficients $a_j(z)$ are holomorphic functions. For $z \in \mathbb{C} \setminus \pi(\overline{\Omega})$, the coefficients $a_j(z) \equiv 0$ for $j < 0$.

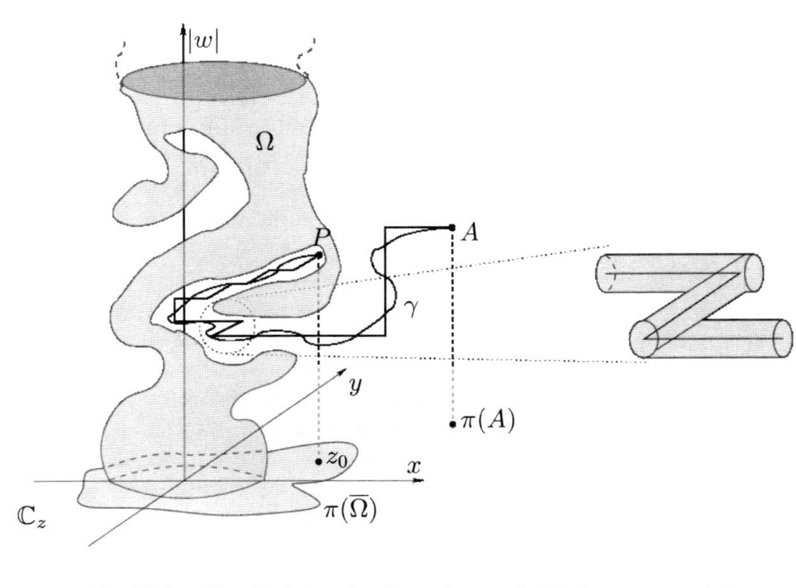

FIGURE 13. Tube-like Reinhardt domains and joining two points by a broken (continuous) line surrounded by tubes.

Next, we go along the broken line, starting at point A. As far as the part of the broken line is outside of $\pi^{-1}(\pi(\overline{\Omega}))$, the series (6.1) is a power series. If the broken line enters $\pi^{-1}(\pi(\overline{\Omega}))$, then we get the Laurent series. However, still the coefficients $a_j(z) \equiv 0$ for $j < 0$.

The nonnegative-index coefficients $a_j(z) \equiv 0$ for $j \geq 0$ are holomorphic for $|z| > R$ and along the projected curve $\pi(\gamma)$ can be holomorphically extended to the disc $|z| < R$. Since the disc is a simply connected domain, such extension is unique and single-valued.

We proved that the function $f^-(z, w)$ is entire, which implies that the CR function f can be holomorphically extended to Ω, namely

$$(f^+ - f^-)|_\Omega = \text{holomorphic function on } \Omega,$$

which completes the proof of the theorem. □

Corollary 6.2. *Suppose that Ω is a Reinhardt tube-like domain which satisfies the conditions* (1) – (3) *from the Conjecture, and additionally there are no complex* 1*-dimensional varieties in its closure. Then the Hartogs tangential type phenomenon holds for this domain.*

The proof immediately follows from Theorem 6.1 and from the equivalence of condition (4) in the Conjecture with nonexistence of 1-dimensional complex variety in $\mathbb{C}^2$ contained entirely in $\overline{\Omega}$.

7. Extension property and complex curve on the boundary of the domain

In Corollary 6.2 we proved that the extension property holds for Reinhardt tube-like domains if there is no complex 1-dimensional variety in its closure. It appears that the latter assumption cannot be relaxed. In this section we will provide an example in $\mathbb{C}^2$, which looks like a tube with concavities in $(x, y, |w|)$ coordinates, as in Figure 14, where the extension property does not hold if such a variety exists which touches the boundary of Ω. The CR function on the boundary which cannot be holomorphically extended to the inside is C^∞ on the boundary except a "small" subset where is only continuous. However, this example can be modified to include C^k-functions, $k \geq 1$, but more technicalities are needed and we do not present it here.

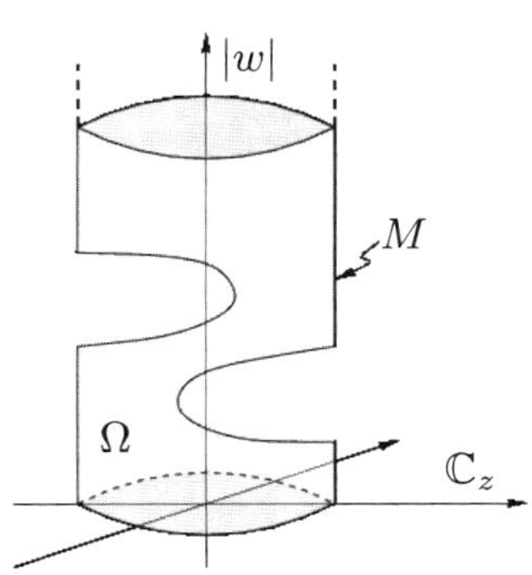

FIGURE 14. Tube with two concavities.

Statement of example. Give $\mathbb{C}^2$ the following coordinates: (z, w) with $z = x + iy$. The goal is to create a domain Ω in $\mathbb{C}^2$ with smooth boundary which has the following properties:

(1) For each $w \in \mathbb{C}$, the slice $\Omega_w = \{z \in \mathbb{C}; (z, w) \in \Omega\}$ is bounded and simply connected.
(2) Ω does not contain any complex varieties of the form $\{z = g(w); w \in \mathbb{C}\}$, where g is entire; in particular, if $L_z = \{(z, w) \in \mathbb{C}^2; w \in \mathbb{C}\}$, then L_z is *not* contained in Ω for any $z \in \mathbb{C}$.
(3) There is a continuous CR function f defined on $b\Omega$ which does *not* analytically continue to the inside of Ω.

It is easy to construct domains which satisfy (1) and (3) — for example, $\Omega = \{(z, w) \in \mathbb{C}^2; |z| < 1\}$ with $f(z, w) = 1/z$. It is harder to find an example which also satisfies property (2) (note that the above example contains the variety $\{z = 0\}$, which is the singular set for f).

As Figure 14 illustrates, the domain Ω looks somewhat like an infinite pipe centered about the w-axis, with two "wedges" removed to ensure property (2).

The CR function f will be constructed using two different branches of a square root which are pasted together. First, Ω will be constructed to satisfy properties (1) and (3), as well as property (2). Here are the basic steps:

Step 1. Defining the square root. Let

$$F(z) = \sqrt{z^2 - 1} = \sqrt{z + 1}\sqrt{z - 1} = r^{1/2}e^{i\phi/2}s^{1/2}e^{i\theta/2}$$

where r, s, ϕ, θ are given in Figure 15 and where $-\pi < \phi, \theta \leq \pi$. Note that F is analytic on the set $\mathbb{C} - [-1, 1]$ where $[-1, 1]$ denotes the line segment from $z = -1$ to $z = +1$. Later on we need the function F restricted to the disc $|z| \leq 2$ only.

The function $F|_{y>0}$ analytically extends across the open interval $(-1, 1)$ to a function denoted by F^+ by allowing θ to assume values that are greater than π, see Figure 17. Likewise, $F|_{y<0}$ analytically extends across the interval $(-1, 1)$ to a function denoted by F^- by allowing θ to assume values which are less than $-\pi$. Of course, F^+ does not agree with F^- on the interval $(-1, 1)$.

The functions $F^{\pm}$ are holomorphic on $\mathbb{C}^2 \setminus ((-\infty, -1] \cup [1, \infty))$, however we need them restricted to the disc $|z| \leq 2$ only. It should be noted that these functions can be holomorphically extended across the interval $(-\infty, -1)$ and $(1, \infty)$ but this extension should be separated from the extension which comes through the "gate" $(-1, 1)$ by a curve starting at -1 and another curve starting at 1 and both going

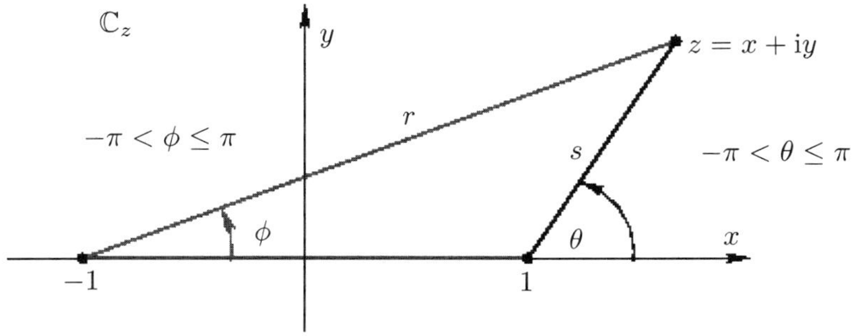

FIGURE 15. Construction of the function F.

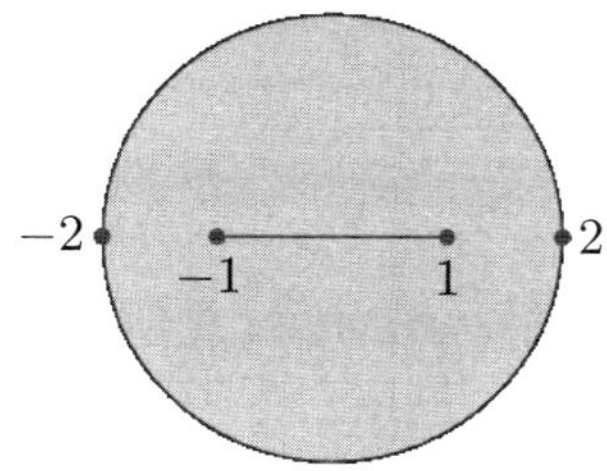

FIGURE 16. $F\colon \mathbb{C}^2 \setminus [-1,1] \to \mathbb{C}$. We need the function F restricted to the disc $|z| \le 2$.

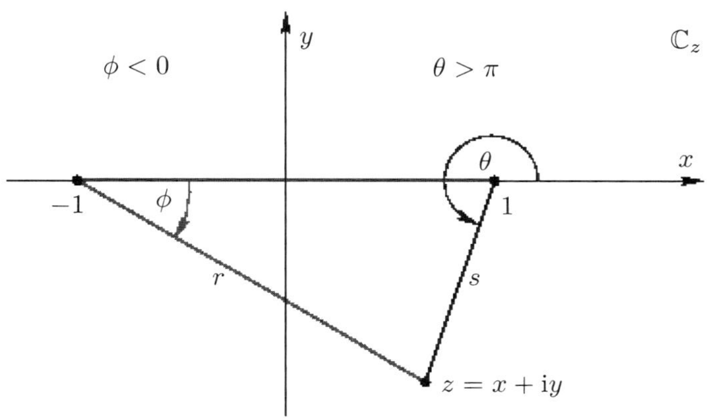

FIGURE 17. Construction of the function F^+.

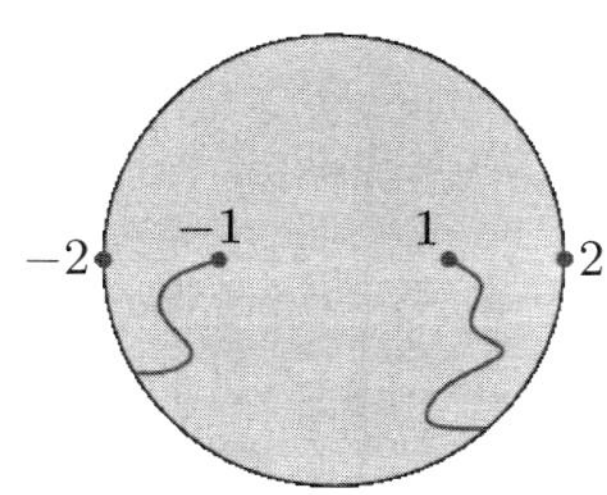

FIGURE 18. $F^+\colon \mathbb{C}^2 \setminus \{\text{curves}\} \to \mathbb{C}$. We need the function F^+ restricted to the disc $|z| \le 2$.

to infinity in a suitable way, like in Figure 18. Moreover, the functions $F^{\pm}$ agree with F on the circle of radius 2.

Step 2. Defining the cross sections of Ω. Let S_0 be the circle of radius 2, centered at the origin in the z-plane. Note F is analytic near S_0. For $0 \le t \le 1$, let S_t^+ be a C^∞ homotopy deformation of S_0 (when $t = 0$) to S_1^+ as shown in Figure 19. Note that S_1^+ lies in $\{y < 0\}$ except that it touches $z = -1, +1$. By definition, each S_t^+ is a closed loop and contains the bottom portion of the circle S_0.

Likewise, define S_t^- as a homotopy of S_0 (when $t = 0$) to the curve S_1^- as given in the figure. Each S_t^- is a closed loop which contains the top portion of the circle S_0.

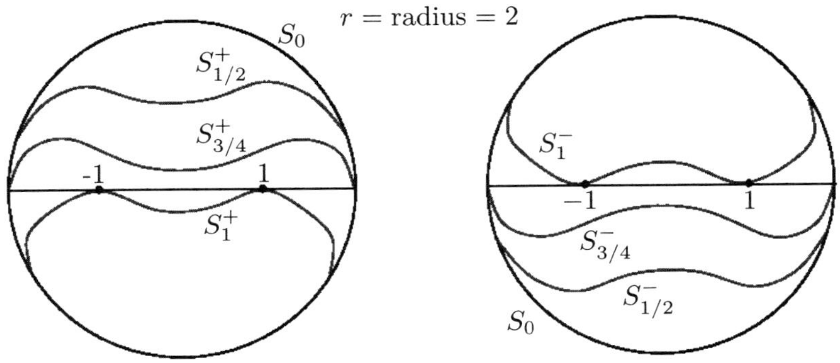

FIGURE 19. Modified circles for positive and negative parameters.

The function F, extended as F^+ across the interval $(-1,1)$, is analytic near each S_t^+ for $0 \leq t \leq 1$, except at $z=-1$ and $z=+1$, when $t=1$. At these two points, $F^+|_{S_1^+}$ is just continuous (and takes the value 0 at $z=-1,\ +1$). Likewise, F, extended as F^- across the interval $(-1,1)$ is analytic near each S_t^- for $0 \leq t \leq 1$ except at $z=-1,+1$ when $t=1$, where $F^-|_{S_1^-}$ is only continuous.

Step 3. Construction of Ω. The set Ω is constructed as the interior of a deformed infinite pipe centered about the w-axis with base bounded by the circle S_0 in the z-plane. The boundary $b\Omega$ will be described in terms of its cross sections $b\Omega_w$ as follows:

- For $0 \leq |w| \leq 1$, let $\mathrm{b}\Omega_w = S_0$.
- For $1 \leq |w| \leq 2$, let $\mathrm{b}\Omega_w = S_t^+$, where $t = |w|-1$ (so $\mathrm{b}\Omega_w$ deforms from S_0 to S_1^+ as $|w|$ increases from 1 to 2). Note that at the level $|w|=2$ in Figure 20, the cross section $b\Omega_w$ is S_1^+ as defined in Step 2.

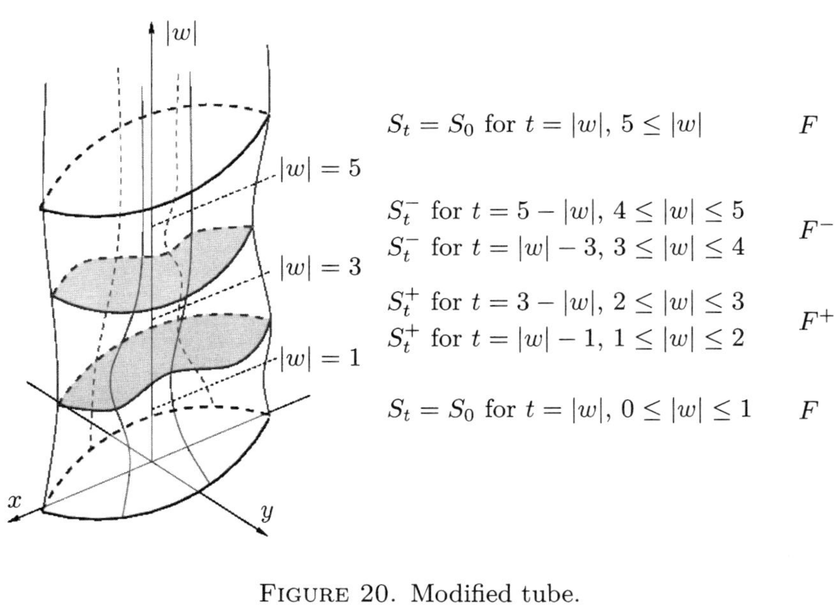

FIGURE 20. Modified tube.

- For $2 \leq |w| \leq 3$, let $b\Omega_w = S_t^+$, where $t = 3 - |w|$ (so $b\Omega_w$ deforms from S_1^+ to S_0).
- For $3 \leq |w| \leq 4$, let $b\Omega_w = S_t^-$ where $t = |w| - 3$. Note that at the level $|w| = 4$ in Figure 20, $b\Omega_w$ is S_1^-.
- For $4 \leq |w| \leq 5$, let $b\Omega_w = S_t^-$ where $t = 5 - |w|$ (so $b\Omega_w$ deforms from S_1^- to S_0).
- For $|w| \geq 5$, let $b\Omega_w = S_0$.

Step 4. Construction of the CR function f. Note that property (1) from the beginning of this section for Ω is obvious. For property (2), note that since $\Omega \subset \{|z| \leq 2\}$, Louiville's Theorem implies that any complex variety which is contained in Ω must be a vertical complex line. If $L_z = \{(z,w) \in \mathbb{C}^2; w \in \mathbb{C}\}$, then L_z is not contained in $\overline{\Omega}$ unless $z = -1$ or $+1$.

Define $f\colon b\Omega \to \mathbb{C}$ as follows:

- On $b\Omega_w$ for $0 \leq |w| \leq 1$, let $f(z,w) = F(z)$.
- On $b\Omega_w$ for $1 \leq |w| \leq 3$, let $f(z,w) = F(z)$, extended as F^+.
- On $b\Omega_w$ for $3 \leq |w| \leq 5$, let $f(z,w) = F(z)$, extended as F^-.
- On $b\Omega_w$ for $|w| \geq 5$, let $f(z,w) = F(z)$.

Note that f is C^∞ and CR on $b\Omega$ except on the set

$$C = \{(z,w) \in b\Omega; z = -1, +1\},$$

which consists of four circles. Also note that f is continuous at C and that C has codimension two in $b\Omega$. An easy approximation argument (with cut off functions that vanish in a neighborhood of C) can be used to show that f is a CR distribution on all of $b\Omega$ (including C). Clearly, f cannot be extended analytically to all of the inside of Ω since such an extension restricted to Ω_w with $|w| \leq 1$ would provide an extension of $F(z) = \sqrt{z^2-1}$ from $S_0 = \{|z| = 2\}$ to its interior $\{|z| < 2\}$.

References

1. A. Andreotti and C. D. Hill, *E. E. Levi convexity and the Hans Lewy problem.* I: *Reduction to vanishing theorems*, Ann. Scuola Norm. Sup. Pisa (3) **26** (1972), 325–363.
2. S. Bochner, *A theorem on analytic continuation of functions in several variables*, Ann. of Math. (2) **39** (1938), no. 1, 14–19.
3. ———, *Analytic and meromorphic continuation by means of Green's formula*, Ann. of Math. (2) **44** (1943), 652–673.
4. A. Boggess, *CR manifolds and the tangential Cauchy–Riemann complex*, Stud. Adv. Math., CRC Press, Boca Raton, FL, 1991.
5. A. Boivin and R. Dwilewicz, *Extension and approximation of CR functions on tube manifolds*, Trans. Amer. Math. Soc. **350** (1998), no. 5, 1945–1956.
6. M. Colţoiu and J. Ruppenthal, *On Hartogs' extension theorem on $(n-1)$-complete complex spaces*, J. Reine Angew. Math. **637** (2009), 41–47.
7. P. Dolbeault and G. Henkin, *Chaînes holomorphes de bord donné dans* $\mathbf{CP}^n$, Bull. Soc. Math. France **125** (1997), no. 3, 383–445.
8. R. J. Dwilewicz, *Additive Riemann–Hilbert problem in line bundles over* $\mathbb{CP}^1$, Canad. Math. Bull. **49** (2006), no. 1, 72–81.
9. ———, *Holomorphic extensions in complex fiber bundles*, J. Math. Anal. Appl. **322** (2006), no. 2, 556–565.
10. R. Dwilewicz and J. Merker, *On the Hartogs–Bochner phenomenon for CR functions in* $\mathrm{P}_2(\mathbb{C})$, Proc. Amer. Math. Soc. **130** (2002), no. 7, 1975–1980 (electronic).
11. L. Ehrenpreis, *A new proof and an extension of Hartogs' theorem*, Bull. Amer. Math. Soc. **67** (1961), 507–509.

12. G. Fichera, *Caratterizzazione della traccia, sulla frontiera di un campo, di una funzione analitica di più variabili complesse*, Atti Accad. Naz. Lincei. Rend. Cl. Sci. Fis. Mat. Nat. (8) **22** (1957), 706–715.
13. R. Fueter, *Über einen Hartogs'schen Satz*, Comment. Math. Helv. **12** (1939), 75–80.
14. F. Hartogs, *Zur Theorie der analytischen Funktionen mehrerer unabhängiger Veränderlichen, insbesondere über die Darstellung derselben durch Reihen, welche nach Potenzen einer Veränderlichen fortschreiten*, Math. Ann. **62** (1906), no. 1, 1–88.
15. F. R. Harvey and H. B. Lawson Jr., *On boundaries of complex analytic varieties.* I, Ann. of Math. (2) **102** (1975), no. 2, 223–290; II **106** (1977), no. 2, 213–238.
16. S. M. Ivashkovich, *The Hartogs-type extension theorem for meromorphic maps into compact Kähler manifolds*, Invent. Math. **109** (1992), no. 1, 47–54.
17. H. Kneser, *Die Randwerte einer analytischen Funktion zweier Veränderlichen*, Monatsh. Math. Phys. **43** (1936), no. 1, 364–380.
18. J. J. Kohn and H. Rossi, *On the extension of holomorphic functions from the boundary of a complex manifold*, Ann. of Math. (2) **81** (1965), 451–472.
19. C. Laurent-Thiébaut, *Phénomène de Hartogs-Bochner dans les variétés CR*, Topics in Complex Analysis (Warsaw, 1992) (P. Jakóbczak and W. Pleśniak, eds.), Banach Center Publ., vol. 31, Polish Acad. Sci., Warsaw, 1995, pp. 233–247.
20. E. Martinelli, *Sopra una dimostrazione di R. Fueter per un teorema di Hartogs*, Comment. Math. Helv. **15** (1943), 340–349.
21. ———, *Sulla determinazione di una funzione analitica di più variabili complesse in un campo, assegnatane la traccia sulla frontiera*, Ann. Mat. Pura Appl. (4) **55** (1961), 191–202.
22. J. Merker and E. Porten, *A Morse-theoretical proof of the Hartogs extension theorem*, J. Geom. Anal. **17** (2007), no. 3, 513–546.
23. J. Merker and E. Porten, *The Hartogs extension theorem on $(n-1)$-complete complex spaces*, J. Reine Angew. Math. **637** (2009), 23–39.
24. R. M. Range, *Extension phenomena in multidimensional complex analysis: correction of the historical record*, Math. Intelligencer **24** (2002), no. 2, 4–12.
25. F. Sarkis, *CR-meromorphic extension and the nonembeddability of the Andreotti – Rossi CR structure in the projective space*, Internat. J. Math. **10** (1999), no. 7, 897–915.
26. ———, *Hartogs – Bochner type theorem in projective space*, Ark. Mat. **41** (2003), no. 1, 151–163.
27. F. Severi, *Risoluzione generale del problema di Dirichlet per le funzioni biarmoniche*, Atti R. Accad. Naz. Lincei Rend. Cl. Sci. Fis. Mat. Nat. (6) **13** (1931), 795–804.

Department of Mathematics, Texas A&M University, College Station, TX 77843, USA

E-mail address: boggess@math.tamu.edu

Department of Mathematics and Statistics, Missouri University of Science and Technology, Rolla, MO 65409, USA and Faculty of Mathematics, Cardinal Stefan Wyszyński University, Wóycickiego 1/3, 01-938 Warsaw, Poland

E-mail address: romand@mst.edu

Department of Mathematics, Statistics and Computer Science, University of Illinois at Chicago, Chicago, IL 60607, USA

E-mail address: zbigniew@uic.edu

Centre de Recherches Mathématiques
CRM Proceedings and Lecture Notes
Volume **55**, 2012

Integration Formulae and Kernels in Singular Subvarieties of $\mathbb{C}^n$

Luis M. Hernández-Pérez and Eduardo S. Zeron

ABSTRACT. The Bochner – Martinelli and Ramirez – Khenkin integration formulae are a pair of cornerstones of the field of several complex variables. They are naturally defined on open domains of $\mathbb{C}^n$. An obvious problem is to produce respective integration formulae on general smooth or singular varieties. We propose in this work a simple technique for producing integration formulae on singular subvarieties defined as the zero locus of special polynomials in $\mathbb{C}^n$. The main idea is to *push down* the integration kernels from a neighborhood of the subvariety into the subvariety itself.

1. Introduction

The Bochner – Martinelli and Ramirez – Khenkin integration formulae are a pair of cornerstones of the field of several complex variables. Let Ω be a bounded domain in $\mathbb{C}^n$ with piecewise smooth boundary $\delta\Omega$, and $\aleph$ be a $(0,q)$-form whose coefficients are continuous functions on the closure $\overline{\Omega}$. If the differential $\bar{\partial}\aleph$ (calculated as a distribution) is also continuous in $\overline{\Omega}$, then the following identity holds in Ω for the Bochner-Martinelli kernels $B_q(z,\xi)$,

$$\aleph(z) = \int_{\delta\Omega} \aleph \wedge B_q - \int_{\Omega} (\bar{\partial}\aleph) \wedge B_q + \bar{\partial}\Big[\int_{\Omega} \aleph \wedge B_{q-1}\Big]; \tag{1.1}$$

see [10, 11]. It is impossible to enumerate all the applications of the integration formulae to complex analysis, geometry and other areas. We may mention for example their use for solving the Neumann $\bar{\partial}$-equation in strictly pseudoconvex domains of $\mathbb{C}^n$. A natural problem is to produce integration formulae on general varieties. Let Ω be an open domain compactly contained in a smooth or singular complex variety Σ. If Ω has a piecewise smooth boundary $\delta\Omega$, the problem is to produce integration formulae similar to (1.1) for differential forms $\aleph$ such that $\aleph$

1991 *Mathematics Subject Classification.* Primary 32F20; Secondary 32W05, 35N15.
Key words and phrases. Bochner Martinelli kernel, singular subvariety.
Research supported by Cinvestav (México) and CONACYT (México).
The authors want to thank one of the referees for his suggestions, they were very important for improving the proof of the main theorem in this paper.

This is the final form of the paper.

and $\bar{\partial}\aleph$ are both continuous on $\overline{\Omega}$. One must be careful when we define $\aleph$ and $\bar{\partial}\aleph$ on the singular part of Σ, but the details are clarified after Definition 1.1.

There is a vast literature in books and papers on integration formulae for smooth complex manifolds; see for example the references [8–10, 12, 16]. Andersson and Samuelsson also produced integration formulae for singular subvarieties Σ of $\mathbb{C}^n$, but they delimited their work to analyse differentials forms $\aleph$ whose restriction to the regular part of Σ extends onto a smooth form well defined on a neighborhood of Σ; see the references [1–3]. We propose in this work a simple technique for producing explicit integration formulae in a family of singular subvarieties of $\mathbb{C}^n$, and for general continuous differential forms $\aleph$. The main idea is to *push down* the integration kernels from a neighborhood of the subvariety into the subvariety itself. We are interested in deducing explicit and simple integral formulae, so as to use them for calculating Hölder and L^p estimates.

We restrict our analysis to those subvarieties Σ in $\mathbb{C}^{n+1}$ generated as the zero locus of a polynomial $s^m - p(z)$ for $s \in \mathbb{C}$ and $z \in \mathbb{C}^n$, because the first entry s can be easily expressed as the m-root of $p(z)$ and several of the main singular subvarieties presented in [4, 5, 7, 17] are the zero locus of such a polynomial. Nevertheless the technique presented in this paper can be applied to analyze other subvarieties Σ of $\mathbb{C}^\kappa$ such that some entries of $z \in \Sigma$ can be easily expressed in terms of the other ones. We may work for example on the subvariety with a simple-elliptic singularity $\widetilde{D}_5$ given in [7, p. 63] as the zero locus of the system of polynomials

$$x^2 + y^2 - \lambda zw \quad \text{and} \quad xy - z^2 - w^2,$$

for we have that $x \pm y = \sqrt{\lambda zw \pm 2(z^2 + w^2)}$. Fix from now on Σ as the zero locus of the polynomial $s^m - p(z)$ for (s, z) in $\mathbb{C} \times \mathbb{C}^n$, $m \geq 1$, and $p \not\equiv 0$. The m-root of $p(z)$ is naturally defined in Σ, but we need to extend this property to a set that contains Σ and has non empty interior in $\mathbb{C}^{n+1}$. Thus define the closed set

$$\Pi = \left\{ (s + \nu, z) \in \mathbb{C} \times \mathbb{C}^n : (s, z) \in \Sigma, |\nu| \leq \frac{|s|}{m} \right\}, \tag{1.2}$$

A really important fact is that for every point (u, z) in Π, there is a unique solution $s \in \mathbb{C}$ to the system

$$s^m = p(z) \quad \text{and} \quad |u - s| \leq |s|/m. \tag{1.3}$$

It is obvious that there is only one solution s when $m = 1$ or $p(z)$ vanishes. Thus assume that $p(z)$ does not vanish, $m \geq 2$, and there are two different solutions s and ρs to the system (1.3). It is easy to see that $\rho \neq 1$ is an m-root of the unity and

$$|1 - \rho| \leq |1 - u/s| + |\rho - u/s| \leq 2/m,$$

but this is impossible. Hence the following natural projection is well defined

$$\pi \colon \Pi \to \Sigma \quad \text{with } \pi(u, z) = (\sqrt[m]{p}(z), z) \tag{1.4}$$

and $\sqrt[m]{p}(z)$ the unique solution s to (1.3). The projection π is only differentiable at those points (u, z) of Π where $p(z)$ is different from zero. On the other hand we will work with general differential forms.

Definition 1.1. Let $\Sigma \subset \mathbb{C}^{n+1}$ be the zero locus of $\tau^m - p(\xi)$ for $\tau \in \mathbb{C}$ and $\xi \in \mathbb{C}^n$. Given any open or closed set X in Σ, a general differential form $\aleph$ of bidegree $(0,q)$ in X is any linear combination

$$\aleph = \sum_{|I|=q} f_I(\tau,\xi)\,\mathrm{d}\bar{\xi}_I + \sum_{|J|=q-1} g_J(\tau,\xi)\,\mathrm{d}\bar{\tau}\wedge \mathrm{d}\bar{\xi}_J, \tag{1.5}$$

where the coefficients f_I and g_J are all arbitrary functions defined on X.

The representation (1.5) is by no means unique. For example the differential form $\mathrm{d}\bar{s}^m$ is obviously equal to $\mathrm{d}\bar{p}(z)$ in Σ. We are specially interested in the case when X is the closure of a domain in Σ and the coefficients f_I and g_J in (1.5) are all continuous functions. In this case the coefficients of the pull-back $\pi^*\aleph$ are not necessarily continuous, because the differential of the m-root of $p(z)$ in (1.4) is involved. However the coefficients of $p(z)\pi^*\aleph$ are indeed continuous. We may now introduce the integration kernels deduced from the Bochner – Martinelli ones.

Definition 1.2. Consider the functions

$$F(\hat{s},x) := \int_{|w|<1/m} \frac{\mathrm{d}\overline{w}\wedge \mathrm{d}w}{(x+|w-\hat{s}|^2)^{n+1}}, \tag{1.6}$$

$$F^*(\hat{s},x) := \int_{|w|=1/m} \frac{\overline{(w-\hat{s})}\,\mathrm{d}w}{(x+|w-\hat{s}|^2)^{n+1}}, \tag{1.7}$$

defined for $\hat{s}\in\mathbb{C}$, the real numbers $x\geq 0$, and the integer numbers n and m greater than or equal to one. These functions are well defined and continuous at those points $(\hat{s},x)$ with $x>0$ or $|\hat{s}|>\frac{1}{m}$. Define as well the following integration kernels for the complex vectors (τ,ξ) and (s,z) in $\mathbb{C}\times\mathbb{C}^n$,

$$K_1(\tau,\xi;s,z) := F\left(\frac{s-\tau}{\tau},\frac{\|\xi-z\|^2}{|\tau|^2}\right)\frac{\omega(\xi;z)}{2\pi\mathrm{i}|\tau|^{2n}}, \tag{1.8}$$

$$K_2(\tau,\xi;s,z) := F^*\left(\frac{s-\tau}{\tau},\frac{\|\xi-z\|^2}{|\tau|^2}\right)\frac{\mathrm{d}\omega(\xi;z)}{2n\pi\mathrm{i}|\tau|^{2n}}, \tag{1.9}$$

where the total differential $\mathrm{d}\omega$ stands for $(\mathrm{d}_\xi+\mathrm{d}_z)\omega$ and

$$\omega(\xi;z) := \sum_{j=1}^{n}\frac{\overline{(\xi_j-z_j)}\,\mathrm{d}\xi_j}{(2\pi\mathrm{i})^n}\wedge\bigwedge_{k\neq j}\big[(\mathrm{d}\bar{\xi}_k-\mathrm{d}\bar{z}_k)\wedge\mathrm{d}\xi_k\big].$$

Finally recall that an open set G in Σ is said to have piecewise-smooth boundary δG if the closure $\overline{G}$ is compact and δG is the union of a finite number of compact smooth submanifolds (with boundary); so that the $(2n-1)$-real dimensional volume of δG is always finite. The following theorem is the main result of this work.

Theorem 1.3 (Main). *Let Σ be a subvariety in $\mathbb{C}^{n+1}$ given as the zero locus of the holomorphic polynomial $s^m-p(z)$ for (s,z) in $\mathbb{C}\times\mathbb{C}^n$, $m\geq 1$, and $p\not\equiv 0$. The singular part of Σ is contained in $H=\{0\}\times\mathbb{C}^n$. Given an open domain G in Σ with piecewise smooth boundary δG, consider any $(0,q)$-form $\aleph$ whose coefficients f_I and g_J are continuous functions on the closure $\overline{G}$ according to the expansion (1.5). Assume that the differential $\bar{\partial}\aleph$ exists in the sense of distributions and is continuous in the difference $G\setminus H$; moreover it has a continuous extension into $\overline{G}$, so that we may say that $\bar{\partial}\aleph$ exists and is continuous in $\overline{G}$. Then the following identity holds for (s,z) in $G\setminus H$ and every fixed integer $\kappa\geq 2$,*

$$\frac{p^\kappa(z)\aleph}{(-1)^q n!} = B_{\partial G}(p^\kappa\pi^*\aleph) - B_G(p^\kappa\pi^*\bar{\partial}\aleph) + \bar{\partial}_z B_G(p^\kappa\pi^*\aleph)$$

where

$$B_G(p^\kappa\pi^*\aleph) = \int_{(\tau,\xi)\in G} \big[p^\kappa(\xi)\pi^*\aleph\big] \wedge K_1(\tau,\xi;s,z), \tag{1.10}$$

$$B_G(p^\kappa\pi^*\bar\partial\aleph) = \int_{(\tau,\xi)\in G} \big[p^\kappa(\xi)\pi^*\bar\partial\aleph\big] \wedge K_1(\tau,\xi;s,z), \tag{1.11}$$

$$\begin{aligned} B_{\delta G}(p^\kappa\pi^*\aleph) = &\int_{(\tau,\xi)\in\delta G} \big[p^\kappa(\xi)\pi^*\aleph\big] \wedge K_1(\tau,\xi;s,z) \\ &+ (-1)^q \int_{(\tau,\xi)\in G} \big[p^\kappa(\xi)\pi^*\aleph\big] \wedge K_2(\tau,\xi;s,z). \end{aligned} \tag{1.12}$$

We must point out that the main difference between the previous theorem and Koppelman formula in [10, p. 57] is the extra integral term in the operator $B_{\delta G}$. However there is a simple technique for removing this extra term: Let (s,z) be any fixed point in Σ with the first entry $s\neq 0$, The kernel K_2 in (1.9) is smooth at every point (τ,ξ) in some neighborhood of Σ, because of the following identity

$$F^*\left(\frac{s-\tau}{\tau}, \frac{\|\xi-z\|^2}{|\tau|^2}\right) = \int_{|u-\tau|=|\tau|/m} \frac{|\tau|^{2n}\overline{(u-s)}\,\mathrm{d}u}{(\|\xi-z\|^2+|u-s|^2)^{n+1}}.$$

Suppose there is a form K_3 smooth on a neighborhood of $\overline{G}\times\overline{G}$ such that

$$(\bar\partial_\xi+\bar\partial_z)K_3(\tau,\xi;s,z) = K_2(\tau,\xi;s,z). \tag{1.13}$$

The kernel K_1 in (1.8) satisfies a similar equation according to (2.17), but K_1 is unfortunately singular when $\xi = z$. We can easily calculate from (1.13) that the second integral term in (1.12) can be rewritten in terms of K_3, i.e.,

$$(-1)^q\int_G \big[p^\kappa\pi^*\aleph\big]\wedge(K_2-\bar\partial_z K_3) + \int_G \big[p^\kappa\pi^*\bar\partial\aleph\big]\wedge K_3 = \int_{\delta G}\big[p^\kappa\pi^*\aleph\big]\wedge K_3.$$

Therefore K_1+K_3 is the integration kernel that we are looking for, because the main result in the main Theorem 1.3 can be rewritten as follows:

$$\begin{aligned} \frac{p^\kappa(z)\aleph}{(-1)^q n!} = &\int_{\xi\in\delta G}\big[p^\kappa\pi^*\aleph\big]\wedge(K_1+K_3) - \int_{\xi\in G}\big[p^\kappa\pi^*\bar\partial\aleph\big]\wedge(K_1+K_3) \\ &+ \bar\partial_z\int_{\xi\in G}\big[p^\kappa\pi^*\aleph\big]\wedge(K_1+K_3). \end{aligned}$$

For example, if $\Sigma\cong\mathbb{C}^n$ is the hyperspace given as the zero locus of the trivial polynomial $s-1$, we automatically have

$$\begin{aligned} F(0,x) &= \int_{|w|<1}\frac{\mathrm{d}\overline{w}\wedge\mathrm{d}w}{(x+|w|^2)^{n+1}} = \frac{2\pi\mathrm{i}}{n}\left[\frac{1}{x^n}-\frac{1}{(x+1)^n}\right], \\ F^*(0,x) &= \int_{|w|=1}\frac{\overline{w}\,\mathrm{d}w}{(x+|w|^2)^{n+1}} = \frac{2\pi\mathrm{i}}{(x+1)^{n+1}}. \end{aligned}$$

The following identity can be easily calculated using (2.16),

$$(\bar\partial_\xi+\bar\partial_z)\frac{\omega(\xi;z)}{(1+\|\xi-z\|^2)^n} = \frac{\mathrm{d}\omega(\xi;z)}{(1+\|\xi-z\|^2)^{n+1}} = nK_2(1,\xi;1,z).$$

Recall the definition (1.8) – (1.9) for the kernels K_1 and K_2. The form that we are looking for is the classical Bochner – Martinelli kernel for $\Sigma \cong \mathbb{C}^n$,

$$K_1(1,\xi,1,z) + \frac{\omega(\xi;z)}{n(1+\|\xi - z\|^2)^n} = \frac{\omega(\xi;z)}{n\|\xi - z\|^{2n}}.$$

Considering again the general result presented in the main Theorem 1.3, one of the remaining problems is to find a simple smooth solution K_3 to equation (1.13), in order to eliminate the extra integral term of the operator $B_{\delta\Omega}$ in (1.11). We will analyze this problem in a future work.

The next section of this paper is devoted to analyze the basic properties of the functions and kernels introduced in Definition 1.2. The main Theorem 1.3 is proved in the third chapter of this paper.

2. Basic properties

We begin defining the positive volume form in the space $\mathbb{C}^n$,

$$\text{(2.1)} \qquad \mathrm{dvol}(\xi) := \bigwedge_{j=1}^{n} \left[\frac{\mathrm{d}\bar{\xi}_j \wedge \mathrm{d}\xi_j}{2\pi \mathrm{i}} \right] = \frac{1}{\pi^n} \bigwedge_{j=1}^{n} [\mathrm{d}x_j \wedge \mathrm{d}y_j],$$

for $\xi = x + iy$ in $\mathbb{C}^n$. Fubini's theorem automatically yields the following identity for every $\mathcal{L}^1$ function h defined on the space $\mathbb{C}^{n+1}$,

$$\text{(2.2)} \qquad \int_{(u,\xi)} h(u,\xi)\,\mathrm{d}\bar{u} \wedge \mathrm{d}u \wedge \mathrm{dvol}(\xi) = \int_{\xi} \left[\int_{u} h(u,\xi)\,\mathrm{d}\bar{u} \wedge \mathrm{d}u \right] \mathrm{dvol}(\xi),$$

where (u,ξ) lies in $\mathbb{C} \times \mathbb{C}^n$; see for example [6, p. 334]. The combination of Fubini and Stoke's theorems yields an interesting result. Let $\{U_\xi\}$ be a collection of open sets in $\mathbb{C}$ parametrized by the points $\xi \in \mathbb{C}^n$, such that U_ξ is empty for $|\xi|$ large enough. Assume that $h(u,\xi)$ is differentiable with respect to the variable u and that each boundary δU_ξ is piecewise smooth and depends smoothly on ξ, then

$$\text{(2.3)} \qquad \begin{aligned} \int_{\substack{\xi \in \mathbb{C}^n \\ u \in \delta U_\xi}} h(u,\xi)\,\mathrm{d}u \wedge \mathrm{dvol}(\xi) &= \int_{\substack{\xi \in \mathbb{C}^n \\ u \in U_\xi}} \frac{\partial h}{\partial \bar{u}}\,\mathrm{d}\bar{u} \wedge \mathrm{d}u \wedge \mathrm{dvol}(\xi) \\ &= \int_{\xi} \left[\int_{u \in U_\xi} \frac{\partial h}{\partial \bar{u}}\,\mathrm{d}\bar{u} \wedge \mathrm{d}u \right] \mathrm{dvol}(\xi) \\ &= \int_{\xi} \left[\int_{u \in \delta U_\xi} h(u,\xi)\,\mathrm{d}u \right] \mathrm{dvol}(\xi). \end{aligned}$$

The volume form dvol is also used for integrating functions on the subvariety Σ. Recall that Σ is the zero locus of $\tau^m - p(\xi)$ for (τ,ξ) in $\mathbb{C} \times \mathbb{C}^n$, $m \geq 1$, and $p \not\equiv 0$. Consider an open set G in Σ with compact closure $\overline{G}$ and a continuous function h defined on $\overline{G}$. It is easy to see that

$$\left| \int_{(\tau,\xi) \in G} h(\tau,\xi)\,\mathrm{dvol}(\xi) \right| \leq m\|h\|_{\overline{G}} \int_{\xi \in \eta(G)} \mathrm{dvol}(\xi)$$

for the projection $\eta(\tau,\xi) = \xi$ and the positive volume form dvol in (2.1). Notice that η is an m-branched covering from Σ onto $\mathbb{C}^n$. The ramification locus of η coincides with the zero locus of p, and so it has codimension one. In particular, if we introduce the open set

$$\text{(2.4)} \qquad G_\beta = \{(\tau,\xi) \in G : |\tau| < \beta\} \subset \Sigma \quad \text{for } \beta > 0,$$

its image $\eta(G_\beta)$ is equal to $\{\xi \in \eta(G) : |p(\xi)| < \beta^m\}$, and so the $2n$-real dimensional volume of this image converges to zero when $\beta \to 0$. Whence

$$\lim_{\beta\to 0} \int_{(\tau,\xi)\in G_\beta} h(\tau,\xi)\,\mathrm{dvol}(\xi) = 0. \tag{2.5}$$

Assume now that the open set $G \subset \Sigma$ has piecewise-smooth boundary δG. It is easy to see that the set G_β given in (2.4) has piecewise-smooth boundary as well. In particular

$$\delta G_\beta = \overline{\{(\tau,\xi) \in \delta G : |\tau| < \beta\}} \cup \{(\tau,\xi) \in G : |\tau| = \beta\}, \tag{2.6}$$

and so the $(2n-1)$-real dimensional volume of every δG_β is finite. Notice that we may use the positive m-root of $|p(\xi)|$ instead of $|\tau|$ in the equation above because every point (τ,ξ) lies in Σ. Consider the differential form ω defined a pair of lines below (1.9). There are exactly n monomials in ω that contain no differential $\mathrm{d}\bar{z}_k$; they are indeed

$$\omega_j^0(\xi; z) = \frac{\overline{(\xi_j - z_j)}\,\mathrm{d}\xi_j}{(2\pi \mathrm{i})^n} \wedge \bigwedge_{k\neq j} (\mathrm{d}\bar{\xi}_k \wedge \mathrm{d}\xi_k) \tag{2.7}$$

for $1 \le j \le n$. Each differential $\mathrm{d}_\xi \omega_j^0$ is equal to $\mathrm{dvol}(\xi)$ according to (2.1). Let h be any continuous function defined on $\overline{G}$. It is easy to prove that the integral of $h\omega_j^0$ over the boundary δG_β in (2.6) satisfies the inequality

$$\begin{aligned}\left|\int_{(\tau,\xi)\in\delta G_\beta} h(\tau,\xi)\omega_j^0(\xi;z)\right| &\le C\|h\|_{\overline{G_\beta}}\,\mathrm{Volume}_{[2n-1]}(\delta G) \\ &\quad + mC\|h\|_{\overline{G_\beta}} \int_{\substack{\xi\in\eta(G)\\ |p(\xi)|=\beta^m}} \mathrm{d}|\xi_j| \wedge \bigwedge_{k\neq j} \left|\frac{\mathrm{d}\bar{\xi}_k \wedge \mathrm{d}\xi_k}{2i}\right|,\end{aligned}$$

where $C > 0$ is a real constant that only depends on the form ω and the size and shape of G. Since $\overline{G}$ is compact, the image $\eta(G)$ is contained in some compact ball. Whence there exists a real constant $M_2 > 0$ such that

$$\left|\int_{(\tau,\xi)\in\delta G_\beta} h(\tau,\xi)\omega_j^0(\xi;z)\right| \le M_2\|h\|_{\overline{G_\beta}} \tag{2.8}$$

for all constants $\beta > 0$, indexes $1 \le j \le n$, and continuous functions h defined on $\overline{G}$. We also need to analyze the functions and kernels introduced in Definition 1.2, so we introduce the functions

$$F(\hat{s}, x) = \int_{|w|<1/m} \frac{\mathrm{d}\overline{w} \wedge \mathrm{d}w}{(x + |w-\hat{s}|^2)^{n+1}}, \tag{2.9}$$

$$F^*(\hat{s}, x) = \int_{|w|=1/m} \frac{\overline{(w-\hat{s})}\,\mathrm{d}w}{(x + |w-\hat{s}|^2)^{n+1}}, \tag{2.10}$$

for $\hat{s} \in \mathbb{C}$, the real number $x \ge 0$, and the integer $n \ge 1$. These functions are well defined and continuous at those points $(\hat{s}, x)$ with $x > 0$ or $|\hat{s}| > 1/m$. We can easily deduce that

$$\left[n + x\frac{\partial}{\partial x}\right] F(\hat{s}, x) = \int_{|w|<\frac{1}{m}} \frac{n|w-\hat{s}|^2 - x}{(x + |w-\hat{s}|^2)^{n+2}}\,\mathrm{d}\overline{w} \wedge \mathrm{d}w.$$

Let d_w be the total differential calculated with respect to the variable w. The identity below easily holds for $\hat{s} \neq w$ or $x > 0$,

$$\mathrm{d}_w\left[\frac{\overline{(w-\hat{s})}\,\mathrm{d}w}{(x+|w-\hat{s}|^2)^{n+1}}\right] = \frac{x-n|w-\hat{s}|^2}{(x+|w-\hat{s}|^2)^{n+2}}\,\mathrm{d}\overline{w}\wedge \mathrm{d}w.$$

The application of Stokes' theorem to (2.10) yields

$$-F^*(\hat{s},x) = nF(\hat{s},x) + xF_x(\hat{s},x), \tag{2.11}$$

whenever $x > 0$ or $|\hat{s}| > 1/m$. Nevertheless one can deduce that $F^*(\hat{s},x)$ is well defined and continuous at those points $(\hat{s},x)$ with $x \geq 0$ and $|\hat{s}| \neq 1/m$. We analyse as well the integration kernel introduced in (1.8) for the complex vectors (τ,ξ) and (s,z) in $\mathbb{C}\times\mathbb{C}^n$,

$$K_1(\tau,\xi;s,z) = F\left(\frac{s-\tau}{\tau}, \frac{\|\xi-z\|^2}{|\tau|^2}\right)\frac{\omega(\xi;z)}{2\pi\mathrm{i}|\tau|^{2n}}, \tag{2.12}$$

where

$$\omega(\xi;z) = \sum_{j=1}^{n}\frac{\overline{(\xi_j-z_j)}\,\mathrm{d}\xi_j}{(2\pi\mathrm{i})^n}\wedge\bigwedge_{k\neq j}\big[(\mathrm{d}\bar{\xi}_k-\mathrm{d}\bar{z}_k)\wedge \mathrm{d}\xi_k\big]. \tag{2.13}$$

We shall prove that the anti-holomorphic differential of the kernel K_1 in (2.12) is equal to minus the integration kernel

$$K_2(\tau,\xi;s,z) = F^*\left(\frac{s-\tau}{\tau}, \frac{\|\xi-z\|^2}{|\tau|^2}\right)\frac{\mathrm{d}\omega(\xi;z)}{2n\pi\mathrm{i}|\tau|^{2n}}, \tag{2.14}$$

where F^* is given in (2.10) and the total differential $d\omega = (\mathrm{d}_\xi + \mathrm{d}_z)\omega$ satisfies

$$\frac{\mathrm{d}\omega(\xi;z)}{n} = \frac{(\bar{\partial}_\xi+\bar{\partial}_z)\omega(\xi;z)}{n} = \bigwedge_{k=1}^{n}\left[\frac{(\mathrm{d}\bar{\xi}_k-\mathrm{d}\bar{z}_k)\wedge\mathrm{d}\xi_k}{2\pi\mathrm{i}}\right]. \tag{2.15}$$

The first equality above holds because $\omega(\xi;z)$ is of bidegree $(n,n-1)$ with respect to the vector (ξ,z). We also have that

$$\big[(\bar{\partial}_\xi+\bar{\partial}_z)\|\xi-z\|^2\big]\wedge\omega(\xi;z) = \|\xi-z\|^2\,\mathrm{d}\omega(\xi;z)/n. \tag{2.16}$$

The previous identity and equations (2.11) – (2.15) can be used for calculating the anti-holomorphic differential of the kernel K_1 in (2.12):

$$K_2(\tau,\xi;s,z) = -\left[nF + \frac{\|\xi-z\|^2}{|\tau|^2}F_x\right]\frac{\mathrm{d}\omega(\xi;z)}{2n\pi\mathrm{i}|\tau|^{2n}} = -(\bar{\partial}_\xi+\bar{\partial}_z)K_1(\tau,\xi;s,z), \tag{2.17}$$

for $\tau\neq 0$ and $z\neq\xi$, or for $\tau\neq 0$ and $|s-\tau| > \frac{|\tau|}{m}$. Recall that the notation F_x in (2.11) and (2.17) stands for the partial derivative of $F(\hat{s},x)$ with respect to the second variable x. Finally we need to express the Bochner – Martinelli – Koppelman formula according to the positive volume form dvol introduced in (2.1). Let Ω be an open set in $\mathbb{C}^n$ with piecewise smooth boundary, and $\aleph$ be a continuous $(0,q)$-form defined on $\overline{\Omega}$. Assume that the differential $\bar{\partial}\aleph$ is continuous on Ω and has a continuous extension onto the closure $\overline{\Omega}$, then the following integral formula holds in Ω,

$$\frac{(-1)^q\aleph(z)}{(n-1)!} = \int_{\delta\Omega}\frac{\aleph\wedge\omega}{\|\xi-z\|^{2n}} - \int_{\Omega}\frac{(\bar{\partial}_\xi\aleph)\wedge\omega}{\|\xi-z\|^{2n}} + \bar{\partial}_z\int_{\Omega}\frac{\aleph\wedge\omega}{\|\xi-z\|^{2n}}, \tag{2.18}$$

where $\omega = \omega(\xi; z)$ is given in (2.13). The integrals in (2.18) are all calculated according to the convention introduced in [10, p. 44]. For example the last two integrals in (2.18) are calculated only over those monomials whose degree is equal to $2n$ with respect to the variable ξ, so that they contain the form $\mathrm{dvol}(\xi)$. Equation (2.18) is easily deduced by rewriting the Bochner – Martinelli – Koppelman formula presented in [10] or [11]. This equation can be directly deduced from Stoke's theorem as well, we only need to observe that the following identity holds,

$$(\bar{\partial}_\xi + \bar{\partial}_z)\left(\frac{\omega(\xi; z)}{\|\xi - z\|^{2n}}\right) = 0$$

according to (2.16). The problem is then reduced to prove (2.18) when $\aleph$ is equal to the smooth form $h(\xi)\,\mathrm{d}\bar{\xi}_J$ with $|J| = q$ and Ω is the ball with centre at z and radius $\varepsilon > 0$ small enough. One may calculate in particular that

$$\begin{aligned}\lim_{\varepsilon\to 0}&\left[\int_{\|\xi - z\| = \varepsilon} \frac{h(\xi) d\bar{\xi}_J \wedge \omega}{\varepsilon^{2n}} + \int_{\|\xi - z\| < \varepsilon} \bar{\partial}_z \frac{h(\xi)\,\mathrm{d}\bar{\xi}_J \wedge \omega}{\|\xi - z\|^{2n}}\right] \\ &= h(z) \lim_{\varepsilon\to 0} \int_{\|\xi - z\| < \varepsilon} \left[(-1)^q \frac{\mathrm{d}\bar{\xi}_J \wedge (\bar{\partial}_\xi \omega)}{\varepsilon^{2n}} + \bar{\partial}_z \frac{\mathrm{d}\bar{\xi}_J \wedge \omega}{\|\xi - z\|^{2n}}\right] \\ &= \frac{(-1)^q h(z)\,\mathrm{d}\bar{z}_J}{(n-1)!}.\end{aligned}$$

We may now prove the main result of this paper.

3. Proof of the Main Theorem 1.3

Let $H_\beta = \{|s| \leq \beta\} \times \mathbb{C}^n$ be a closed set in $\mathbb{C}^{n+1}$ for $\beta \geq 0$. Assume from now on that Σ is a singular subvariety in $\mathbb{C}^{n+1}$ given as the zero locus of the holomorphic polynomial $\tau^m - p(\xi)$ for (τ, ξ) in $\mathbb{C} \times \mathbb{C}^n$, $m \geq 1$, and $p \not\equiv 0$. The singular part of Σ is obviously contained in H_0. Recall the closed set Π of $\mathbb{C}^{n+1}$ and the natural projection π from Π onto Σ defined in the formulae (1.2) – (1.4). Given any bounded domain G in Σ with piecewise smooth boundary δG, we define the bounded open set Ω in $\mathbb{C}^{n+1}$ as the interior of the inverse image $\pi^{-1}(G \setminus H_\beta)$, i.e.,

$$\Omega = \left\{(\tau + \nu, \xi) \in \mathbb{C} \times \mathbb{C}^n : (\tau, \xi) \in G \setminus H_\beta, |\nu| < \frac{|\tau|}{m}\right\}. \tag{3.1}$$

The difference $G \setminus H_\beta$ is obviously contained in Ω. Let $\aleph$ be a differential form of bidegree $(0, q)$ and with continuous coefficients in the closure $\overline{G}$; i.e., $\aleph$ is a linear combination

$$\aleph = \sum_{|I|=q} f_I(\tau, \xi)\,\mathrm{d}\bar{\xi}_I + \sum_{|J|=q-1} g_J(\tau, \xi)\,\mathrm{d}\bar{\tau} \wedge \mathrm{d}\bar{\xi}_J, \tag{3.2}$$

where the coefficients f_I and g_J are all continuous functions on $\overline{G}$. We assume that $\bar{\partial}\aleph$ exists in the sense of distributions and is continuous on $\Sigma \setminus H_0$. Recall that the singular part of Σ is contained in H_0. Moreover the given hypotheses imply that the differential $\bar{\partial}\aleph$ has a continuous extension onto $\overline{G}$, so that we may say that $\bar{\partial}\aleph$ exists and is continuous in $\overline{G}$. The pull-back of $\aleph$ with respect to the projection π given in (1.4) can be easily calculated in a simple formula,

$$\pi^*\aleph = \sum_I f_I(\sigma(\xi), \xi)\,\mathrm{d}\bar{\xi}_I + \sum_J g_J(\sigma(\xi), \xi)\overline{\left[\frac{\sigma(\xi)\partial p(\xi)}{m\,p(\xi)}\right]} \wedge \mathrm{d}\bar{\xi}_J, \tag{3.3}$$

where $\pi(u,\xi) = (\sigma(\xi),\xi)$ and the function $\sigma(\xi) = \sqrt[m]{p}(\xi)$ is the unique solution to (1.3). It is easy to see that the $(0,q)$-forms $\aleph$ and $\pi^*\aleph$ are equal in $G \setminus H_0$. One of the main problems is that the coefficients of the pull-back $\pi^*\aleph$ are not necessarily continuous in the closure $\overline{G}$, because we have to deal with the differential of $\sigma(\xi)$. The same can be said about the coefficients of $\pi^*\bar{\partial}\aleph$. Nevertheless this problem can be easily solved by multiplying the pull-backs $\pi^*\aleph$ and $\pi^*\bar{\partial}\aleph$ times the holomorphic polynomial $p^\kappa(\xi)$ for any integer exponent $\kappa \geq 2$.

The main idea is to apply the Bochner – Martinelli – Koppelman integration formulae on the open set Ω and the form $p^\kappa(\xi)\pi^*\aleph$, and then to rewrite it in terms of integrals over the set $\overline{G}$ in Σ. We suppose that the domain G has piecewise smooth boundary, i.e. δG is the union of a finite number of smooth real manifolds. It is then easy to deduce that Ω in (3.1) has piecewise boundary for any $\beta > 0$. In particular

$$(3.4)\quad \delta\Omega = \left\{(\tau+\nu,\xi) \in \mathbb{C}\times\mathbb{C}^n : (\tau,\xi) \in G \setminus H_\beta,\ |\nu| = \frac{|\tau|}{m}\right\} \cup \left\{(\tau+\nu,\xi) \in \mathbb{C}\times\mathbb{C}^n : (\tau,\xi) \in \delta(G \setminus H_\beta),\ |\nu| \leq \frac{|\tau|}{m}\right\}.$$

Recall that H_β is equal to $\{|s| \leq \beta\} \times \mathbb{C}^n$, so that $|\tau| > \beta$ in (3.4); moreover the singular part of Σ is contained in H_0. We also need the Bochner – Martinelli kernel written in terms of the vectors (u,ξ) and (s,z) in $\mathbb{C}\times\mathbb{C}^n$; see equations (2.13) and (2.18), or references [10, 11]:

$$(3.5)\qquad B(u,\xi;s,z) = \frac{\overline{(u-s)\,\mathrm{d}u \wedge (\mathrm{d}\omega(\xi;z)/n) + (\mathrm{d}\bar{u}-\mathrm{d}\bar{s}) \wedge \mathrm{d}u \wedge \omega(\xi;z)}}{2\pi\mathrm{i}(|u-s|^2 + \|\xi - z\|^2)^{n+1}},$$

where $\omega(\xi;z)$ and $\mathrm{d}\omega(\xi;z)$ are respectively defined in formulae (2.13)) and (2.15)). Whence, since the form $p^\kappa\pi^*\aleph$ in (3.3) and its $\bar{\partial}$-differential are both continuous in the closure $\overline{\Omega}$ for any fixed integer $\kappa \geq 2$, the Bochner – Martinelli – Koppelman formula (2.18) yields the following identity for (s,z) in Ω,

$$(3.6)\qquad \frac{p^\kappa\pi^*\aleph}{(-1)^q n!} = B_{\delta\Omega}(p^\kappa\pi^*\aleph) - B_\Omega(p^\kappa\pi^*\bar{\partial}\aleph) + (\bar{\partial}_s + \bar{\partial}_z)B_\Omega(p^\kappa\pi^*\aleph),$$

The open set Ω has piecewise smooth boundary and the operators B_Ω and $B_{\delta\Omega}$ are naturally given by the formula

$$(3.7)\qquad B_M(p^\kappa\pi^*\aleph) = \int_{(u,\xi)\in M} \left[p^\kappa(\xi)\pi^*\aleph\right] \wedge B(u,\xi;s,z),$$

where M is equal to Ω or $\delta\Omega$. The integral above is calculated according to the convention introduced in [10, p. 44], so that only those terms whose degree is equal to the real dimension of M are integrated. We have that the forms $\pi^*\aleph$ and $\aleph$ are equal in $G \setminus H_0$, because of (3.2) – (3.3) and since $G \setminus H_\beta$ is naturally contained in Ω. Hence we only need to rewrite the operators B_Ω and $B_{\delta\Omega}$ as integrals over the open set G or its boundary δG, in order to deduce the desired integration formula for the singular subvariety Σ. Recall the functions F and F^* given in (2.9) – (2.10), and rewrite them using the change of variable w equal to $(u-\tau)/\tau$,

$$F\Big(\frac{s-\tau}{\tau},\frac{\|\xi-z\|^2}{|\tau|^2}\Big)=\int_{|u-\tau|<|\tau|/m}\frac{|\tau|^{2n}\,d\bar u\wedge du}{(\|\xi-z\|^2+|u-s|^2)^{n+1}},\tag{3.8}$$

$$F^*\Big(\frac{s-\tau}{\tau},\frac{\|\xi-z\|^2}{|\tau|^2}\Big)=\int_{|u-\tau|=|\tau|/m}\frac{|\tau|^{2n}\,\overline{(u-s)}\,du}{(\|\xi-z\|^2+|u-s|^2)^{n+1}}.\tag{3.9}$$

Now we may analyze the operator B_Ω defined according to (3.5) – (3.7). Since the real dimension of Ω is equal to $2n+2$, the integrals on Ω must be calculated only over those monomials that contain all the differentials $d\bar u$, du, $d\bar\xi$, and $d\xi$. The coefficients of $\pi^*\aleph$ in (3.3) are locally constant with respect to the variable u; moreover the form $\pi^*\aleph$ contains neither the differential du nor $d\bar u$. Hence

$$\begin{aligned}B_\Omega(p^\kappa\pi^*\aleph)&=\int_{(u,\xi)\in\Omega}\frac{[p^\kappa\pi^*\aleph]\wedge d\bar u\wedge du\wedge\omega(\xi;z)}{2\pi i(|u-s|^2+\|\xi-z\|^2)^{n+1}}\\&=\int_{(\tau,\xi)\in G\setminus H_\beta}\int_{|u-\tau|<|\tau|/m}\frac{d\bar u\wedge du\wedge[p^\kappa\pi^*\aleph]\wedge\omega(\xi;z)}{2\pi i(|u-s|^2+\|\xi-z\|^2)^{n+1}}\\&=\int_{(\tau,\xi)\in G\setminus H_\beta}F\Big(\frac{s-\tau}{\tau},\frac{\|\xi-z\|^2}{|\tau|^2}\Big)\big[p^\kappa(\xi)\pi^*\aleph\big]\wedge\frac{\omega(\xi;z)}{2\pi i|\tau|^{2n}}.\end{aligned}\tag{3.10}$$

The last two equalities above follows from Fubini's theorem (2.2) and the definition of Ω in (3.1). The integral with respect to the variable u yields in particular the function F in (3.8). Moreover, since the variable u in (3.10) runs along an open disc in $\mathbb{C}$, we can easily understand why only those monomials in (3.5) that contain the volume form $d\bar u\wedge du$ can be integrated in (3.10). We can simplify the final identity in (3.10) by using the kernel K_1 defined in (2.12),

$$B_\Omega(p^\kappa\pi^*\aleph)=\int_{(\tau,\xi)\in G\setminus H_\beta}\big[p^\kappa(\xi)\pi^*\aleph\big]\wedge K_1(\tau,\xi;s,z).\tag{3.11}$$

A similar formula can be deduced for $B_\Omega(p^\kappa\pi^*\bar\partial\aleph)$ given in (3.6) – (3.7). Nevertheless the operator $B_{\delta\Omega}(p^\kappa\pi^*\aleph)$ in (3.6) must be calculated carefully, because the boundary $\delta\Omega$ the union of two sets; see equation (3.4). Whence

$$\begin{aligned}&B_{\delta\Omega}(p^\kappa\pi^*\aleph)\\&\quad=\int_{(\tau,\xi)\in\delta(G\setminus H_\beta)}\int_{|u-\tau|<|\tau|/m}\frac{d\bar u\wedge du\wedge[p^\kappa\pi^*\aleph]\wedge\omega(\xi;z)}{2\pi i(|u-s|^2+\|\xi-z\|^2)^{n+1}}\\&\qquad+(-1)^q\int_{(\tau,\xi)\in G\setminus H_\beta}\int_{|u-\tau|=|\tau|/m}\frac{\overline{(u-s)}\,du\wedge[p^\kappa\pi^*\aleph]\wedge d\omega(\xi;z)}{2n\pi i(|u-s|^2+\|\xi-z\|^2)^{n+1}}\\&\qquad-\int_{(\tau,\xi)\in G\setminus H_\beta}\int_{|u-\tau|=|\tau|/m}\frac{d\bar s\wedge du\wedge[p^\kappa\pi^*\aleph]\wedge\omega(\xi;z)}{2\pi i(|u-s|^2+\|\xi-z\|^2)^{n+1}}.\end{aligned}\tag{3.12}$$

The integrals in the first line of (3.12) are calculated over those monomials in (3.5) that contain the volume form $d\bar u\wedge du$, because the variable u runs along an open disc in $\mathbb{C}$; while the integrals in the second and third lines are calculated over those monomials in (3.5) that contain either du or $d\bar u$, because the variable u runs along a circumference in $\mathbb{C}$. A direct application of Fubini's theorems (2.2) – (2.3) with the kernels K_ℓ and functions given in (2.12), (2.14), (3.8), and (3.9) yields the following formula for (3.12) minus the term in third line,

$$(3.13)\quad B^*_{\delta\Omega}(p^\kappa\pi^*\aleph) := \int_{(\tau,\xi)\in\delta(G\setminus H_\beta)} \left[p^\kappa\pi^*\aleph\right]\wedge K_1(\tau,\xi;s,z) + (-1)^q\int_{(\tau,\xi)\in G\setminus H_\beta}\left[p^\kappa\pi^*\aleph\right]\wedge K_2(\tau,\xi;s,z).$$

We do not include the term in the third line of (3.12) because it is cancelled by the antiholomorphic differential $\bar\partial_s B_\Omega(p^\kappa\pi^*\aleph)$ in (3.6), i.e.

$$\begin{aligned}\int_{(\tau,\xi)\in G\setminus H_\beta}\int_{|u-\tau|=|\tau|/m} &\frac{\mathrm{d}\bar s\wedge\mathrm{d}u\wedge[p^\kappa\pi^*\aleph]\wedge\omega(\xi;z)}{2\pi\mathrm{i}(|u-s|^2+\|\xi-z\|^2)^{n+1}}\\ &= (n+1)\int_{\substack{(\tau,\xi)\in G\setminus H_\beta\\ \xi\neq z}}\int_{|u-\tau|<|\tau|/m}\frac{(u-s)\,\mathrm{d}\bar s\wedge\mathrm{d}\bar u\wedge\mathrm{d}u\wedge[p^\kappa\pi^*\aleph]\wedge\omega}{2\pi\mathrm{i}(|u-s|^2+\|\xi-z\|^2)^{n+2}}\\ &= \bar\partial_s B_\Omega(p^\kappa\pi^*\aleph).\end{aligned}$$

Recall the expression of $B_\Omega(p^\kappa\pi^*\aleph)$ given in the second line of (3.10). Bochner–Martinelli–Koppelman formula (3.6) produces then the desired result on $G\setminus H_\beta$ for any $\beta>0$, recall that the form $\aleph$ and its pull-back $\pi^*\aleph$ coincide on $G\setminus H_0$ because of the system of equations (3.2)–(3.3),

$$(3.14)\qquad \frac{p^\kappa(z)\aleph}{(-1)^q n!} = B^*_{\delta\Omega}(p^\kappa\pi^*\aleph) - B_\Omega(p^\kappa\pi^*\bar\partial\aleph) + \bar\partial_z B_\Omega(p^\kappa\pi^*\aleph).$$

Finally we need to calculate the limit of the previous equation as β converges to zero. Let (s,z) be any fixed point in $G\setminus H_0$, so that neither s nor $p(z)$ vanishes. Functions F and F^* in (3.8)–(3.9) are obviously continuous with respect to (τ,ξ) for $\xi\neq z$; moreover

$$(3.15)\qquad \lim_{\tau\to 0}\frac{1}{|\tau|^{2n}}F\left(\frac{s-\tau}{\tau},\frac{\|\xi-z\|^2}{|\tau|^2}\right)=0 \quad \text{if } \xi\neq z.$$

The same equality holds for F^* instead of F. On the other hand the coefficients of the forms ω in (2.13) and $p^\kappa\pi^*\aleph$ in (3.3) are all continuous on the closure $\overline{G}$, because we have chosen the exponent $\kappa\geq 2$. Therefore equations (3.10)–(3.11) can be rewritten as follows

$$(3.16)\qquad B_\Omega(p^\kappa\pi^*\aleph) = \sum_{|J|=q-1}\mathrm{d}\bar z_J\int_{(\tau,\xi)\in G\setminus H_\beta} h_J(\tau,\xi;s,z)\,\mathrm{dvol}(\xi)$$

for some coefficients h_J that are continuous at each (τ,ξ) in $\overline{G}$ with $\xi\neq z$. The term $d\bar z_J$ appears in the formula above because $\pi^*\aleph$ is a $(0,q)$-form and the integral in the third line of (3.10) is calculated over an open set in the regular part of the $2n$-real dimensional subvariety Σ, so that we integrate only those monomials that contain the volume form $\mathrm{dvol}(\xi)$ defined in (2.1). Consider the set G_β defined in equation (2.4), so that

$$(3.17)\qquad \left[G\cap H_{\beta/2}\right]\subset G_\beta\subset\left[G\cap H_\beta\right].$$

Recall that $H_\beta=\{|s|\leq\beta\}\times\mathbb{C}^n$ and that $s^m=p(z)$ for every (s,z) in the compact set $\overline{G}$. The polynomial $p(z)$ does not vanish because (s,z) lies in $G\setminus H_0$. Hence there exists a real number $\theta>0$ small enough such that $p(\xi)$ and $p(z)$ are different for every point (τ,ξ) in $\overline{G_\theta}$, and so each coefficient h_J in (3.16) is continuous on $\overline{G_\theta}$. Equations (2.5) and (3.17) yields a simple way to calculate the limit as β goes to zero of the operator $B_\Omega(p^\kappa\pi^*\aleph)$ expressed in (3.11) or (3.16). We can define in particular

$$(3.18)\qquad B_G(p^\kappa\pi^*\aleph) := \int_{(\tau,\xi)\in G} \big[p^\kappa(\xi)\pi^*\aleph\big]\wedge K_1(\tau,\xi;s,z) = \lim_{\beta\to 0}\int_{(\tau,\xi)\in G\setminus H_\beta} \big[p^\kappa(\xi)\pi^*\aleph\big]\wedge K_1(\tau,\xi;s,z).$$

The operator $B_G(p^\kappa\pi^*\bar\partial\aleph)$ can be defined in a similar way,

$$(3.19)\qquad B_G(p^\kappa\pi^*\bar\partial\aleph) := \int_{(\tau,\xi)\in G} \big[p^\kappa(\xi)\pi^*\bar\partial\aleph\big]\wedge K_1(\tau,\xi;s,z).$$

Let (s,z) be again a fixed point in $G\setminus H_0$, so that neither s nor $p(z)$ vanishes. The operator $B^*_{\delta\Omega}(p^\kappa\pi^*\aleph)$ in (3.13) can be expanded as follows

$$(3.20)\qquad B^*_{\delta\Omega}(p^\kappa\pi^*\aleph) = \sum_{|I|=q} \mathrm{d}\bar z_I \int_{(\tau,\xi)\in G\setminus H_\beta} h^*_I(\tau,\xi;s,z)\,\mathrm{dvol}(\xi) + \sum_{|I|=q} \mathrm{d}\bar z_I \sum_{j=1}^{n} \int_{(\tau,\xi)\in\delta(G\setminus H_\beta)} h_{j,I}(\tau,\xi;s,z)\omega^0_j(\xi;z).$$

for some coefficients h^*_I and $h_{j,I}$ that are continuous at (τ,ξ) in $\overline{G}$ with $\xi\neq z$. The form ω^0_j was defined in (2.7) and appears in connection to ω in (2.13) because the real dimension of the boundary of $G\setminus H_\beta$ is equal to $2n-1$. The coefficients h^*_I and $h_{j,I}$ are obviously determined by the form $p^\kappa\pi^*\aleph$ in (3.3) and the functions F and F^* given in (3.8) – (3.9); moreover they are all continuous on the compact set $\overline{G_\theta}$ for the parameter $\theta>0$ deduced some paragraphs above and the set G_β defined in (2.4). Finally observe that all the coefficients of $p^\kappa\pi^*\aleph$ in (3.3) contain the factor $p(\xi)$ because of the exponent $\kappa\geq 2$; moreover F is continuous on $\overline{G_\theta}$ and satisfies equation (3.15). Whence there exists a real constants $M_3>0$ such that

$$(3.21)\qquad |h_{j,I}(\tau,\xi;s,z)| \leq M_3|p(\xi)| \quad\text{and}\quad \delta(G_\beta)=\delta(G\cap H_\beta)$$

for all points (τ,ξ) in $\overline{G_\theta}$, indexes $\{j,I\}$, and real numbers $\beta>0$. Since $|p(\xi)|$ is less than or equal to β^m for every point (τ,ξ) in H_β, we easily have that equations (2.8) and (3.21) yields that

$$\lim_{\beta\to 0}\left|\int_{(\tau,\xi)\in\delta(G\cap H_\beta)} h_{j,I}(\tau,\xi;s,z)\omega^0_j(\xi;z)\right| = 0.$$

Thus the operator $B_{\delta G}(p^\kappa\pi^*\aleph)$ may be defined as the limit when β goes to zero of the operator $B^*_{\delta\Omega}(p^\kappa\pi^*\aleph)$ expressed in (3.13) and (3.20),

$$(3.22)\qquad B_{\delta G}(p^\kappa\pi^*\aleph) := \int_{(\tau,\xi)\in\delta G} \big[p^\kappa\pi^*\aleph\big]\wedge K_1(\tau,\xi;s,z) + (-1)^q\int_{(\tau,\xi)\in G} \big[p^\kappa\pi^*\aleph\big]\wedge K_2(\tau,\xi;s,z).$$

The desired result on $G\setminus H_0$ is obtained by applying the limit $\beta\to 0$ to the Bochner – Martinelli – Koppelman formula (3.14), recall that the form $\aleph$ and its pull-back $\pi^*\aleph$ coincide on $G\setminus H_0$ because of (3.2) – (3.3),

$$\frac{p^\kappa(z)\aleph}{(-1)^q n!} = B_{\delta G}(p^\kappa\pi^*\aleph) - B_G(p^\kappa\pi^*\bar\partial\aleph) + \bar\partial_z B_G(p^\kappa\pi^*\aleph).$$

References

1. M. Andersson, *Integral representation with weights*. I, Math. Ann. **326** (2003), no. 1, 1–18.
2. ———, *Integral representation with weights*. II: *Division and interpolation*, Math. Z. **254** (2006), no. 2, 315–332.
3. M. Andersson and H. Samuelsson, *Weighted Koppelman formulas and the $\bar{\partial}$-equation on an analytic space*, J. Funct. Anal. **261** (2011), no. 3, 777–802.
4. V. I. Arnol′d, S. M. Guseĭn-Zade, and A. N. Varchenko, *Singularities of differentiable maps*. Vol. I: *The classification of critical points, caustics and wave fronts*, Monogr. Math., vol. 82, Birkhäuser, Boston, MA, 1985.
5. V. I. Arnol′d, V. V. Goryunov, O. V. Lyashko, and V. A. Vasil′ev, *Singularity theory*. I, Springer, Berlin, 1998.
6. E. M. Chirka, *Complex analytic sets*, Math. Appl. (Soviet Ser.), vol. 46, Kluwer, Dordrecht, 1989.
7. A. Dimca, *Singularities and topology of hypersurfaces*, Universitext, Springer, New York, 1992.
8. T. E. Hatziafratis, *An explicit Koppelman type integral formula on analytic varieties*, Michigan Math. J. **33** (1986), no. 3, 335–341.
9. ———, *Integral representation formulas on analytic varieties*, Pacific J. Math. **123** (1986), no. 1, 71–91.
10. G. Henkin and J. Leiterer, *Theory of functions on complex manifolds*, Monogr. Math., vol. 79, Birkhäuser, Basel, 1984.
11. C. Laurent-Thiébaut, *Holomorphic function theory in several variables: An introduction*, Universitext, Springer, London, 2011.
12. A. Palm, *Integral representation formulas on strictly pseudoconvex domains in Stein manifolds*, Duke Math. J. **43** (1976), no. 1, 41–62.
13. J. Ruppenthal, *The $\bar{\partial}$-equation on homogeneous varieties with an isolated singularity*, Math. Z. **263** (2009), no. 2, 447–472.
14. J. Ruppenthal and E. S. Zeron, *An explicit $\bar{\partial}$-integration formula for weighted homogeneous varieties*, Michigan Math. J. **58** (2009), no. 2, 441–457.
15. ———, *An explicit $\bar{\partial}$-integration formula for weighted homogeneous varieties*. II: *Forms of higher degree*, Michigan Math. J. **59** (2010), no. 2, 283–295.
16. E. L. Stout, *An integral formula for holomorphic functions on strictly pseudoconvex hypersurfaces*, Duke Math. J. **42** (1975), 347–356.
17. S. S.-T. Yau and Y. Yu, *Gorenstein quotient singularities in dimension three*, Mem. Amer. Math. Soc. **105** (1993), no. 505, viii+88.

Departamento de Matemáticas, Centro de Investigación y de Estudios Avanzados del Instituto Politécnico Nacional, Apartado Postal 14-740, México D.F., 07000, México

E-mail address, L. M. Hernández-Pérez: `lhperez@math.cinvestav.mx`

E-mail address, E. S. Zeron: `eszeron@math.cinvestav.mx`

Centre de Recherches Mathématiques
CRM Proceedings and Lecture Notes
Volume **55**, 2012

Invariant Potential Theory, Derivatives of Inner Functions, and $B^{p,q}$ Spaces in the Unit Ball of $\mathbb{C}^n$

Manfred Stoll

Abstract. Let $\mathbb{B}_n$ be the unit ball in $\mathbb{C}^n$ with boundary S. A bounded holomorphic function f on $\mathbb{B}_n$ is called an inner function if

$$\lim_{r\to 1^-} |f(rt)| = 1 \quad \text{for a.e. } t \in S.$$

For $0 < p < 1$ and $0 < q < \infty$, the space $B^{p,q}$ is defined as the space of holomorphic functions f on $\mathbb{B}_n$ for which

$$\|f\|^q_{B^{p,q}} = \int_{\mathbb{B}_n} (1-|z|^2)^{n/p-n-1} |f(z)|^q \,\mathrm{d}\nu(z) < \infty,$$

where ν is normalized Lebesgue measure on $\mathbb{B}_n$. The case $q = 1$ gives the usual B^p spaces of holomorphic functions on $\mathbb{B}_n$. Using techniques and results of potential theory with respect to the Laplace Beltrami operator on $\mathbb{B}_n$, we prove that for $n/(n+q) < p < \min\{n/q, 1\}$, the radial derivative Rf of an inner function f satisfies

$$Rf \in B^{p,q} \iff \int_{\mathbb{B}_n} (1-|z|^2)^{n/p-q-n-1} |\widetilde{\nabla} f(z)|^r \,\mathrm{d}\nu(z) < \infty$$

for all r, q, with $1 \le q \le r \le 2$, where $\widetilde{\nabla}$ denotes the gradient with respect to the Bergman metric on $\mathbb{B}_n$. From this it now follows that for $n \ge 2$ and $q \ge 1$, if f is an inner function on $\mathbb{B}_n$ with $Rf \in B^{p,q}$ with $p \ge 2n/(2n+q)$, then f is constant. In particular, if f is an inner function with $Rf \in B^{2n/(2n+1)}$ and $n \ge 2$, then f is a constant.

1. Introduction

In this survey paper we present some recent results concerning membership of the radial derivative of an inner function on the unit ball in some classical function spaces. Let $\mathbb{B}_n$ denote the unit ball in $\mathbb{C}^n$ $(n \ge 1)$ with boundary S. A bounded holomorphic function f on $\mathbb{B}_n$ is an **inner function** if

$$\lim_{r\to 1^-} |f(rt)| = 1 \quad \text{for almost every } t \in S$$

2010 *Mathematics Subject Classification.* 32A36, 32A17, 32A25.

Key words and phrases. Bergman metric, Bergman space, Dirichlet space, Hardy space, inner functions, radial derivative.

This is the final form of the paper.

When $n = 1$, every inner function f can be written as

$$f(z) = z^m B(z) S_\nu(z),$$

where $B(z)$ is a Blaschke product whose zeros $\{a_n\}$ $a_n \neq 0$ satisfy

$$\sum_{n=1}^{\infty} (1 - |a_n|) < \infty,$$

and S_ν is a singular inner function given by

$$S_\nu(z) = \exp\biggl[-\int_0^{2\pi} \frac{e^{it} + z}{e^{it} - z}\, d\nu(t)\biggr], \tag{1.1}$$

where ν is a singular measure.

For $n \geq 2$, the existence of inner functions on $\mathbb{B}_n$ was a long standing open problem until 1982 when A. B. Aleksandrov [5] (see also [6]) proved the existence of a positive singular measure μ on S such that

$$P[\mu](z) = \operatorname{Re} h(z),$$

where h is a non constant holomorphic function on $\mathbb{B}_n$, and $P[\mu]$ is the Poisson integral of μ.[1] Hence

$$F_\mu(z) = \exp[-h(z)] = \exp\biggl\{-\int_S \biggl[\frac{2}{(1 - \langle z, t\rangle)^n} - 1\biggr] d\mu(t)\biggr\}$$

is a nonconstant singular inner function on $\mathbb{B}_n$. The existence of nonconstant inner functions on $\mathbb{B}_n$ was proved independently by Erik Løw [18].

In the late 1960s and early seventies, when $n = 1$, there was a flurry of activity involving the existence of inner functions whose derivatives lie in certain classical function spaces. For $0 < p \leq \infty$, we denote by H^p the classical **Hardy space** of holomorphic functions on $\mathbb{B}_n$. Also, for $0 < p < 1$,

$$B^p = \biggl\{f \text{ holomorphic} : \|f\|_{B^p} = \int_{\mathbb{B}_n} (1 - |z|^2)^{n/p-n-1} |f(z)|\, d\nu(z) < \infty\biggr\}.$$

where ν is normalized Lebesgue measure on $\mathbb{B}_n$. It is well known that for $0 < p < 1$ the space B^p is the containing Banach space of the Hardy space H^p ([14] for $n = 1$ and [11, 19] for $n > 1$). More generally, as in [1], for $0 < p < 1$, $0 < q < \infty$, we define the spaces $B^{p,q}$ as follows:

$$B^{p,q} = \biggl\{f \text{ holomorphic} : \|f\|^q_{B^{p,q}} = \int_{\mathbb{B}_n} (1 - |z|^2)^{n/p-n-1} |f(z)|^q\, d\nu(z) < \infty\biggr\}.$$

The weighted **Bergman spaces** A^q_α, $\alpha > -1$, $0 < q < \infty$, are defined by

$$A^q_\alpha = \biggl\{f \text{ holomorphic} : \|f\|^q_{A^q_\alpha} = \int_{\mathbb{B}_n} (1 - |z|^2)^\alpha |f(z)|^q\, d\nu(z) < \infty\biggr\}.$$

The special case A^q_0 is denoted by A^q. We also have that $B^{p,q} = A^q_{\frac{n}{p}-n-1}$. Finally, $L^p(\mathbb{B}_n)$ denotes the usual Lebesgue space of measurable functions f for which $\int_{\mathbb{B}_n} |f|^p d\nu < \infty$.

As we will see in Section 3, if f is an inner function on $\mathbb{B}_n$, then for $0 < q < \infty$, the radial derivative Rf of f is in the space $B^{p,q}$ for all p, $0 < p < n/(n+q)$. In particular, for $q = 1$, $Rf \in B^p$ for all p, $0 < p < n/(n+1)$. By a result of Y. Dupain

[1] Here P can be either the Euclidean Poisson kernel or the invariant Poisson kernel.

[13] there exists an inner function f on $\mathbb{B}_n$ with $Rf \in B^{n/(n+1)}$. In the paper, as a consequence of the main result (Theorem 5.4), we obtain the following.

Theorem A (Corollary 5.6). *Suppose $n \geq 2$. If f is an inner function on $\mathbb{B}_n$ with $Rf \in B^{p,q}$ with $q \geq 1$ and $p \geq 2n/(2n+q)$, then f is a constant. In particular, if $Rf \in B^{2n/(2n+1)}$, then f is a constant.*

As another application of Theorem 5.4 we obtain the following result which connects membership of Rf in certain B^p spaces to membership in weighted Bergman spaces.

Theorem B (Theorem 5.9). *Let f be an inner function on $\mathbb{B}_n$, $n \geq 1$. If $1 \leq q \leq 2$ and $0 < (1-\alpha) < \frac{1}{2}q$, then*

$$Rf \in B^{n/(n+\alpha)} \quad \textit{if and only if} \quad Rf \in A^q_{q+\alpha-2}.$$

When $n = 1$, it is known that if f is an inner function on the unit disk $\mathbb{D}$ with $f' \in A^2$, then f is a finite Blaschke product [20]. For $n \geq 2$, by a result of E. Bedford and A. Taylor [10], if f is an inner function on $\mathbb{B}_n$ with $Rf \in A^2$ then f is a constant. Combining Theorems A and B, as well as known results when $n = 1$, we obtain the following.

Theorem C (Corollary 5.10). *Let f be an inner function on $\mathbb{B}_n$, $n \geq 1$.*

(a) *If $n = 1$ and $f' \in A^2_{1/2}$ or $f' \in A^{3/2}$, then f is a Blaschke product.*

(b) *If $n \geq 2$ and $Rf \in A^2_{1/2}$ or $Rf \in A^{3/2}$, then f is constant.*

Part (b) of Theorem C is a generalization of the Bedford and Taylor result.

In Section 2 we provide a brief summary of some known results concerning membership of derivatives of inner functions in either H^p or B^p when $n = 1$, and in Section 3 we provide a summary of the few known results when $n \geq 2$. In Section 4 we provide a brief review of potential theory with respect to the Laplace – Beltrami operator. Finally, Section 5 contains the proofs of the main results of the paper.

Acknowledgments.

(1) The author would like to thank the organizers André Boivin and Javad Mashreghi for inviting him to participate in this conference in honor of his long time friends Paul Gauthier and Kohur Gowrisankaran.

(2) The author would also like to acknowledge that the main results of this paper in the case $q = 1$, namely Theorems 5.4 and 5.9 and their corollaries, were proved by the author's doctoral student Matt Gamel in his dissertation [15] and resulting publication [16].

2. Some known results when $n = 1$

It is easily proved (see Proposition 3.2) that if f is an inner function in $\mathbb{D}$, then $f' \in B^p$ for all $p < \frac{1}{2}$. In 1955, W. Rudin [22] proved that if B is a Blaschke product whose zeroes $\{a_n\}$ satisfy

$$\sum (1 - |a_n|) \log \frac{1}{(1 - |a_n|)} < \infty,$$

then $B' \in B^{1/2}$. Subsequently in 1973, D. Protas [21] proved that if

$$\sum (1 - |a_n|)^\alpha < \infty$$

for some α, $0 < \alpha < 1$, then $B' \in B^{1/(1+\alpha)}$. As a consequence, for every p, $\frac{1}{2} \le p < 1$, there exists a Blaschke product B such that $B' \in B^p$. In the same paper Protas also proved that if

$$\sum (1 - |a_n|)^\alpha < \infty$$

for some α, $0 < \alpha < \frac{1}{2}$, then $B' \in H^{1-\alpha}$. Furthermore, if

$$\sum (1 - |a_n|)^{1/2} \log \frac{1}{(1 - |a_n|)} < \infty,$$

then $B' \in H^{1/2}$. On the other hand, P. Ahern and D. Clark proved in 1974 ([3]) that if f is an inner function on $\mathbb{D}$ with $f' \in H^{1/2}$, then f is a Blaschke product. In a subsequent paper, [4], the same authors proved in 1976 that if f is an inner function on $\mathbb{D}$ with $f' \in B^{2/3}$, then f is a Blaschke product.

For singular inner functions, it is easily proved that if ν is a finite sum of point mass measures on $\partial\mathbb{D}$, then the singular inner function S_ν given by (1.1) satisfies $S_\nu' \in H^p$ for all $p < \frac{1}{2}$. In 1970, M. Cullen [12] provided sufficient conditions on a singular measure ν such that the singular inner function S_ν satisfies $S_\nu' \in H^p$ for $p < \frac{1}{2}$. Also, H. Allen and C. Belna [7] proved that if $\nu = \sum \delta_{a_n}$ with $\{a_n\} \in \ell^{1/q}$, $1 < q < \infty$, then $S_\nu' \in B^\gamma$ for all $\gamma < 2q/(3q+1)$. In particular, for every p, $\frac{1}{2} \le p < \frac{2}{3}$, there exists a singular inner function S_ν with $S_\nu' \in B^p$.

3. Some known results when $n \ge 2$

In this section we provide a brief summary of some of the known results when $n \ge 2$. For this we introduce the following notation. For f holomorphic on $\mathbb{B}_n$, the complex gradient ∂f of a holomorphic function f is defined by

$$\partial f = \left(\frac{\partial f}{\partial z_1}, \ldots, \frac{\partial f}{\partial z_n}\right),$$

and the radial derivative Rf of f is given by

$$Rf(z) = \sum_{j=1}^{n} z_j \frac{\partial f}{\partial z_j}.$$

Proposition 3.1 ([10]). *If f is an inner function on $\mathbb{B}_n$, $n \ge 2$, with $|\partial f| \in L^2$, then f is constant.*

Proposition 3.2. *If f is an inner function (or bounded holomorphic function) on $\mathbb{B}_n$, $n \ge 1$, and $0 < q < \infty$, then $Rf \in B^{p,q}$ for all p, $0 < p < n/(n+q)$.*

PROOF. If f is a bounded holomorphic function on $\mathbb{B}_n$, then as a consequence of the Szegő integral formula one has

$$f(z) = \int_S \frac{\hat{f}(\zeta)}{(1 - \langle z, \zeta\rangle)^n} \, d\sigma(\zeta)$$

where σ denotes normalized surface area measure on S and $\hat{f}$ is the boundary function of f which exists a.e. on S. Thus

$$Rf(z) = n \int_S \frac{\langle z, \zeta\rangle \hat{f}(\zeta)}{(1 - \langle z, \zeta\rangle)^{n+1}} \, d\sigma(\zeta).$$

Therefore

$$|Rf(r\eta)| \le C \int_S \frac{1}{|1-\langle r\eta, \zeta\rangle|^{n+1}}\, d\sigma(\zeta),$$

which by [23, Proposition 1.4.10]

$$\le (1-r^2)^{-1}.$$

Thus

$$\|Rf\|^q_{B^{p,q}} \le M \int_0^1 r^{2n-1}(1-r^2)^{n/p-n-q-1}\, dr$$

which is finite for all $p < n/(n+q)$. □

Proposition 3.3 ([13]). *For $n \ge 2$, there exists an inner function f on $\mathbb{B}_n$ such that $|\partial f| \in L^1(\mathbb{B}_n)$, i.e., $Rf \in B^{n/(n+1)}$.*

4. Invariant potential theory on $\mathbb{B}_n$

We denote by $\mathcal{M}$ the group of Möbius transformations on $\mathbb{B}_n$. For $z \in \mathbb{B}_n$, let φ_z denote the Möbius transformation on $\mathbb{B}_n$ satisfying

- $\varphi_z(0) = z$ and
- $\varphi_z \circ \varphi_z = I$, the identity map.

Then each $\psi \in \mathcal{M}$ is of the form $\psi = \mathcal{U} \circ \varphi_z$ for some φ_z and unitary transformation $\mathcal{U}$. It is known that the automorphism φ_z satisfies

$$1 - |\varphi_z(w)|^2 = \frac{(1-|z|^2)(1-|w|^2)}{|1-\langle z, w\rangle|^2}. \tag{4.1}$$

Also, we denote by $\tilde{\Delta}, \widetilde{\nabla}$, and λ the Laplacian, gradient, and measure with respect to the Bergman metric on $\mathbb{B}_n$. Then for a C^2 function f

$$\tilde{\Delta} f(z) = \Delta(f \circ \varphi_z)(0) = 4(1-|z|^2) \sum_{i,j=1}^{n} (\delta_{i,j} - \bar{z}_i z_j) \frac{\partial^2 f(z)}{\partial \bar{z}_i \partial z_j},$$

where Δ is the usual Laplacian on $\mathbb{R}^{2n}$. For f holomorphic we have

$$|\widetilde{\nabla} f(z)|^2 = |\partial(f \circ \varphi_z)(0)|^2 = (1-|z|^2)\big[|\partial f(z)|^2 - |Rf(z)|^2\big].$$

The measure $d\lambda$ is invariant under $\mathcal{M}$ and is given by

$$d\lambda(z) = \frac{d\nu(z)}{(1-|z|^2)^{n+1}}.$$

For a holomorphic function f we have the following inequality:

$$(1-|z|^2)^2|Rf(z)|^2 \le (1-|z|^2)^2|\partial f(z)|^2 \le |\widetilde{\nabla} f(z)|^2 \le (1-|z|^2)|\partial f(z)|^2. \tag{4.2}$$

The C^∞ functions that are annihilated by $\tilde{\Delta}$ on $\mathbb{B}_n$ are called **$\mathcal{M}$-harmonic** or **invariant harmonic** in $\mathbb{B}_n$. Any continuous function h satisfying

$$h\big(\psi(0)\big) = \int_S h\big(\psi(rt)\big)\, d\sigma(t) \tag{4.3}$$

for all $\psi \in \mathcal{M}$ and $0 < r < 1$ is C^∞ on $\mathbb{B}_n$ and satisfies $\tilde{\Delta} h = 0$. An upper semi continuous function h is **$\mathcal{M}$-subharmonic** on $\mathbb{B}_n$ if one has

$$h\big(\psi(0)\big) \le \int_S h\big(\psi(rt)\big)\, d\sigma(t)$$

for all $\psi \in \mathcal{M}$ and $0 < r < 1$. For C^2 functions h this is equivalent to $\tilde{\Delta} h \geq 0$.

For an $\mathcal{M}$-subharmonic function f, the **Riesz measure** of f is the nonnegative Borel measure μ_f on $\mathbb{B}_n$ satisfying

$$\int_{\mathbb{B}_n} \psi \, \mathrm{d}\mu_f = \int_{\mathbb{B}_n} f \, \tilde{\Delta} \psi \, \mathrm{d}\lambda$$

for all $\psi \in C_c^2(\mathbb{B}_n)$, the class of twice continuously differentiable functions on $\mathbb{B}_n$ with compact support. If f is C^2, then by Green's identity,

$$\mathrm{d}\mu_f = \tilde{\Delta} f \, \mathrm{d}\lambda.$$

The **Green's function** G for the operator $\tilde{\Delta}$ is given by $G(z,w) = g(\varphi_z(w))$, where

$$g(z) = c_n \int_{|z|}^{1} t^{-2n+1}(1-t^2)^{n-1} \, \mathrm{d}t.$$

The Green's function g satisfies

$$\text{(4.4)} \qquad g(z) \leq \begin{cases} C_n \, |z|^{-2n+2} & \text{for all } z \in \mathbb{B}_n, \\ C_{n,\delta}(1-|z|^2)^n & \text{for all } z \in \mathbb{B}_n, \ |z| \geq \delta \ (0 < \delta < 1). \end{cases}$$

For a Borel measure μ on $\mathbb{B}_n$, the **Green potential** G_μ of μ is defined by

$$G_\mu(z) = \int_{\mathbb{B}_n} G(z,w) \, \mathrm{d}\mu(w), \quad z \in \mathbb{B}_n.$$

If the measure μ satisfies

$$\text{(4.5)} \qquad \int_{\mathbb{B}_n} (1-|z|^2)^n \, \mathrm{d}\mu < \infty,$$

then G_μ is a nonnegative $\mathcal{M}$-superharmonic function on $\mathbb{B}_n$ which is finite a.e. on $\mathbb{B}_n$.

The **Riesz decomposition theorem** for $\mathcal{M}$-superharmonic functions is as follows: *If V is any nonnegative $\mathcal{M}$-superharmonic function on $\mathbb{B}_n$ satisfying*

$$\text{(4.6)} \qquad \lim_{r \to 1} \int_S V(rt) \, \mathrm{d}\sigma(t) = 0,$$

then V is the Green potential of a nonnegative measure μ on $\mathbb{B}$ satisfying (4.5). *If V is C^2 then* $\mathrm{d}\mu = -\tilde{\Delta} V \, \mathrm{d}\lambda$. The Riesz decomposition theorem for $\mathcal{M}$-subharmonic functions was originally proved by D. Ullrich in [26]. For proofs of the above the reader is also referred to [24].

The **invariant Dirichlet space** $\widetilde{\mathcal{D}}_\gamma^p$ is defined as follows: for $\gamma \in \mathbb{R}$, $0 < p < \infty$,

$$\widetilde{\mathcal{D}}_\gamma^p = \left\{ f \text{ holomorphic} : \widetilde{D}_\gamma^p(f) := \int_{\mathbb{B}_n} (1-|z|^2)^\gamma |\widetilde{\nabla} f(z)|^p d\lambda(z) < \infty \right\}.$$

For $f \in \widetilde{\mathcal{D}}_\gamma^p$ set

$$\|f\|_{\widetilde{\mathcal{D}}_\gamma^p} = |f(0)| + \big(\widetilde{D}_\gamma^p(f)\big)^{1/p}.$$

When $n = 1$ and f is holomorphic, $|\widetilde{\nabla} f(z)| = (1-|z|^2)|f'(z)|$. Thus $f \in \widetilde{\mathcal{D}}_\gamma^p$ if and only if

$$\int_{\mathbb{D}} (1-|z|^2)^{\gamma+p-2} |f'(z)|^p \, \mathrm{d}A(z),$$

where A is area measure on the unit disk $\mathbb{D}$. If f is holomorphic (or harmonic) on $\mathbb{D}$, then $|f'(z)|^p$ ($|\nabla f|^p$) is subharmonic for all $p > 0$. As a consequence, for

$p > 0$, $\widetilde{\mathcal{D}}^p_\gamma$ contains nonconstant harmonic or holomorphic functions if and only if $\gamma > 1 - p$. For $n \geq 2$ we have the following.

Theorem 4.1 ([25] **and** [8] **in the case $\lambda = 0$).** *For $n \geq 2$, $0 < p < \infty$, $\widetilde{\mathcal{D}}^p_\gamma$ contains nonconstant holomorphic functions if and only if $\gamma > (n - \frac{1}{2}p)$.*

PROOF [25, Theorem 3.3]. Without loss of generality we assume that $\partial_1 f(0) \neq 0$. For $\lambda \in \mathbb{D}$, set $s_\lambda = \sqrt{1 - |\lambda|^2}$. If $z' = (z_2, \ldots, z_n) \in \mathbb{B}_{n-1}$, then $(\lambda, s_\lambda z') \in \mathbb{B}_n$. From the definition of $\widetilde{\nabla} f$, one can show that

$$|\widetilde{\nabla} f(\lambda, s_\lambda z')| \geq (1 - |\lambda|^2)(1 - |z'|^2)|h_\lambda(z')|^2,$$

where

$$h_\lambda(z') = -s_\lambda \partial_1 f(\lambda, s_\lambda z') + \overline{\lambda} \sum_{j=2}^{n} z_j \partial_j f(\lambda, s_\lambda z').$$

Therefore

$$\int_S |\widetilde{\nabla} f(r\zeta)|^p \, \mathrm{d}\sigma(\zeta) \geq (1-r^2)^{p/2} \int_{\mathbb{D}} (1-|\lambda|^2)^{n-2+p/2} \left[\int_{S_{n-1}} |h_\lambda(r\zeta')|^p \, \mathrm{d}\sigma(\zeta') \right] \mathrm{d}A(\lambda).$$

Since h_λ is holomorphic on $\mathbb{B}_{n-1}$,

$$\int_{S_{n-1}} |h_\lambda(r\zeta')|^p \, \mathrm{d}\sigma(\zeta') \geq |h_\lambda(0')|^p = (1 - |\lambda|^2)^{p/2} |\partial_1 f(r\lambda, 0')|^p.$$

Finally, since $\lambda \to \partial f(r\lambda, 0')$ is holomorphic in $\mathbb{D}$,

$$\int_S |\widetilde{\nabla} f(r\zeta)|^p \, \mathrm{d}\sigma(\zeta) \geq c_n (1 - r^2)^{p/2} |\partial_1 f(0)|^2.$$

Therefore

$$\widetilde{D}^p_\gamma(f) \geq C |\partial_1 f(0)|^p \int_0^1 r^{2n-1}(1 - r^2)^{\gamma + p/2 - n - 1} \, \mathrm{d}r,$$

and the above integral is finite if and only if $\gamma > n - \frac{1}{2}p$.

Finally, if $f(z) = z_1$, then $\widetilde{D}^p_\gamma(f) < \infty$ for all $\gamma > n - \frac{1}{2}p$, proving that $\widetilde{\mathcal{D}}^p_\gamma$ contains a nonconstant holomorphic function. □

The proof used by Arazy, Fisher, Jansen, and Peetre in the case $\lambda = 0$ is based on the following lemma.

Lemma 4.2 ([8]). *Let V be a linear subspace of $\mathcal{H}(\mathbb{B}_n)$ such that*

(V1) *if $f \in V$ and $\psi \in \mathrm{Aut}(\mathbb{B}_n)$, then $f \circ \psi \in V$,*
(V2) *if $f \in V$, then $g(z) = \int_0^{2\pi} \mathrm{e}^{-\mathrm{i}\theta} f(\mathrm{e}^{\mathrm{i}\theta} z) \, \mathrm{d}\theta \in V$.*

Then either V contains only constant functions, or V contains the linear function z_1.

The proof that $\widetilde{\mathcal{D}}^p_\gamma$ satisfies (V1) is straightforward. The same methods as used in [8] can be used to show that $\widetilde{\mathcal{D}}^p_\gamma$ satisfies (V2).

5. Derivatives of Inner Functions

Lemma 5.1. *If $f\colon \mathbb{B}_n \to \mathbb{D}$ is holomorphic, then for all $z \in \mathbb{B}_n$,*

$$|\widetilde{\nabla} f(z)| \le (1 - |f(z)|^2).$$

PROOF. Fix $z \in \mathbb{B}_n$. By the $\mathcal{M}$-invariance of $\widetilde{\nabla}$,

$$|\widetilde{\nabla} f(z)| = \left|\widetilde{\nabla}\Big(f(\varphi_z(0))\Big)\right| = |\widetilde{\nabla}(f\circ\varphi_z)(0)| = |\partial(f\circ\varphi_z)(0)|.$$

Let $u = f(z)$ and consider

$$h(w) = \varphi_u \circ f \circ \varphi_z(w).$$

Then $h\colon \mathbb{B}_n \to \mathbb{D}$ with $h(0) = 0$. Thus $|\partial h(0)| \le 1$ ([23, Theorem 8.13]). But

$$|\partial h(0)|^2 = \sum_{i=1}^n \left|\frac{\partial h}{\partial w_i}(0)\right|^2 = \frac{1}{(1-|f(z)|^2)^2}\sum_{i=1}^n \left|\frac{\partial(f\circ\varphi_z)}{\partial w_i}(0)\right| = \frac{|\partial(f\circ\varphi_z)(0)|^2}{(1-|f(z)|^2)^2},$$

from which the conclusion now follows. □

For f an inner function on $\mathbb{B}_n$, $0 < q < \infty$ and $0 < p < 1$, set

$$N_{p,q}(f) = \int_{\mathbb{B}_n} (1-|z|^2)^{n/p-n-q-1}(1-|f(z)|^2)^q \,\mathrm{d}\nu(z).$$

For the proof of Theorem 5.3 we require the following lemma which follows as a consequence of Hardy's inequality. For completeness we include the details

Lemma 5.2. *For $q \ge 1$, $\alpha > 0$ and f a nonnegative function in $L^q([0,1])$,*

$$\int_0^1 (1-r)^{-\alpha-1}\left[\int_r^1 f(s)\,\mathrm{d}s\right]^q \mathrm{d}r \le C_{\alpha,q}\int_0^1 (1-r)^{q-\alpha-1} f^q(r)\,\mathrm{d}r,$$

where $C_{\alpha,q}$ is a constant depending only on α and q.

PROOF. By the change of variable $s = 1-t$

$$\int_r^1 f(s)\,\mathrm{d}s = \int_0^{1-r} f(1-t)\,\mathrm{d}t = \int_0^{1-r} g(t)\,\mathrm{d}t,$$

where $g(t) = f(1-t)$. Thus if we let $\rho = 1-r$,

$$\int_0^1 (1-r)^{-\alpha-1}\left[\int_r^1 f(s)\,\mathrm{d}s\right]^q \mathrm{d}r = \int_0^1 \rho^{-\alpha-1}\left[\int_0^\rho g(s)\,\mathrm{d}s\right]^q \mathrm{d}\rho.$$

Extend g to all of $[0,\infty)$ by setting $g(s) = 0$ for $s > 1$. Then

$$\int_0^1 \rho^{-\alpha-1}\left[\int_0^\rho g(s)\,\mathrm{d}s\right]^q \mathrm{d}\rho \le \int_0^\infty \rho^{-\alpha-1}\left[\int_0^\rho g(s)\,\mathrm{d}s\right]^q \mathrm{d}\rho$$

which by Hardy's inequality [17, 9.9.10]

$$\le C_{\alpha,q}\int_0^\infty \rho^{q-\alpha-1} g(\rho)^q\,\mathrm{d}\rho$$
$$= C_{\alpha,q}\int_0^1 \rho^{q-\alpha-1} g(\rho)^q\,\mathrm{d}\rho,$$

from which the result now follows. □

Theorem 5.3. *Let f be an inner function on $\mathbb{B}_n$. Then*

(a) *$\|Rf\|^q_{B^{p,q}} \le N_{p,q}(f)$ for all p, $0 < p < 1$ and all q, $0 < q < \infty$.*

Conversely,

(b) *for* $1 \le q < \infty$ *and* $n/(n+q) < p < 1$, $N_{p,q}(f) \le C_{p,q}\|Rf\|^q_{B^{p,q}}$.

PROOF. The conclusion (a) follows immediately from the fact that by Lemma 5.1 and inequality (4.2), for an inner function f we have

$$|Rf(z)| \le \frac{(1-|f(z)|^2)}{(1-|z|^2)}.$$

(b) For $\gamma > n$, $q \ge 1$, and $\zeta \in S$, set

$$I_\gamma(f,\zeta) = 2n\int_0^1 r^{2n-1}(1-r^2)^{\gamma-n-q-1}(1-|f(r\zeta)|^2)^q \, dr.$$

By the fundamental theorem of calculus, for a.e. $\zeta \in S$,

$$1-|f(t\zeta)|^2 \le 2|f(\zeta)-f(t\zeta)| \le 2\int_t^1 |\langle \partial f(s\zeta),\zeta\rangle \, ds \le 2\int_t^1 |\partial f(s\zeta)| \, ds.$$

Therefore, for a.e. $\zeta \in S$,

$$\begin{aligned} I_\gamma(f,\zeta) &\le \int_0^1 r^{2n-1}(1-r^2)^{\gamma-n-q-1}\left[\int_r^1 |\partial f(s\zeta)| \, ds\right]^q dr, \\ &\le C_{n,q}\int_0^1 (1-r)^{\gamma-n-q-1}\left[\int_r^1 |\partial f(s\zeta)| \, ds\right]^q dr, \end{aligned}$$

where $C_{n,q}$ is a constant depending on n and q. Thus by Lemma 5.2

$$I_\gamma(f,\zeta) \le C_{n,q,\gamma}\int_0^1 (1-r)^{\gamma-n-1}|\partial f(r\zeta)|^q \, dr,$$

provided $\gamma < n+q$. Integrating over S and using Fubini's theorem gives

$$\int_S I_\gamma(f,\zeta)\, d\sigma(\zeta) \le C_{n,q,\gamma}\int_0^1 (1-r)^{\gamma-n-1}\int_S |\partial f(r\zeta)|^q \, d\sigma(\zeta)\, dr.$$

Since $|\partial f(z)|$ is plurisubharmonic on $\mathbb{B}_n$, it is easily shown that for $\gamma > n$,

$$\int_0^1 (1-r)^{\gamma-n-1}\int_S |\partial f(r\zeta)|^q \, d\sigma(\zeta)\, dr \le C\int_{\mathbb{B}_n} (1-|z|^2)^{\gamma-n-1}|\partial f(z)|^q \, d\nu(z).$$

Finally, it is well known ([9]) that

$$\int_{\mathbb{B}_n} (1-|z|^2)^{\gamma-n-1}|\partial f(z)|^q \, d\nu(z) \le C\int_{\mathbb{B}_n} (1-|z|^2)^{\gamma-n-1}|Rf(z)|^q \, d\nu(z).$$

The conclusion now follows by taking $\gamma = n/p$. □

Theorem 5.4. *Let* f *be an inner function on* $\mathbb{B}_n$.

(a) *If* $n/(n+q) < p < 1$ *and* $q \ge 1$, *then for all* $r \ge q$,

$$\int_{\mathbb{B}_n} (1-|z|^2)^{n/p-q}|\widetilde{\nabla} f(z)|^r \, d\lambda(z) \le C\|Rf\|^q_{B^{p,q}},$$

where C *is a constant independent of* f.

(b) *If* $n/(n+q) < p < \min\{n/q, 1\}$, $q \ge 1$, *then for all* r, $0 < r \le 2$,

$$\|Rf\|^q_{B^{p,q}} \le C\int_{\mathbb{B}_n} (1-|z|^2)^{n/p-q}|\widetilde{\nabla} f(z)|^r \, d\lambda(z),$$

where C *is a constant independent of* f.

Remark 5.5. The special case $q = 1$ of the above theorem was proved by Matt Gamel in his dissertation [15]. See also [16] for a published version of the result.

Proof. (a) By Lemma 5.1,

$$\int_{\mathbb{B}_n} (1-|z|^2)^{n/p-q}|\widetilde{\nabla} f(z)|^r \, d\lambda(z) \le \int_{\mathbb{B}_n} (1-|z|^2)^{n/p-q-n-1}(1-|f(z)|^2)^r \, d\nu(z),$$

which since $r \ge q$,

$$\begin{aligned} &\le \int_{\mathbb{B}_n} (1-|z|^2)^{n/p-q-n-1}(1-|f(z)|^2)^q \, d\nu(z) \\ &\le C\|Rf\|^q_{B^{p,q}}. \end{aligned}$$

The last inequality follows by Theorem 5.3(b).

(b) For $0 < p < 1$,

$$\|Rf\|^q_{B^{p,q}} \le \int_{\mathbb{B}_n} (1-|z|^2)^{n/p-q-n-1}(1-|f(z)|^2)^q \, d\nu(z),$$

which for $q \ge 1$

$$\le \int_{\mathbb{B}_n} (1-|z|^2)^{n/p-q-n-1}(1-|f(z)|^2) \, d\nu(z).$$

Set $V(z) := 1 - |f(z)|^2$. Then V is $\mathcal{M}$-superharmonic with

$$\lim_{r\to 1^-} \int_S V(r\zeta)\, d\sigma(\zeta) = 0.$$

By the Riesz decomposition theorem, there exists a regular Borel measure $\mu = \mu_V$ on $\mathbb{B}_n$ such that

$$V(z) = \int_{\mathbb{B}_n} G(z,w) d\mu(w).$$

By the Fubini – Tonelli theorem,

$$\begin{aligned} &\int_{\mathbb{B}_n} (1-|z|^2)^{n/p-q-n-1}(1-|f(z)|^2) \, d\nu(z) \\ &\qquad = \int_{\mathbb{B}_n} (1-|z|^2)^{n/p-q} \int_{\mathbb{B}_n} G(z,w)\, d\mu(w)\, d\lambda(z) \\ &\qquad = \int_{\mathbb{B}_n}\int_{\mathbb{B}_n} (1-|z|^2)^{n/p-q} g\big(\varphi_w(z)\big)\, d\lambda(z)\, d\mu(w). \end{aligned}$$

By the invariance property of $d\lambda(z)$,

$$\begin{aligned} &\int_{\mathbb{B}_n} (1-|z|^2)^{n/p-q} g\big(\varphi_w(z)\big)\, d\lambda(z) \\ &\qquad = \int_{\mathbb{B}_n} (1-|\varphi_w(z)|^2)^{n/p-q} g(z)\, d\lambda(z) \\ &\qquad = (1-|w|^2)^{n/p-q} \int_{\mathbb{B}_n} \frac{(1-|z|^2)^{n/p-q}}{|1-\langle z,w\rangle|^{2(n/p-q)}} g(z)\, d\lambda(z) \\ &\qquad = (1-|w|^2)^{n/p-q} I(w), \end{aligned}$$

where

$$I(w) := \int_{\mathbb{B}_n \setminus B_{1/2}} + \int_{B_{1/2}} \frac{(1-|z|^2)^{n/p-q-n-1}}{|1-\langle z,w\rangle|^{2(n/p-q)}} g(z)\, d\nu(z) = I_1(w) + I_2(w).$$

Since $g(z) \le c_n(1-|z|^2)^n$ for $|z| \ge \frac{1}{2}$,

$$I_1(w) \le C \int_{\mathbb{B}_n} \frac{(1-|z|^2)^{n/p-q-1}}{|1-\langle z,w\rangle|^{2n/p-2q}} d\nu(z) = C \int_{\mathbb{B}_n} \frac{(1-|z|^2)^t}{|1-\langle z,w\rangle|^{n+1+t+c}}\, \mathrm{d}\nu(z),$$

where $t = n/p - q - 1$ and $c = n/p - n - q$. Since $p < n/q$, we have $t > -1$. Also, since $p > n/(n+q)$ we have $c < 0$. Thus by [23, Proposition 1.4.10], $I_1(w) \le C$ for all $w \in \mathbb{B}_n$.

Since $g(z) \le c_n|z|^{-2n+2}$ for all $z \in \mathbb{B}_n$,

$$\begin{aligned} I_2(w) &\le C_{n,q,p} \int_0^{1/2} r(1-r^2)^{n/p-q-n-1} \int_S \frac{1}{|1-r\langle \zeta,w\rangle|^{2n/p-2q}}\, \mathrm{d}\sigma(\zeta)\, \mathrm{d}r \\ &\le C_n \int_0^{1/2} r(1-r^2)^{n/p-q-n-1}\, \mathrm{d}r \le C. \end{aligned}$$

Consequently, $I(w) \le C$ for the given values of p, q. Thus

$$\|Rf\|^q_{B^{p,q}} \le C \int_{\mathbb{B}_n} (1-|w|^2)^{n/p-q}\, \mathrm{d}\mu(w) < \infty.$$

Since $|f(z)|^2$ is C^2 we have

$$\mathrm{d}\mu_V(z) = -\tilde{\Delta} V(z)\, \mathrm{d}\lambda(z).$$

Because f is holomorphic, a simple computation gives

$$\tilde{\Delta}|f(z)|^2 = 4|\widetilde{\nabla} f(z)|^2.$$

Consequently,

$$d\mu_V(z) = 4|\widetilde{\nabla} f(z)|^2 d\lambda(z).$$

Note also that by Lemma 5.1, $|\widetilde{\nabla} f(z)| < 1$. Thus for $0 < r \le 2$,

$$\begin{aligned} \int_{\mathbb{B}_n} (1-|w|^2)^{n/p-q}\, \mathrm{d}\mu(w) &= 4 \int_{\mathbb{B}_n} (1-|w|^2)^{n/p-q} |\widetilde{\nabla} f(w)|^2\, \mathrm{d}\lambda(w) \\ &= 4 \int_{\mathbb{B}_n} (1-|w|^2)^{n/p-q} |\widetilde{\nabla} f(w)|^r |\widetilde{\nabla} f(w)|^{2-r}\, \mathrm{d}\lambda(w) \\ &\le 4 \int_{\mathbb{B}_n} (1-|w|^2)^{n/p-q} |\widetilde{\nabla} f(w)|^r\, \mathrm{d}\lambda(w). \end{aligned}$$

This yields the desired result. □

Corollary 5.6. *Suppose $n \ge 2$, $q \ge 1$, and $n/(n+q) < p < 1$. If f is an inner function on $\mathbb{B}_n$ with $Rf \in B^{p,q}$ with $p \ge 2n/(2n+q)$, then f is a constant. In particular, if $Rf \in B^{2n/(2n+1)}$, then f is a constant.*

PROOF. By part (a) of Theorem 5.4, with $r = q$, if $Rf \in B^{p,q}$, $q \ge 1$, then $f \in \widetilde{\mathcal{D}}^q_{n/p-q}$. But by Theorem 4.1, if $n/p - q \le n - \frac{1}{2}q$, i.e., $p \ge 2n/(2n+q)$, then f is a constant. □

Another immediate consequence of Theorem 5.4 is the following:

Corollary 5.7. *Suppose $\frac{n}{n+q} < p < \min\{n/q, 1\}$ and f is an inner function on $\mathbb{B}_n$. Then*

$$Rf \in B^{p,q} \iff f \in \widetilde{\mathcal{D}}^r_{n/p-q} \quad \textit{for all } r, q \textit{ with } 1 \le q \le r \le 2.$$

The special case $q = 1$ of Theorem 5.4 gives the following.

Corollary 5.8 ([15,16]). *Suppose $n/(n+1) < p < 1$ and f is an inner function on $\mathbb{B}_n$. Then*

$$Rf \in B^p \iff f \in \widetilde{\mathcal{D}}^r_{n/p-1} \quad \textit{for all } r \textit{ with } 1 \le r \le 2.$$

Theorem 5.9 ([15, 16]). *Suppose f is an inner function on $\mathbb{B}_n$, $n \ge 1$, $1 \le q \le 2$, and $0 < (1-\alpha) < \frac{1}{2}q$. Then*

$$Rf \in B^{n/(n+\alpha)} \iff Rf \in A^q_{q+\alpha-2}.$$

Proof. Let $p = n/(n+\alpha)$ and $1 \le q \le 2$. Then by Corollary 5.8, $Rf \in B^{n/(n+\alpha)}$ if and only if $f \in \widetilde{\mathcal{D}}^q_{n+\alpha-1}$, i.e.,

$$(1-|z|^2)^{(\alpha-1)/q}|\widetilde{\nabla} f(z)| \in L^q\big((1-|z|^2)^{-1}\,d\nu(z)\big).$$

By Remark 5.2 of [8], if $(1-\alpha) < \frac{1}{2}q$, then this is equivalent to

$$(1-|z|^2)^{(q+\alpha-1)/q}|Rf(z)| \in L^q\big((1-|z|^2)^{-1}\,d\nu(z)\big),$$

i.e., $Rf \in A^q_{q+\alpha-2}$. □

In the following corollary we indicate some special results that are consequences of the previous theorem.

Corollary 5.10 ([15,16]). *Let f be an inner function on $\mathbb{B}_n$, $n \ge 1$. Then the following hold.*

(1) $Rf \in B^{n/(n+\alpha)} \Leftrightarrow Rf \in A^2_\alpha$, $0 < \alpha < 1$.
(2) *If $n \ge 2$ and $Rf \in A^2_{1/2}$, then f is constant.*
(3) *If $n = 1$ and $Rf \in A^2_{1/2}$, then f is a Blaschke product.*
(4) *If $1 < p < 2$, then $Rf \in A^p$ if and only if $Rf \in B^{n/(n+2-p)}$.*
(5) *If $n \ge 2$ and $Rf \in A^{3/2}$, then f is constant.*
(6) *If $n = 1$ and $f' \in A^{3/2}$, then f is a Blaschke product.*

Proof. The result (1) follows immediately by taking $q = 2$. With $q = 2$ and $\alpha = \frac{1}{2}$ we have that $Rf \in A^2_{1/2}$ implies that $Rf \in B^{2n/(2n+1)}$, which by Corollary 5.6 implies that for $n \ge 2$ the function f is a constant. When $n = 1$, $Rf \in A^2_{1/2}$ implies that $Rf \in B^{2/3}$, which by a result of P. Ahern and D. Clark [4] implies that f is a Blaschke product.

The result (4) follows from Theorem 5.9 by taking $\alpha = 2-p$. Then for $1 < p < 2$ we have $0 < (1-p) < \frac{1}{2}p$. With $p = \frac{3}{2}$ we now have that $Rf \in A^{3/2}$ if and only if $Rf \in B^{2n/(2n+1)}$ from which (5) and (6) now follow. □

Remark 5.11. (a) The results (2) and (5) are an improvement on Proposition 3.1 previously proved by E. Bedford and A. Taylor in [10]. The results (3) and (6) are new in the case $n = 1$, whereas (5) is new for $n \ge 2$.

(b) With regard to the H^p spaces, in [15, 16] Matt Gamel proved that if f is an inner function on $\mathbb{B}_n$, $n \ge 2$, with $Rf \in H^{\frac{1}{2}}$, then f is a constant.

In view of the results proved to date, there remain numerous open questions concerning derivatives of inner functions on $\mathbb{B}_n$, $n \ge 2$. The most obvious of these are as follows.

Open questions. Does there exist an inner function f on $\mathbb{B}_n$, $n \ge 2$, with

(1) $Rf \in B^p$ for some p, $n/(n+1) < p < 2n/(2n+1)$?

(2) $Rf \in A^p$ for some p, $1 < p < \frac{3}{2}$? In view of (4) of Corollary 5.10, we have that (1) and (2) are equivalent.
(3) $Rf \in H^p$ for some p, $0 < p < \frac{1}{2}$?
(4) $Rf \in N$, the Nevanlinna class?

Remark 5.12. When $n > 1$ it is easy to find bounded holomorphic functions F on $\mathbb{B}_n$ with $|F(z)| \le 1$ for all $z \in \mathbb{B}_n$ for which RF is in either H^p, $B^{p,q}$, or A^q_α. For example, if we set

$$\Pi(z) = n^{n/2} z_1 \cdots z_n,$$

then by the geometric-arithmetic mean inequality $|\Pi(z)| \le |z|^n$ with $\Pi(\mathbb{B}_n) = \mathbb{D}$ and $\Pi(\overline{\mathbb{B}}_n) = \overline{\mathbb{D}}$ [2]. Clearly $R\Pi(z) = n\Pi(z)$ and thus $R\Pi$ is in H^p for all p and A^q_α for all q, $0 < q < \infty$ and all $\alpha > -1$. More complicated examples can be obtained as follows. If f is any holomorphic function on $\mathbb{D}$ with $|f(z)| \le 1$ for all $z \in \mathbb{D}$, setting $F(z) = f\big(\Pi(z)\big)$ gives a holomorphic function on $\mathbb{B}_n$ with $|F(z)| \le 1$ for all $z \in \mathbb{B}_n$. Furthermore, $RF(z) = n\Pi(z) f'\big(\Pi(z)\big)$. Thus for $n \ge 2$,

$$\int_S |Rf(r\zeta)|^p \, \mathrm{d}\sigma(\zeta) \le n^p r^{np} \int_S |f'(r^n \Pi(\zeta)|^p \, \mathrm{d}\sigma(\zeta)$$

which by [2, Theorem 1]

$$\le c_n n^p \int_0^1 (1-\rho)^{(n-3)/2} \int_0^{2\pi} \big|f'\big(r^n \rho \mathrm{e}^{\mathrm{i}\theta}\big)\big|^p \, \mathrm{d}\theta \, \mathrm{d}\rho.$$

Finally, since $|f'|^p$ is subharmonic on $\mathbb{D}$, we have

$$\int_S |RF(r\zeta)|^p \, \mathrm{d}\sigma(\zeta) \le C_{n,p} \int_0^1 (1-\rho)^{(n-3)/2} \int_0^{2\pi} \big|f'\big(\rho \mathrm{e}^{\mathrm{i}\theta}\big)\big|^p \, \mathrm{d}\theta \, \mathrm{d}\rho,$$

for all r, $0 < r < 1$. Thus by an appropriate choice of f we can obtain $RF \in H^p$. Variations of this can also be used to obtain functions mapping $\mathbb{B}_n$ into or onto $\mathbb{D}$ for which $RF \in B^{p,q}$ or A^q_α. These examples also show that inner is required for many of the above results when $n \ge 2$.

References

1. P. Ahern, *The mean modulus and the derivative of an inner function*, Indiana Univ. Math. J. **28** (1979), no. 2, 311–347.
2. ———, *On the behavior near a torus of functions holomorphic in the ball*, Pacific J. Math. **107** (1983), no. 2, 267–278.
3. P. R. Ahern and D. N. Clark, *On inner functions with H^p-derivative*, Michigan Math. J. **21** (1974), 115–127.
4. ———, *On inner functions with B^p derivative*, Michigan Math. J. **23** (1976), no. 2, 107–118.
5. A. B. Aleksandrov, *The existence of inner functions in a ball*, Mat. Sb. (N.S.) **118(160)** (1982), no. 2, 147–163, 287 (Russian); English transl., Math. USSR-Sb. **46** (1983), 1143–159.
6. ———, *Function theory in the ball*, Current Problems in Mathematics. Fundamental Directions, Vol. 8 (R. V. Gamkrelidze, ed.), Itogi Nauki i Tekhniki, Akad. Nauk SSSR Vsesoyuz. Inst. Nauchn. i Tekhn. Inform., Moscow, 1985, pp. 115–190, 274 (Russian); English transl., Several Complex Variables. II, Encyclopaedia Math. Sci., vol. 8, Springer, Berlin, 1994, pp. 107–178.
7. H. A. Allen and C. L. Belna, *Singular inner functions with derivative in B^p*, Michigan Math. J. **19** (1972), 185–188.
8. J. Arazy, S. D. Fisher, S. Janson, and J. Peetre, *Membership of Hankel operators on the ball in unitary ideals*, J. London Math. Soc. (2) **43** (1991), no. 3, 485–508.
9. F. Beatrous and J. Burbea, *Holomorphic Sobolev spaces on the ball*, Dissertationes Math. (Rozprawy Mat.) **276** (1989), 60.

10. E. Bedford and B. A. Taylor, *Two applications of a nonlinear integral formula to analytic functions*, Indiana Univ. Math. J. **29** (1980), no. 3, 463–465.
11. W. Y. Chen, *Lipschitz spaces of holomorphic functions over bounded symmetric domains in* $\mathbf{C}^N$, J. Math. Anal. Appl. **81** (1981), no. 1, 63–87.
12. M. R. Cullen, *Derivatives of singular inner functions*, Michigan Math. J. **18** (1971), 283–287.
13. Y. Dupain, *Gradients des fonctions intérieures dans la boule unité de* $\mathbf{C}^n$, Math. Z. **193** (1986), no. 1, 85–94.
14. P. L. Duren, B. W. Romberg, and A. L. Shields, *Linear functionals on* H^p *spaces with* $0 < p < 1$, J. Reine Angew. Math. **238** (1969), 32–60.
15. M. R. Gamel, *On the integrability of derivatives of holomorphic and* $\mathcal{M}$*-subharmonic inner functions in the unit ball of* $\mathbb{C}^n$, Ph.D. Dissertation, University of South Carolina, 2011.
16. ———, *On inner functions with* H^p *and* B^p *derivatives in the unit ball of* $\mathbb{C}^n$, J. Math. Anal. Appl. **378** (2011), no. 1, 295–305.
17. G. H. Hardy, J. E. Littlewood, and G. Pólya, *Inequalities*, 2nd ed., Cambridge Univ. Press, Cambridge, 1959.
18. E. Løw, *A construction of inner functions on the unit ball in* $\mathbf{C}^p$, Invent. Math. **67** (1982), no. 2, 223–229.
19. J. Mitchell and K. T. Hahn, *Representation of linear functionals in* H^p *spaces over bounded symmetric domains in* C^N, J. Math. Anal. Appl. **56** (1976), no. 2, 379–396.
20. D. J. Newman and H. S. Shapiro, *The Taylor coefficients of inner functions*, Michigan Math. J. **9** (1962), 249–255.
21. D. Protas, *Blaschke products with derivative in* H^p *and* B^p, Michigan Math. J. **20** (1973), 393–396.
22. W. Rudin, *The radial variation of analytic functions*, Duke Math. J. **22** (1955), 235–242.
23. ———, *Function theory in the unit ball of* $\mathbf{C}^n$, Grundlehren Math. Wiss., vol. 241, Springer, New York, 1980.
24. M. Stoll, *Invariant potential theory in the unit ball of* $\mathbf{C}^n$, London Math. Soc. Lecture Note Ser., vol. 199, Cambridge Univ. Press, Cambridge, 1994.
25. ———, *On the integrability of eigenfunctions of the Laplace – Beltrami operator in the unit ball of* $\mathbb{C}^n$, Potential Anal. **16** (2002), no. 3, 205–220.
26. D. Ullrich, *Radial limits of* M*-subharmonic functions*, Trans. Amer. Math. Soc. **292** (1985), no. 2, 501–518.

Department of Mathematics, University of South Carolina, Columbia, SC 29208, USA

E-mail address: stoll@math.sc.edu

Centre de Recherches Mathématiques
CRM Proceedings and Lecture Notes
Volume **55**, 2012

Logarithmic Hölder Estimates of p-Harmonic Extension Operators in a Metric Measure Space

Tsubasa Itoh

Abstract. Let $1 < p < \infty$ and let X be a metric measure space with a doubling measure and a $(1,p)$-Poincaré inequality. Let Ω be a bounded domain in X. For a function f on $\partial\Omega$ we denote by $\mathcal{P}_\Omega f$ the p-harmonic extension of f over Ω. It is well known that if Ω is p-regular and $f \in C(\partial\Omega)$, then $\mathcal{P}_\Omega f$ is continuous in $\overline{\Omega}$. We characterize the family of domains such that logarithmic Hölder continuity of boundary functions f ensures logarithmic Hölder continuity of $\mathcal{P}_\Omega f$.

1. Introduction

Let $X = (X, d, \mu)$ be a complete connected metric measure space endowed with a metric d and a positive complete Borel measure μ such that $0 < \mu(U) < \infty$ for all nonempty bounded open sets U.

By the symbol C we denote an absolute positive constant whose value is unimportant and may change from line to line. Let $B(x,r) = \{y \in X : d(x,y) < r\}$ denote the open ball centered at x with radius r. We assume that μ is doubling, i.e., there is a constant $C \geq 1$ such that $\mu(B(x,2r)) \leq C\mu(B(x,r))$ for every $x \in X$ and $r > 0$. Let $1 < p < \infty$. We assume that X admits a $(1,p)$-Poincaré inequality (see [5]).

We denote by Cap_p the p-capacity defined on X (see [2]). Let $\Omega \subset X$ be a bounded domain with $\mathrm{Cap}_p(X \setminus \Omega) > 0$. For a function f on $\partial\Omega$ we denote by $\mathcal{P}_\Omega f$ the p-Perron solution of f over Ω (see [3]). A point $\xi \in \partial\Omega$ is said to be a *p-regular point* (with respect to the p-Dirichlet problem) if

$$\lim_{\Omega \ni x \to \xi} \mathcal{P}_\Omega f(x) = f(\xi)$$

for every $f \in C(\partial\Omega)$. If every boundary point is a p-regular point, then Ω is called *p-regular.* It is well known that if Ω is p-regular and $f \in C(\partial\Omega)$, then $\mathcal{P}_\Omega f$ is p-harmonic in Ω and continuous in $\overline{\Omega}$. It is natural to raise the following question:

2010 *Mathematics Subject Classification.* 31E05, 31C45, 35J60.

Key words and phrases. Modulus of continuity, p-harmonic, p-Dirichlet solution, metric measure space, p-capacity.

This is the final form of the paper.

Question 1.1. Does improved continuity of a boundary function f guarantee improved continuity of $\mathcal{P}_\Omega f$?

Aikawa and Shanmugalingam [2] studied this question in the context of Hölder continuity. Aikawa [1] investigated this question in the context of general modulus of continuity for the classical setting, i.e., for harmonic functions in a Euclidean domain. The purpose of this paper is to study this question in the context of logarithmic Hölder continuity in a metric measure space.

We consider the function $\psi_{\alpha\beta}$ defined by

$$\psi_{\alpha\beta}(t) = \begin{cases} t^\alpha(-\log t)^{-\beta} & \text{for } 0 < t < t_0, \\ t_0^\alpha(-\log t_0)^{-\beta} & \text{for } t \geq t_0. \end{cases}$$

where either $0 < \alpha < 1$ and $\beta \in \mathbb{R}$ or $\alpha = 0$ and $\beta > 0$; and t_0 is so small that $\psi_{\alpha\beta}$ is concave. Let E be a subset of X and let f be a function on X. We say that f is $\psi_{\alpha\beta}$-Hölder continuous if $|f(x) - f(y)| \leq C\psi_{\alpha\beta}(d(x,y))$ for $x, y \in E$. If f is $\psi_{\alpha 0}$-Hölder continuous, then f is α-Hölder continuous in the classical sense. If f is $\psi_{0\beta}$-Hölder continuous, then f is considered to be logarithmic Hölder continuous. In general, $\psi_{\alpha\beta}$-Hölder continuity is a mixture of Hölder continuity and logarithmic Hölder continuity.

Let E be a subset in X. We consider the family $\Lambda_{\psi_{\alpha\beta}}(E)$ of all bounded continuous functions f on E with norm

$$\|f\|_{\psi_{\alpha\beta},E} = \sup_{x\in E}|f(x)| + \sup_{\substack{x,y\in E\\ x\neq y}} \frac{|f(x) - f(y)|}{\psi_{\alpha\beta}(d(x,y))} < \infty.$$

We define the operator norm

$$\|\mathcal{P}_\Omega\|_{\psi_{\alpha\beta}} = \sup_{\substack{f\in\Lambda_{\psi_{\alpha\beta}}(\partial\Omega)\\ \|f\|_{\psi_{\alpha\beta},\partial\Omega}\neq 0}} \frac{\|\mathcal{P}_\Omega f\|_{\psi_{\alpha\beta},\Omega}}{\|f\|_{\psi_{\alpha\beta},\partial\Omega}}.$$

Observe that logarithmic Hölder-continuity of a boundary function f ensures logarithmic Hölder-continuity of $\mathcal{P}_\Omega f$ if and only if $\|\mathcal{P}_\Omega\|_{\psi_{\alpha\beta}} < \infty$. Therefore we characterize the family of domains Ω for which $\|\mathcal{P}_\Omega\|_{\psi_{\alpha\beta}} < \infty$.

In this paper, we state the results obtained in [6]. In Section 2, we give the characterizations of the family of domains Ω for which $\|\mathcal{P}_\Omega\|_{\psi_{\alpha\beta}} < \infty$. In Section 3, we characterize the family of domains such that improved continuity of a boundary function f ensures improved continuity of $\mathcal{P}_\Omega f$ in the context of general modulus of continuity. See [6] for their proofs.

2. Results

In this section, we give some characterizations of the family of domains Ω for which $\|\mathcal{P}_\Omega\|_{\psi_{\alpha\beta}} < \infty$.

The interior regularity of p-harmonic functions is known.

Theorem A ([8, Theorem 5.2]). *There exists $\alpha_0 \in (0,1]$ depending only on X and p such that if u is a p-harmonic function in a domain Ω in X, then u is α_0-Hölder continuous on every compact subset of Ω.*

Aikawa [1] estimated Dirichlet solutions by the Poisson integral representation of harmonic functions on balls. Since we are dealing with (nonlinear) p-harmonic functions, we do not have the Poisson integral representation. We instead use the local Hölder continuity of p-harmonic functions, so that we restrict ourselves to $\alpha < \alpha_0$.

Let U be an open set in X and let E be a Borel set in ∂U. We denote by $\omega_p(x, E, U)$ the p-harmonic measure evaluated at x of E in U (see [2]). Note that the p-harmonic *measure* is not a measure, i.e., the p-harmonic measure is not additive. We define two decay properties for p-harmonic measures. We say that Ω enjoys the *Local Harmonic Measure Decay property* with $\psi_{\alpha\beta}$ (abbreviated to the LHMD($\psi_{\alpha\beta}$) property) if there are positive constants C and r_0 depending only on Ω and $\psi_{\alpha\beta}$ such that

$$(2.1)\qquad \omega_p\big(x, \Omega \cap \partial B(a,r), \Omega \cap B(a,r)\big) \le C\frac{\psi_{\alpha\beta}\big(d(x,a)\big)}{\psi_{\alpha\beta}(r)} \quad \text{for } x \in \Omega \cap B(a,r),$$

whenever a $\in \partial\Omega$ and $0 < r < r_0$. We say that Ω enjoys the *Global Harmonic Measure Decay property* with $\psi_{\alpha\beta}$ (abbreviated to the GHMD($\psi_{\alpha\beta}$) property) if there are positive constants C and r_0 depending only on Ω and $\psi_{\alpha\beta}$ such that

$$(2.2)\qquad \omega_p(x, \partial\Omega \setminus B(a,r), \Omega) \le C\frac{\psi_{\alpha\beta}\big(d(x,a)\big)}{\psi_{\alpha\beta}(r)} \quad \text{for } x \in \Omega \cap B(a,r),$$

whenever a $\in \partial\Omega$ and $0 < r < r_0$. By the comparison principle (see [7, Theorem 7.2]) it is easy to see that (2.1) implies (2.2).

Without loss of generality, we may assume that Ω is a bounded p-regular domain (see [2, Proposition 2.1]). For $a \in \partial\Omega$ we define a test function $\tau_{a,\psi_{\alpha\beta}}$ on $\partial\Omega$ by

$$\tau_{a,\psi_{\alpha\beta}}(\xi) = \psi_{\alpha\beta}\big(d(a,\xi)\big) \quad \text{for } \xi \in \partial\Omega.$$

Then we have the following theorem.

Theorem 2.1. *Let Ω be a bounded p-regular domain. Suppose that α and β satisfy either $0 < \alpha < \alpha_0$ and $\beta \in \mathbb{R}$ or $\alpha = 0$ and $\beta > 0$, where α_0 is as in Theorem* A. *Consider the following four conditions*:

(i) $\|\mathcal{P}_\Omega\|_{\psi_{\alpha\beta}} < \infty$.

(ii) *There is a constant C such that*

$$\mathcal{P}_\Omega \tau_{a,\psi_{\alpha\beta}}(x) \le C\psi_{\alpha\beta}\big(d(x,a)\big) \quad \textit{for } x \in \Omega,$$

whenever $a \in \partial\Omega$.

(iii) *Ω has the* GHMD($\psi_{\alpha\beta}$) *property.*

(iv) *Ω has the* LHMD($\psi_{\alpha\beta}$) *property.*

Then we have

$$\text{(i)} \iff \text{(ii)} \implies \text{(iii)} \impliedby \text{(iv)}.$$

Moreover, if $\alpha > 0$ and if X is Ahlfors Q-regular, i.e.,

$$C^{-1}r^Q \le \mu(B(x,r)) \le Cr^Q \quad \textit{for every } B(x,r),$$

then (iii) $\Leftrightarrow$ (iv).

See [6, Theorems 1.2 and 4.1] for the proof of this theorem. Aikawa and Shanmugalingam [2] showed the case $\beta = 0$ of Theorem 2.1.

The implication (iv) $\Rightarrow$ (i) with the same exponent α and β does not necessarily hold in above theorem (see [2, Remark 2.4]). However, we obtain the following theorem.

Theorem 2.2 ([6, Theorem 1.3]). *Let Ω be a bounded p-regular domain. Consider the following conditions:*

(i) $0 < \alpha < \alpha' < \alpha_0$ *and* $\beta, \beta' \in \mathbb{R}$.
(ii) $0 = \alpha < \alpha' < \alpha_0$ *and* $\beta > 0$, $\beta' \in \mathbb{R}$.
(iii) $\alpha = \alpha' = 0$ *and* $0 < \beta < \beta'$.

Assume that either (i), (ii), *or* (iii) *holds. If Ω has the* LHMD$(\psi_{\alpha'\beta'})$ *property, then* $\|\mathcal{P}_\Omega\|_{\psi_{\alpha\beta}} < \infty$.

PROOF. Let $a \in \partial\Omega$ and $u = \mathcal{P}_\Omega \tau_{a,\psi_{\alpha\beta}}$. We will show (ii) in Theorem 2.1 holds, i.e., $u(x) \le C\psi_{\alpha\beta}\big(d(x,a)\big)$. For $\rho > 0$ we define a function $f(\rho)$ by

$$f(\rho) = \sup_{\Omega\cap S(a,\rho)} u(x).$$

It is sufficient to show that

$$f(\rho) \le C\psi_{\alpha\beta}(\rho) \tag{2.3}$$

for small $\rho > 0$.

Let $0 < \rho < r < \operatorname{diam}(\Omega)$. By definition of $\tau_{a,\psi_{\alpha\beta}}$ and the comparison principle yields

$$u(x) \le \psi_{\alpha\beta}(r) + f(r)\omega_p(x, \Omega \cap \partial B(a,r), \Omega \cap B(a,r))$$

for all $x \in \Omega \cap B(a,r)$. Let $\psi(r) = \psi_{\alpha'\beta'}(r)/\psi_{\alpha\beta}(r)$. Since Ω has the LHMD$(\psi_{\alpha'\beta'})$ property,

$$f(\rho) \le \psi_{\alpha\beta}(r) + C_1 f(r)\frac{\psi_{\alpha'\beta'}(\rho)}{\psi_{\alpha'\beta'}(r)} \le \psi_{\alpha\beta}(r) + C_1 f(r)\frac{\psi_{\alpha\beta}(\rho)}{\psi_{\alpha\beta}(r)}\frac{\psi(\rho)}{\psi(r)} \tag{2.4}$$

for some C_1. Let r_0 be a small positive number. Since

$$\psi(r) = \psi_{\alpha''\beta''} \quad \text{for } 0 < r < r_0$$

where $\alpha'' = \alpha' - \alpha$ and $\beta'' = \beta' - \beta$, we have

$$M = \sup_{0<\rho<r\le r_0} \left\{ \frac{\psi_{\alpha\beta}(r)}{\psi_{\alpha\beta}(\rho)} : \frac{\psi(\rho)}{\psi(r)} = \frac{1}{2C_1} \right\} < \infty. \tag{2.5}$$

Let $0 < r < r_0$. Then we can find a sequence $\{\rho_j\}_{j=0}^n$ such that $r = \rho_0 < \rho_1 < \cdots < \rho_{n-1} < r_0 \le \rho_n < \operatorname{diam}(\Omega)$,

$$\frac{\psi(\rho_j)}{\psi(\rho_{j+1})} = \frac{1}{2C_1} \quad \text{for } j = 0, 1, \dots, n-1,$$

and

$$\frac{\psi_{\alpha\beta}(\rho_{j+1})}{\psi_{\alpha\beta}(\rho_j)} \le M \quad \text{for } j = 0, 1, \dots, n-1.$$

Hence, by (2.4) we obtain

$$f(\rho_j) \le \psi_{\alpha\beta}(\rho_{j+1}) + \frac{1}{2} f(\rho_{j+1}) \frac{\psi_{\alpha\beta}(\rho_j)}{\psi_{\alpha\beta}(\rho_{j+1})} \quad \text{for } j = 0, 1, \dots, n-1.$$

These inequalities imply that

$$\begin{aligned}f(r) = f(\rho_0) &\le \psi_{\alpha\beta}(\rho_1) + \psi_{\alpha\beta}(\rho_0)\sum_{j=1}^{n-1}\frac{1}{2^j}\frac{\psi_{\alpha\beta}(\rho_{j+1})}{\psi_{\alpha\beta}(\rho_j)} + \frac{1}{2^n}f(\rho_n)\frac{\psi_{\alpha\beta}(\rho_0)}{\psi_{\alpha\beta}(\rho_n)}\\ &\le M\psi_{\alpha\beta}(\rho_0) + M\psi_{\alpha\beta}(\rho_0)\sum_{j=1}^{n-1}\frac{1}{2^j} + f(\rho_n)\frac{\psi_{\alpha\beta}(\rho_0)}{\psi_{\alpha\beta}(\rho_n)}\\ &\le M\psi_{\alpha\beta}(\rho_0) + M\psi_{\alpha\beta}(\rho_0) + f(\rho_n)\frac{\psi_{\alpha\beta}(\rho_0)}{\psi_{\alpha\beta}(\rho_n)}\\ &\le (2M + \frac{\psi_{\alpha\beta}(\mathrm{diam}(\Omega))}{\psi_{\alpha\beta}(r_0)})\psi_{\alpha\beta}(r),\end{aligned}$$

where $f \le \psi_{\alpha\beta}(\mathrm{diam}(\Omega))$ and $r_0 \le \rho_n$ are used in the last inequality. Thus (2.3) follows, and so (ii) in Theorem 2.1 holds. Hence $\|\mathcal{P}_\Omega\|_{\psi_{\alpha\beta}} < \infty$ by Theorem 2.1. $\square$

We give more geometrical characterizations of domains Ω for which $\|\mathcal{P}_\Omega\|_{\psi_{\alpha\beta}} < \infty$. We say that $E \subset X$ is *uniformly p-fat* or satisfies the *p-capacity density condition* if there are constants $C > 0$ and $r_0 > 0$ such that

$$\frac{\mathrm{Cap}_p\big(E \cap B(a,r), B(a,2r)\big)}{\mathrm{Cap}_p\big(B(a,r), B(a,2r)\big)} \ge C, \tag{2.6}$$

whenever $a \in E$ and $0 < r < r_0$.

For $\alpha > 0$ we obtain the following theorem.

Theorem 2.3 ([6, Theorem 1.4]). *Let Ω be a bounded p-regular domain. If $X \setminus \Omega$ is uniformly p-fat, then there is a constant $0 < \alpha_1 \le \alpha_0$ such that $\|\mathcal{P}_\Omega\|_{\psi_{\alpha\beta}} < \infty$ for $0 < \alpha < \alpha_1$ and $\beta \in \mathbb{R}$. Conversely, if $\|\mathcal{P}_\Omega\|_{\psi_{\alpha\beta}} < \infty$ for some $0 < \alpha < \alpha_0$ and $\beta \in \mathbb{R}$, then $X \setminus \Omega$ is uniformly p-fat, provided that there is a constant $Q \ge p$ such that X is Ahlfors Q-regular.*

To prove Theorem 2.3 we observe the following lemma (see [2, Lemma 6.1]).

Lemma 2.4. *A domain Ω has the* LHMD(φ_{α_2}) *property for some $\alpha_2 > 0$ if and only if $X \setminus \Omega$ is uniformly p-fat.*

PROOF OF THEOREM 2.3. First suppose that $X \setminus \Omega$ is uniformly p-fat. It follows from Lemma 2.4 that there is a constant $\alpha_2 > 0$ such that Ω has the LHMD(φ_{α_2}) property. Let $\alpha_1 = \min\{\alpha_0, \alpha_2\}$. Then Ω has the LHMD(φ_{α_1}) property. Theorem 2.2 yields that $\|\mathcal{P}_\Omega\|_{\psi_{\alpha\beta}} < \infty$ for $0 < \alpha < \alpha_1$ and $\beta \in \mathbb{R}$.

Conversely, suppose that $\|\mathcal{P}_\Omega\|_{\psi_{\alpha\beta}} < \infty$ for some $0 < \alpha < \alpha_0$ and $\beta \in \mathbb{R}$. Assume that X is Ahlfors Q-regular with $Q \ge p$. By Theorem 2.1 Ω has the LHMD($\psi_{\alpha\beta}$) property. Let $0 < \alpha' < \alpha$. By Theorem 2.2 we obtain that $\|\mathcal{P}_\Omega\|_{\varphi_{\alpha'}} < \infty$. Theorem 2.1 implies that Ω has the LHMD($\varphi_{\alpha'}$) property. Lemma 2.4 yields that $X \setminus \Omega$ is uniformly p-fat. $\square$

Aikawa and Shanmugalingam [2] showed the case $\beta = 0$ of Theorem 2.3. Moreover, for $\alpha = 0$ we obtain the following theorem.

Theorem 2.5 ([6, Theorem 1.5]). *If $X \setminus \Omega$ is uniformly p-fat, then $\|\mathcal{P}_\Omega\|_{\psi_{0\beta}} < \infty$ for every $\beta > 0$.*

PROOF. Suppose that $X \setminus \Omega$ is uniformly p-fat. It follows from Lemma 2.4 that there is a constant $\alpha_2 > 0$ such that Ω has the LHMD(φ_{α_2}) property. Let $\alpha_1 = \min\{\alpha_0, \alpha_2\}$. Then Ω has the LHMD(φ_{α_1}) property. Theorem 2.2 yields that $\|\mathcal{P}_\Omega\|_{\psi_{0\beta}} < \infty$ for every $\beta > 0$. □

3. Modulus of continuity

In this section, we consider general modulus of continuity of p-Perron solutions $\mathcal{P}_\Omega f$. See [6] for the proof of theorems and the corollary in this section. Note that the case of $p = 2$ and Euclidean domains was studied by Aikawa [1].

Let $\mathcal{M}$ be the family of all positive nondecreasing concave functions ψ on $(0, \infty)$ with $\psi(0) = \lim_{t\to 0} \psi(t) = 0$. In particular, if either $0 < \alpha < 1$ and $\beta \in \mathbb{R}$ or $\alpha = 0$ and $\beta > 0$, then $\psi_{\alpha\beta} \in \mathcal{M}$. For $\psi \in \mathcal{M}$, we say that f is *ψ-Hölder continuous* if $|f(x) - f(y)| \le C\psi\big(d(x, y)\big)$. The modulus of continuity of a uniformly continuous function on any geodesic space is comparable to a certain concave function. See [4, Chapter 2, §6]. Therefore, we have only to check ψ-Hölder continuity for $\psi \in \mathcal{M}$ to study Question 1.1 in the context of modulus of continuity.

Next we define the operator norm for $\psi \in \mathcal{M}$. Let E be a subset in X. We consider the family $\Lambda_\psi(E)$ of all bounded continuous functions f on E with norm

$$\|f\|_{\psi,E} = \sup_{x\in E}|f(x)| + \sup_{\substack{x,y\in E\\ x\neq y}} \frac{|f(x) - f(y)|}{\psi\big(d(x, y)\big)} < \infty.$$

We define the operator norm

$$\|\mathcal{P}_\Omega\|_\psi = \sup_{\substack{f\in\Lambda_\psi(\partial\Omega)\\ \|f\|_{\psi,\partial\Omega}\neq 0}} \frac{\|\mathcal{P}_\Omega f\|_{\psi,\Omega}}{\|f\|_{\psi,\partial\Omega}}.$$

Let ψ, $\varphi \in \mathcal{M}$. We say that $\varphi \precsim \psi$ if there are $r_0 > 0$ and $C > 0$ such that

$$\frac{\varphi(s)}{\varphi(r)} \le C\frac{\psi(s)}{\psi(r)} \quad \text{for } 0 < s < r < r_0.$$

Let $\mathcal{M}_0$ be the family of all $\psi \in \mathcal{M}$ with $t^{\alpha_0} \precsim \psi(t)$, where α_0 is a positive constant such that every p-harmonic function in Ω is locally α_0-Hölder continuous in Ω as explained in Section 2 (see [8]). For example, if either $0 < \alpha < \alpha_0$ and $\beta \in \mathbb{R}$ or $\alpha = 0$ and $\beta > 0$, then $\psi_{\alpha\beta} \in \mathcal{M}_0$. But if $\alpha = \alpha_0$ and $\beta < 0$, then $\psi_{\alpha_0\beta} \notin \mathcal{M}_0$. Hence we see that $\mathcal{M}_0 \subsetneq \mathcal{M}$. We use the locally Hölder continuity of p-harmonic functions as Section 2, so that we restrict ourselves to the case $\psi \in \mathcal{M}_0$.

Let $\psi \in \mathcal{M}$. We say that Ω enjoys the *Local Harmonic Measure Decay property* with ψ (abbreviated to the LHMD(ψ) property) if there are positive constants C and r_0 depending only on Ω and ψ such that

$$\text{(3.1)} \qquad \omega_p(x, \Omega \cap \partial B(a, r), \Omega \cap B(a, r)) \le C\frac{\psi\big(d(x, a)\big)}{\psi(r)} \quad \text{for } x \in \Omega \cap B(a, r),$$

whenever a $\in \partial\Omega$ and $0 < r < r_0$. We say that Ω enjoys the *Global Harmonic Measure Decay property* with ψ (abbreviated to the GHMD(ψ) property) if there are positive constants C and r_0 depending only on Ω and ψ such that

$$\text{(3.2)} \qquad \omega_p(x, \partial\Omega \setminus B(a, r), \Omega) \le C\frac{\psi\big(d(x, a)\big)}{\psi(r)} \quad \text{for } x \in \Omega \cap B(a, r),$$

whenever a $\in \partial\Omega$ and $0 < r < r_0$.

For $a \in \partial\Omega$ we define a test function $\tau_{a,\psi}$ on $\partial\Omega$ by

$$\tau_{a,\psi}(\xi) = \psi\big(d(a,\xi)\big) \quad \text{for } \xi \in \partial\Omega.$$

Then we obtain the generalization of Theorem 2.1.

Theorem 3.1 ([6, Theorem 1.2]). *Let $\psi \in \mathcal{M}_0$ and let Ω be a bounded p-regular domain. Consider the following conditions:*

(i) $\|\mathcal{P}_\Omega\|_\psi < \infty$.

(ii) *There is a constant C such that*

$$\mathcal{P}_\Omega \tau_{a,\psi} \le C\psi\big(d(x,a)\big) \quad \textit{for } x \in \Omega,$$

whenever $a \in \partial\Omega$.

(iii) *Ω has the* GHMD(ψ) *property.*

(iv) *Ω has the* LHMD(ψ) *property.*

Then we have

$$\text{(i)} \iff \text{(ii)} \implies \text{(iii)} \impliedby \text{(iv)}.$$

Moreover, we have the generalizations of Theorem 2.2.

Theorem 3.2 ([6, Theorem 5.1]). *Let $\psi, \psi_1 \in \mathcal{M}_0$. Let $\psi_2 = \psi_1/\psi$. Suppose that $\lim_{r\to 0}\psi_2(r) = 0$ and there are constants $0 < C < 1$ and $r_0 > 0$ such that ψ_2 is increasing on $(0, r_0)$ and*

$$\sup_{0<\rho<r\le r_0} \left\{ \frac{\psi(r)}{\psi(\rho)} : \frac{\psi_2(\rho)}{\psi_2(r)} = C \right\} < \infty. \tag{3.3}$$

If Ω has the LHMD(ψ_1) *property, then $\|\mathcal{P}_\Omega\|_\psi < \infty$.*

Corollary 3.3 ([6, Corollary 5.2]). *Let $\psi, \psi_1 \in \mathcal{M}_0$. Let $\psi_2 = \psi_1/\psi$. Suppose that there are constants $0 < C < 1$ and $r_0 > 0$ such that ψ is increasing on $(0, r_0)$ and*

$$\inf_{0<r\le r_0} \frac{\psi_2(r)}{\psi_2(Cr)} > 1. \tag{3.4}$$

If Ω has the LHMD(ψ_1) *property, then $\|\mathcal{P}_\Omega\|_\psi < \infty$.*

References

1. H. Aikawa, *Modulus of continuity of the Dirichlet solutions*, Bull. Lond. Math. Soc. **42** (2010), no. 5, 857–867.
2. H. Aikawa and N. Shanmugalingam, *Hölder estimates of p-harmonic extension operators*, J. Differential Equations **220** (2006), no. 1, 18–45.
3. A. Björn, J. Björn, and N. Shanmugalingam, *The Perron method for p-harmonic functions in metric spaces*, J. Differential Equations **195** (2003), no. 2, 398–429.
4. R. A. DeVore and G. G. Lorentz, *Constructive approximation*, Grundlehren Math. Wiss., vol. 303, Springer, Berlin, 1993.
5. J. Heinonen and P. Koskela, *Quasiconformal maps in metric spaces with controlled geometry*, Acta Math. **181** (1998), no. 1, 1–61.
6. T. Itoh, *Modulus of continuity of p-Dirichlet solutions in a metric measure space*, preprint.
7. J. Kinnunen and O. Martio, *Nonlinear potential theory on metric spaces*, Illinois J. Math. **46** (2002), no. 3, 857–883.
8. J. Kinnunen and N. Shanmugalingam, *Regularity of quasi-minimizers on metric spaces*, Manuscripta Math. **105** (2001), no. 3, 401–423.

Department of Mathematics, Hokkaido University, Kita 10, Nishi 8, Kita-Ku, Sapporo, Hokkaido, 060-0810, Japan

E-mail address: tsubasa@math.sci.hokudai.ac.jp

Centre de Recherches Mathématiques
CRM Proceedings and Lecture Notes
Volume **55**, 2012

Meromorphic Approximation on Noncompact Riemann Surfaces

Nadya Askaripour and André Boivin

ABSTRACT. This paper is based on the work of G. Schmieder on holomorphic and meromorphic uniform approximation on closed subsets of noncompact Riemann surfaces. Some conditions have been added to the original (but incorrect) statement of Schmieder's extension theorem and its proof has been clarified. These added conditions slightly affect the definition of a class of closed sets introduced by Schmieder which he called "weakly of infinite genus." Following Schmieder, we then show that the sets in this "new" class are sets of meromorphic approximation. The class contains all the previously known sets of meromorphic approximation.

1. Introduction

A closed subset E of a noncompact Riemannn surface R is called a set of holomorphic (respectively meromorphic) approximation if every function f holomorphic on a neighborhood (depending on f) of E can be approximated uniformly on E by functions holomorphic (respectively meromorphic) on all of R. In this generality, a complete characterization of sets of approximation still eludes us, though many special cases are known, for example when $R = \mathbb{C}$, or when $E^0 = \varnothing$, or when E is compact (see [2, 4–7, 9, 12, 13, 19, 23–25, 28, 32] and the references therein). In 1984, G. Schmieder [28] introduced a new "larger" class of sets which he called "weakly of infinite genus" ("schwach von unendlichem Geschlecht" in German) and proved that they were sets of approximation.

Unfortunately, it was soon realized that the proof of Schmieder was incorrect, or at least that it contained some gaps. More precisely, it was realized that the so-called "extension theorem" [28, Satz 1] could not hold in the full generality that it was stated. This, in turn, left some doubt about how large this new class of sets really was.

In this paper, we have added some extra hypotheses to the original statement of the extension theorem of Schmieder and corrected and clarified its proof. We then

2010 *Mathematics Subject Classification.* 30F35, 46E15.
Key words and phrases. Riemann surfaces, approximation on closed sets, uniform approximation, meromorphic approximation.
The research of the second author is partially supported by a grant from NSERC of Canada.
This is the final form of the paper.

accordingly modified the definition of sets "weakly of infinite genus" (but kept the same terminology). Following the original paper of Schmieder, we then show that these sets are sets of meromorphic approximation. We also show that this "new" version of the class still includes all the previously known sets of meromorphic approximation and is, in general, larger.

We note that Schmieder also claimed results on sets of holomorphic approximation in [28, §5], but some of the statements and proofs in that section of his paper also contain errors and gaps. We hope to address these issues in a subsequent paper.

2. The extension theorem

2.1. A special exhaustion. Let R be a noncompact Riemann surface which in this paper we always assume to be connected. In [21], I. Richards has shown that every Riemann surface can be represented *topologically* as a sphere with handles attached and points removed on the equator. As part of this representation of R, we have a domain $\widetilde{R}$ of the *Riemann sphere* with pairs $\{\mathcal{D}_j^+, \mathcal{D}_j^-\}$ of open disjoint discs removed. The surface R is obtained from $\widetilde{R}$ by identifying the points of the two boundary circles $\{C_j^+, C_j^-\}$. For a more precise statement, see [21]. If we now let $\pi\colon \widetilde{R} \to R$ be the natural projection onto this quotient space, then each $\pi(C_j^+ \cup C_j^-)$ is a single "*handle circle*" C_j in R to which we associate a *handle* in R. We say that a handle is contained in a set A if A contains the "handle circle" C associated with that handle.

Let E be a closed subset of R. From this topological representation of R, it follows easily that we can find subsets of R which we denote $G, D_0, D_1, D_2, \ldots$ and $W_1, W_2, \ldots$ with the following properties:

(1) The set G is an open and connected subset (that is, a *domain*) of R which contains E, and whose boundary consists of at most a countable number of Jordan curves and "open" Jordan arcs. See Definition 2.1 and Remark 2.2 below.

(2) Each compact subset of R meets at most a finite number of components of ∂G.

(3) All the handles contained in G are also contained in E.

(4) The family $\{D_n\}$ is a normal exhaustion of R by smoothly bounded domains [29, p. 141]. We also require that D_0 be *planar* (recall that a set M is planar if each simple closed curve that *locally separates* M also *separates* M). The sets D_n can easily be constructed by taking a suitable exhaustion $\{\widetilde{D}_n\}$ of the domain $\widetilde{R}$ mentioned above, and putting $D_n = \pi(\widetilde{D}_n)$. We moreover assume that if one of the circles C_j^+ or C_j^- mentioned above is contained in $\widetilde{D}_n$, then both of them are. That is, each handle of R is either contained in D_n or in its complement.

(5) For each $n \geq 0$, $G \setminus D_n$ has only a finite number of components.

(6) The W_j's ($j \geq 1$) are open connected neighborhoods around each handle contained in G. The W_j's are chosen such that their closure $\overline{W}_j$ are disjoint and also contained in G. It thus follows that $G \setminus \cup \overline{W}_j$ is planar. Without loss of generality we also assume that $W_j = \pi(\widetilde{R} \cap \widehat{W})$ where $\widehat{W}$ is a topological disc containing $C_j^+ \cup C_j^-$. We can further assume that the exhaustion $\{D_n\}$ has been chosen such that for each $n > 0$, $D_n \setminus \overline{D}_{n-1}$ contains exactly one of these neighborhoods, that is, for each n, $W_n \subset\subset D_n \setminus \overline{D}_{n-1}$. Note that this implies that *we are thus assuming* G (*and thus* E) *to be of infinite genus.* For more on this, see Remarks 2.25 and 5.2.

(7) We also assume that the boundary of each D_n does not intersect with the "holes" of G, (by a *hole* of G, we mean a component of $R \setminus G$ whose closure is compact in R).

Definition 2.1. By a *Jordan curve*, we mean, as usual, a simple closed curve in R; that is, the image of a continuous injective mapping of the circle $S^1 = \{z \in \mathbb{C} : |z| = 1\}$ into R. By an *open Jordan arc*, we mean the image of an injective continuous mapping of an open interval into R, or equivalently, but maybe more suggestively here, the homeomorphic image of the real line $-\infty < t < \infty$ into R, since the open Jordan arcs under consideration form the unbounded components of the boundary of G

Remark 2.2. Maybe the simplest way to construct a domain G as described above is as follows. Let Γ_j be the handle circles contained in E and C_j the handle circles not contained in the set E. For each such C_j, choose a point $p_j \in C_j \setminus E$ and let K_j be a closed parametric disc "centered" at p_j, contained in $R \setminus E$ and chosen inductively to be so small that $\{K_j\}$ is a disjoint locally finite family. Put $G = R \setminus \bigcup_j K_j$. We may moreover assume that each $\widetilde{D}_n$ (see Property 4) is disjoint from every $\widetilde{K}_j = \pi^{-1}(K_j)$, assuring that Property 7 above is satisfied.

This construction is provided to illustrate that it is not difficult to obtain the sets described above. It should be noted though that it leads to a set G that will be very large. In particular, as constructed, the complement of G will always fail to contain an unbounded open Jordan arc, a further property that we later require of our set G (and thus of E). See Lemma 2.8 and Section 2.4.

2.2. Stitching up. Let V be a domain of the extended plane $\overline{\mathbb{C}} = \mathbb{C} \cup \{\infty\}$ (i.e. the Riemann sphere) which contains ∞ in its interior and such that every component of the complement of V is either a point or a horizontal line segment. Such a region V will be called a horizontal slit domain. We recall the well-known Koebe's planarity theorem.

Theorem 2.3 (see [1, Chap. III, §4; 30, §9.1]). *Every (connected) planar Riemann surface can be mapped one-to-one and conformally onto a horizontal slit domain. Moreover, the area of the complement is zero.*

We say that a domain V in the extended plane is *n-connected* if $\overline{\mathbb{C}} \setminus V$ has n components. Thus the only 0-connected domain is $\overline{\mathbb{C}}$ itself and a n-connected domain is simply connected if and only if $n \leq 1$. We say that V is a *nondegenerate n-connected* domain if it is an n-connected domain with $n > 0$, and no component of its complement in $\overline{\mathbb{C}}$ is a point. Thus, according to this definition, $\mathbb{C}$ and $\overline{\mathbb{C}}$ are considered degenerate, but the unit disc is not.

A domain Ω contained in the Riemann sphere $\overline{\mathbb{C}}$ is called a circular domain if its boundary consists of a finite number of disjoint nondegenerate circles. It is known that every nondegenerate n-connected domain in $\overline{\mathbb{C}}$ is conformally equivalent to a circular domain bounded by n circles. Note also that if Ω and Λ are two circular domains and $f \colon \Omega \to \Lambda$ is a conformal equivalence, then f must be a Möbius transformation. For details, see [8, §15.7] but notice that the definition of n-connected given in [8, §15.1] differ slightly from the one given above.

Definition 2.4. A planar Riemann surface will be called *n-connected* if it is conformally equivalent to an n-connected domain and *nondegenerate* if it is conformally equivalent to a circular domain.

Let r be an noncompact planar Riemann surface which is n-connected and nondegenerate, and assume $n \geq 2$. It follows from the discussion above that r can be represented by a circular domain $\Omega = \overline{\mathbb{C}} \setminus \{K_1, \dots, K_n\}$ under a bijective meromorphic map f, where each closed disc K_i is contained in $\mathbb{C}$, $i = 1, \dots, n$

Let $K_1 = \{z \mid |z - z_1| \leq \rho_1\}$. Set r_1 to be $r \sqcup K_1$, and define charts on r_1 as follows. Let $p_0 \in r_1$. If $p_0 \in r \cap r_1$, we use the charts in r containing p_0. If $p_0 \in K_1^\circ \cap r_1$, we use a small neighborhood of p_0 in $\mathbb{C}$ (small enough to be contained in K_1°) and the identity map. If $p_0 \in \partial K_1 \cap r_1$, let U be a neighborhood of p_0 in $\mathbb{C}$, and define $U' := f^{-1}((U \setminus K_1) \cap \Omega) \sqcup (U \cap K_1)$ to be a neighborhood of p_0 in r_1. A chart is then defined as follows:

$$F \colon U' \to \mathbb{C}$$

$$p \longmapsto \begin{cases} f(p) & \text{if } p \in f^{-1}(U \setminus K_1 \cap \Omega) \\ p & \text{if } p \in U \cap K_1 \end{cases}$$

Following [28, §6.1], we now show that these charts are compatible with each other and with the original structure.

Suppose $p_0, p_1 \in \partial K_1$, let U_{p_0} and U_{p_1} be neighborhoods of p_0 and p_1 in r_1, and F_0 and F_1 be the related charts, as defined above.

Suppose $p \in U_{p_0} \cap U_{p_1}$. We then have either $p \in (U_{p_0} \cap U_{p_1}) \cap K_1$ or $p \in (U_{p_0} \cap U_{p_1}) \cap K_1^c$.

If $p \in (U_{p_0} \cap U_{p_1}) \cap K_1$, then $F_1 F_0^{-1}(p) = F_1(p) = p$, and if $p \in (U_{p_0} \cap U_{p_1}) \cap K_1^c$, then $F_1 F_0(p) = F_1(f^{-1}(p)) = f(f^{-1}(p)) = p$.

So, in either case, $F_1 {F_0}^{-1}(p) = p$, and the new charts are compatible with the original charts of r and of K_1 (as a subset of $\mathbb{C}$). Thus r_1 preserve the analytic structure of r and K_1.

If $n > 2$, the same process can be repeated for $K_2, \dots, K_{n-1}$, to obtain a noncompact simply connected Riemann surface $\tilde{r}$. We have thus obtained the following:

Lemma 2.5 (see [28, §6.1]**).** *Let r be an noncompact planar Riemann surface which is n-connected and nondegenerate, then there exists a noncompact simply connected Riemann surface $\tilde{r}$ which contains r and $\tilde{r}$ preserves the analytic structure of r.*

2.3. Constructing planar Riemann surfaces and simply connected domains.

Definition 2.6 (see [28]**).** The analytic structure of a sequence of Riemann surfaces R_n is said to *encompass* the analytic structure of R if, for each compact $K \subset R$, there exists $n_0 \in \mathbb{N}$ such that $K \subset R_n$, for every $n \geq n_0$. The inclusion indicates not only the inclusion of sets, but also that the analytic structures of K as a subset of R and of K as a subset of R_n are the same. We will use the notation $R_n\}_{n\to\infty} R$.

Note that the limit R is not unique in the above definition. For example if $R_n := \mathbb{C} \setminus [n, \infty)$, then $R_n\}_{n\to\infty} B$, for each domain $B \subset \mathbb{C}$.

The following was stated and used in [28] without proof.

Proposition 2.7. *Suppose $R_{n,k}\}_{k\to\infty} R_n$ and $R_n\}_{n\to\infty} R$, then there exists a sequence $k(n)$ such that $R_{n,k(n)}\}_{n\to\infty} R$.*

PROOF. We can write $R = \bigcup_{j=0}^{\infty} D_j$, where $\{D_j\}$ is an exhaustion of R by precompact domains.

Since $R_n\}_{n\to\infty}R$ there exists a strictly increasing sequence $\{N_j\}_{j=0}^{\infty}$, $N_j \in \mathbb{N}$, such that $\overline{D}_j \subset R_n$ whenever $n \geq N_j$. Because $\overline{D}_j \subset R_{N_j}$ and $R_{N_j,k}\}_{k\to\infty}R_{N_j}$, there exists $k(N_j) \in \mathbb{N}$ such that, $\overline{D}_j \subset R_{N_j,k}$ whenever $k \geq k(N_j)$. Similarly, for every $\ell > N_j$, there is a $k_j(\ell)$ such that $\overline{D}_j \subset R_{\ell,k}$ whenever $k \geq k_j(\ell)$. Define

$$r_\ell := \begin{cases} R_{\ell,1}, & \text{if } 1 \leq \ell < N_0; \\ R_{N_j,k(N_j)}, & \text{if } \ell = N_0, N_1, N_2, \ldots; \\ R_{\ell,k_j(\ell)}, & \text{if } N_j < \ell < N_{j+1}\ (j = 0,1,2,\ldots). \end{cases}$$

We now have r_ℓ defined for all $\ell \in \mathbb{N}$ and in the form required by the proposition. We claim that $r_\ell\}_{\ell\to\infty}R$.

Indeed suppose K is a compact subset of R. Since $\{D_j\}$ is a normal exhaustion of R, we have $K \subset D_j$, for j large enough, say $j \geq n_K$. By construction, $\overline{D}_j \subset r_\ell$ for every $\ell \geq N_j$, and thus $K \subset r_\ell$ for every $\ell \geq N_{n_K}$. This completes the proof. □

Lemma 2.8. *Let R be a noncompact Riemann surface, E be a closed subset of R, and G, D_n $(n \geq 0)$ and W_n $(n \geq 0)$ be subsets of R as described in Section 2.1. For each $n \geq 0$, let X_n denote the closure of D_n in R. Then there exists a sequence of Riemann surfaces R_n, with $X_n \subset R_n$ (and thus $R_n\}_{n\to\infty}R$), and a sequence of domains $\mathcal{G}_n \subset R_n$ whose boundary in R_n consist of at most a countable number of Jordan curves and open Jordan arcs and such that:*

(a) *$\mathcal{G}_n \setminus X_n$ is planar and $\mathcal{G}_n\}_{n\to\infty}G$.*
(b) *The set $E_n = E \setminus (\bigcup_{j>n} W_j)$. is a closed subset of R_n, and $\mathcal{G}_n \supset E_n$.*
(c) *If it is possible to find the domain G in such a way that $R\setminus\overline{G}$ also contains an unbounded open Jordan arc, then the sequence $\mathcal{G}_n$ can be chosen such that, for n sufficiently large, say $n \geq n_0$, no component of the complement of $\mathcal{G}_n \setminus X_n$* relative to *$R_n$ is bounded (i.e., compact).*

Remark 2.9. Note that we are assuming Property 6 from Section 2.1, and thus that G is of infinite genus. It should also be noted that our goal here is the construction of the sequences $\mathcal{G}_n \subset R_n$ satisfying (a), (b) and (c). This may explain our (and thus Schmieder's) slightly overcomplicated proof of (a).

PROOF. For ease of notation, we will assume $n \geq 0$ to be fixed and we will drop the dependence on n from the notation unless absolutely necessary. We let $G_n := G \setminus (\bigcup_{j>n} \overline{W}_j)$, and for $k \geq 0$, we let A_k $(= A_k^n) := D_k \cap G_n$. It follows that when $k \geq n+1$, $A_k \setminus \bar{A}_{k-1}$ is planar, and consists of a finite number of components, $\mathcal{O}^{k,1}, \ldots, \mathcal{O}^{k,N_k}$, each component having finite connectivity (this follows from the the way that G, the D_n's and the W_n's were selected).

Each $\mathcal{O}^{k,j}$ can be represented under a bijective meromorphic map $\theta^{k,j}$ by a circular domain $\overline{\mathbb{C}} \setminus (K_1^{k,j} \cup \cdots \cup K_{m_j}^{k,j})$ where each $K_i^{k,j}$ is a closed disc in $\mathbb{C}$ (see Section 2.2). Note that if we assume, as we did, that ∂W_k is a Jordan curve, then the W_k's hole corresponds to a single disc $K_i^{k,j}$ in this representation.

The boundaries of the holes (i.e., the compact components of the complement) of G_n in R which are contained between D_{k-1} and D_k for some $k \geq n+1$, including the W_k's hole, will correspond exactly with the boundaries of some of these discs $K_i^{k,j}$ under the bijective map $\theta^{k,j}$. For each fix k, $k \geq n+1$ and each fix j,

$1 \leq j \leq N_k$, using the technique of Section 2.2, we "stitch" all these discs (and only those discs) to $\mathcal{O}^{k,j}$. Denote by $\widetilde{\mathcal{O}}^{k,j}$ the Riemann surface obtained this way, and let X_n denote the closure of D_n in R. Set

$$\widetilde{d_k} := \left(\bigcup_{j=1}^{N_k} \widetilde{\mathcal{O}}^{k,j}\right) \cup \left((X_k \setminus X_{k-1}) \setminus (A_k \setminus \bar{A}_{k-1})\right),$$

with the obvious identification of the corresponding boundaries of $\widetilde{\mathcal{O}}^{k,j}$ and $(X_k \setminus X_{k-1}) \setminus (A_k \setminus \bar{A}_{k-1})$.

Recalling that we have assumed $n \geq 0$ to be fixed, we now define R_n for each such n to be $R_n := X_n \cup (\bigcup_{k=n+1}^{\infty} \tilde{d}_k)$, again with the obvious identification of the boundaries. It follows easily that $R_n\}_{n\to\infty}R$. We note also that $G_n \subset R_n$. Now we let

$$\mathcal{G}_n := (G \cap X_n) \cup \left(\bigcup_{k=n+1}^{\infty} \left(\bigcup_{j=1}^{N_k} \widetilde{\mathcal{O}}^{k,j}\right)\right),$$

which, as a subset of R_n, is G_n with its holes (in R) filled outside of D_n. Obviously $\mathcal{G}_n\}_{n\to\infty}G$ and $\mathcal{G}_n \setminus X_n$ is planar for each $n \geq 0$. It also follows from Property 1 in Section 2.1 and this construction that the boundary of $\mathcal{G}_n$ in R_n consists of at most a countable number of Jordan curves and open Jordan arcs. This proves the first part.

Let $E_n = E \setminus (\bigcup_{j>n} W_j)$. It follows from our construction and Property 3 in Section 2.1 that E_n is a subset of both R and of R_n and that it is *closed* in both R and R_n. Also E_n (as a subset of R_n) is contained in $\mathcal{G}_n$.

Now we prove (c). Unless specifically mentioned, all sets in the rest of this proof should now be considered as subsets of R_n. Suppose B is a component of $(\mathcal{G}_n \setminus X_n)^c$. Note that $(\mathcal{G}_n \setminus X_n)^c = (\mathcal{G}_n \cap (X_n)^c)^c = (\mathcal{G}_n)^c \cup X_n$.

Now first assume $B \cap X_n = \varnothing$. By the construction of $\mathcal{G}_n$, B cannot be contained in a precompact subset of $R \setminus G_n$. The set B must thus contained points which are contained in some noncompact (i.e., unbounded) component of $R \setminus G_n$. By construction, this component also belongs to $R_n \setminus \mathcal{G}_n$ and remains unbounded as a subset of R_n. Thus B must be unbounded.

If $B \cap X_n \neq \varnothing$, then $X_n \subset B$ since X_n is a connected set contained in R_n. Under the extra assumption that it is possible to find a connected and unbounded open Jordan arc Γ in $R \setminus \overline{G}$, we have that for large enough n, say $n \geq n_0$, $X_n \cap \Gamma \neq \varnothing$ and, as explained above, Γ must then also be contained in B as a subset of R_n, and so B is unbounded. It thus follows that, as a subset of R_n, the complement of $\mathcal{G}_n \setminus X_n$ does not have bounded components if $n \geq n_0$. □

Remark 2.10. In [28], Schmieder's intented $\mathcal{G}_n \setminus X_n$ to be (for each n) a planar and simply connected Riemann surface conformally equivalent to the unit disc (see Lemma 2.15 and Section 2.4 below). Unfortunately, his instructions often led (unintentionally) to the construction of (surfaces conformally equivalent to) the complex plane or the Riemann sphere. To avoid this, we only stitch those discs that corresponds (via boundary components) to holes of G_n in R, and we assume the existence on an unbounded open Jordan arc in the complement of $\overline{G}$. This eventually prevents the construction of either the complex plane or the Riemann sphere but still does not ensure that $\mathcal{G}_n \setminus X_n$ is simply connected, as shown by the following example. We start with $\mathbb{C} \setminus \{0\}$ from which we remove, for each $n \geq 1$, the pair of open disjoint discs $\{\mathcal{D}_n^+, \mathcal{D}_n^-\}$, centred at $(2n-1)+i$ and $(2n-1)-i$ respectively and of radius 1/4. Denote by $\widetilde{R}$ the domain in $\mathbb{C}$ thus constructed. Let $\widetilde{G} = \{z \in \widetilde{R} : |z| > -1\}$, $\widetilde{D}_0 = \{z \in \widetilde{R} : |z-1| < \frac{1}{8}\}$ and, for $n \geq 1$, let $\widetilde{D}_n = \{z \in \widetilde{R} : 1/2n < |z| < 2n\}$. Now let R be the surface obtained from $\widetilde{R}$ by identifying the points on the pair of boundary circles $\{C_n^+, C_n^-\}$ corresponding to $\{\mathcal{D}_n^+, \mathcal{D}_n^-\}$, $n = 1, 2, \ldots$, and let $\pi\colon \widetilde{R} \to R$ be the natural projection into this quotient space. Now set $G = \pi(\widetilde{G})$ and $D_n = \pi(\widetilde{D}_n)$, $n \geq 0$. Following the steps in the proof of Lemma 2.8, we get that $\pi(|z| = \frac{3}{4}) \subset \mathcal{G}_n \setminus X_n$ for all $n \geq 1$, but this curve is not homotopic to zero in R. We thus must make an additional assumption on the set G (and thus on E):

Condition 2.11. *We assume that for each n, $\mathcal{G}_n \setminus X_n$ (as constructed in the proof of Lemma 2.8 above) is simply connected.*

Condition 2.11 will fail whenever $\widetilde{R}$ consist of the Riemann sphere with at least two points removed, say z_1 and z_2, and $\widetilde{G}$ contains a closed curve containing one and only one of these points in its interior. For example, in the definition of $\widetilde{G}$, if we replace $\{z \in \widetilde{R} : |z| > -1\}$ by $\{z \in \widetilde{R} : |z| > 0\}$, then the construction will lead to simply connected sets. We could say that in the first case, G is "too fat" near the ideal boundary point $\{z = 0\}$.

Definition 2.12 (Ideal and real boundary points). Let A be a closed, unbounded subset of a noncompact Riemann surface R and let R' be another Riemann surface containing A such that the analytic structure of R and R' are compatible on $R \cap R'$ and such that A is bounded in R'. The nonempty set $\bar{A} \setminus A$ in R' (where $\bar{A}$ denote the closure of A in R') is called the ideal boundary of A with respect to R'. The boundary points of A which are in R are called real boundary points.

Lemma 2.13. *Assume all the hypotheses of Lemma 2.8, including that $R \setminus \overline{G}$ also contains an unbounded open Jordan arc, and assume that $\mathcal{G}_n$ and R_n have been constructed as described above. Then, for each fixed $n \in \mathbb{N}$, $n \geq n_0$, there exists a sequence of Riemann surfaces $R_{n,k}$, $k \in \mathbb{N}$, such that:*

(i) $\mathcal{G}_n \subset R_{n,k} \subset R_{n,k+1} \subset R_n$, $k = 1, 2, \ldots$

(ii) *$\mathcal{G}_n$ is finitely connected in $R_{n,k}$.*

(iii) *The real boundary of the closure of $\mathcal{G}_n$ in $R_{n,k}$ consists of finite number of Jordan curves and open Jordan arcs.*

(iv) $R_{n,k}\}_{k\to\infty} R_n$.

Proof. Fix $n \in \mathbb{N}$, $n \geq n_0$. Let $\Gamma_1, \Gamma_2, \ldots$ be the unbounded components of $\partial\mathcal{G}_n$, and let $R_{n,k}$ be the component of $R_n \setminus (\bigcup\{\Gamma_j; j > k\})$ which contains $\mathcal{G}_n$. It follows immediately that $R_{n,k} \subset R_n$, that $\mathcal{G}_n$ is finitely connected in $R_{n,k}$, and that $R_{n,k} \setminus \mathcal{G}_n$ consists of only a finite number of precompact sets, none of them outside of D_n.

Fix a compact set K in R. Only a finite number of the Γ_j's can intersect with K, so for k_0 large enough, K is contained in the component of $R_n \setminus (\bigcup \Gamma_j, j > k_0)$ that contains G, i.e., R_{n,k_0}, and so $K \subset R_{n,k}$, for $k > k_0$. It thus follows that $R_{n,k}\}_{k\to\infty} R_n$. □

Now suppose that $\mathcal{R}$ is a noncompact Riemann surface and that $B \subset \mathcal{R}$ is an unbounded bordered domain. Thus the boundary of of B consists of open Jordan arcs and Jordan curves. Now we assume that B is simply connected. Thus the boundary of B consists only of open Jordan arcs and we shall assume they are finite (> 0) in number. By the Riemann mapping theorem, there exists a conformal map $f\colon B \to \mathbb{D}$ from B onto the unit disc $\mathbb{D} = \{z \in \mathbb{C} : |z| < 1\}$. Because the boundary of B is composed of open Jordan arcs, it follows from Carathéodory's theorem that f has a continuous extension to the boundary ∂B of B (see [20, §2.1]). Note that $M := f(\partial B)$ is an open subset of the boundary of the unit disc, consisting of a finite number of open arcs of the circle. The complement of M with respect to $\{|z| = 1\}$ is thus a finite number of closed arcs $A_j = \{\mathrm{e}^{\mathrm{i}t} : \alpha_j \leq t \leq \beta_j\}$.

We now must make an assumption on the set B.

Condition 2.14. *We assume that each of the arcs A_j degenerate to a single point a_j.*

That condition 2.14 is not always satisfied can be seen by the following example. Let $A = \{\mathrm{e}^{\mathrm{i}t} : |t| \leq \pi/2\}$, $\mathcal{R} = \overline{\mathbb{C}} \setminus A$, and $B = \mathbb{D}$. Then B is an unbounded simply connected domain in $\mathcal{R}$ with a single (real) boundary component which is the Jordan arc $\partial B = \{\mathrm{e}^{\mathrm{i}t} : \pi/2 < t \leq 3\pi/2\}$. The conformal map f is the identity function $f(z) = z$ so $f(\partial B) = \partial B$. Thus the complement of M with respect to $\partial\mathbb{D}$ is the closed arc A.

In this example, for condition 2.14 to hold on $\mathcal{R}$, we would need our set B to be much "thinner" at infinity, that is near the set A.

Lemma 2.15. *Suppose B is an unbounded simply connected domain with a finite number of boundary components in $\mathcal{R}$, each of which is an open Jordan arc. Moreover, suppose that B satisfies condition* 2.14, *and that $\mathcal{F}$ is a closed subset of $\mathcal{R}$ contained in B. Then there exists a sequence R^ν such that, for each ν, $\mathcal{F} \subset B \subset R^\nu$, $\mathcal{F}$ is precompact in R^ν, and $R^\nu\}_{\nu\to\infty}\mathcal{R}$.*

PROOF. As above, let $f\colon B \to \mathbb{D}$ be a conformal map from B onto the unit disc $\mathbb{D} = \{z \in \mathbb{C} : |z| < 1\}$, and let $M := f(\partial B)$. By assumption, the complement of M with respect to $\{|z| = 1\}$ consists in a finite number of points $a_j = e^{it_j}$, $j = 1, \dots, \mu$. Let $A_j^\nu := \{\mathrm{e}^{\mathrm{i}t} : t_j - 1/\nu \leq t \leq t_j + 1/\nu\}$ for $\nu \geq \nu_0$, where ν_0 is taken large enough to have $A_j^{\nu_0} \cap A_i^{\nu_0} = \varnothing$ whenever $i \neq j$.

Now, let R' be $R \setminus (\bigcup_{j=1}^{\mu} f^{-1}(A_j^\nu \setminus \{a_j\}))$. Take μ copies $\overline{\mathbb{C}_j}$ of $\overline{\mathbb{C}}$ and "stitch up" R' to the circular domains $\overline{\mathbb{C}_j} \setminus \{z \in \overline{\mathbb{C}_j} : |z| < 1\}$ along the arcs $\{\mathrm{e}^{\mathrm{i}t} : t_j - 1/\nu < t < t_j + 1/\nu\}$, using the technique described in Section 2.2. Denote the new surface by R^ν.

From the way the conformal structure has been selected for R^ν, R and R^ν are compatible on $R \cap R^\nu$, and from the way R^ν is defined it contains the set B (and thus the set $\mathcal{F}$). We also note that, since the closure of $f(\mathcal{F})$ in $\mathbb{C}$ can intersect the unit circle $\{|z| = 1\}$ only at the points a_j's, all the points of the ideal boundary of $\mathcal{F}$ with respect to R^ν now form a bounded subset of R^ν, and thus $\mathcal{F}$, as a subset of R^ν, becomes precompact.

Finally, if K is a fixed compact subset of R, then for large enough ν we must have $K \cap \left(\bigcup_{j=1}^{\mu} f^{-1}(A_j^\nu \setminus \{a_j\})\right) = \varnothing$. It follows that $K \subset R^\nu$ for large enough ν, and thus $R^\nu\}_{\nu\to\infty}R$. □

Remark 2.16. The condition in Lemma 2.15 on the set $\mathcal{F}$ can be relaxed. Indeed it follows easily from the proof that if $\mathcal{R}$ is one of the surfaces $R_{n,k}$ from Lemma 2.13, B is a connected component of $\mathcal{G}_n \setminus X_n$ and $E_n = E \setminus \bigcup_{j>n} W_j$, then

$(E \cap X_n) \cup (E_n \cap B)$ becomes precompact, as a subset of the surface R^ν constructed in the proof above, though it may not satisfy all the conditions of the lemma.

2.4. Extension theorem: Statement and proof. Let R be a noncompact Riemann surface, E be a closed subset of R, and G, D_n $(n \geq 0)$ and W_n $(n \geq 1)$ be subsets of R as described in Section 2.1. We recall that, in particular, this means that

Condition 2.17. *The family* $\{D_n\}$ *is a normal exhaustion of* R.

Condition 2.18. *For each* $n > 0$, W_n *is of genus one.*

Condition 2.19. $G \setminus (\bigcup_{n\in\mathbb{N}} \overline{W}_n)$ *is planar.*

Note that it follows from condition 2.18 that we are assuming that

Condition 2.20. G *(and* E*) are of infinite genus.*

Moreover, assume that

Condition 2.21. $R\setminus\overline{G}$ *contains some connected neighborhood of an unbounded open Jordan arc*

Let X_n denote the closure of D_n in R. Following the proof of Lemma 2.8, we construct a sequence of Riemann surfaces R_n, with $X_n \subset R_n$, and a sequence of domains $\mathcal{G}_n \subset R_n$ whose boundary in R_n consist of at most a countable number of Jordan curves and open Jordan arcs such that $R_n\}_{n\to\infty}R$, $\mathcal{G}_n \setminus X_n$ is planar, $\mathcal{G}_n\}_{n\to\infty}G$, $E_n = E \setminus (\bigcup_{j>n} W_j)$ is a closed subset of R_n, and $\mathcal{G}_n \supset E_n$.

Condition 2.22. *We now also assume condition* 2.11.

Fix $n \geq n_0$. Then by Lemma 2.13 there exist sequences $R_{n,k}$ such that $R_{n,k}\}_{k\to\infty}R_n$, $\mathcal{G}_n \subset R_{n,k}$, and $\mathcal{G}_n \setminus X_n$ consist of a finite number of simply connected components $B^1_{n,k}, \ldots, B^{N(n,k)}_{n,k}$ in $R_{n,k}$, such that the boundary of each $B^i_{n,k}$ in $R_{n,k}$ consists of finite number of unbounded open Jordan arcs. Assume that

Condition 2.23. *Each* $B^i_{n,k}$ *satisfies condition* 2.14.

Now fix k. By the method of Lemma 2.15 and Remark 2.16, since none of the B^i share the same side of a boundary curve, we may simultaneously stich along the boundaries of the $B^i_{n,k}$, $i = 1, \ldots, N(n,k)$ to obtain noncompact Riemann surfaces $R^\nu_{n,k}$ such that $R^\nu_{n,k}\}_{\nu\to R_{n,k}}$ and, for each $i = 1, \ldots, N(n,k)$, $(E\cap X_n)\cup(E_n \cap B^i_{n,k})$ becomes precompact in $R^\nu_{n,k}$.

Note that $E_n = \bigcup_{i=1}^{N(n,k)}(E \cap X_n) \cup (E_n \cap B^i_{n,k})) = E \cap \mathcal{G}_n$. Note also that so far we have:

$$R^\nu_{n,k}\}_{\nu\to\infty}R_{n,k}, \qquad R_{n,k}\}_{k\to\infty}R_n \qquad \textit{and} \qquad R_n\}_{n\to\infty}R.$$

Applying Proposition 2.7, we thus get

Theorem 2.24 (Extension theorem). *Under the above hypotheses,*

(1) *for each* $n \in \mathbb{N}$, *there exists a noncompact Riemann surface* r_n *on which* $E_n := E \setminus (\bigcup_{j>n} \overline{W}_j)$ *is relatively compact;*
(2) *for each* $n \in \mathbb{N}$, $X_n \subset r_n$, *and thus* $r_n\}_{n\to\infty}R$.

Remark 2.25. In [28], G. Schmieder did not assume that G (and thus R) was of infinite genus, he did not assume that $R \setminus G$ contained a connected unbounded open Jordan arc, and he did not assume explicitly conditions 2.22 and 2.23. For his (and thus our) purpose, the first condition is not really restrictive, as the main goal is to obtain "new" sets of approximation, and a complete characterization of those sets is known on Riemann surfaces of finite genus (see [7] and the references therein). It is not hard to see that when $R \setminus E$ is precompact, the extension theorem in its original version does not hold in general, the simplest example being the complement of the unit disc in $\mathbb{C}$. The second condition is introduced in part to avoid this situation. Note that when $R \setminus E$ is precompact, the set E may, or may not be a set of approximation (see Remark 5.3). Conditions 2.22 and 2.23 were used by Schmieder in his proof, but only implicitly. They are the most restrictive of all the conditions mentioned. Roughly speaking, they can be interpreted as meaning that the set E (and its neighborhood G) must be "thin at infinity."

Remark 2.26. We would like to remind the reader at this point of the existence of *maximal* noncompact Riemann surfaces. If there exists a conformal mapping ψ of R into a Riemann surface R_1, then the pair (R_1, ψ) is called an *extension* of R. Note that R itself is an extension of R according to this definition. An extension (R_1, ψ) is called *proper* if $R_1 \setminus \psi(R) \neq \varnothing$. A Riemann surface is called *maximal* if it has no proper extension. Conditions are known that characterize maximal and nonmaximal Riemann surfaces (see, for example, [16] and the references therein). It would be interesting to revisit the extension theorem of Schmieder above in view of these conditions.

3. The fusion lemma on noncompact Riemann surfaces

3.1. Introduction. The fusion lemma of Alice Roth [9, 23] allows for the simultaneous approximation of two meromorphic functions in the complex plane. This lemma says the following: Let K_1 and K_2 be disjoint compact sets in $\overline{\mathbb{C}}$ and let m_1 and m_2 be meromorphic functions on $\mathbb{C}$ whose values are close to each other on a compact set K. Then one can find a third meromorphic function which is close to m_1 on $K_1 \cup K$ and close to m_2 on $K_2 \cup K$; how close depends (surprisingly) neither on the functions m_1 and m_2 nor on K, but only on K_1, K_2 and the maximum of $|m_1 - m_2|$ on K.

This phenomenon has assumed a central position in complex approximation theory (see [12]). In particular P. M. Gauthier has shown that the proof of A. Roth can also be carried over to noncompact Riemann surfaces [11, p. 143]. In these earlier works, one always only found the uniform estimate $|m_j - m| < a\alpha$ on $K_j \cup K$ with $a > 2$. But in the planar case, it can be shown that $|m_2(z) - m(z)| \to 0$ as $|z| \to \infty$, $z \in K_2$. Here, without loss of generality, we assumed that ∞ belongs to K_2, as in the proof of A. Roth.

In Section 3.3, we shall for completeness give a proof (due to G. Schmieder [27]) of a stronger version of the fusion lemma on Riemann surfaces, where the question of how well the function m approximate (or can approximate) the function m_1 on K_1, respectively m_2 on K_2 is considered. However, the basic idea remains the proof of A. Roth for the planar case.

A further question in this connection is whether one can omit the assumption that $K_1 \cap K_2 = \varnothing$ and instead suppose that m_1 and m_2 are close on $K_1 \cap K_2$ thus dropping K completely. This is in general not possible, in fact not even when K_1

and K_2 are rectangles as D. Gaier has shown [10]. From this fact, it is not hard to see that for numbers $a = a(K_1, K_2)$ having the above properties, it is not possible to give a common estimate valid for all disjoint K_1 and K_2.

3.2. Notations and Pompeiu formula. For a closed subset T of a Riemann surface R, we denote by $M(T)$ (respectively $H(T)$) the family of functions f meromorphic (respectively holomorphic) in a neighborhood (depending on f) of T. Let $M_R(T)$ be the class of (continuous) functions $f\colon T \to \overline{\mathbb{C}}$ for which there exists a sequence $f_n \in M(R)$ such that all f_n have the same set P_f of poles on T and $|f_n - f|$ converges uniformly to zero on $T \setminus P_f$. Let $H_R(T)$ be the class of (continuous) functions $f\colon T \to \overline{\mathbb{C}}$ for which there exists a sequence $f_n \in H(R)$ such that $|f_n - f|$ converges uniformly to zero on T. Now let R be a noncompact Riemann surface. According to Behnke and Stein [4], there exists a Cauchy kernel ω on R (see for example [4, 11, 24]); let one such kernel be fixed.

Lemma 3.1 (Pompeiu formula). *If $\sigma\colon R \to \mathbb{C}$ is a C^1-function with compact support, then*

$$\sigma(Q) = \frac{1}{2\pi \mathrm{i}} \iint_R \omega(P, Q) \wedge \bar{\partial}\sigma$$

for all $Q \in R$, where $\bar{\partial}\sigma$ is the 1-form given in the local parameters by $(\partial\sigma/\partial\bar{\zeta})\,\mathrm{d}\bar{\zeta}$ and $\wedge$ denotes the exterior product.

This lemma follows from a more general theorem obtained by S. Scheinberg [26, Proposition 7].

3.3. Formulation and proof of the fusion lemma.

Theorem 3.2 (Fusion lemma). *Let K_1 and K be compact subsets and K_2 a closed subset of an open Riemann surface R with $K_1 \cap K_2 = \varnothing$, and suppose*

$$M(K_1 \cup K_2 \cup K) \subset M_R(K_1 \cup K_2 \cup K).$$

Then, there exists a continuous function

$$C\colon R \to \mathbb{R}_+$$

such that for each pair $m_1, m_2 \in M(R)$ with

$$\|m_1 - m_2\|_K < \varepsilon,$$

there exists a function $m \in M(R)$ which satisfies the inequality

$$|m(p) - m_j(p)| < C(p) \cdot \varepsilon \qquad (p \in K_j \cup K, j = 1, 2).$$

The function C depends only on K_1 and K_2.

Remark 3.3. If K_2 is also compact then $M(K_1 \cup K_2 \cup K) \subset M_R(K_1 \cup K_2 \cup K)$ always by [4, Theorem 13]. Note that in [4], the holomorphy which is assumed in a neighborhood of $K_1 \cup K_2 \cup K$ can be circumvented. Indeed we first subtract a meromorphic function on R having the same poles and principal parts on $K_1 \cup K_2 \cup K$ and after approximation we add it again. If for a compact set K_2, we replace the function $C(p)$ by the constant $c = \|C\|_{K_1 \cup K_2 \cup K}$, then we obtain the weaker form of the fusion lemma (see [9, 11, 23]).

Remark 3.4. $M(X) \subset M_{\mathbb{C}}(X)$ holds for every closed set $X \subset \mathbb{C}$. This follows from the approximation theorem of Roth [9, 22], taking Remark 3.3 into consideration.

In order to prove the fusion lemma (following [27]), we choose pairwise smoothly bordered neighborhoods U_1 and U_2 of K_1 and K_2 on R with $\overline{U}_1 \cap \overline{U}_2 = \varnothing$, and we may choose U_2 so large that $R \setminus U_2$ is compact. Further, let $E = R \setminus (U_1 \cup U_2)$, and $\chi\colon R \to [0,1]$ be a C^1 function with $\chi|_{U_1} = 1$ and $\chi|_{U_2} = 0$ (see, e.g., [18, Corollary 2.2.15]). Let ω be some Cauchy-kernel on R (see [4, Theorem 12]). The function

$$B(Q) = \frac{1}{2\pi} \iint_E |\omega(P,Q) \wedge \bar{\partial}\chi(P)| \tag{3.1}$$

is well-defined and continuous on R as can be seen by passing to a local coordinate z and then changing to polar coordinate $\zeta - z = \rho e^{it}$.

Now let $q = m_1 - m_2$ be fixed. By assumption there is a neighborhood U_3 of K with $\|q\|_{\overline{U}_3} < \varepsilon$.

Now we define the function q_1 on R as follows: On $U_1 \cup U_2 \cup U_3$ we set $q_1 = q$; then we extend $q_1|_{U_3}$ continuously by Tietze's theorem (see, e.g., [17, p. 242]) to $E \setminus U_3$ (with respect to the relative topology of E) so that $\|q_1\|_E < \varepsilon$. Note that q_1 need not be continuous on R. By passing to local coordinates we see that

$$g(Q) = -\frac{1}{2\pi i} \iint_E q_1(P)\omega(P,Q) \wedge \bar{\partial}\chi(P)$$

represents a function that is holomorphic on $R \setminus E$.

From our choice of χ it follows that the function $f = \chi q_1 + g$ is holomorphic on U_2 and meromorphic on U_1 with the same poles as q. From the Pompeiu formula (Lemma 3.1), we have

$$f(Q) = \frac{1}{2\pi i} \iint_E (q_1(Q) - q_1(P))\omega(P,Q) \wedge \bar{\partial}\chi(P)$$

for all $Q \in R$ with $q_1 \neq \infty$.

From the properties of the Cauchy kernel (see [4]), it follows that f is holomorphic on U_3. Indeed, on U_3, we have $q_1 = q$ which is holomorphic on U_3. Since the differential $\omega(P,Q)$ is holomorphic in Q, the differential $\big(q_1(P) - q_1(Q)\big)\omega(P,Q)$ is also holomorphic on U_3, and consequently f is holomorphic on U_3.

It thus follows that f is meromorphic on $U_1 \cup U_2 \cup U_3$. Now set $X = K_1 \cup K_2 \cup K$ and let $h \in H(X)$ be bounded and nonvanishing (h may be constant). From the hypothesis that $M(X) \subset \overline{M}(X)$, it follows that there exists a function $m_3 \in M(R)$ such that

$$|m_3 - f| \leq \varepsilon |h| \qquad \text{on } X.$$

Since $|q_1| < \varepsilon$ on E, we have $|g(Q)| < \varepsilon . B(Q)$, for all $Q \in R$. With $m = m_2 + m_3$ we have the following estimates:

On K_1:

$$\begin{aligned} |m - m_1| &\leq |f - (m_1 - m_2)| + |m_3 - f| \\ &= |f - q| + |m_3 - f| \\ &\leq |\chi - 1|\,|q| + |g| + |m_3 - f| < \varepsilon \cdot B + \varepsilon \cdot |h| = (B + |h|)\varepsilon. \end{aligned}$$

On K_2:

$$\begin{aligned} |m - m_2| &\leq |m_3 - f| + |f| \\ &\leq |m_3 - f| + |\chi|\,|q| + |g| < \varepsilon \cdot |h| + \varepsilon \cdot B = (B + |h|)\varepsilon. \end{aligned}$$

Taking into consideration that $0 \le \chi \le 1$, we have on K:

$$|m - m_j| \le 2|q| + |g| + |m_3 - f| < 2\varepsilon + \varepsilon \cdot B + \varepsilon \cdot |h|$$

for $j = 1, 2$. Now let $C\colon R \to \mathbb{R}_+$ be any continuous function with

$$C(P) > \begin{cases} B(P) + |h(P)|, & \text{for } P \in K_1 \cup K_2; \\ B(P) + 2 + |h(P)|, & \text{for } P \in K \setminus (K_1 \cup K_2). \end{cases}$$

This function satisfies the conclusion of the fusion lemma. □

Suppose $Y(P,Q)$ is a meromorphic function on $R \times R$ and $\Omega = Y(P,Q)\,\mathrm{d}P$ represents a Cauchy-kernel except on a set of the form

$$P = \bigcup_{j=1}^{M} (P_j \times R) \cup \bigcup_{i=1}^{L} (R \times Q_i)$$

for finitely many points $P_1, \dots, P_M, Q_1, \dots, Q_L \in R$. Then Ω is called a *pseudo-Cauchy kernel*.

Remark 3.5. If Ω is a pseudo-Cauchy kernel with exceptional set P as above where $P_1, \dots, P_M, Q_1, \dots, Q_L \notin K_1 \cup K_2 \cup K$, then the above proof goes through with ω replaced by Ω. The function f can then have additional poles. However, these lie outside of $K_1 \cup K_2 \cup K$. Since Ω has an influence on the function $C(P)$, such a choice could be advantageous.

3.4. The Newtonian potential of an annulus. It will now be shown (following [27]) that the function $C(P)$, in the fusion lemma can be chosen near 1 on K_1 and K_2 but can be smaller in the special case where, $R = \mathbb{C}$ and $E = \mathbb{C} \setminus (U_1 \cup U_2)$ can be chosen to be an annulus.

The Newtonian (surface) potential of the disc $|\zeta| \le r$ is the function

$$F_r(Z) = F(r,z) = \iint_{|\zeta| \le r} \frac{1}{|\zeta - z|}\,\mathrm{d}\xi\,\mathrm{d}\eta$$

where $\zeta = \xi + \mathrm{i}\eta$. If we set, as usual,

$$K(k) = \int_0^{\pi/2} (1 - k^2 \sin^2 t)^{-1/2}\,\mathrm{d}t$$

$$E(k) = \int_0^{\pi/2} (1 - k^2 \sin^2 t)^{1/2}\,\mathrm{d}t,$$

then, using [15, (233,5e)], we obtain by an elementary calculation

$$F(r,z) = \begin{cases} 4rE(|z|/r) & \text{for } |z| \le r \\ 4|z|(E(r/|z|) - (1 - r^2/|z|^2)K(r/|z|)) & \text{for } |z| > r. \end{cases} \tag{3.2}$$

Now let $U_1 = \{|\zeta| < r\}$ and $U_2 = \{|\zeta| > r + \Delta r\}$, so that $E = \{r \le |\zeta| \le r + \Delta r\}$. Let ω be the usual Cauchy kernel.

For every $\alpha > 0$, there exists a C^1-function $\sigma : \mathbb{R}_{>0} \to [0,1]$ with $\sigma|_{[0,r]} = 1$, $\sigma|_{[r+\Delta r,\infty)} = 0$ and $|\sigma'| \le r/(\Delta r) + \alpha$ on $\mathbb{R}_{>0}$. Then $\chi(\zeta) = \sigma(|\zeta|)$ has the properties in the fusion lemma, and for $B(z)$ (see (3.1)) we have:

$$B(z) \le \frac{1}{2\pi}\left(\frac{1}{\Delta r} + \alpha\right)\bigl(F(r + \Delta r, z) - F(r,z)\bigr), \qquad z \in \mathbb{C}.$$

For $B(r,\Delta r,z) = (2\pi\Delta r)^{-1}\big(F(r+\Delta r,z) - F(r,z)\big)$, we have the estimate $B(r,\Delta r,z) \le r/(\Delta r)+1$, where for $|z|\le r+\Delta r$ follows from (3.2), and otherwise from elementary geometry after writing F in polar coordinates centered at z.

The number α was chosen arbitrarily. Hence from the fusion lemma together with Remark 3.4, we have the following:

Corollary 3.6. *If K_1 is a compact subset of the disc $|\zeta| < r+\Delta r$, K_2 closed subset of $|z| > r+\Delta r$ $(r, \Delta r > 0)$, and $\delta > 0$, and K a compact set, then we have:*

If $m_1, m_2 \in M(\mathbb{C})$ with $\|m_1 - m_2\|_K < \varepsilon$. Then there exists a function $m \in M(\mathbb{C})$ such that

$$\|m - m_j\|_{K_j} \le \left(\frac{r}{\Delta r} + 1 + \delta\right)\varepsilon,$$

$$\|m - m_j\|_{K} \le \left(\frac{r}{\Delta r} + 2 + \delta\right)\varepsilon, \qquad j = 1,2.$$

In [27], an illustration is provided showing the effect on $B(r,\Delta r,z)$ for various choices of r and Δr.

4. The localization theorem — Part 1

The localization theorem has been proved by A. Roth [23, Theorem 1] for planar surfaces. It allows one to give a positive answer to the question: Is it possible to approximate uniformly a function f on a closed set E by global meromorphic functions when approximable on each compact subset of E?

The generalization of the theorem of A. Roth to noncompact Riemann surfaces was given by P. M. Gauthier [12, Theorem 1], for the case when E is contained in a neighborhood which consist of disjoint components, each of finite genus. As in [28], we will now show that this localization theorem is valid for a larger class of closed sets E. Since the proof requires the extension theorem, we will need to impose the extra conditions introduced in Section 2. The definition of the larger class uses Cauchy kernels which are themselves determined by the analytic structure of the Riemann surface R; to determine if a given set E belongs to this class is not as easy as in the case of Gauthier, as it is the analytic structure that manifests itself deeply through the Cauchy kernels, and not anymore simply some topological conditions. Furthermore, we must take into consideration the analytic structure if we want to improve on the result of Gauthier, as many examples show [11, 26].

4.1. Special localization on the surfaces $r_n \cap R$. Let $E \subset R$ be a closed set which satisfies the hypothesis of the extension theorem in Section 2. Let $D_0 \subset\subset D_1 \subset\subset \dots$ be an exhaustion of R with $X_n := \overline{D_n} \subset r_n$, as in the extension theorem. Let us choose the Jordan domains W_n and define $E_n := E \setminus (\bigcup_{j>n} W_j)$ as before. We now prove the following:

Proposition 4.1 (see [28]). *For the Riemann surface $\mathcal{R} = r_n \cap R$ and for a function $h \in M(X_{n-1} \cup E_n)$ (i.e., h is meromorphic in a neighborhood U of $X_{n-1} \cup E_n$ and $U \subset \mathcal{R}$), the following are equivalent:*

(i) $h \in M_{\mathcal{R}}(X_{n-1} \cup E_n)$

(ii) $h|_{K\cap(X_{n-1}\cup E_n)} \in M_{\mathcal{R}}\big(K \cap (X_{n-1} \cup E_n)\big)$ *for each compact $K \subset \mathcal{R}$*

Proof. It is trivial that (i) $\Rightarrow$ (ii).

To prove (ii) $\Rightarrow$ (i), let n be fixed, $h \in M(X_{n-1} \cup E_n)$, assume that (ii) holds and let δ be a positive number.

The argument follows closely that of A. Roth [23]. The goal is to approach h on $X_{n-1} \cup E_n$. Let d_j be an exhaustion of $r_n \cap R$. For each fixed j and for the following three bounded subsets of r_n

$$K_1^j = \bar{d}_j, \qquad K_2^j = (X_{n-1} \cup E_n) \setminus d_{j+1}, \qquad \text{and} \qquad K^j = (X_{n-1} \cup E_n) \cap \bar{d}_{j+1},$$

we choose $a_j \in \mathbb{R}_+$ such that the fusion lemma holds for these three sets with a_j replacing $C(P)$ (that is $a_j \geq \|C\|_{K_1^j \cup K_2^j \cup K^j}$). Moreover, we assume that the a_j's are chosen such that $1 < a_1 < a_2 < \cdots$.

Let $\delta_1, \delta_2, \ldots$ be a decreasing sequence of positive numbers such that $\sum_1^\infty \delta_j = \delta/2$. From the hypothesis on h, there exist functions $q_j \in M(r_n \cap R)$ such that $|q_j - h| < \delta_j/2a_j$ on K_j. From which it follows that

$$|q_{j+1} - q_j| < \frac{\delta_j}{a_j} \qquad \text{on } K^j \text{ for all } j.$$

Using the theorem of Behnke and Stein (see [4]) we can approximate q_j uniformly on K_j by functions in $M(r_n)$; in the case where q_j has poles on K^j, we first subtract a function in $M(r_n)$ which has the same poles on K^j and we add this function after having done the approximation. We can thus suppose in what follows that $q_j \in M(r_n)$.

By the fusion lemma (Lemma 3.2), there exist functions $\varphi_j \in M(r_n)$ such that

$$|\varphi_j - q_j| < \delta_j \qquad \text{on } \bar{d}_j \cup K^j$$

and

$$|\varphi_j - q_{j+1}| < \delta_j \qquad \text{on } D_{n-1} \cup E_n = K_2^j \cup K^j.$$

This implies

$$\sum_{\nu=j}^{\infty} |\varphi_\nu - q_\nu| < \sum_{\nu=j}^{\infty} \delta_\nu < \frac{\delta}{2} \qquad \text{on } \bar{d}_j.$$

Consequently, the function

$$H := q_1 + \sum_{\nu=1}^{\infty} (\varphi_\nu - q_\nu)$$

is meromorphic on $\bigcup d_j = r_n \cap R$. On K^1, we have

$$|H - h| \leq |q_1 - h| + \sum_{\nu=1}^{\infty} |\varphi_\nu - q_\nu| \leq \frac{\delta_1}{2a_1} + \sum_{\nu=1}^{\infty} \delta_\nu < \delta,$$

and on $K^j \setminus K^{j-1}$, $(j > 1)$, we get

$$\begin{aligned}
|H - h| &= \left| q_1 + \sum_{\nu=1}^{j-1} (\varphi_\nu - q_\nu) - h + \sum_{\nu=j}^{\infty} (\varphi_\nu - q_\nu) \right| \\
&= \left| \sum_{\nu=1}^{j-2} (\varphi_\nu - q_{\nu+1}) + (\varphi_{j-1} - h) + \sum_{\nu=j}^{\infty} (\varphi_\nu - q_\nu) \right| \\
&\leq \sum_{\nu=1}^{j-2} \delta_\nu + |\varphi_{j-1} - q_j| + |q_j - h| + \sum_{\nu=j}^{\infty} \delta_j \leq 2\delta.
\end{aligned}$$

We thus obtain $|H-h|<2\delta$ on $X_{n-1}\cup E_n$ which is what we needed to prove. □

Proposition 4.2. *Each function g meromorphic in a neighborhood U of $X_{n-1}\cup E_n$ with respect to $r_n\cap R$ can be uniformly approximated on $X_{n-1}\cup E_n$ by functions in $M(r_n\cap R)$.*

This is known, since $X_{n-1}\cup E_n$ is of finite genus (see [12,25]). This also follows from Proposition 4.1 and the theorem of Behnke and Stein [4, Theorem 13]. The eventual poles of g in $X_{n-1}\cup E_n$ are treated as in the previous proof.

5. Definition of the sets "weakly of infinite genus" (wig)

Let R, E, D_n, X_n, W_n, E_n and r_n $(n>0)$ be defined as before. For each $W_n\neq\varnothing$, we fix a domain V_n with $W_n\subset\subset V_n$, such that the $\overline{V}_n$'s are mutually disjoint and

(1) $\overline{V}_n\subset D_n\setminus X_{n-1}$ from which it follows that
(2) $\overline{V}_n\subset r_n$

For each of the Riemann surfaces $r_n\cap R$, let w_n be a (given) Cauchy kernel (or only a pseudo Cauchy kernel (see [28, §7.2] and Section 3.2). For each n, let

$$K_1=\overline{W}_n\cap E,\qquad K_2=X_{n-1}\cup(E_n\setminus V_n),\qquad\text{and}\qquad K=(\overline{V}_n\setminus W_n)\cap E.$$

For these sets, we now choose open neighborhoods U_1 and U_2 of K_1 and K_2 respectively with $U_1\cap U_2=\varnothing$, and a C^1-function with compact support $\chi_n\colon r_n\cap R\to[0,1]$ with $\chi_n\,|_{U_1}=1$, $\chi_n\,|_{U_2}=0$. Set $\mathcal{D}_n=(r_n\cap R)\setminus(U_1\cup U_2)$ and

$$B_n(Q)=\frac{1}{2\pi}\iint_{\mathcal{D}_n}|w_n(P,Q)\wedge\bar{\partial}\chi_n(P)|,\qquad Q\in r_n\cap R,$$

where $\bar{\partial}\chi_n$ is the 1-form which in the local coordinate ζ is given by $(\partial\chi(\zeta)/\partial\bar{\zeta})\,d\bar{\zeta}$.

We have that $K_1\cup K_2\cup K=X_{n-1}\cup E_n$. From Proposition 4.2, we obtain

$$M(K_1\cup K_2\cup K)\subset M_{r_n\cap R}(K_1\cup K_2\cup K).$$

We can thus apply the fusion lemma with K_1, K_2 and K on the Riemann surface $r_n\cap R$ with a function $C_n(P)$ satisfying

$$C_n(P)\geq\begin{cases}B_n(P)+|\delta(P)| & \text{for } P\in K_1\cup K_2;\\ B_n(P)+2+|\delta(P)| & \text{for } P\in K\setminus(K_1\cup K_2),\end{cases}$$

where δ is an arbitrary function in $H(K_1\cup K_2\cup K)$ (see [28]). Fix such a function C_n. Moreover, let h_0 be a nonvanishing bounded (say by $\lambda_1>0$) holomorphic function on a neighborhood (with respect to R) of the closed planar set E_0. Now set

$$\varepsilon_0(P):=|h_0(P)|\qquad\text{and}\qquad e_0:=\|\varepsilon_0\|_{(\overline{V_1}\setminus W_1)\cap E}.$$

We define inductively the sequence of functions ε_n and a sequence of numbers e_n as follows:

$$\varepsilon_n(P)=\begin{cases}e_{n-1}(2C_n(P)+1) & \text{for } P\in W_n\cap E;\\ 2e_{n-1}C_n(P)+\varepsilon_{n-1}(P) & \text{for } P\in E_{n-1}.\end{cases}$$

$$e_n=\|\varepsilon_n\|_{(\overline{V}_{n+1}\setminus W_{n+1})\cap E}.$$

We note that each ε_n is defined on E_n respectively.

Definition 5.1 (see [28]). (a) A closed subset E of a noncompact Riemann surface R is said to be a *wig*1*-set* (*weakly of infinite genus, type* 1) if it is of finite genus or if it is possible to carry the above construction in such a way that the following inequalities are satisfied for all $n = 0, 1, 2, \ldots$

$$e_n \leq \frac{\lambda_{n+2}}{2\|C_{n+1}\|_{E_n \cup D_n}} \tag{5.1}$$

$$e_n \leq \frac{\sum_{j=1}^{n+2} \lambda_j}{\|2C_{n+1} + 1\|_{W_{n+1} \cap E}} \tag{5.2}$$

where λ_n are arbitrary positive numbers satisfying $\sum \lambda_n < \infty$.

(b) A closed subset $E \subset R$ is said to be a *wig*2*-set* (*weakly of infinite genus, type* 2) if there exists a decomposition of E by mutually disjoint closed sets A_j such that each compact subset of R intersect with only a finite number of A_j and each set $\bigcup_{j=1}^{n} A_j$ is a wig1-set $n \in \mathbb{N}$.

(c) $E \subset \mathbb{R}$ is said to be *weakly of infinite genus* (or a *wig-set*) if E is a wig1-set or a wig2-set.

Remark 5.2. Note that, this definition was given by G. Schmieder in [28]. It depends on $R, E, G, D_n, W_n, V_n, E_n$ $(n > 0)$ as defined above and on the Riemann surfaces r_n $(n > 0)$ as obtained in his extension theorem. The statement and the proof of this theorem needed to be fixed. This was done in Section 2. By doing so, additional requirements were imposed on the set G (and thus E) and on the D_n's and W_n's (see Section 2.4). This in turn, has changed the class of wig-sets.

Remark 5.3. After introducing this definition, Schmieder proceeded to show that wig-sets were sets of approximation and that this new class included all previously known such sets. Closed sets contained in an open set of finite genus are known to be sets of approximation (see [12]) and thus should be included in the class of wig-sets. The original definition of G. Schmieder for wig1-sets, did not need a separate clause for sets of finite genus because, based on his incorrect formulation of the extension theorem, it already contained that class. But if we take for example $R = \mathbb{C}$ and let E be the complement of an open precompact set (such as the open unit disc), then E is contained in an open set of finite genus (in fact, of genus zero) and is a set of (meromorphic) approximation. Notice that there exists no open neighborhood G for E such that $R \setminus G$ contains an unbounded and connected open Jordan arc. So a definition solely based on our formulation of the extension theorem would miss this set.

We also note that there are known examples of Riemann surfaces R, and of closed subsets $E \subset R$ defined as the complement of open precompact sets in R that failed to be sets of (meromorphic) approximation (see for example [6, §1.11]). From our previous discussion, it follows that the surfaces R must necessarily be of infinite genus.

In the case when there exist a covering of E by mutually disjoint open sets, each of finite genus, then, following Gauthier [12], E is said to be "*essentially of finite genus*," which we denote by the abbreviation *efg*. If E can be covered by one open set of finite genus, E will be called a set of finite genus, or in short, a *fg-set*.

Remark 5.4. Each efg-set is a wig2-set.

PROOF. If E is an efg-set, then E can be decomposed in a sequence A_j of mutually disjoint fg sets, such that each compact intersect with only a finite number of sets A_j as it follows from the definition of efg-sets. □

Proposition 5.5. *Let $E \subset R$ be a wig1-set and $A \subset R$ be compact. Then $A \cup E$ is a wig1-set.*

PROOF. Let $\{\lambda_n\}$ be a sequence such that $\sum \lambda_n < \infty$. Suppose D_n, W_n, V_n, E_n and r_n be as in Section 2, and assume $A \subset D_n$ for $n \geq n_0$. Let $E' = E \cup A$, let W'_n and D'_n be as in Section 2 for E'.

Note that E' might contain more handles than E (but only finitely many more). We can select W_n's in a way that $\{W_1, W_2, \dots\} \subset \{W'_1, W'_2, \dots\}$ and also D'_ns in a way that $\{D_1, D_2, \dots\} \subset \{D'_1, D'_2, \dots\}$. So we can choose the indices in a way to have $D'_{n_1+k} = D_{n_0+k}$ and $W'_{n_1+k} = W_{n_0+k}$, for any $k \in \mathbb{N} \cup \{0\}$.

If we let h'_0 to be the same function as h_0 all the inequalities for $n \geq n_1$ would be the same as for the set E up to indexing. It easy to show that if $|h'_0(P)| < |h_0(P)|$ then $e'_k \leq e_k$ for any k.

We can select h'_0 in a way to have $e'_0, \dots, e'_{n_1-1}$ small enough to satisfy all the related inequalities. Also we can start with h'_0 to have $|h_0(P)| \leq |h'_0(P)|$. for any $P \in E_0$. □

We will give a short heuristic interpretation of wig-sets. The reader is referred to [28, §7] for more precise statements.

If one thinks of the Riemann surface as the result of gluing handles (caps with handles) onto planar domains (suitably punctured), then we can think of $\mathcal{D}_j$ as the gluing areas (common to the domain and the lid).

Let E be a closed subset of the surface, and assume for simplicity that all $\mathcal{D}_j$ are contained in E° (in any case, the $\mathcal{D}_j$'s contained in the complement of E are irrelevant for the following discussion).

If ω_j is the usual Cauchy kernel (see [28, §7.2]), then B_j denotes the electric potential obtained by assigning the charge density $|\partial\chi_j/\partial\bar{\zeta}|$ to $\mathcal{D}_j$ (Newton potential); the charge density depends only on $\mathcal{D}_j$ (χ) and so C_j is an upper bound for the potential; ε_n majorize the potential of a weighted superposition of simultaneous charge on the first n sets $\mathcal{D}_j$, so ε_n becomes arbitrary small if one gets "far away" from those already charged $\mathcal{D}_j$ (i.e., when one approach the ideal boundary).

So the inequalities (i) and (ii) in the Definition 5.1 of wig-sets can be interpreted as follows: The numbers λ_j are given; for (5.1) to be satisfied, the larger C_{n+1} is on $E_n \cup D_n$ the smaller ε_n must be in a neighborhood of $(n+1)$th handle. The $(n+1)$th handle must therefore be at a sufficiently large distance of its predecessors for E to be a wig1-set.

6. The localization theorem — Part 2

Using the localization theorem below, Schmieder [28] has obtained nice approximation theorems for wig-sets. Recalling that there was problem with his definition of wig-sets, we now re-state and prove this Theorem for the new class of wig-sets introduced in Section 5. The proof follows the proof found in [28].

Theorem 6.1 (Localization theorem). *Let E be wig-set on a given non-compact Riemann surface R. Given a function $f \in M(E)$, then the following are equivalent*

(i) $f \in M_R(E)$
(ii) $f|_{K\cap E} \in M_R(K \cap E)$ *for all compact sets* $K \subset R$.

PROOF (See [28, §3.3]). That (i) implies (ii) is trivial. To prove (ii) implies (i), assume first that E is a wig1-set of infinite genus (see [12] for the proof when E is of finite genus). Let $D_n, W_n, V_n, r_n, e_n, C_n$ and ε_n be as defined in Sections 2 and 5. The proof will be established by constructing a sequence of functions g_n with the following properties:

(1) $g_n \in M(r_n \cap R)$.
(2) $\|g_n - g_{n-1}\|_{D_{n-1}} \leq \lambda_{n+1}$.
(3) $|g_n - F| \leq \varepsilon_n$ pointwise on E_n, where $F = (f/\varepsilon)\sum_{j=1}^{\infty} \lambda_j$ for an arbitrary but fixed ε and the λ_n's are the positive numbers given in the definition of wig1-sets (and whose sum converges).

Moreover, from (5.2) we also have,

(4) $\varepsilon_n \leq \sum_{j=1}^{n+1} \lambda_j$ on E_n.

As given in the definition of a wig-set, let h_0 be holomorphic on a neighborhood of E_0, without zeroes and such that $\|h_0\| < \lambda_1$. According to Proposition 4.2 (with $D_{-1} = \varnothing$), there exists $\tilde{g}_0 \in M(r_0 \cap R)$ with the property:

$$\|\tilde{g}_0 - \frac{F}{h_0}\|_{E_0} < \frac{1}{2}.$$

Let $g_0 := h_0\tilde{g}_0 \in M(E_0)$, also according to Proposition 4.2, the function g_0 can be approximated by functions in $M(r_0 \cap R)$. Let $g_0' \in M(r_0 \cap R)$ and $|g_0 - g_0'| < \frac{1}{2}\varepsilon_0$, we have,

$$|g_0' - F| < |g_0 - g_0'| + |g_0 - F| < \frac{1}{2}\varepsilon_0 + \frac{1}{2}|h_0| = \varepsilon_0 \qquad \text{pointwise on } E_0.$$

Just for simplicity in the notation we use g_0 instead of g_0' in the following:

Assuming now that we have already constructed $g_0, \ldots, g_{n-1}, g_n$ can be constructed as follows:

Because of our assumption, there exists a meromorphic function h on R satisfying

$$\|h - F\|_{V_n \cap E} < e_{n-1}.$$

Of course, since $r_n \cap R$ is open in R, we also have that $h \in M(r_n \cap R)$. We now get:

$$\|h - g_{n-1}\|_{(\overline{V}_n \setminus W_n)\cap E} \leq \|h - F\| + \|g_{n-1} - F\| \leq e_{n-1} + \|\varepsilon_{n-1}\|_{(\overline{V}_n \setminus W_n)\cap E} = 2e_{n-1}.$$

We now apply the fusion lemma with sets K_1, K_2 and K as given in the previous Section; this yields a function $g_n \in M(r_n \cap R)$ with the properties,

$$|g_n - g_{n-1}| < 2\varepsilon_{n-1}C_n \qquad \text{pointwise on } D_{n-1} \subset K_2$$

and hence because of the assumption on the e_n, we have

$$\|g_n - g_{n-1}\| < \lambda_{n+1}. \tag{6.1}$$

On $E_{n-1} \subset K_2 \cup K$, we also have

$$\begin{aligned} |g_n - F| &\leq |g_n - g_{n-1}| + |g_{n-1} - F| \\ &\leq 2e_{n-1}C_n + \varepsilon_{n-1} = \varepsilon_n \leq \lambda_{n+1} + \sum_{j=1}^{n} \lambda_j = \sum_{j=1}^{n+1} \lambda_j. \end{aligned} \tag{6.2}$$

On $\overline{W}_n \cap E = K_1$, we have:

$$|g_n - F| \leq |g_n - h| + |h - F| \leq 2e_{n-1}C_n + e_{n-1} = e_{n-1}(2C_n + 1) = \varepsilon_n \leq \sum_{j=1}^{n+1} \lambda_j. \tag{6.3}$$

Therefore we have shown that $|g_n - F| \leq \varepsilon_n \leq \sum_{j=1}^{n+1} \lambda_j$ on all of E_n, and the induction is complete.

Because (6.1) holds, the sequence $\{g_n\}$ is uniformly convergent to a function g which is meromorphic (at least) on $\bigcup D_n = R$. From (6.2) and (6.3), we get $\|g - F\|_E \leq \sum_{j=1}^{\infty} \lambda_j$. Moreover for $h := \varepsilon(\sum_1^{\infty} \lambda_j)^{-1} g \in M(R)$, because $F = (f/\varepsilon) \sum_1^{\infty} \lambda_j$, we have $\|h - F\|_E < \varepsilon$.

Since ε was an arbitrary positive number, it follows that $f \in M_R(E)$. This proves the localization theorem for wig1-sets.

Now let E be a wig2-set, and $E = A_1 \cup A_2 \cup \cdots$ a decomposition of E as in the definition of a wig2-set. Let $G_1 \subset G_2 \subset \cdots \subset R$ be an exhaustion of R by relatively compact sets G_j. Because of the local finiteness of the A_j's, we may assume that G_j meets the sets $A_1, \dots, A_j$ and has empty intersections with the remaining A_k's. Now let f be a function on E satisfying (ii) and let δ be a positive number.

A_1 is a wig1-set; let $f_1 \in M(R)$ such that $\|f - f_1\|_{A_1} < \delta/2$. Set $G_0 = \varnothing$ and $f_0 = f_1$. Assume that $f_1, \dots f_{n-1} \in M(R)$ have already been constructed with the properties

$$\|f_j - f_{j-1}\|_{G_{j-1}} \leq \delta 2^{-j} \tag{6.4}$$

and

$$\|f_j - f\|_{A_1 \cup \cdots \cup A_j} \leq \delta(2^{-1} + \cdots + 2^{-j}) \qquad \text{for } j = 1, \dots, n-1. \tag{6.5}$$

Thus $g_n|_{\overline{G}_{n-1} \cup A_1 \cup \cdots \cup A_{n-1}} = f_{n-1}$ and $g_n|_{A_n} = f$ defines a function g_n on the wig1-set $Y_n = G_{n-1} \cup A_1 \cup \cdots \cup A_n$ and this function satisfies (ii) with E replaced by Y_n.

Moreover, there exists $f_n \in M(R)$ satisfying $\|f_n - g_n\|_{X_n} \leq \delta 2^{-n}$. This implies

$$\|f_n - f_{n-1}\|_{G_{n-1}} \leq \delta 2^{-n}$$

$$\|f_n - f\|_{A_1 \cup \cdots \cup A_{n-1}} \leq \delta(2^{-1} + \cdots + 2^{-n})$$

and

$$\|f_n - f\|_{A_n} \leq \delta 2^{-n} \leq \delta(2^{-1} + \cdots + 2^{-n}).$$

Because (6.4) holds, the sequence of functions f_n constructed in this manner converges to a function $F \in M(R)$ that satisfies $\|F - f\|_E < \delta$, in light of (6.5). This completes the proof of the theorem for wig-sets. □

7. Approximation by meromorphic function

Using the localization theorem above, G. Schmieder [28] proved that wig-sets are sets of meromorphic approximation. Recalling that there was a problem with his definition of wig-sets, we now state and prove these theorems for the class of wig-sets as defined in Section 5. The proofs follow verbatim the proofs found in [28].

Theorem 7.1 (see [28, §4]). *Let E be a subset of a noncompact Riemann surface R. If E is a wig-set, then $M(E) \subset M_R(E)$.*

PROOF. This follows from the localization theorem in an analogous way that Proposition 4.2 follows from Proposition 4.1. □

Corollary 7.2 (see [28, §4]). *Let E be a subset of a noncompact Riemann surface R. If E is a wig-set, $f \in M(E)$ and $g \in M(R)$, $g \not\equiv 0$, then there exist a function $F \in M(R)$ such that*

$$|f(p) - F(p)| \leq |g(p)| \qquad \text{for } p \in E.$$

PROOF. It is suffices to approximate uniformly on E the function f/g. □

Corollary 7.3 (see [28, §4]). *Let E be a subset of a noncompact Riemann surface R. Assume that E is a wig-set, $f \in M(E)$, $g \in M(R)$ is nonconstant and bounded on E, and let $Z := \{p \in E : g(p) = 0\}$. Then for each $\delta > 0$, there exist a function $F \in M(R)$ such that*

(i) $\|f - F\|_E < \delta$.
(ii) $f(p) = F(p)$ *for* $p \in Z$.

PROOF. It suffices to approximate the function $(f/g)\|g\|_E$ uniformly on E within the constant δ. □

Acknowledgements. We would like to express our deepest gratitude to the referee who read our manuscript with infinite care and suggested numerous changes and corrections that greatly improved this paper.

References

1. L. V. Ahlfors and L. Sario, *Riemann surfaces*, Princeton Math. Ser., vol. 26, Princeton Univ. Press, Princeton, NJ, 1960.
2. N. U. Arakeljan, *Uniform approximation on closed sets by entire functions*, Izv. Akad. Nauk SSSR Ser. Mat. **28** (1964), 1187–1206 (Russian).
3. H. Behnke and F. Sommer, *Theorie der analytischen Funktionen einer komplexen Veränderlichen*, Die Grundlehren der mathematischen Wissenschaften in Einzeldarstellungen mit besonderer Berücksichtigung der Anwendungsgebiete, vol. 77, Springer, Berlin, 1955.
4. H. Behnke and K. Stein, *Entwicklung analytischer Funktionen auf Riemannschen Flächen*, Math. Ann. **120** (1949), 430–461.
5. E. Bishop, *Subalgebras of functions on a Riemann surface*, Pacific J. Math. **8** (1958), 29–50.
6. A. Boivin and P. M. Gauthier, *Holomorphic and harmonic approximation on Riemann surfaces*, Approximation, Complex Analysis, and Potential Theory (Montréal, QC, 2000) (N. Arakelian, P. M. Gauthier, and G. Sabidussi, eds.), NATO Sci. Ser. II Math. Phys. Chem., vol. 37, Kluwer, Dordrecht, 2001, pp. 107–128.
7. A. Boivin and B. Jiang, *Uniform approximation by meromorphic functions on Riemann surfaces*, J. Anal. Math. **93** (2004), 199–214.
8. J. B. Conway, *Functions of one complex variable.* II, Grad. Texts in Math., vol. 159, Springer, New York, 1995.
9. D. Gaier, *Vorlesungen über Approximation im Komplexen*, Birkhäuser, Basel, 1980 (German); English transl., *Lectures on complex approximation*, translated by R. McLaughlin, Birkhäuser, Boston, MA, 1987.
10. ______, *Remarks on Alice Roth's fusion lemma*, J. Approx. Theory **37** (1983), no. 3, 246–250.
11. P. M. Gauthier, *On the possibility of rational approximation*, Padé and Rational Approximation (Tampa, FL, 1976) (E. B. Saff and R. S. Varga, eds.), Academic Press, New York, 1977, pp. 261–264.
12. ______, *Meromorphic uniform approximation on closed subsets of open Riemann surfaces*, Approximation Theory and Functional Analysis (Campinas, 1977) (J. B. Prolla, ed.), North-Holland Math. Stud., vol. 35, North-Holland, Amsterdam, 1979, pp. 139–158.
13. P. M. Gauthier and W. Hengartner, *Uniform approximation on closed sets by functions analytic on a Riemann surface*, Approximation Theory (Poznań, 1972) (Z. Ciesielski and J. Musielak, eds.), Reidel, Dordrecht, 1975, pp. 63–69.

14. R. E. Greene and S. G. Krantz, *Function theory of one complex variable*, 3rd ed., Grad. Stud. Math., vol. 40, Amer. Math. Soc., Providence, RI, 2006.
15. W. Gröbner and N. Hofreiter, *Integraltafel.* II, 2nd ed., Springer, Wien-Innsbruck, 1958.
16. N. Jin, *On maximal Riemann surfaces*, Hiroshima Math. J. **26** (1996), no. 2, 385–404.
17. J. L. Kelley, *General topology*, Grad. Texts in Math., vol. 27, Springer, New York, 1975. Reprint of the 1955 edition [Van Nostrand, Toronto, ON].
18. R. Narasimhan, *Analysis on real and complex manifolds*, North-Holland Math. Library, vol. 35, North-Holland, Amsterdam, 1985. Reprint of the 1973 edition.
19. A. A. Nersesjan, *Uniform and tangential approximation by meromorphic functions*, Izv. Akad. Nauk Armjan. SSR Ser. Mat. **7** (1972), no. 6, 405–412, 478 (Russian).
20. Ch. Pommerenke, *Boundary behaviour of conformal maps*, Grundlehren Math. Wiss., vol. 299, Springer-Verlag, Berlin, 1992.
21. I. Richards, *On the classification of noncompact surfaces*, Trans. Amer. Math. Soc. **106** (1963), 259–269.
22. A. Roth, *Approximationseigenschaften und Strahlengrenzwerte meromorpher und ganzer Funktionen*, Ph.D. Thesis, Eidgenössische Technische Hochschule (ETH), Zürich, 1938; Comment. Math. Helv. **11** (1938), no. 1, 77–125
23. ———, *Uniform and tangential approximations by meromorphic functions on closed sets*, Canad. J. Math. **28** (1976), no. 1, 104–111.
24. S. Scheinberg, *Uniform approximation by functions analytic on a Riemann surface*, Ann. of Math. (2) **108** (1978), no. 2, 257–298.
25. ———, *Uniform approximation by meromorphic functions having prescribed poles*, Math. Ann. **243** (1979), no. 1, 83–93.
26. ———, *Noninvariance of an approximation property for closed subsets of Riemann surfaces*, Trans. Amer. Math. Soc. **262** (1980), no. 1, 245–258.
27. G. Schmieder, *Über die simultane Approximation zweier meromorpher Funktionen*, Bayer. Akad. Wiss. Math.-Natur. Kl. Sitzungsber. (1983), 133–139 (1984).
28. ———, *Approximation auf abgeschlossenen Teilen Riemannscher Flächen*, Manuscripta Math. **46** (1984), no. 1-3, 165–192.
29. H. Seifert and W. Threlfall, *Seifert and Threlfall: a textbook of topology*, translated by M. A. Goldman, with a preface by J. S. Birman and *Topology of 3-dimensional fibered spaces* by H. Seifert, translated from German by W. Heil, Pure Appl. Math., vol. 89, Academic Press, New York, 1980. Translated from the German edition of 1934.
30. G. Springer, *Introduction to Riemann surfaces*, Addison-Wesley, Reading, MA, 1957; 2nd ed., Chelsea, New York, NY, 1981
31. H. Tietz, *Zur Realisierung Riemannscher Flächen.* II, Math. Ann. **136** (1958), 41–45.
32. A. G. Vituškin, *Analytic capacity of sets in problems of approximation theory*, Uspehi Mat. Nauk **22** (1967), no. 6 (138), 141–199 (Russian); English transl., Russian Math. Surveys **22** (1967), 139–200.

Department of Mathematics and Statistics, University of Calgary, 2500 University Drive NW, Calgary AB T2N 1N4, Canada
E-mail address: `naskarip@ucalgary.ca`

Department of Mathematics, Middlesex College, The University of Western Ontario, London, ON N6A 5B7, Canada
E-mail address: `boivin@uwo.ca`

Centre de Recherches Mathématiques
CRM Proceedings and Lecture Notes
Volume **55**, 2012

On a Family of Outer Functions

Javad Mashreghi

ABSTRACT. We introduce a class of positive functions φ, with $\log\varphi$ integrable, such that the Hilbert transform of $\log\varphi$ is a step function with jumps which are integer multiples of 2π. Such a function creates an outer function Φ which is positive on the unit circle $\mathbb{T}$.

1. Introduction

Let $\varphi \geq 0$ be a measurable function on the unit circle $\mathbb{T}$ such that

$$\log\varphi \in L^1(\mathbb{T}).$$

Then the formula

$$\Phi(z) = \exp\left\{\int_{\mathbb{T}} \frac{\zeta + z}{\zeta - z} \log\varphi(\zeta)\,\mathrm{d}m(\zeta)\right\} \tag{1.1}$$

gives an analytic function on $\mathbb{C} \setminus \mathbb{T}$. We usually restrict z to the open unit disc $\mathbb{D}$ and call Φ the *outer function* created by φ. This function satisfies

$$\lim_{r\to 1^-} |\Phi(r\zeta)| = \varphi(\zeta) \quad (\text{a.e. on } \mathbb{T}), \tag{1.2}$$

and, for $0 < p \leq \infty$,

$$\Phi \in \mathrm{H}^p \iff \varphi \in L^p(\mathbb{T}), \tag{1.3}$$

where H^p denotes the classical Hardy space on $\mathbb{D}$, consisting of all analytic functions f for which

$$\|f\|_p = \sup_{0\leq r<1} \left(\frac{1}{2\pi}\int_0^{2\pi} \left|f\left(r e^{\mathrm{i}\theta}\right)\right|^p \mathrm{d}\theta\right)^{1/p} < \infty.$$

A more precise version of (1.2) is

$$\lim_{r\to 1^-} \Phi(r\zeta) = \varphi(\zeta)\exp\left\{\mathrm{i}\,\widetilde{\log\varphi}(\zeta)\right\} \quad (\text{a.e. on } \mathbb{T}), \tag{1.4}$$

2010 *Mathematics Subject Classification.* Primary 30D55; Secondary 44A15.
Key words and phrases. Outer function, Hilbert transform, extreme points.
This work was supported by NSERC (Canada) and FQRNT (Quebec).
This is the final form of the paper.

where $\widetilde{\log \varphi}$ represents the Hilbert transform of $\log \varphi$, i.e.,

$$\widetilde{\log \varphi}(\mathrm{e}^{\mathrm{i}\theta}) = \lim_{\varepsilon \to 0} \int_{\varepsilon}^{\pi} \frac{\log \varphi(\mathrm{e}^{\mathrm{i}(\theta - t)}) - \log \varphi(\mathrm{e}^{\mathrm{i}(\theta + t)})}{2 \tan(t/2)} \, dt.$$

See [3, p. 109].

In this short note, our goal is to introduce a class of functions φ for which

$$\exp\Big\{ \mathrm{i} \, \widetilde{\log \varphi}(\zeta) \Big\} = 1 \quad \text{(a.e. on } \mathbb{T}).$$

In other words, $\widetilde{\log \varphi}$ just assumes integral multiples of 2π almost everywhere on $\mathbb{T}$. Since the function φ that we introduce is $\mathcal{C}^\infty$ on $\mathbb{T} \setminus E$, where the exceptional set E is the union of the sequence of zeros on $\mathbb{T}$ (which tends to 1) and the point 1 itself, then $\widetilde{\log \varphi}$ is a step function on complementary intervals of E. See [3, Theorem 5.6]. At the end, we exploit this construction to generalize a result of J. Inoue [2] about the existence of outer nonexposed elements in H^1.

2. An outer function

Let $(\alpha_n)_{n \geq 1}$ be a sequence on the unit circle $\mathbb{T}$ which converges to the point 1. Since

$$\left| 1 - \frac{\alpha_n - z}{1 - z} \right| = \frac{|1 - \alpha_n|}{|1 - z|} \qquad (n \geq 1),$$

the condition

$$\sum_{n \geq 1} |1 - \alpha_n| < \infty \tag{2.1}$$

ensures that the product

$$f(z) = \prod_{n \geq 1} \frac{\alpha_n - z}{1 - z}$$

defines an analytic function on $\mathbb{C} \setminus \{1\}$ with zeros precisely at α_n, $n \geq 1$. If we write $\alpha_n = \mathrm{e}^{\mathrm{i}\theta_n}$, where $0 < |\theta_n| \leq \pi$, then (2.1) is equivalent to

$$\sum_{n \geq 1} |\theta_n| < \infty. \tag{2.2}$$

Even though each factor $(\alpha_n - z)/(1 - z)$ is an outer function, the condition (2.1) is not enough to ensure that f is also outer. However, a slightly stronger growth restriction does the job.

Lemma 2.1. *Let $(\theta_n)_{n \geq 1}$ be a sequence of real numbers such that $\theta_n \to 0$ and*

$$\sum_{n \geq 1} |\theta_n| \log(1/|\theta_n|) < \infty. \tag{2.3}$$

Put

$$f(z) = \prod_{n \geq 1} \frac{\mathrm{e}^{\mathrm{i}\theta_n} - z}{1 - z} \qquad (z \in \mathbb{D}).$$

Then f is an outer function.

PROOF. Let

$$\varphi_n(z) = \frac{1-e^{-i\theta_n}z}{1-z} \quad (n \ge 1).$$

Since this factor is the quotient of two outer functions, we can write

$$\varphi_n(z) = \exp\left\{\int_{\mathbb{T}} \frac{\zeta+z}{\zeta-z} \log|\varphi_n(\zeta)|\, dm(\zeta)\right\} \quad (z \in \mathbb{D}).$$

Since $\varphi_n(0) = 1$, we have

$$\int_{\mathbb{T}} \log|\varphi_n(\zeta)|\, dm(\zeta) = 0,$$

a fact which can also be directly verified. In return, the above identity implies

$$\int_{\mathbb{T}} \log^{+}|\varphi_n(\zeta)|\, dm(\zeta) = \int_{\mathbb{T}} \log^{-}|\varphi_n(\zeta)|\, dm(\zeta),$$

and thus

$$\int_{\mathbb{T}} \big|\log|\varphi_n(\zeta)|\big|\, dm(\zeta) = 2\int_{\mathbb{T}} \log^{-}|\varphi_n(\zeta)|\, dm(\zeta). \tag{2.4}$$

But, we can estimate the last integral. Without loss of generality, assume that $0 < \theta_n < 1$. Since

$$|\varphi_n(e^{i\theta})| = \left|\frac{\sin\big((\theta-\theta_n)/2\big)}{\sin(\theta/2)}\right|,$$

the inequality $|\varphi_n| \le 1$ holds on the interval $[\theta_n/2, \pi+\theta_n/2]$. Thus,

$$\begin{aligned}
\int_{\mathbb{T}} \log^{-}|\varphi_n(\zeta)|\, dm(\zeta) &= -\frac{1}{2\pi}\int_{\theta_n/2}^{\pi+\theta_n/2} \log\left|\frac{\sin((\theta-\theta_n)/2)}{\sin(\theta/2)}\right| d\theta \\
&= -\frac{1}{\pi}\int_{-\theta_n/4}^{\pi/2-\theta_n/4} \log|\sin\vartheta|\, d\vartheta + \frac{1}{\pi}\int_{\theta_n/4}^{\pi/2+\theta_n/4} \log|\sin\vartheta|\, d\vartheta \\
&= -\frac{2}{\pi}\int_{0}^{\theta_n/4} \log|\tan\vartheta|\, d\vartheta.
\end{aligned}$$

Finally, integration by parts gives

$$\int_{\mathbb{T}} \log^{-}|\varphi_n(\zeta)|\, dm(\zeta) \asymp \theta_n \log 1/\theta_n \quad (n \ge 1).$$

The constants involved in the above estimation do not depend on n. Therefore, the condition (2.3) and (2.4) ensure that the series

$$\sum_{n\ge 1} \log|\varphi_n|$$

converges in $L^1(\mathbb{T})$-norm. This allows us to write

$$f(z)/f(0) = \prod_{n\ge 1} \frac{1-e^{-i\theta_n}z}{1-z} = \exp\left\{\int_{\mathbb{T}} \frac{\zeta+z}{\zeta-z}\left(\sum_{n\ge 1}\log|\varphi_n(\zeta)|\right) dm(\zeta)\right\} \quad (z \in \mathbb{D}).$$

This representation means that f is an outer function. □

It is interesting to know to what extent Lemma 2.1 is reversible. More precisely, if f is an outer function, what can we say about $(\theta_n)_{n\ge 1}$? For example, can we say the condition (2.3) is also necessary?

Theorem 2.2. *Let $(\theta_n)_{n\geq 1}$ be a sequence of real numbers such that*

$$\lim_{n\to\infty} \theta_n = 0$$

and

$$\sum_{n\geq 1} |\theta_n| \log(1/|\theta_n|) < \infty.$$

Put

$$F(z) = \prod_{n\geq 1} \frac{(e^{i\theta_n} - z)(1 - e^{-i\theta_n} z)}{(1-z)^2}. \tag{2.5}$$

The F has the following properties:

(1) *F is an analytic function on $\mathbb{C} \setminus \{1\}$;*
(2) *F is outer on $\mathbb{D}$;*
(3) *F has zeros of order 2 at each $e^{i\theta_n}$, $n \geq 1$;*
(4) *For each $e^{i\theta} \in \mathbb{T} \setminus \{1\}$,*

$$F(e^{i\theta}) \geq 0.$$

(5) *$F \notin H^{1/2}$.*
(6) *On each complementary interval in $\mathbb{T} \setminus \{1, e^{i\theta_n} : n \geq 1\}$, the Hilbert transform of $\log|F|$ exists and is a constant function. The constant is an integer multiple of 2π.*

PROOF. We have

$$F(z) = \prod_{n\geq 1} \frac{(z-\alpha_n)(-1+\bar{\alpha}_n z)}{(1-z)^2} = \prod_{n\geq 1} \bar{\alpha}_n \times \left(\prod_{n\geq 1} \frac{z-\alpha_n}{z-1} \right)^2.$$

Hence, by Lemma 2.1, F is an analytic function on $\mathbb{C} \setminus \{1\}$, with zeros of order 2 at α_n, and outer on $\mathbb{D}$. Moreover, for each $e^{i\theta} \in \mathbb{T} \setminus \{1\}$, we have

$$\frac{(e^{i\theta} - e^{i\theta_n})(1 - e^{-i\theta_n} e^{i\theta})}{(e^{i\theta} - 1)(1 - e^{i\theta})} = \frac{(1 - e^{i(\theta_n - \theta)})(1 - e^{i(\theta - \theta_n)})}{(1 - e^{-i\theta})(1 - e^{i\theta})} = \left| \frac{1 - e^{i(\theta - \theta_n)}}{1 - e^{i\theta}} \right|^2 \geq 0.$$

Hence, $F(e^{i\theta}) \geq 0$. The last claim now follows from (1.4) an that, on each complementary interval, $\widetilde{\log|F|}$ is a $\mathcal{C}^\infty$ function. Hence, it has to be a constant function which is an integer multiple of 2π. A uniqueness theorem, which is a consequence of the canonical factorization theorem, says that if $g \in H^{1/2}$ and g is real on $\mathbb{T}$, then g is constant. Hence, F cannot be in $H^{1/2}$. □

3. A nonexposed extreme function

Let $\mathcal{X}$ be a Banach space, and let Ω be a closed convex subset of $\mathcal{X}$. A point $p \in \Omega$ is called an *extreme point* of Ω if $\Omega \setminus \{p\}$ is convex. The set of all extreme points of $\mathfrak{B}(\mathcal{X})$, the closed unit ball $\mathcal{X}$, is denoted by $\text{Ext}(\mathcal{X})$. For Hardy spaces a complete description of the extreme points of the closed unit ball is available. For $1 < p < +\infty$, we have

$$\text{Ext}(H^p) = \{f \in H^p : \|f\|_p = 1\}.$$

As a matter of fact, this result is a simple consequence of the fact that

$$\text{Ext}(L^p) = \{f \in L^p : \|f\|_p = 1\}.$$

However, the story is completely different for H^1 and H^∞. According to a result of de Leeuw – Rudin [1],

$$\mathrm{Ext}(H^1) = \{f \in H^1 : \|f\|_1 = 1, f \text{ is outer}\}.$$

In their paper, they also show that $f \in H^\infty$ is an extreme point of the closed unit ball of H^∞ if and only if

$$\int_0^{2\pi} \log\left(1 - |f(e^{it})|\right) dt = -\infty.$$

In particular, any inner function is an extreme point. But, as they pointed out, this result have been found jointly by Arens, Buck, Carleson, Hoffman, and Royden, during the Princeton Conference on Functions of a Complex Variable in September 1957.

A nonzero function $f \in H^1$ is called *rigid* if the functional equation

$$\frac{g}{|g|} = \frac{f}{|f|}, \qquad (g \not\equiv 0, g \in H^1),$$

holds if and only if $g = \lambda f$, where λ is a positive constant. A rigid function is surely an extreme point of H^1. However, the converse is not true. The first construction of this type was given by J. Inoue [2]. He constructed the outer function in Theorem 2.2 with the specific sequence $\alpha_n = e^{i/n^2}$. But, to verify that F is outer, he used another method by appealing to Zygmund's LlogL theorem. In the following, we generalize his construction.

For simplicity, assume that $1 > \theta_1 > \theta_2 > \cdots > 0$ and $(\theta_n)_{n\geq 1}$ fulfills (2.3). Consider a small arc centered at $e^{i\theta_n}$, say $I_n = \left(e^{\theta_n - \varepsilon_n}, e^{\theta_n + \varepsilon_n}\right)$, such that the arcs are disjoint and

$$\left|F(e^{i\theta})\right| \leq 1 \quad (e^{i\theta} \in I_n).$$

For disjointness of I_n, we need to assume that

$$\theta_n - \varepsilon_n > \theta_{n+1} + \varepsilon_{n+1} \quad (n \geq 1). \tag{3.1}$$

Note that this restriction implies

$$0 < \varepsilon_n < \theta_n \quad (n \geq 1). \tag{3.2}$$

We have certain liberty in choosing ε_n as small as we wish. But, the price is paid below in the definition of g.

Define

$$g(e^{i\theta}) = \begin{cases} \min\left\{1, \left|F(e^{i\theta})\right|^{-1}\right\}, & \text{if } e^{i\theta} \in \mathbb{T} \setminus \bigcup_{n\geq 1} I_n, \\ t_n/\varepsilon_n, & \text{if } e^{i\theta} \in I_n, \end{cases}$$

where $t_n \in (0, 1)$ is chosen such that

$$\sum_{n\geq 1} t_n < \infty, \tag{3.3}$$

and

$$\sum_{n\geq 1} \theta_n \log \frac{1}{t_n} < \infty, \tag{3.4}$$

For example, (3.3) and (3.4) are satisfied if we take $t_n = \theta_n^k$, where $k \geq 1$ is a fixed exponent. The reason for the above assumptions is clarified below.

Firstly, by (3.3), we have

$$\begin{aligned}\int_{\mathbb{T}} g(\mathrm{e}^{\mathrm{i}\theta})\,\mathrm{d}\theta &= \int_{\mathbb{T}\setminus\bigcup_{n\geq 1} I_n} g(\mathrm{e}^{\mathrm{i}\theta})\,\mathrm{d}\theta + \sum_{n\geq 1}\int_{I_n} g(\mathrm{e}^{\mathrm{i}\theta})\,\mathrm{d}\theta\\ &\leq \int_{\mathbb{T}\setminus\bigcup_{n\geq 1} I_n} \mathrm{d}\theta + 2\sum_{n\geq 1} t_n < \infty.\end{aligned}$$

In other words,

$$g \in L^1(\mathbb{T}). \tag{3.5}$$

Secondly, by (3.2) and (3.4), we have

$$\begin{aligned}\int_{\mathbb{T}} \big|\log g(\mathrm{e}^{\mathrm{i}\theta})\big|\, dd\theta &= \int_{\mathbb{T}\setminus\bigcup_{n\geq 1} I_n} \big|\log g(\mathrm{e}^{\mathrm{i}\theta})\big|\,\mathrm{d}\theta + \sum_{n\geq 1}\int_{I_n} \big|\log g(\mathrm{e}^{\mathrm{i}\theta})\big|\,\mathrm{d}\theta\\ &\leq \int_{\mathbb{T}} \Big|\log\big|F(\mathrm{e}^{\mathrm{i}\theta})\big|\Big|\,\mathrm{d}\theta + \sum_{n\geq 1} 2\varepsilon_n|\log t_n/\varepsilon_n|\\ &\leq \|\log|F|\|_1 + 2\sum_{n\geq 1}\varepsilon_n \log\frac{1}{\varepsilon_n} + 2\sum_{n\geq 1}\varepsilon_n\log\frac{1}{t_n}\\ &\leq \|\log|F|\|_1 + 2\sum_{n\geq 1}\theta_n \log\frac{1}{\theta_n} + 2\sum_{n\geq 1}\theta_n\log\frac{1}{t_n} < \infty.\end{aligned}$$

Hence, we can construct the unique outer function G such that $|G| = g$, almost everywhere on $\mathbb{T}$, and $G(0) > 0$, i.e.,

$$G(z) = \exp\left\{\int_{\mathbb{T}} \frac{\zeta + z}{\zeta - z}\log g(\zeta)\,\mathrm{d}m(\zeta)\right\} \quad (z \in \mathbb{D}).$$

In the light of (1.3), the condition (3.5) ensures that $G \in \mathrm{H}^1$. Moreover, the construction of G is such that

$$\big|F(\mathrm{e}^{\mathrm{i}\theta})G(\mathrm{e}^{\mathrm{i}\theta})\big| \leq \begin{cases} 1, & \text{if } \mathrm{e}^{\mathrm{i}\theta} \in \mathbb{T}\setminus\bigcup_{n\geq 1} I_n,\\ t_n/\varepsilon_n, & \text{if } \mathrm{e}^{\mathrm{i}\theta} \in I_n.\end{cases}$$

This observation shows that

$$\int_{\mathbb{T}} \big|F(\mathrm{e}^{\mathrm{i}\theta})G(\mathrm{e}^{\mathrm{i}\theta})\big|\,\mathrm{d}\theta \leq \int_{\mathbb{T}\setminus\bigcup_{n\geq 1} I_n} \mathrm{d}\theta + 2\sum_{n\geq 1} t_n < \infty.$$

Therefore, the outer function FG is also in H^1.

By part (iv) of Theorem 2.2, the normalized outer function $FG/\|FG\|_1$ has the same argument as the normalized outer function $G/\|G\|_1$ at almost all points of $\mathbb{T}$. Therefore, $G/\|G\|_1$ and $FG/\|FG\|_1$ are not rigid functions. This means that $G/\|G\|_1$ is an extreme point of the unit ball of H^1 which is not an exposed point. The same statement holds for $FG/\|FG\|_1$.

References

1. K. de Leeuw and W. Rudin, *Extreme points and extremum problems in H_1*, Pacific J. Math. **8** (1958), 467–485.
2. J. Inoue, *An example of a nonexposed extreme function in the unit ball of H^1*, Proc. Edinburgh Math. Soc. (2) **37** (1994), no. 1, 47–51.
3. J. Mashreghi, *Representation theorems in Hardy spaces*, London Math. Soc. Stud. Texts, vol. 74, Cambridge Univ. Press, Cambridge, 2009.

4. D. Sarason, *Exposed points in* H^1. I, The Gohberg Anniversary Collection, Vol. II (Calgary, AB, 1988) (H. Dym, S. Goldberg, M. A. Kaashoek, and P. Lancaster, eds.), Oper. Theory Adv. Appl., vol. 41, Birkhäuser, Basel, 1989, pp. 485–496.

Département de mathématiques et de statistique, Université Laval, Québec, QC G1K 7P4, Canada

E-mail address: Javad.Mashreghi@mat.ulaval.ca

Centre de Recherches Mathématiques
CRM Proceedings and Lecture Notes
Volume **55**, 2012

On C^m-Subharmonic Extension Sets of Walsh-Type

P. V. Paramonov

ABSTRACT. This survey is an extended version of the lecture given at the Conference "Complex Analysis and Potential Theory" (Montréal, June 20 – 23, 2011). The possibility of C^m-extension of subharmonic functions from compact sets onto $\mathbb{R}^N$ is discussed.

1. Introduction

We refer the reader to the papers [2, 3] and literature given in them for results on continuous extension of subharmonic functions. Several different settings of the problem can be found there mostly connected with *Runge*-type extensions, meaning that the functions we wish to extend (subharmonically) from a given closed set X are assumed subharmonic in a *neighbourhood* (depending on the function) of X.

In this survey we discuss the main results of the papers [4, 6–8]. The last theorem of the present paper (Theorem 4.1) is new. Here we deal with the following C^m-subharmonic extension problem of *Walsh*-type, that is where the functions to be (properly) extended from a given compact set X onto $\mathbb{R}^N$ are assumed to be subharmonic only in the *interior* of X. This type of extension problem was also mentioned in [3, p. 52]. A geophysical motivation for this problem is given at the end of Section 2 below.

Problem 1.1. Given $N \geq 2$ and $m \geq 0$, for which compact sets X in $\mathbb{R}^N$ can every function $f \in C^m(X)$, subharmonic in X°, be extended to a function F subharmonic and C^m on all of $\mathbb{R}^N$ such that

$$\|F\|_{BC^m(\mathbb{R}^N)} \leqslant A\|f\|_{C^m(X)}, \quad N \geq 3, \tag{1.1}$$

with $A > 0$ depending only on N, m, X? For $N = 2$ estimate (1.1) should be slightly changed (see later).

A compact set X satisfying this property is called a C^m*-subharmonic extension set*. Briefly speaking, it is shown ([4, 6–8, 10]), that for all $m \in [1, 3)$ any closed Lyapunov ($C^{1+\delta}$-smooth, $\delta > 0$) domain X in $\mathbb{R}^N$ with connected complement is a C^m-subharmonic extension set. In Theorem 4.1 below it is proved that, under

2010 *Mathematics Subject Classification.* Primary 31B05; Secondary 30E10.

The research was partially supported by the Program "Leading Scientific Schools of the Russian Federation" (grant NSh-3476.2010.1).

This is the final form of the paper.

some natural restrictions, a finite union of disjoint C^m-subharmonic extension sets is also a C^m-subharmonic extension set.

The following two examples show that the cases $m \in [0,1)$ and $X^\circ = \varnothing$ are not of primary interest in the context of our problem. The first example is based just on the Poisson formula.

Example 1.2. For $t \in [-\frac{1}{2}, \frac{1}{2}]$ define the C^1-function $f(t) = -|t|/\big|\log|t|\big|$, $f(0) = 0$. Then, there do not exist $\delta > 0$ and a function F, continuous and subharmonic on $\{\mathbf{x} \in \mathbb{R}^2 \mid |\mathbf{x}| < \delta\}$, such that $F(x_1, 0) = f(x_1)$ for $|x_1| < \delta$.

The second example then can be obtained using conformal mappings.

Example 1.3. Let B be an open disc in $\mathbb{R}^2$. Then one can find $g \in C^1(\partial B)$ such that the solution G of the Dirichlet problem in B with boundary data g is not in $C^1(\overline{B})$. Moreover, $G \in C^m(\overline{B})$ for all $m \in [0,1)$, but G cannot be extended continuously to a subharmonic function in any neighbourhood of $\overline{B}$.

Thus, it is natural to begin with the following case.

2. C^1-subharmonic extension from Lyapunov-Dini domains onto $\mathbb{R}^N$

A function $\varepsilon(\cdot) \in C\big([0,+\infty)\big)$ with the properties $\varepsilon(0) = 0$, $\varepsilon\colon (0,+\infty) \to (0,+\infty)$, $\varepsilon(t)$ nondecreasing and $\varepsilon(t)/t$ decreasing on $(0,+\infty)$, and

$$\int_0^1 \frac{\varepsilon(t)}{t}\, \mathrm{d}t < +\infty,$$

is called a function of *Dini-class.*

A bounded domain D in $\mathbb{R}^N$ ($N \in \{2,3,\dots\}$) with C^1-smooth boundary $S = \partial D$ is called a *Lyapunov – Dini* domain (briefly, (L-D)-domain), if there exists a function of Dini-class $\varepsilon(\cdot)$ (called a *Dini-function* for D) such that for each $\mathbf{x}$ and $\mathbf{y}$ on S one has

$$|\mathbf{n_x} - \mathbf{n_y}| \le \varepsilon(|\mathbf{x} - \mathbf{y}|),$$

where $\mathbf{n_x}$ is the *inward* (with respect to D) unit normal to S at the point $\mathbf{x} \in S$.

As usual, $\overline{E}$, E° and ∂E denote respectively the closure, the interior and the boundary of the set $E \neq \varnothing$ in $\mathbb{R}^N$ and $\|f\|_E = \sup_{\mathbf{x}\in E}|f(\mathbf{x})|$ denotes the uniform norm of the function f on E (with $\|\cdot\| = \|\cdot\|_{\mathbb{R}^N}$). For an open set Ω in $\mathbb{R}^N$ we denote by $H(\Omega)$ (respectively, $SH(\Omega)$) the class of all real functions *harmonic* (respectively, *subharmonic*) on Ω.

Recall that for a compact set X in $\mathbb{R}^N$ and $m \in \{0,1,2,\dots\}$, one denotes by $C^m(X)$ the space $C^m(\mathbb{R}^N)|_X$ with the norm

$$\|f\|_{m,X} = \inf\|F\|_m,$$

where the latter infimum is taken over all functions $F \in C^m(\mathbb{R}^N)$ with the property $F|_X = f$ and $\|F\|_m := \max_{|\beta|\le m}\{\|\partial^\beta F\|\} < +\infty$. Notice that, if $X = \overline{X^\circ}$ then for each $f \in C^m(X)$ the derivatives

$$\partial^\beta f(\mathbf{x}) = \frac{\partial^{|\beta|} f(\mathbf{x})}{\partial x_1^{\beta_1} \cdots \partial x_N^{\beta_N}}$$

with $|\beta| := \beta_1 + \dots + \beta_N \le m$ ($\beta = (\beta_1,\dots,\beta_N)$, $\beta_n \in \{0,1,2,\dots\}$, $\mathbf{x} = (x_1,\dots,x_N)$) are uniquely defined for all $\mathbf{x} \in X$, and so in this case the space $C^m(X)$ can be identified with the space $C^m_{\mathrm{jet}}(X)$ of Whitney-jets (see [13]). If $m = 0$, we omit the indexes m in the notations $C^m(X)$ and $\|\cdot\|_{m,X}$.

For an open set Ω in $\mathbb{R}^N$ and $m \in \{0, 1, 2, \dots\}$, one denotes by $BC^m(\Omega)$ the space of all functions f in $C^m(\Omega)$ (which is the space of all real continuous functions in Ω having continuous derivatives $\partial^\beta f(\mathbf{x})$ in Ω for all β with $|\beta| \le m$) that have finite norm

$$\|f\|_{m,\Omega} = \max_{|\beta|\le m} \{\|\partial^\beta f\|_\Omega\}.$$

Definitions of the *normed* spaces $BC^m(X)$ (and *Fréchet* spaces $C^m(X)$) for all $m \ge 0$ and open or *closed* sets $X \subset \mathbb{R}^N$ can be found in Section 5 below. For *compact* sets X the spaces $BC^m(X)$ and $C^m(X)$ can be identified.

In what follows we fix $N \in \{2, 3, \dots\}$, an arbitrary function of Dini-class $\varepsilon(\cdot)$ and $d \in (0, +\infty)$. Let D be an (L-D)-domain in $\mathbb{R}^N$ with Dini-function $\varepsilon(\cdot)$ and $\operatorname{diam} D < d$ (presumably, a domain D with such parameters exists). Set $D_\mathrm{o} = \mathbb{R}^N \setminus \overline{D}$. The constant $A \in (0, +\infty)$ in the following Theorems 2.1 – 2.7 (see [4, 7]) depends only on N, ε and d.

Theorem 2.1. *Let $u \in H(D) \cap C^1(\overline{D})$, and let u_o be the (unique) solution of the Dirichlet problem in D_o with boundary data $u_\mathrm{o}|_{\partial D_\mathrm{o}} = u|_{\partial D_\mathrm{o}}$ (in the unbounded component of the set D_o we additionally require that $u_\mathrm{o}(\infty) = 0$ for $N \ge 3$ or $|u_\mathrm{o}(\infty)| < +\infty$ for $N = 2$, where $u_\mathrm{o}(\infty) = \lim_{|\mathbf{x}|\to+\infty} u_\mathrm{o}(\mathbf{x})$ must exist). Then $u_\mathrm{o} \in BC^1(\overline{D_\mathrm{o}})$ and*

$$\|u_\mathrm{o}\|_{1,\overline{D_\mathrm{o}}} \le A\|u\|_{1,\overline{D}}.$$

We shall say that u_o is the C^1-*reflection* of the function u over the boundary S of the domain D. So one sees that, in contrast to the Dirichlet problem (which solutions in general do not preserve class C^1 even in balls, see Example 1.3 above), the *reflection operator* $u \to u_\mathrm{o}$ preserves class C^1 (in conditions of Theorem 2.1). In Theorem 2.7 below we give a rather useful refinement of Theorem 2.1. From Theorem 2.1 we obtain also the following result on C^1-subharmonic extension.

Theorem 2.2. *Suppose that D has connected complement and $u \in H(D) \cap C^1(\overline{D})$. Then one can find a function $F \in SH(\mathbb{R}^N) \cap BC^1(\mathbb{R}^N)$ for $N \ge 3$ (or $F \in SH(\mathbb{R}^N) \cap C^1(\mathbb{R}^N)$ for $N = 2$) such that $F|_{\overline{D}} = u$ and*

$$\|F\|_1 \le A\|u\|_{1,\overline{D}}, \qquad N \ge 3,$$
$$\|\nabla F\| \le A\|\nabla u\|_{\overline{D}}, \quad N = 2.$$

It is well known that (in our context) $\mathbb{R}^N \setminus D$ (and consequently D_o) is connected if and only if $S = \partial D$ is connected. Also, it can be easily seen that the requirement of connectedness of $\mathbb{R}^N \setminus D$ is *necessary* in Theorems 2.2, 2.4, 3.1 and 3.2 (see [3, p. 48]). The following localization property for $N = 2$ holds even for every Jordan domain (see [4, Theorem 2.5]).

Corollary 2.3. *Suppose that D has connected complement and $f \in SH(D) \cap C^1(\overline{D})$. If for each $\mathbf{a} \in \partial D$ there are a ball $B_\mathbf{a}$ centered at $\mathbf{a}$ and a function $g_\mathbf{a} \in SH(B_\mathbf{a}) \cap C^1(\overline{B_\mathbf{a}})$ with $g_\mathbf{a}|_{B_\mathbf{a}\cap D} = f$, then there exists a function $F \in SH(\mathbb{R}^N) \cap BC^1(\mathbb{R}^N)$ for $N \ge 3$ (respectively, $F \in SH(\mathbb{R}^N) \cap C^1(\mathbb{R}^N)$ for $N = 2$) such that $F|_{\overline{D}} = f$ and $\|F\|_1 < +\infty$ (respectively, $\|\nabla F\| < +\infty$ for $N = 2$).*

Our first main result is the following.

Theorem 2.4. *Suppose D has connected complement and $\varepsilon(\cdot)$ satisfies the following "logarithmic" Dini-condition*:

$$\int_0^1 \frac{\varepsilon(t)}{t} \log\left(\frac{1}{t}\right) \mathrm{d}t < +\infty. \tag{2.1}$$

Then, for each function $f \in SH(D) \cap C^1(\overline{D})$ one can find $F \in SH(\mathbb{R}^N) \cap BC^1(\mathbb{R}^N)$ for $N \geq 3$ (or $F \in SH(\mathbb{R}^N) \cap C^1(\mathbb{R}^N)$ for $N = 2$) such that $F|_{\overline{D}} = f$ and

$$\|F\|_1 \leq A\|f\|_{1,\overline{D}}, \qquad N \geq 3,$$
$$\|\nabla F\| \leq A\|\nabla f\|_{\overline{D}}, \quad N = 2.$$

Moreover, for any $\delta > 0$ the function F can be chosen to be harmonic outside the δ-neighbourhood of the set $\overline{D}$ (then, of course, the constant A additionally depends on δ).

Remark 2.5. The function $\varepsilon(t) = t^\alpha$ satisfies (2.1) for each $\alpha \in (0, 1]$, and so each closed Lyapunov domain with connected complement is a C^1-subharmonic extension set.

Remark 2.6. The functions $\varepsilon(t) = \varepsilon_p(t) = 1/\big(\log(1/t)\big)^p$ $(0 < t \leq \mathrm{e}^{-p-1})$, $\varepsilon(t) = \varepsilon(\mathrm{e}^{-p-1})$ $(t > \mathrm{e}^{-p-1})$, satisfy (2.1) if and only if $p > 2$. In [4, Theorem 3.1] it was proved that the (L-D)-condition is in some sense necessary in Theorem 2.4. For instance, in the scale of functions $\varepsilon_p(\cdot)$ the extension property (in the sense of Theorem 2.4) fails for $p \in (0, 1]$. For $p > 1$, domains with the Dini-function $\varepsilon_p(\cdot)$ are (L-D)-domains. The cases $p > 2$ are covered by Theorem 2.4. The cases $p \in (1, 2]$ still remain unsolved.

Theorem 2.4 is based on a constructive, but rather complicated technique called "C^1-reflection of subharmonic functions." The first main step in this direction is based on the following theorem.

Theorem 2.7. *Let D be an (L-D)-domain in $\mathbb{R}^N$ with Dini-function $\varepsilon(\cdot)$ satisfying the logarithmic Dini-condition*

$$\int_0^1 \frac{\varepsilon(t)}{t} \log\left(\frac{1}{t}\right) \mathrm{d}t < +\infty.$$

Let $\psi \in C^1(S)$, $S = \partial D$, and u (respectively, u_{o}) be the solution of the Dirichlet problem in D (respectively, in $D_{\mathrm{o}} = \mathbb{R}^N \setminus \overline{D}$) with boundary data ψ. Let $\mathbf{a} \in S$ and $\mathbf{x} \in D$ be such that $\mathbf{x} - \mathbf{a} = |\mathbf{x} - \mathbf{a}|\mathbf{n}_{\mathbf{a}}$ with $|\mathbf{x} - \mathbf{a}| < r_0 = r_0(N, d, \varepsilon)$ (small enough). Then

$$\big|\nabla u(\mathbf{x}) - \big(\nabla u_{\mathrm{o}}(\mathbf{x}^*_{\mathbf{a}})\big)^*_{\mathbf{a}}\big| \leq A\|\psi\|_{1,S}. \tag{2.2}$$

In the latter estimate $(\cdot)^*_{\mathbf{a}}$ denotes the symmetrization (applied to the corresponding point or vector) with respect to the hyperplane $T_{\mathbf{a}}$ tangent to S at the point $\mathbf{a} \in S$.

The second main step in the proof of Theorem 2.4 is a special construction of C^1-reflection of the *Riesz measure* of f, $\mu = \Delta f$, from D onto D_{o} (let μ^* be the resulting measure in D_{o}, see [7, Section 3]). Between the *Green potentials* of the measures μ in D and μ^* in D_{o} there is a connection analogous to (2.2).

It is worth mentioning that in $\mathbb{R}^3$ our extension results have a very natural physical interpretation. If, in the notations of Theorem 2.4, we consider the positive measure $\sigma = \Delta F$ (in the distributional sense) as the "mass distribution" on the "rigid body" $\Sigma = \operatorname{Supp}\sigma$ then, by Newton's classical law of gravitation, the vector field $-4\pi\gamma\nabla(\Phi * \sigma)$ is precisely the *gravitational field*, induced by σ. Here $\Phi(x) = -1/(4\pi|x|)$ is the standard fundamental solution for the Laplacian Δ in $\mathbb{R}^3$, and γ is the universal gravitational constant. Therefore, under the hypotheses of Theorem 2.4, given $f \in SH(D) \cap C^1(\overline{D})$, one can find a positive compactly supported measure σ such that its (globally defined) *gravitation potential* $4\pi\gamma F = 4\pi\gamma\Phi * \sigma$ is equal to $4\pi\gamma f + \text{const}$ on $\overline{D}$. In particular, ∇F represents a continuous extension of the *vector field* ∇f. In Theorem 3.1 below, in the case $m = 2$, we also preserve the continuity of the *density* $\mathrm{d}\mu/\,\mathrm{d}\mathbf{x}$ of the Riesz measure $\mu = \Delta f$ with respect to the Lebesque measure $\mathrm{d}\mathbf{x}$. See [1] and references therein for some concrete applications of potential theory to the theory of gravitation.

3. C^m-subharmonic extension from Lyapunov – Dini domains onto $\mathbb{R}^N$

Our second main result concerns the cases $m \in (1,3)$ (see [6, 8]).

Theorem 3.1. *Suppose D is an (L-D)-domain in $\mathbb{R}^N$ with connected complement, $N \geq 2$, and let $m \in (1,3)$. There is a constant $A = A(N,m,d,\varepsilon) > 0$ such that for every $f \in C^m(\overline{D}) \cap SH(D)$ one can find $F \in BC^m(\mathbb{R}^N) \cap SH(\mathbb{R}^N)$ for $N \geq 3$ (or $F \in C^m(\mathbb{R}^N) \cap SH(\mathbb{R}^N)$ for $N = 2$) with $F|_{\overline{D}} = f$ and*

$$\|F\|_m \leqslant A\|f\|_{m,\overline{D}}, \qquad N \geq 3,$$
$$\|\nabla F\|_{m-1} \leqslant A\|\nabla f\|_{m-1,\overline{D}}, \quad N = 2.$$

For $m \in [3, +\infty)$ an analog of Theorem 3.1 does not hold even for balls in $\mathbb{R}^N$, but the statement of Theorem 3.1 remains valid for functions $f \in C^m(\overline{D}) \cap H(D)$. Jointly with Theorem 2.2 we obtain the following result.

Theorem 3.2. *Let D be an (L-D)-domain in $\mathbb{R}^N$ with connected complement, $N \geq 2$, and let $m \geq 1$. There is a constant $A = A(N,m,d,\varepsilon) > 0$ such that for every function $u \in H(D) \cap C^m(\overline{D})$ one can find $F \in SH(\mathbb{R}^N) \cap BC^m(\mathbb{R}^N)$ for $N \geq 3$ (or $F \in SH(\mathbb{R}^N) \cap C^m(\mathbb{R}^N)$ for $N = 2$) with $F|_{\overline{D}} = u$ and*

$$\|F\|_m \leqslant A\|u\|_{m,\overline{D}}, \qquad N \geq 3,$$
$$\|\nabla F\|_{m-1} \leqslant A\|\nabla u\|_{m-1,\overline{D}}, \quad N = 2.$$

The proofs of the above formulated results are based on: Special properties of Green functions in (L-D)-domains (see [14] and some refinements of these results in [7]); Vitushkin's localization technique [11] (see also [5, 9]); Warschawski's regularity theorems for conformal mappings [12]; and a priori estimates.

In connection with Theorem 3.2 we say that a compact set X in $\mathbb{R}^N$ is a C^mH-*subharmonic extension set* ($m \geq 1$) if there is a constant $A > 0$ such that for each $f \in C^m(X) \cap H(X^\circ)$ one can find $F \in SH(\mathbb{R}^N) \cap BC^m(\mathbb{R}^N)$ for $N \geq 3$ (or $F \in SH(\mathbb{R}^N) \cap C^m(\mathbb{R}^N)$ for $N = 2$) with $F|_X = f$ and

$$\|F\|_m \leqslant A\|f\|_{m,X}, \qquad N \geq 3,$$
$$\|\nabla F\|_{m-1} \leqslant A\|\nabla f\|_{m-1,X}, \quad N = 2.$$

Remark 3.3. It can be proved that in $\mathbb{R}^2$ for $m \in (1,2) \cup (2,3)$ (respectively, for $m \in (1,+\infty)\setminus\mathbb{Z}$) every C^∞-smooth *closed Jordan curve* X is a C^m-subharmonic extension set (respectively, is a C^mH-subharmonic extension set), and this is not so for other m. In particular, in these cases the set $\mathbb{R}^2 \setminus X$ is not connected.

It is not yet clear whether the assumptions in Theorems 3.1 and/or 3.2 are most appropriate (sharp). For instance, it can be proved that any closed parallelepiped in $\mathbb{R}^N$ is a C^m-subharmonic extension set for $m \in [1,3)$, and is a $C^m H$-subharmonic extension set for all $m \geq 1$.

Problem 3.4. Given $N \geq 2$ and $m \in (1,3)$ (respectively, $m \geq 1$), does there exist a C^1-smooth *closed* Jordan domain in $\mathbb{R}^N$ which is not C^m-subharmonic extension set (respectively, is not $C^m H$-subharmonic extension set)?

Another interesting problem can be formulated as follows.

Problem 3.5. In notations of Theorem 2.1, given $m > 1$, under what additional conditions on (L-D)-domain D the reflection operator $u \to u_\circ$ preserves the C^m-class:

$$\|u_\circ\|_{m,\overline{D_\circ}} \leq A\|u\|_{m,\overline{D}}?$$

4. The union of disjoint C^m-subharmonic extension sets

The last (and only new) main result of this paper gives an additional copious variety of C^m- and $C^m H$-subharmonic extension sets. For a compact set K, let $\widehat{K}$ be the union of K with all bounded components of the set $\mathbb{R}^N \setminus K$.

Theorem 4.1. *Let $N \geq 2$ and $m \in [1,3)$ (respectively, $m \geq 1$). Suppose that $X_1, \dots, X_J$ is a finite family of C^m-subharmonic extension sets (respectively, $C^m H$-subharmonic extension sets) such that the compacts $\widehat{X_j}$ ($j \in \{1, \dots, J\}$) are pairwise disjoint. Then the union $X = X_1 \cup \dots \cup X_J$ is also a C^m-subharmonic extension set (respectively, $C^m H$-subharmonic extension set).*

PROOF. In this proof we take $N \geq 3$ (the dimension $N = 2$ needs some simple appropriate changes). It is well known topological fact that, if $K_1 = \widehat{K_1}$, $K_2 = \widehat{K_2}$ and $K_1 \cap K_2 = \varnothing$, then the compact set $K = K_1 \cup K_2$ also has connected complement, that is $K = \widehat{K}$. Therefore, it is enough to consider the case $J = 2$. Fix $f \in C^m(X) \cap SH(X^\circ)$ (respectively, $f \in C^m(X) \cap H(X^\circ)$) with $M = \|f\|_{m,X}$. Take $\delta = \operatorname{dist}(\widehat{X_1}, \widehat{X_2})/8$ and let W_j be 3δ-neighbourhoods of $\widehat{X_j}$ ($j \in \{1,2\}$). Put $Y_j = \overline{W_j}$. By the hypothesis, for each j there exists $F_j \in BC^m(\mathbb{R}^N) \cap SH(\mathbb{R}^N)$ such that $F_j|_{X_j} = f|_{X_j}$ and $\|F_j\|_m \leq A\|f\|_{m,X_j} \leq AM$ (here and below the constant $A \geq 1$ can be different in different occurrences). Since Y_j ($j \in \{1,2\}$) are disjoint, by the Whitney extension theorem [13] we can find a function $g \in C^m(\mathbb{R}^N)$ with compact support $\operatorname{Supp}(g) = S_g$ such that $g|_{Y_j} = F_j|_{Y_j}$ ($j \in \{1,2\}$) and $\|g\|_m \leq AM$. Let $\{\varphi_i\}_{i \in I}$ be a finite δ-partition of unity on a neighbourhood of S_g, which means that there are some open balls B_i of radius δ such that $\varphi_i \in C_0^\infty(B_i)$, $0 \leq \varphi_i \leq 1$ and $\sum_{i \in I} \varphi_i = 1$ on some neighbourhood of S_g.

Denote by $\Phi(\mathbf{x})$ the standard fundamental solution for the Laplacian Δ in $\mathbb{R}^N$ ($N \geq 3$):

$$\Phi(\mathbf{x}) = -\frac{\alpha_N}{|\mathbf{x}|^{N-2}}, \quad \alpha_N > 0.$$

Define the Vitushkin localized functions [11] (see also [5, 9]):

$$g_i = \Phi * (\varphi_i \Delta g).$$

By [5, 9] we have $g_i \in BC^m(\mathbb{R}^N)$, $\|g_i\|_m \leq A\|g\|_m \leq A^2 M$ and $g \equiv \sum_{i \in I} g_i$. Let I_1 be subset in I such that $B_i \subset W_1 \cup W_2$ for each $i \in I_1$, and let $I_2 = I \setminus I_1$. For $i \in I_1$

(say $B_i \subset W_j$) we have $\Delta g_i = \varphi_i \Delta g = \varphi_i \Delta F_j \geq 0$, so that $g_i \in SH(\mathbb{R}^N)$. For $i \in I_2$ we have $\Delta g_i = \varphi_i \Delta g$, so that $g_i \in H(\mathbb{R}^N \setminus \overline{B_i})$ and then g_i is harmonic on δ-neighbourhood of $\widehat{X_1} \cup \widehat{X_2}$. Since $f = g$ on X, it remains to properly extend each g_i, $i \in I_2$. Recall that the set $Y := \widehat{X_1} \cup \widehat{X_2}$ also has connected complement. Fix $i \in I_2$ and take an open ball B which contains Y and $\overline{B_i}$. There exists a smooth Jordan curve Γ_i outside Y, that connects ∂B_i and ∂B. Then, clearly, one can find an (L-D)-domain D_i with connected complement and such that $Y \subset D_i \subset B \setminus (\overline{B_i} \cup \Gamma_i)$. It remains to apply Theorem 3.2 to g_i and D_i. □

Remark 4.2. By Remark 3.3, the unit circle $X_1 = \{\mathbf{x} \in \mathbb{R}^2 \mid |\mathbf{x}| = 1\}$ for $m \in (1,2) \cup (2,3)$ (respectively, for $m \in (1,+\infty) \setminus \mathbb{Z}$) is a C^m-subharmonic extension set (respectively, is a $C^m H$-subharmonic extension set). Since, clearly, the disc $X_2 = \{|\mathbf{x}| \leq \frac{1}{2}\}$ is a C^m-subharmonic extension set (respectively, a $C^m H$-subharmonic extension set for the considered m), but the set $X_1 \cup X_2$ is not, one sees that the requirements on $\widehat{X_j}$ to be pairwise disjoint *cannot be omitted* in Theorem 4.1 (at least for $N = 2$ and $m \in (1,+\infty) \setminus \mathbb{Z}$).

It is also worth mentioning that Theorems 2.4, 3.1, 3.2 and 4.1 have word for word analogs for the spaces $B\mathrm{Lip}^m$ ($N \geq 3$) and Lip^m ($N = 2$), which are defined below (see [8, Section 3]).

5. Notations

For completeness and consistency of notations we now present definitions for the spaces of smooth functions which appeared above.

For a real function g on a (more than one-point) set $E \subset \mathbb{R}^N$ and $\mu \in [0,1]$, define the so-called μ-modulus of continuity of the function g on E as follows:

$$\omega_E^\mu(f,\delta) = \sup \frac{|g(x) - g(y)|}{|x-y|^\mu}, \quad \delta \in (0,+\infty],$$

where the supremum (not necessarily finite) is taken over all $x \in E$ and $y \in E$ with $0 < |x - y| < \delta$.

For $m \in \mathbb{Z}_+ = \{0,1,2,\dots\}$, $\mu \in (0,1]$ and open $\Omega \subset \mathbb{R}^N$, denote

$$B\mathrm{Lip}^{m+\mu}(\Omega) = \{f \in BC^m(\Omega) \mid \omega_\Omega^{m,\mu}(f,\infty) < \infty\},$$

where, for $E \subset \Omega$,

$$\omega_E^{m,\mu}(f,\delta) = \max\{\omega_E^\mu(\partial^\alpha f,\delta)\},$$

and the maximum is taken over all N-indices α such that $|\alpha| = m$. For $m \in \mathbb{Z}_+$ and $\mu \in (0,1)$ denote

$$BC^{m+\mu}(\Omega) = \{f \in B\mathrm{Lip}^{m+\mu}(\Omega) \mid \omega_K^{m,\mu}(f,\delta) \to 0 \text{ as } \delta \to 0\}$$

(the latter condition is for every compact subset K of Ω).

The norm in $B\mathrm{Lip}^{m+\mu}(\Omega)$ (for $\mu \in (0,1]$) and in $BC^{m+\mu}(\Omega)$ (for $\mu \in (0,1)$) is defined as

$$\|f\|_{m+\mu,\Omega} = \max\{\|f\|_{m,\Omega}, \omega_\Omega^{m,\mu}(f,\infty)\}.$$

For every $m \geq 0$ and open Ω denote by $C^m(\Omega)$ (respectively, $\mathrm{Lip}^m(\Omega)$, $m > 0$) the space of all functions f in Ω such that $f \in BC^m(E)$ (respectively, $f \in B\mathrm{Lip}^m(E)$) for each bounded open set E with $\overline{E} \subset \Omega$. Clearly, $C^{m_1}(\Omega) \subset C^{m_2}(\Omega)$ (and $\mathrm{Lip}^{m_1}(\Omega) \subset \mathrm{Lip}^{m_2}(\Omega)$) for every $m_1 \geq m_2 \geq 0$. Define also $C^\infty(\Omega) = \cap_{m \geq 0} C^m(\Omega)$. The spaces $C^m(\Omega)$ and $\mathrm{Lip}^m(\Omega)$ are endowed with Fréchet metrics in a standard way. It is easily seen that (for each $m > 0$) the space $BC^m(\Omega)$ is the closure in $B\mathrm{Lip}^m(\Omega)$ of the space $C^\infty(\Omega) \cap B\mathrm{Lip}^m(\Omega)$. Therefore, the spaces $BC^m(\Omega)$ (and the spaces $BC^m(X)$ defined below) naturally appear, for instance, in problems of approximations by harmonic functions on Ω (respectively on a closed set X) in the norms $\|\cdot\|_{m,\Omega}$ (respectively, $\|\cdot\|_{m,X}$). We omit the *index* m and (or) Ω in the above notations for the norms whenever $m = 0$ and (or) $\Omega = \mathbb{R}^N$.

Let now X be a *closed* set in $\mathbb{R}^N$ and $m \geq 0$ arbitrary. Denote by $BC^m(X)$ (respectively, $B\mathrm{Lip}^m(X)$, $m > 0$) the space of functions f on X such that f can be extended to a function in the space $BC^m(\mathbb{R}^N)$ (respectively, $B\mathrm{Lip}^m(\mathbb{R}^N)$). For $f \in BC^m(X)$ (respectively, $f \in B\mathrm{Lip}^m(X)$) define

$$\|f\|_{m,X} = \inf \|F\|_m,$$

where the infimum is taken over all possible extensions $F \in BC^m(\mathbb{R}^N)$ (respectively, $F \in B\mathrm{Lip}^m(\mathbb{R}^N)$), $F|_X = f$.

The corresponding Fréchet spaces $C^m(X)$ (respectively, $\mathrm{Lip}^m(X)$, $m > 0$) for closed sets X can now be defined by analogy with the ones for open sets. For a compact set X the spaces $C^m(X)$ can be identified with $BC^m(X)$ (respectively, $\mathrm{Lip}^m(X)$ with $B\mathrm{Lip}^m(X)$).

References

1. W. Freeden, *Basis systems and their role in the approximation of the Earth's gravitational field*, Gerlands Beitr. Geophysik **94** (1985), no. 1, 19–34.
2. S. J. Gardiner, *Superharmonic extension and harmonic approximation*, Ann. Inst. Fourier (Grenoble) **44** (1994), no. 1, 65–91 (English, with English and French summaries).
3. P. M. Gauthier, *Subharmonic extensions and approximations*, Canad. Math. Bull. **37** (1994), no. 1, 46–53.
4. M. S. Mel′nikov and P. V. Paramonov, *C^1-extension of subharmonic functions from closed Jordan domains in $\mathbb{R}^2$*, Izv. Ross. Akad. Nauk Ser. Mat. **68** (2004), no. 6, 105–118 (Russian); English transl., Izv. Math. **68** (2004), no. 6, 1165–1178.
5. A. G. O'Farrell, *Metaharmonic approximation in Lipschitz norms*, Proc. Roy. Irish Acad. Sect. A **75** (1975), no. 24, 317–330.
6. P. V. Paramonov, *On the C^m-extension of subharmonic functions*, Izv. Ross. Akad. Nauk Ser. Mat. **69** (2005), no. 6, 139–152 (Russian); English transl., Izv. Math. **69** (2005), no. 6, 1211–1223.
7. ———, *On the C^1-extension and C^1-reflection of subharmonic functions from Lyapunov–Dini domains into $\mathbb{R}^N$*, Mat. Sb. **199** (2008), no. 12, 79–116 (Russian); English transl., Sb. Math. **199** (2008), no. 11-12, 1809–1846.
8. ———, *On the C^m-extension of subharmonic functions from Lyapunov–Dini domains to $\mathbb{R}^N$*, Mat. Zametki **89** (2011), no. 1, 149–152 (Russian); English transl., Math. Notes **89** (2011), no. 1-2, 160–164.
9. J. Verdera, *C^m approximation by solutions of elliptic equations, and Calderón–Zygmund operators*, Duke Math. J. **55** (1987), no. 1, 157–187.
10. Dzh. Verdera, M. S. Mel′nikov, and P. V. Paramonov, *C^1-approximation and the extension of subharmonic functions*, Mat. Sb. **192** (2001), no. 4, 37–58 (Russian); English transl., Sb. Math. **192** (2001), no. 3-4, 515–535.
11. A. G. Vitushkin, *Analytic capacity of sets in problems of approximation theory*, Uspehi Mat. Nauk **22** (1967), no. 6 (138), 141–199 (Russian); English transl., Russian Math. Surveys **22** (1967), no. 6, 139–200.
12. S. E. Warschawski, *On differentiability at the boundary in conformal mapping*, Proc. Amer. Math. Soc. **12** (1961), 614–620.
13. H. Whitney, *Analytic extensions of differentiable functions defined in closed sets*, Trans. Amer. Math. Soc. **36** (1934), no. 1, 63–89.
14. K.-O. Widman, *Inequalities for the Green function and boundary continuity of the gradient of solutions of elliptic differential equations*, Math. Scand. **21** (1967), 17–37 (1968).

MECHANICS AND MATHEMATICS FACULTY, MOSCOW STATE UNIVERSITY, 119991 MOSCOW, RUSSIAN FEDERATION.

E-mail address: petr.paramonov@list.ru

Centre de Recherches Mathématiques
CRM Proceedings and Lecture Notes
Volume **55**, 2012

On Maximal Plurisubharmonic Functions

A. Sadullaev

Abstract. In this paper we consider a class of maximal plurisubharmonic functions and prove some of its properties. As an application, we prove maximality of some extremal functions.

1. Introduction

It is well-known that the equation $\Delta u = 0$ defines harmonic functions on the complex plane $\mathbb{C}$. They have the maximality property in the class of subharmonic (sh) functions. Namely, if u is harmonic in a domain $\Omega \subset \mathbb{C}$ then $u(z) \geq v(z)$ $\forall\, z \in \Omega$ for any subharmonic function $v \in \mathrm{sh}(\Omega)$ such, that $\underline{\lim}_{z\to\partial\Omega}\big(u(z) - v(z)\big) \geq 0$. The maximality property of harmonic function direct us to define maximal plurisubharmonic (psh) function in multidimensional case $\Omega \subset \mathbb{C}^n$.

2. Maximal functions

Definition ([10])**.** We say that a function $u \in \mathrm{psh}(\Omega)$ is maximal in the domain Ω if the maximality principle holds: if $v \in \mathrm{psh}(\Omega) : \underline{\lim}_{z\to\partial\Omega}\big(u(z) - v(z)\big) \geq 0$, then $u(z) \geq v(z)$, $\forall\, z \in \Omega$.

Unlike the classical case $n = 1$, where every harmonic function is C^∞ smooth, the maximal psh function in $\mathbb{C}^n$, $n > 1$, is not always belong to class C^∞. For example, the function $\ln|z_1|$, which is maximal in $\mathbb{C}^2(z_1, z_2)$, shows that there exist maximal functions, which are not bounded from below.

The condition $\underline{\lim}_{z\to\partial\Omega}\big(u(z) - v(z)\big) \geq 0$ in the above definition means that for arbitrary fixed $\varepsilon > 0$ there exist a compact $K \subset \Omega$ such that $v(z) \leq u(z) + \varepsilon$ outside K. In particular, $v(z) = -\infty$ if $u(z) = -\infty$.

Next proposition is useful in many applications.

Proposition 2.1. *The followings are equivalent*:

(1) *u is maximal in Ω*;

2010 *Mathematics Subject Classification.* Primary 32U05, 32U15, 32U35; Secondary 31B15.

Key words and phrases. Plurisubharmonic function, Green function, maximal function, extremal function.

Partially supported by the fundamental research of Khorezm Mamun Academy, Grant 1-024.

This is the final form of the paper.

(2) *for any domain* $G \subset\subset \Omega$ *and for any function*

$$v \in \operatorname{psh}(G): \varliminf_{z \to \partial G} \big(u(z) - v(z)\big) \geq 0$$

implies $u(z) \geq v(z)$, $\forall\, z \in G$;

(3) *for any domain* $G \subset\subset \Omega$ *and for any function* $v \in \operatorname{psh}(\Omega) : u|_{\partial G} \geq v|_{\partial G}$ *implies* $u(z) \geq v(z)$, $\forall\, z \in G$.

PROOF. The implication (3) $\Rightarrow$ (1) is clear. For implications (1) $\Rightarrow$ (2) $\Rightarrow$ (3) we note that the function

$$w(z) = \begin{cases} \max\{u(z), v(z)\}, & \text{if } z \in G \\ u(z), & \text{if } z \in \Omega \setminus G. \end{cases} \tag{1}$$

is psh in Ω and $\varliminf_{z \to \partial\Omega} \big(u(z) - w(z)\big) = 0$. Hence $u(z) \geq w(z)$, $\forall\, z \in \Omega$ and $u(z) \geq v(z)$, $\forall\, z \in G$. □

The Monge – Ampère operator $(\mathrm{dd^c} u)^n$ play big role in the theory of maximal psh functions. We recall the following standard notations:

$$\mathrm{d} = \partial + \bar\partial, \quad \mathrm{d^c} = \mathrm{i}(\bar\partial - \partial),$$

where

$$\partial = \frac{\partial}{\partial z_1}\,\mathrm{d}z_1 + \cdots + \frac{\partial}{\partial z_n}\,\mathrm{d}z_n, \quad \bar\partial = \frac{\partial}{\partial \bar z_1}\,\mathrm{d}\bar z_1 + \cdots + \frac{\partial}{\partial \bar z_n}\,\mathrm{d}\bar z_n$$

so, that

$$\mathrm{dd^c} u = 2\mathrm{i}\partial\bar\partial u,$$

$$\mathrm{d}u \wedge \mathrm{d^c} u = 2\mathrm{i}\partial u \wedge \bar\partial u$$

and

$$(\mathrm{dd^c} u)^n = \mathrm{dd^c} u \wedge \cdots \wedge \mathrm{dd^c} u = \mathrm{const}\det\left(\frac{\partial^2 u}{\partial z_j \partial \bar z_k}\right) \mathrm{d}V$$

In classical case of $n = 1$ the operator $(\mathrm{dd^c} u)^n$ is the Laplace operator, which is linear: $\mathrm{dd^c} u$ is easily defined and it is a positive Borel measure for every subharmonic function u.

The definition of the Monge – Ampère operator $(\mathrm{dd^c} u)^n$ in general case of arbitrary psh functions is more complicated. Bremermann [4] and Kerzman [6] noted, that a function $u \in C^2(\Omega) \cap \operatorname{psh}(\Omega)$ is maximal if and only if $(\mathrm{dd^c} u)^n = 0$. Bedford and Taylor [1] have defined the Monge – Ampère operator for $u \in \operatorname{psh}(\Omega) \cap L^\infty_{\mathrm{loc}}(\Omega)$ by the recurrent relation as a positive current:

$$\int (\mathrm{dd^c} u)^k \wedge \varphi = \int u(\mathrm{dd^c} u)^{k-1} \wedge \mathrm{dd^c}\varphi, \quad \varphi \in D^{(n-k,n-k)}, k = 1, 2, \ldots, n-1. \tag{2}$$

Here $D^{(n-k,n-k)}$ is the space of test forms, i.e. , space of all differential forms of bi-degree $(n-k, n-k)$, with C^∞ coefficients. Moreover, later in [2] it was proved that $(\mathrm{dd^c} u)^k$ defined as above is *well-defined*, i.e. , $(\mathrm{dd^c} u_j)^k \mapsto (\mathrm{dd^c} u)^k$ for any approximation $u_j \downarrow u : u \in \operatorname{psh}(\Omega) \cap L^\infty_{\mathrm{loc}}(\Omega)$. In the class of bounded psh functions, a function $u(z) \in \operatorname{psh}(\Omega) \cap L^\infty_{\mathrm{loc}}(\Omega)$ is maximal if and only if $(\mathrm{dd^c} u)^n = 0$ and the following comparison principle holds : if $u, v \in \operatorname{psh}(\Omega) \cap L^\infty_{\mathrm{loc}}(\Omega)$ and the set $F = \{z \in \Omega : u(z) < v(z)\} \subset\subset \Omega$, then

$$\int_F (\mathrm{dd^c} u)^n \geq \int_F (\mathrm{dd^c} v)^n. \tag{3}$$

But the definition of $(dd^c u)^n$ for arbitrary $u \in \text{psh}(\Omega)$ is still a challenging problem. Namely, first, in 1975 Shiffman and Taylor showed that there is $u \in \text{psh}(C^n)$ such that $\int (dd^c u_j)^n \to \infty$, where $u_j \downarrow u$. Kiselman [7] constructed a simple example that function $u(z) = (-\ln|z_1|)^{1/n}(|z_2|^2 + \cdots + |z_n|^2 - 1)$, which is psh near the origin, has this property and its Monge – Ampère mass is unbounded near $z_1 = 0$.

Second, U. Cegrell [5] suggested the following example. For psh function $u(z) = \ln|z_1|^2 + \cdots + \ln|z_n|^2$ if we take the approximation

$$u_j(z) = \ln\left(|z_1 \cdots z_n|^2 + \frac{1}{j}\right) \downarrow u(z),$$

then $(dd^c u_j)^n \mapsto 0$; if we take the approximation

$$v_j(z) = \ln\left(|z_1|^2 + \frac{1}{j}\right) + \cdots + \ln\left(|z_n|^2 + \frac{1}{j}\right) \downarrow u(z),$$

then $(dd^c u_j)^n \mapsto n!4^n \delta_0$, where δ_0 — the Dirac measure.

This examples show that for arbitrary psh functions the operator $(dd^c u)^n$ cannot be well-defined (by approximation $u_j \downarrow u$) and we should not define maximal functions by Monge – Ampère operator $(dd^c u)^n$ or by arbitrary approximation $u_j \downarrow u$.

On the other hand, Z. Błocki [3] showed that, the function

$$u(z, w) = -(\ln|z| \ln|w|)^{1/2}$$

is maximal in the $U^2 \setminus (0,0)$, where $U^2 = \{|z| < 1, |w| < 1\}$ is the unit be-disk, but for the special approximation $u_j = \max\{u, -j\}$ the $(dd^c u_j)^n$ does not tends 0 in $U^2 \setminus (0,0)$. This example shows that the special approximation $\max\{u, -j\} \downarrow u$ is also not suitable for establishing the criteria for maximality. Nevrtheless, it is true

Theorem 2.2. *If for $u(z) \in \text{psh}(\Omega)$ there exist a sequence $u_j(z) \in \text{psh}(\Omega) \cap L^\infty_{\text{loc}}(\Omega)$: $u_j \downarrow u$, $(dd^c u_j)^n \mapsto 0$ then u is maximal. Conversely, if u is maximal, then there exist an approximation $\{u_j(z)\}$ such that*

$$(4) \qquad \begin{gathered} u_j \in \text{psh}(\Omega_j) \cap L^\infty_{\text{loc}}(\Omega_j), \quad (dd^c u_j)^n = 0, \\ \Omega_j \subset\subset \Omega_{j+1} \subset\subset \Omega, \quad \Omega = \bigcup_{j=1}^{\infty} \Omega_j, \quad u_j(z) \downarrow u(z). \end{gathered}$$

Theorem 2.2 was first proved in [10] under the assumption, that the sequence is continuous. Similar property of maximal psh functions was considered also by Błocki and Cegrell.

PROOF. Let $u(z) \in \text{psh}(\Omega)$ and suppose there exist a sequence $u_j(z) \in \text{psh}(\Omega) \cap L^\infty_{\text{loc}}(\Omega) : u_j \downarrow u$, $(dd^c u_j)^n \mapsto 0$. Assume by contrary that u is not maximal. Then there exist a domain $G \subset\subset \Omega$ and a function $v \in \text{psh}(\Omega) : u|_{\partial G} \geq v|_{\partial G}$ such that $u(z^0) < v(z^0)$, $z^0 \in G$.

We fix a small $\varepsilon > 0$ such that $u(z^0) + \varepsilon < v(z^0)$. Then there exist $\delta = \delta(\varepsilon) > 0$ such that the $\text{psh}(\Omega)$ function $\tilde{v} = v + \delta(|z - z^0|^2/\rho^2 - 1)$, where $\rho = max\{|z^0 - w| : w \in \partial G\}$, satisfies the conditions $u|_{\partial G} + \varepsilon > \tilde{v}|_{\partial G}$, $u(z^0) + \varepsilon < \tilde{v}(z^0)$. We can choose $j_0 \in N$ big enough that

$$(5) \qquad u_{j_0}|_{\partial G} + \varepsilon > \tilde{v}|_{\partial G}, \quad u_{j_0}(z^0) + \varepsilon < \tilde{v}(z^0).$$

We note, that if we take $v' = \max\{v(z), -\|u_{j_0}\|\}$ instead v, then the function $\tilde{v}' = v' + \delta(|z - z^0|^2/\rho^2 - 1)$, which is now in $L^\infty(G)$ also satisfies (5). Using the comparison principle we have

$$(6)\qquad \int_F (dd^c u_{j_0})^n \geq \int_F (dd^c \tilde{v}')^n, \quad F = \{u_{j_0} + \varepsilon < \tilde{v}'\} \subset\subset G.$$

By (5)

$$(7)\qquad \operatorname{mes}\{u_{j_0} + \varepsilon < \tilde{v}\} = c > 0.$$

It is clear that $(dd^c \tilde{v}')^n \geq (dd^c(\delta/\rho^2)|z - z^0|^2)^n = \text{const} > 0$. This together with (6) and (7) gives the contradiction to the claim $\lim_{j\to\infty}(dd^c u_j)^n = 0$.

Let now $u(z)$ be maximal. For fixed domain $G \subset\subset \Omega$ we take an approximation $u_j \downarrow u$, $u_j \in \operatorname{psh}(\overline{G}) \cap C(\overline{G})$, $j = 1, 2, \ldots$. It is well-known that the regularization v_j^* of

$$(8)\qquad v_j = \sup\{w(z) \in \operatorname{psh}(G) \cap (\overline{G}) : w|_{\partial G} \leq u_j|_{\partial G}\}$$

is bounded and psh function in G with vanishing Monge–Ampère operator, $(dd^c v_j^*)^n = 0$. Moreover, since u is maximal, we have $v_j(z) \downarrow u(z)$, $z \in G$.

Now by applying this process for arbitrary compact $G \subset\subset \Omega$, it is not difficult, to construct a sequence of domains $\Omega_j \subset \Omega$ and approximation $u_j(z) \downarrow u(z)$, $u_j \in \operatorname{psh}(\Omega_j) \cap L^\infty_{\text{loc}}(\Omega_j)$, $(dd^c u_j)^n = 0$, where $\Omega_j \subset\subset \Omega_{j+1} \subset\subset \Omega$, $\Omega = \bigcup_{j=1}^\infty \Omega_j$. □

Remark. If the domain $G \subset\subset \Omega$ above is strongly pseudoconvex, then the upper envelop (8) is continuous in $\overline{G}$ by Bremermann–Walsh theorem.

3. Extremal psh functions

The extremal Green function $V(z, K)$, $K \subset \mathbb{C}^n$, or $\mathcal{P}$-measure $\omega(z, K, D)$, where $K \subset D \subset \mathbb{C}^n$, play an important role in the pluripotential theory and in the approximation theory of multidimensional complex analysis (For more details see [11]). They are defined by locally bounded psh functions:

$$V(z, K) = \sup\{u(z) \in \operatorname{psh}(\mathbb{C}^n) : u|_K = 0, u(z) \leq \ln\|z\| + c_u\},$$
$$\omega(z, K, D) = \sup\{u \in \operatorname{psh}(D) : u|_D < 0, u|_K = -1\}.$$

The regularizations V^*, ω^* of the functions V, ω are maximal psh functions off K (we except the trivial case, when K is pluripolar set). In general, extremal psh functions in a domain D can be considered as a solution of Dirichlet problem:

$$(9)\qquad \begin{gathered} u \in \operatorname{psh}(D) \cap L^\infty(D), \quad (dd^c u)^n = 0 \text{ in } D \\ u|_{\partial D} = \varphi(\xi) \text{ in } \partial D. \end{gathered}$$

If D is strictly pseudoconvex and $\varphi \in C(\partial D)$ then the (9) has unique solution (see [1]). The solution can be fined also by Perron method: let $\aleph(\varphi, D)$ class of function $u \in \operatorname{psh}(D)$ such that $u^*(\xi) = \overline{\lim}_{z\to\xi} u(z) \leq \varphi(\xi)$. Then the regularization $\omega^*(z, \varphi, D)$ of $\omega(z, \varphi, D) = \sup\{u(z) : u \in \aleph\}$ is maximal and is solution of Dirichlet problem (9): $\omega^* \in \operatorname{psh}(D) \cap C(\overline{D})$, $(dd^c \omega^*)^n = 0$, $\omega|_{\partial D} = \varphi$.

For arbitrary function $\varphi(\xi)$, which is bounded from above on ∂D, we put $\varphi_j(\xi) = \max\{\varphi(\xi), -j\}$ and take $\omega^*(z, \varphi_j, D)$ to be the solution of (9). As $j \uparrow \infty$ the sequence is decreasing. The limit $\omega^*(z, \varphi, D) = \lim_{j\to\infty} \omega^*(z, \varphi_j, D)$ may, generally speaking, be identically equal $-\infty$. For example, $\omega^*(z, \varphi, D) \equiv -\infty$

if $u \notin L^1(\partial D)$. But this is not the case if $\aleph(\varphi, D) \neq \varnothing$ (there is at least one function $u \in \operatorname{psh}(D) : u^*(\xi) \leq \varphi(\xi)$ and $u \not\equiv -\infty$). Then $\omega^*(z, \varphi, D) \in \operatorname{psh}(D)$ and this function is said to be the solution of (9). However, we can not say on $(dd^c\omega^*)^n = 0$, because the ω^* may be nonbounded and therefore $(dd^c\omega^*)^n$ may not be defined. Nevertheless, we can prove the following fact

Theorem 3.1. *If $\aleph(\varphi, D) \neq \varnothing$, then the $\omega^*(z, \varphi, D) \not\equiv -\infty$ and is maximal in D. Moreover if the domain D is strictly pseudoconvex and $\exp\varphi$ is continuous on ∂D, then $\overline{\lim}_{z\to\xi} \omega^*(z, \varphi, D) = \varphi(\xi)$ on ∂D.*

PROOF. The maximality of $\omega^*(z, \varphi, D)$ follows from Theorem 2.2 and definition of $\omega^*(z, \varphi, D)$. Since $\exp\varphi$ is continuous on ∂D, we have $\varphi_j(\xi) = \max\{\varphi(\xi), -j\} \in C(\partial D)$ and the solution ω_j of (9) satify $\omega_j \in \operatorname{psh}(D) \cap C(\overline{D})$, $\omega_j|_{\partial D} = \varphi_j$, $j = 1, 2, \ldots$. It is clear that $\omega_j \downarrow \omega^*(z, \varphi, D)$ and that

$$\overline{\lim_{z\to\xi}}\, \omega^*(z, \varphi, D) \leq \overline{\lim_{z\to\xi}}\, \omega_j(z) = \varphi_j(\xi) \quad \forall \xi \in \partial D,$$

which yield the following inequality

$$\overline{\lim_{z\to\xi}}\, \omega^*(z, \varphi, D) \leq \varphi(\xi) \quad \forall \xi \in \partial D. \tag{10}$$

On the other hand, fix $u(z) \in \aleph(\varphi, D)$, $u \not\equiv -\infty$. Since $u(z)$ is bounded from above psh function, the set $S = \{\xi \in \partial D : u(\xi) = -\infty\}$ has zero Lebesgue measure on ∂D. Let $\xi^0 \in \partial D \setminus S$ to be a fixed point and let $\chi(z)$ be a chosen psh function of strictly pseudoconvex domain D at $\xi^0 : \chi(z) \in \operatorname{psh}(D) \cap C(\overline{D})$, $\chi(\xi^0) = 0$ and $\chi(z) < 0\ \forall z \in \overline{D} \setminus \{\xi^0\}$. Then for every $\varepsilon > 0$ there exists $M > 0$ such that

$$M\chi(\xi) + u(\xi) - u(\xi^0) + \varphi(\xi^0) - \varepsilon \leq \varphi(\xi) \quad \forall \xi \in \partial D.$$

It follows that the function $\nu(z) = M\chi(z) + u(z) - u(\xi^0) + \varphi(\xi^0) - \varepsilon \in \aleph(\varphi, D)$ and that $\nu(z) \leq \omega^*(z, \varphi, D)\ \forall z \in D$. Hence $\overline{\lim}_{z\to\xi^0} \omega^*(z, \varphi, D) \geq \overline{\lim}_{z\to\xi^0} \nu(z) = \varphi(\xi^0) - \varepsilon$ and since $\varepsilon > 0$ is arbitrary, we have $\overline{\lim}_{z\to\xi^0} \omega^*(z, \varphi, D) \geq \varphi(\xi^0)$. This in combination with (10) gives $\overline{\lim}_{z\to\xi} \omega^*(z, \varphi, D) = \varphi(\xi)$ for every $\xi \in \partial D \setminus S$ and therefore for every $\xi \in \partial D$. □

Problem 3.2. If $\aleph(\varphi, D) \neq \varnothing$, then is it true that $\overline{\lim}_{z\to\xi} \omega^*(z, \varphi, D) = \varphi(\xi)$ for arbitrary, bounded from above function $\varphi(\xi)$?

Example. Let $L = \{z_1 = 0, |z_2| = 1\}$ is to be a circle on the unit ball $B = B(0, 1) \subset \mathbb{C}^2$ and $l = \{z_1 = 0, z_2 = e^{i\alpha}\}$, $\alpha \in [\pi/4, 3\pi/4]$ is an arc on L. We take a function φ on $\partial B : \exp\varphi(\xi) \in C(\partial B)$, $\varphi(\xi) = \varphi(\xi_1, \xi_2) = 1$ if $\operatorname{Im}\xi_2 < 0$, $\varphi(\xi) = -\infty$ if $\xi \in l$ and $\ln|\xi_1| \leq \varphi(\xi)$ on ∂B. Then it is not difficult to see, that $\omega^*(z, \varphi, B) = -\infty$ on L. Therefore, $\underline{\lim}_{z\to\xi} \omega^*(z, \varphi, B) \neq \varphi(\xi)$ for $\{\xi_1 = 0, \xi_2 = e^{i\alpha}, \alpha \in [\pi, 2\pi]\}$.

Let now $K \subset \mathbb{C}^n$ fixed nonpluripolar compact and $\varphi(z)$ arbitrary bounded above function on K. Put $\varphi_j = \max\{\varphi, -j\}$ and

$$V(z, \varphi_j) = \sup\{u(z) \in \operatorname{psh}(\mathbb{C}^n) : u|_K \leq \varphi_j, u(z) \leq \ln\|z\| + c_u\}.$$

Then the regularization $V^*(z, \varphi_j)$ of $V(z, \varphi_j)$ is psh in $\mathbb{C}^n$ and $V^*(z, \varphi_j)$ is decreasing in $j \uparrow \infty$. If there is at least one function

$$u(z) \in \operatorname{psh}(\mathbb{C}^n) : \quad u \not\equiv -\infty, \quad u|_K \leq \varphi, \quad u(z) \leq \ln\|z\| + c_u$$

then the $\lim_{j\to\infty} V^*(z,\varphi_j) = V^*(z,\varphi) \in \mathrm{psh}(\mathbb{C}^n)$, $V^*(z,\varphi) \le \ln\|z\| + C$. We note that if the set $\{z \in K : \varphi(z) = -\infty\}$ is not pluripolar, then $V^*(z,\varphi) \equiv -\infty$. Moreover, if $z^0 \in K$ and $\varphi(z) = -1/\|z - z^0\|$, then we also have $V^*(z,\varphi) \equiv -\infty$. For more details on weighted Green functions see [12, Appendix B] and a survey paper of N. Levenberg [8].

As an immediate consequence of Theorem 3.1 we obtain the following corollary.

Corollary 3.3. *If there is at least one function*

$$u(z) \in \mathrm{psh}(\mathbb{C}^n): \quad u \not\equiv -\infty, \quad u|_K \le \varphi, \quad u(z) \le \ln\|z\| + c_u,$$

then $V^*(z,\varphi)$ is maximal in $\mathbb{C}^n \setminus K$.

References

1. E. Bedford and B. A. Taylor, *The Dirichlet problem for a complex Monge-Ampère equation*, Invent. Math. **37** (1976), no. 1, 1–44.
2. ______, *A new capacity for plurisubharmonic functions*, Acta Math. **149** (1982), no. 1-2, 1–40.
3. Z. Błocki, *A note on maximal plurisubharmonic functions*, Uzbek Math. J. **1** (2009), 28–32.
4. H. J. Bremermann, *On a generalized Dirichlet problem for plurisubharmonic functions and pseudo-convex domains. Characterization of Šilov boundaries*, Trans. Amer. Math. Soc. **91** (1959), 246–276.
5. U. Cegrell, *Sums of continuous plurisubharmonic functions and the complex Monge – Ampère operator in* $\mathbb{C}^n$, Math. Z. **193** (1986), no. 3, 373–380.
6. N. Kerzman, *A Monge – Ampère equation in complex analysis*, Several Complex Variables (Williamstown, MA, 1975) (R. O. Wells Jr., ed.), Proc. Sympos. Pure Math., vol. 30, Amer. Math. Soc., Providence, RI, 1977, pp. 161–167.
7. C. O. Kiselman, *Sur la définition de l'opérateur de Monge-Ampère complexe*, Complex Analysis (Toulouse, 1983) (É. Amar, R. Gay, and Nguyen Thanh Van, eds.), Lecture Notes in Math., vol. 1094, Springer, Berlin, 1984, pp. 139–150.
8. N. Levenberg, *Weighter pluripotential theory results of Bermann – Boucksom*, available at `arXiv:1010.4035`.
9. A. Sadullaev, *The operator* $(dd^c u)^n$ *and condenser capacities*, Dokl. Akad. Nauk SSSR **251** (1980), no. 1, 44–47 (Russian); English transl., Soviet Math. Dokl. **21** (1980), no. 2, 387–391.
10. ______, *Plurisubharmonic measures and capacities on complex manifolds*, Uspekhi Mat. Nauk **36** (1981), no. 4(220), 53–105, 247 (Russian); English transl., Russian Math. Surveys **36** (1981), no. 4, 61–119.
11. ______, *Plurisubharmonic functions*, Current Problems in Mathematics. Fundamental Directions, Vol. 8 (R. V. Gamkrelidze, ed.), Itogi Nauki i Tekhniki, Akad. Nauk SSSR Vsesoyuz. Inst. Nauchn. i Tekhn. Inform., Moscow, 1985, pp. 65–113, 274 (Russian); English transl., Several Complex Variables. II, Encyclopaedia Math. Sci., vol. 8, Springer, Berlin, 1994, pp. 59–106.
12. E. B. Saff and V. Totik, *Logarithmic potentials with external fields*, with Appendix B by T. Bloom, Grundlehren Math. Wiss., vol. 316, Springer, Berlin, 1997.

Mathematics Department, National University of Uzbekistan, 700174 Tashkent, Uzbekistan

E-mail address: `sadullaev@mail.ru`

Centre de Recherches Mathématiques
CRM Proceedings and Lecture Notes
Volume **55**, 2012

On Universality of Series in Banach Spaces

Richard Fournier and Jérôme-Melville Giguère

Abstract. We study a notion of universality for series in a Banach space linked to the famous theorem of Riemann on rearrangements.

Introduction and statement of the results

Let $(X, \|\cdot\|)$ be a Banach space; we say that X is real (complex) if the underlying field of scalars is $\mathbb{R}(\mathbb{C})$. Let also $X^{\mathbb{N}}$ be the set of formal series $S = \sum_{n\in\mathbb{N}} x_n$ endowed with the standard product topology; we denote by $s := \lim_{n\to\infty} \sum_{k=1}^{n} x_k$ the sum of the convergent series S. Finally, let $\mathbb{P}$ denote the set of all permutations (i.e., bijections) $\sigma\colon \mathbb{N} \to \mathbb{N}$.

In this note, we introduce a new type of universality and say that an element $S = \sum_{n\in\mathbb{N}} x_n \in X^{\mathbb{N}}$ is universal if there exists, for any $x \in X$, a permutation $\sigma \in \mathbb{P}$ such that $x = \sum_{n=1}^{\infty} x_{\sigma(n)}$. Inspired by recent papers [1, 4, 5] we shall be concerned by the mere existence and abundance (in various contexts) of universal series (the set, possibly empty, of universal elements in $X^{\mathbb{N}}$, will be denoted by U_X). We shall prove

Theorem 1. *Let X be a separable Banach space and $X_{\mathrm{c}}^{\mathbb{N}}$ be the set of convergent elements of $X^{\mathbb{N}}$. Then U_X is dense in $X^{\mathbb{N}}$. Moreover, $U_X \cup \{0\}$ is spaceable in $X_{\mathrm{c}}^{\mathbb{N}}$ (i.e., it contains a closed infinite-dimensional vector subspace of $X_{\mathrm{c}}^{\mathbb{N}}$) and also dense-lineable in $X^{\mathbb{N}}$ (i.e., it contains a vector subspace of $X^{\mathbb{N}}$ which is dense in $X^{\mathbb{N}}$).*

The above statement is also a generalization of an older result of Talaljan [10], concerning a similar universality in the context of L_p spaces, to the case of a separable Banach space. The notions of density, lineability and spaceability are standard ways of measuring the abundance of universal elements in a given context or another. We refer to the survey of Grosse – Erdmann [7] and to the slightly more recent book of Bayart and Matheron [3] concerning these matters. The abundance of universal elements is also sometimes measured in terms of topological genericity

---

2010 *Mathematics Subject Classification.* 46B25, 40A05, 30K05.

Key words and phrases. Universal series in Banach spaces, density, topological genericity, lineability, spaceability.

We wish to thank Vassili Nestoridis for his help and advice.

This is the final form of the paper.

(here the book of Oxtoby [9] is the standard reference). In this direction we have only a partial result.

Theorem 2. *Let X be a finite-dimensional Banach space. Then $U_X \cap X_c^{\mathbb{N}}$ is a dense G_δ subset of $X_c^{\mathbb{N}}$ (i.e., it is the dense and countable intersection of open subsets of $X_c^{\mathbb{N}}$).*

Here our definition of universality is not a standard one (compare with [2]). However our last result shows that from some point of view our universal series behave in a standard manner :

Theorem 3. *Let X be a separable Banach space, $\sum_{n\in\mathbb{N}} x_n \in U_X$ and $\mathcal{U}$ an infinite subset of $\mathbb{N}$. There exists a permutation $\sigma \in \mathbb{P}$ and, for each $x \in X$, an increasing subsequence $\{n_k\} \subseteq \mathcal{U}$ such that*

$$x = \lim_{k\to\infty} \sum_{n=1}^{n_k} x_{\sigma(n)}.$$

Preliminary results

Of course the foundation of our work is the famous theorem of Riemann on conditionaly convergent real series. Indeed, let $\sum_{n\in\mathbb{N}} a_n$ be such a series and X a finite-dimensional real Banach space with basis $\{e_1, e_2, \ldots, e_n\}$. We claim that the series

$$(1) \qquad a_1e_1 + \cdots + a_1e_n + a_2e_1 + \cdots + a_2e_n + \cdots + a_me_1 + \cdots + a_me_n + \cdots$$

is universal: let $x = \sum_{j=1}^{n} x_je_j$ with $x_j \in \mathbb{R}$; there exists for each j, by Riemann's theorem, a permutation σ_j of $\mathbb{N}$ such that $x_j = \sum_{n\in\mathbb{N}} a_{\sigma_j(n)}$. It is then clear that the series

$$a_{\sigma_1(1)}e_1 + \cdots + a_{\sigma_n(1)}e_n + a_{\sigma_1(2)}e_1 + \cdots + a_{\sigma_n(2)}e_n + \cdots$$

is a rearrangement of (1) which converges to x (just check components separately!). This means that $U_X \neq \varnothing$. However, the result is not that straightforward for general Banach spaces and we shall need several results to be found for example in [8]. For a Banach space X and a series $\sum_{n\in\mathbb{N}} x_n$ in $X^{\mathbb{N}}$, we define the domain of sums of the series by

$$D\left(\sum_{n\in\mathbb{N}} x_n\right) := \left\{x \in X \;\middle|\; \exists\sigma \in \mathbb{P} : \sum_{n\in\mathbb{N}} x_{\sigma(n)} = x\right\}.$$

We also define

$$\Gamma = \Gamma\left(\sum_{n\in\mathbb{N}} x_n\right) := \left\{f \in X^* \;\middle|\; \sum_{n\in\mathbb{N}} |f(x_n)| < \infty\right\},$$

the set of "absolutely convergent" functionals f in the dual X^* for the series $\sum_{n\in\mathbb{N}} x_n$. The following theorem and its corollary are essentially due to Steinitz:

Theorem A. *Given a finite-dimensional Banach space X and a series $\sum_{n\in\mathbb{N}} x_n$ in $X_c^{\mathbb{N}}$ with sums s, we have*

$$D\left(\sum_{n\in\mathbb{N}} x_n\right) = s + \Gamma_0 := \{s + x \mid x \in \Gamma_0\}$$

with $\Gamma_0 = \{x \in X \mid f(x) = 0$ for all $f \in \Gamma\}$.

In particular $\sum_{n\in\mathbb{N}} x_n$ belongs to U_X if and only if $\Gamma = \{0\}$ and we have

Corollary B. *A series $\sum_{n\in\mathbb{N}} x_n$ in $X_c^{\mathbb{N}}$ where X is a finite-dimensional Banach space is universal if and only if for any $f \neq 0$ in X^* we have $\sum_{n\in\mathbb{N}} |f(x_n)| = \infty$.*

It is well-known that the theorem of Steinitz is false for any infinite-dimensional Banach space [8] and indeed we have

Theorem C. *For any infinite-dimensional Banach space X,*

$$U_X \neq \varnothing \iff X \text{ is separable.}$$

Theorem C is given as an exercise in [8, p. 25]; a detailed proof can be found in [6]. Finally we shall make repeated use of the following lemmas, stated here without proof.

Lemma D. *If a series $\sum_{n\in\mathbb{N}} x_n$ is universal and another series $\sum_{n\in\mathbb{N}} y_n$ has only a finite number of nonzero terms, then the series $\sum_{n\in\mathbb{N}}(x_n + y_n)$ is universal. In other words, perturbing or removing a finite number of terms of a universal series yields another universal series.*

Lemma E. *If a series $\sum_{n\in\mathbb{N}} x_n$ is universal and another series $\sum_{n\in\mathbb{N}} y_n$ is convergent then the series*

$$x_1 + y_1 + x_2 + y_2 + \cdots + x_n + y_n + \cdots$$

and any of its rearrangements are still universal.

Proof of Theorem 1

Let Z be a dense and countable subset of X and $P \subset X_c^{\mathbb{N}}$ be the set of finite sums with terms in $Z \cup \{0\}$. Then P is countable and clearly dense in $X^{\mathbb{N}}$. Since X is separable, it follows from Theorem C that U_X is not empty. If we consider any element of U_X, we can find a rearrangement of it, $\sum_{n\in\mathbb{N}} u_n$, belonging to $U_X \cap X_c^{\mathbb{N}}$; then

$$\left\{ \sum_{n\in\mathbb{N}} u_n + \sum_{n\in\mathbb{N}} x_n \;\middle|\; \sum_{n\in\mathbb{N}} x_n \in P \right\} \tag{2}$$

is countable and dense in $X^{\mathbb{N}}$. Further, by Lemma D, each element of the set in (2) is a universal series. Thus, U_X is dense in $X^{\mathbb{N}}$.

Let $\{N_n\}_{n\in\mathbb{N}}$ be an infinite partition of $\mathbb{N}$ into infinite and naturally ordered sets $N_n \subset \mathbb{N}$. We call $N_n(i)$ the ith smallest element of N_n and define a sequence $\{\alpha_m\} \subset X^{\mathbb{N}}$ as follows : given a fixed universal series $\sum_{n\in\mathbb{N}} u_n$, we set

$$\alpha_m = \{\alpha_m(i)\} \quad \text{with } \alpha_m(i) = \begin{cases} 0 & \text{if } i \notin N_m, \\ u_n & \text{if } i = N_m(n) \in N_m. \end{cases}$$

We shall prove that the closure $\overline{V}$ in $X_c^{\mathbb{N}}$ of the vector space $V = \operatorname{span}\{\alpha_m\}$ has the desired properties. It is rather clear that V is infinite-dimensional and therefore $\overline{V}$ is an infinite-dimensional closed vector subspace of $X_c^{\mathbb{N}}$. Let

$$\alpha = \sum_{j=1}^{N} c_j \alpha_{m_j} \in V, \quad 0 \neq c_j \in \mathbb{C}.$$

We have $\alpha(i) = c_j u_n$ if $i = N_{m_j}(n) \in N_{m_j}$ and setting $\alpha' = \sum_{j=2}^{N} c_j \alpha_{m_j}$, this last series is convergent (its sum being of course $(\sum_{j=2}^{N} c_j) \sum_{n\in\mathbb{N}} u_n$) and we may now look at α as a rearrangement of the series

$$c_1 u_1 + \alpha'(1) + c_1 u_2 + \alpha'(2) + c_1 u_3 + \alpha'(3) \cdots \tag{3}$$

which belongs to U_X by Lemma E because the series $c_1 \sum_{n\in\mathbb{N}} u_n$ is universal and $\sum_{n\in\mathbb{N}} \alpha'(n)$ is convergent. We therefore have $V \subset U_X \cup \{0\}$ and what remains to show is that also $\overline{V} \subset U_X \cup \{0\}$; let $\{\beta_j\}_{j\in\mathbb{N}} \subset V$ with $\lim_{j\to\infty} \beta_j := \beta \in X_{\mathrm{c}}^{\mathbb{N}} \setminus \{0\}$. Let $i_0 \in N$ be such that $\beta(i_0) \neq 0$; let also $i_0 = N_{m_0}(n_0)$. We have for each $i \in N_{m_0}$ and j and some $c_{j,m_0} \in \mathbb{C}$

$$\beta_j(i) = c_{j,m_0} u_n \quad \text{with } i = N_{m_0}(n) \in N_{m_0}$$

and clearly $\beta(i) = (\lim_{j\to\infty} c_{j,m_0}) u_n := c_{m_0} u_n$ for some nonzero complex number c_{m_0}. Therefore

$$\sum_{k\in N_{m_0}} \beta(k) = \sum_{n\in\mathbb{N}} c_{m_0} u_n$$

is a universal series while

$$\sum_{k\in\mathbb{N}\setminus N_{m_0}} \beta(k) = \sum_{k\in\mathbb{N}} \beta(k) - \sum_{k\in N_{m_0}} \beta(k)$$

for sure belongs to $X_{\mathrm{c}}^{\mathbb{N}}$. We therefore have $\overline{V} \subset U_X \cup \{0\}$ by Lemma E.

Let us consider again the set $P := \{p_m\}_{m\in\mathbb{N}}$ introduced above and define $\{Y_m\}_{m\in\mathbb{N}} \subset X^{\mathbb{N}}$ by

$$Y_m(i) = \begin{cases} p_m(i) & \text{if } i \notin N_m, \\ p_m(i) + u_n & \text{if } i = N_m(n) \in N_m. \end{cases}$$

By Lemma E, each series Y_m is universal because they are rearrangements of series of the type $\sum_{i\in\mathbb{N}} (p_m(i) + X_i)$ where $p_m \in X_{\mathrm{c}}^{\mathbb{N}}$ and the two series $\sum_{i\in\mathbb{N}} X_i$ and $\sum_{n\in\mathbb{N}} u_n$ belong to U_X. Just as above we also have $\operatorname{span}\{Y_m\} \subset U_X \cup \{0\}$ and the density of $\operatorname{span}\{Y_m\}$ in $X^{\mathbb{N}}$ is a trivial consequence of the density of P in $X^{\mathbb{N}}$.

Proof of Theorem 2

Define $E(s, N)$ as

$$\left\{ \sum_{n\in\mathbb{N}} x_n \in X_{\mathrm{c}}^{\mathbb{N}} \;\middle|\; \sum_{n=1}^{N} |f(x_n)| > s \text{ for all } f \in X^* \text{ with } \|f\| = 1 \right\}.$$

We start this proof by showing that $U_X \cap X_{\mathrm{c}}^{\mathbb{N}} = \bigcap_{s\in\mathbb{N}} \bigcup_{N\in\mathbb{N}} E(s, n)$. Let $\sum_{n\in\mathbb{N}} x_n$ belong to the later and $g \in X^*$, $g \neq 0$, and $f = g/\|g\|$. Then for any $s \in \mathbb{N}$ there must exist $N \in \mathbb{N}$ such that $\sum_{n=1}^{N} |f(x_n)| > s$. Therefore

$$\sum_{n=1}^{\infty} |g(x_n)| = \sum_{n=1}^{\infty} |f(x_n)| = \infty$$

and by Corollary B we have $\sum_{n\in\mathbb{N}} x_n \in U_X$.

Let conversely $\sum_{n\in\mathbb{N}} x_n \in U_X \cap X_c^{\mathbb{N}}$ and $f \in \mathcal{B}$, the unit ball of X_1^*; by the same corollary we have, for a given $s \in \mathbb{N}$

$$\sum_{n=1}^{N_f} |f(x_n)| > s \tag{4}$$

for some $N_f \in \mathbb{N}$. We claim that N_f can be chosen independently of f; otherwise there exists a sequence $\{f_m\}_{m\in\mathbb{N}} \subset \mathcal{B}$ and a sequence of integers N_{f_m} tending to infinity such that

$$\sum_{n=1}^{N_{f_m}} \|f_n(x_n)\| \leq s. \tag{5}$$

By compacity of $\mathcal{B}$, the sequence $\{f_m\}$ has a convergent subsequence $\{f_{m_j}\}$ tending to say $f \in \mathcal{B}$; but then by (4) and (5)

$$s < \sum_{n=1}^{N_f} |f(x_n)| = \lim_{j\to\infty} \sum_{n=1}^{N_f} |f_{m_j}(x_n)| \leq \limsup_{j\to\infty} \sum_{n=1}^{N_{f_{m_j}}} |f_{m_j}(x_n)| \leq s$$

which is plainly absurd! Therefore

$$\forall s \in \mathbb{N}\ \exists N \in \mathbb{N}\ \forall f \in \mathcal{B} \quad \sum_{n=1}^{N} |f(x_n)| > s$$

and this amounts to $\sum_{n\in\mathbb{N}} x_n \in \bigcup_{N\in\mathbb{N}} E(s,N)$ holding for all $s \in \mathbb{N}$.

We now prove that $E(s,N)$ is open. Let $\sum_{n\in\mathbb{N}} x_n \in E(s,N)$. Because the unit ball $\mathcal{B}$ is compact, we have

$$\delta := \frac{\min_{f\in\mathcal{B}} \sum_{n=1}^{N} |f(x_n)| - s}{N} > 0$$

and if $\sum_{n\in\mathbb{N}} y_n \in X^{\mathbb{N}}$ with $\|x_n - y_n\| < \delta$ for $1 \leq n \leq N$, we obtain for any $f \in \mathcal{B}$

$$\begin{aligned}
\left| \sum_{n=1}^{N} |f(x_n)| - \sum_{n=1}^{N} |f(y_n)| \right| &\leq \sum_{n=1}^{N} |f(x_n) - f(y_n)| \\
&\leq \sum_{n=1}^{N} \|x_n - y_n\| \\
&< N\delta \\
&\leq \sum_{n=1}^{N} |f(x_n)| - s
\end{aligned}$$

which of course implies that $\sum_{n=1}^{N} |f(y_n)| > s$ and by definition of the product topology we have established that $E(s,N)$ is open. It follows that $U_X \cap X_c^{\mathbb{N}}$ is G_δ in $X_c^{\mathbb{N}}$. Finally we will show that $U_X \cap X_c^{\mathbb{N}}$ is dense in $X_c^{\mathbb{N}}$. Let $\beta \in X_c^{\mathbb{N}}$ and F a finite subset of $\mathbb{N}$. By Theorem C, there exists γ both in U_X and $X_c^{\mathbb{N}}$. We set $\delta(i)$ equal to $\gamma(i)$ when i does not belong to F and $\delta(i)$ equal to $\beta(i)$ otherwise. Then δ belongs to U_X and $X_c^{\mathbb{N}}$ and coincides with β on F which is an arbitrary finite set. It follows that $U_X \cap X_c^{\mathbb{N}}$ is dense in $X_c^{\mathbb{N}}$. The proof is complete and we notice that we have not used Baire's theorem.

We are unable to decide, for any Banach space, if the statement

$$U_X \neq \varnothing \iff U_X \text{ is a dense } G_\delta \text{ subset of } X^{\mathbb{N}}$$

is true or false. We can however obtain the following variant of Theorem 2 (given here without proof):

Theorem 2′. *Let X be a finite-dimensional vector space and V a subspace of $X^{\mathbb{N}}$, endowed with the natural topology and containing all finite sums, such that for any $\sum_{n\in\mathbb{N}} x_n \in V$ there exists a permutation $\sigma \in \mathbb{P}$ such that*

$$D\left(\sum_{n\in\mathbb{N}} x_n\right) = \sum_{n\in\mathbb{N}} x_{\sigma(n)} + \Gamma_0.$$

Then

$$U_X \cap V \neq \varnothing \iff U_X \cap V \text{ is a dense } G_\delta \text{ subset of } V.$$

The above result holds in particular for $V = X_c^{\mathbb{N}}$.

Proof of Theorem 3

Let $\{v_n\}_{n\in\mathbb{N}}$ be dense in X, $\mathcal{U}$ an infinite subset of $\mathbb{N}$ and $\sum_{n\in\mathbb{N}} x_n \in U_X$. There exists a permutation $\sigma_1 \in \mathbb{P}$ such that $\sum_{n\in\mathbb{N}} x_{\sigma_1(n)} = v_1$. We choose $n_1 \in \mathcal{U}$ such that

$$\left\| \sum_{n=1}^{n_1} x_{\sigma_1(n)} - v_1 \right\| < \frac{1}{2^1} \quad \text{and} \quad 1 \in \{\sigma_1(n)\}_{n=1}^{n_1} \tag{6}$$

and we define $\sigma(n) := \sigma_1(n)$, $1 \leq n \leq n_1$ and $N_1 := \mathbb{N}\backslash\{\sigma(n)\}_{n=1}^{n_1}$. By Lemma D, the series $\sum_{n\in N_1} x_n$ is also universal. There exists a bijection $\sigma_2 = \mathbb{N} \to N_1$ such that $\sum_{n\in\mathbb{N}} x_{\sigma_2(n)} = v_2 - \sum_{n=1}^{n_1} x_{\sigma(n)}$. We choose m_2 such that $n_1 + m_2 \in \mathcal{U}$ and

$$\left\| \sum_{n=1}^{m_2} x_{\sigma_2(n)} - \left(v_2 - \sum_{n=1}^{n_1} x_{\sigma(n)}\right) \right\| < \frac{1}{2^2} \quad \text{and} \quad 2 \in \{\sigma(n)\}_{n=1}^{n_1} \cup \{\sigma_2(n)\}_{n=1}^{n_2}. \tag{7}$$

We define $\sigma(n_1 + k) := \sigma_2(k)$ for $1 \leq k \leq m_2$, $n_2 = n_1 + m_2$. It follows from (7) that

$$\left\| \sum_{n=1}^{n_2} x_{\sigma(n)} - v_2 \right\| < \frac{1}{2^2} \quad \text{and} \quad \{1, 2\} \subset \{\sigma(n)\}_{n=1}^{n_2}. \tag{8}$$

Clearly (compare with (6) and (8)) we can define inductively a permutation σ in $\mathbb{P}$ and an infinite sequence $\{n_k\} \subset \mathcal{U}$ of integers such that

$$\left\| \sum_{n=1}^{n_k} x_{\sigma(n)} - v_k \right\| < \frac{1}{2^k} \quad \text{and} \quad \{1, 2, \ldots, k\} \subset \{\sigma(n)\}_{n=1}^{n_k},$$

for $k = 1, 2, \ldots$. Let now $x \in X$; there exists a subsequence $\{n_{k_j}\}_{j\in\mathbb{N}}$ of integers such that $\lim_j v_{k_j} = x$ and since

$$\left\| \sum_{n=1}^{n_{k_j}} x_{\sigma(n)} - x \right\| \leq \left\| \sum_{n=1}^{n_{k_j}} x_{\sigma(n)} - v_{k_j} \right\| + \|v_{k_j} - x\|$$

the result follows. Theorem 3 admits, as suggested by Vassili Nestoridis, some sort of sharpness. Let X be a separable Banach space with $\{v_n\}_{n\geq 1}$ as a dense subset. Fix $0 \neq x \in X$ and a subsequence $\{n_k\}_{k\geq 1}$ of $\mathbb{N}$ with $n_{k+1} - n_k > k$,

$k \geq 1$. We define $\{\tilde{v}_n\}_{n\geq 0}$ by $\tilde{v}_0 = 0$, $\tilde{v}_{n_k} = x$, $\tilde{v}_{n_k+1} = 2x$ for $k \geq 1$ and for $n \in \mathbb{N} \setminus \{n_k\}_{k\geq 1} \cup \{n_k + 1\}_{k\geq 1}$, we let $\{\tilde{v}_n\}$ to be an enumeration of $\{v_n\}_{n\geq 1}$. We also let $x_n = \tilde{v}_n - \tilde{v}_{n-1}$; clearly $\sum_{k=1}^{j} x_k = \tilde{v}_j$ and since $\{\tilde{v}_n\}_{n\geq 1}$ is clearly also dense in X, the set of partial sums of the formal series $\sum_{k=1}^{\infty} x_k$ is dense in X. Given on the other hand any permutation σ of $\mathbb{N}$, the series $\sum_{k=1}^{\infty} x_{\sigma(k)}$ cannot be convergent because for any $k \in \mathbb{N}$

$$x_{n_k+1} = \tilde{v}_{n_k+1} - \tilde{v}_{n_k} = x \neq 0$$

and $\lim_{k\to\infty} x_{\sigma(k)} \neq 0$; this means that the series $\sum_{k=1}^{\infty} x_k$ cannot be universal.

References

1. F. Bayart, *Topological and algebraic genericity of divergence and universality*, Studia Math. **167** (2005), no. 2, 161–181.
2. F. Bayart, K.-G. Grosse-Erdmann, V. Nestoridis, and C. Papadimitropoulos, *Abstract theory of universal series and applications*, Proc. Lond. Math. Soc. (3) **96** (2008), no. 2, 417–463.
3. F. Bayart and É. Matheron, *Dynamics of linear operators*, Cambridge Tracts in Math., vol. 179, Cambridge Univ. Press, Cambridge, 2009.
4. L. Bernal-González, *Dense-lineability in spaces of continuous functions*, Proc. Amer. Math. Soc. **136** (2008), no. 9, 3163–3169.
5. P. M. Gauthier and R. Clouâtre, *Approximation by translates of Taylor polynomials of the Riemann zeta function*, Comput. Methods Funct. Theory **8** (2008), no. 1-2, 15–19.
6. J.-M. Giguêre, *Universalité, variables complexes et réarrangement*, Master's thesis, Université de Montréal, 2011.
7. K.-G. Grosse-Erdmann, *Universal families and hypercyclic operators*, Bull. Amer. Math. Soc. (N.S.) **36** (1999), no. 3, 345–381.
8. V. M. Kadets and M. I. Kadets, *Rearrangements of series in Banach spaces*, Tartu. Gos. Univ., Tartu, 1988 (Russian); English transl. in Transl. Math. Monogr., vol. 86, Amer. Math. Soc., Providence, RI, 1991.
9. J. C. Oxtoby, *Measure and category: A survey of the analogies between topological and measure spaces*, 2nd ed., Grad. Texts in Math., vol. 2, Springer, New York, 1980.
10. A. A. Talaljan, *Series universal with respect to rearrangement*, Izv. Akad. Nauk SSSR. Ser. Mat. **24** (1960), 567–604 (Russian).

Département de mathématiques et de statistique and Centre de recherches mathématiques, Université de Montréal, C.P. 6128, succ. Centre-ville, Montréal, QC H3J 3J7, Canada

E-mail address: fournier@dms.umontreal.ca

Département de mathématiques et de statistique, Université de Montréal, C.P. 6128, succ. Centre-ville, Montréal, QC H3J 3J7, Canada

E-mail address: jerome-melville.giguere@hotmail.com

Centre de Recherches Mathématiques
CRM Proceedings and Lecture Notes
Volume **55**, 2012

Orlicz Capacity of Balls

Yoshihiro Mizuta and Takao Ohno

ABSTRACT. The notion of classical Newton capacity has been generalized to various forms. Among others, Meyers introduced a general notion of L^p-capacity, which is defined by general potentials of functions in the Lebesgue space L^p and such notion of capacity has been proved to provide rich results in the nonlinear potential theory as well as in the study of various function spaces and partial differential equations; see e.g., Adams-Hedberg [1]. The most useful L^p-capacity is Riesz capacity. The aim in this note is to estimate the Riesz capacity of balls $B(x,r)$ centered at x of radius r in the Orlicz setting.

1. Introduction and statement of results

For $0 < \alpha < n$ and a locally integrable function f on $\mathbf{R}^n$, we define the Riesz potential $I_\alpha f$ of order α by

$$I_\alpha f(x) = \int_{\mathbf{R}^n} |x-y|^{\alpha-n} f(y)\,\mathrm{d}y.$$

In the present note, we treat functions f satisfying an Orlicz condition :

$$\int_{\mathbf{R}^n} |f(y)|^p \varphi(|f(y)|)\,\mathrm{d}y < \infty, \tag{1.1}$$

where $p > 1$ and $\varphi(r)$ is a positive monotone function on the interval $(0,\infty)$ which is of logarithmic type; that is, there exists $c_1 > 0$ such that

$$c_1^{-1}\varphi(r) \le \varphi(r^2) \le c_1\varphi(r) \quad \text{whenever } r > 0. \tag{$\varphi 1$}$$

We set $\Phi_{p,\varphi}(r) = r^p\varphi(r)$ and

$$\Phi_{p,\varphi}(0) = 0,$$

because we will see from ($\varphi 4$) below that

$$\lim_{r\to 0+} \Phi_{p,\varphi}(r) = 0;$$

2010 *Mathematics Subject Classification.* Primary 46E30; Secondary 31B15.

Key words and phrases. Orlicz space, Morrey space, capacity, Riesz potential.

The first author was supported in part by Grant-in-Aid for Scientific Research (C), No. 22540192, Japan Society for the Promotion of Science.

The second author was supported in part by Grant-in-Aid for Young Scientists (B), No. 23740108, Japan Society for the Promotion of Science.

This is the final form of the paper.

see [17, p. 205].

We denote by $L^{\Phi_{p,\varphi}}(\mathbf{R}^n)$ the family of all locally integrable functions g on $\mathbf{R}^n$ such that

$$\Phi_{p,\varphi}(g) \equiv \int_{\mathbf{R}^n} \Phi_{p,\varphi}(|g(x)|)\, dx < \infty.$$

Let G be a bounded open set in $\mathbf{R}^n$. For $E \subset G$, the relative $(\alpha, \Phi_{p,\varphi})$-capacity is defined by

$$C_{\alpha,\Phi_{p,\varphi}}(E;G) = \inf_f \Phi_{p,\varphi}(f),$$

where the infimum is taken over all functions f such that $f = 0$ outside G and

$$I_\alpha f(x) \geq 1 \quad \text{for all } x \in E$$

(cf. Adams – Hedberg [1], Meyers [12], Ziemer [22] and the first author [13,14]).

Our first aim in the present note is to give an estimate of $(\alpha, \Phi_{p,\varphi})$-capacity of balls. We denote by $B(x,r)$ the open ball centered at x of radius r. For fixed α, set

$$\varphi_p(r) = \int_r^1 [t^{n-\alpha p}\varphi(t^{-1})]^{-1/(p-1)}\, dt/t$$

when $0 \leq r < 1$.

As an extension of Adams – Hurri-Syrjänen [3], Joensuu [11] and Mizuta [14, Section 8.3, Lemma 3.1], we state our theorem in the following.

Theorem A. *There exists a constant $A_1 > 0$ such that*

$$A_1^{-1}\varphi_p(r)^{-(p-1)} \leq C_{\alpha,\Phi_{p,\varphi}}\big(B(0,r); B(0,1)\big) \leq A_1\varphi_p(r)^{-(p-1)}$$

whenever $0 < r < \frac{1}{2}$.

Theorem A was proved by Adams – Hurri-Syrjänen [3, Theorem 2.11] in the case when $\varphi(t) = (\log(\mathrm{e}+t))^\beta$ with $p = n/\alpha > 1$ and $0 \leq \beta \leq p-1$. Recently Joensuu [9–11] treated the case when φ is nondecreasing. Their main idea was to use the rearrangement equivalent norm for $\|f\|_{L^{\Phi_{p,\varphi}}(\mathbf{R}^n)}$ ([citeBS ,9,10]). Our proof will be done straightforward from the definition of capacity, and several technical assumptions posed in [11] are removed. If $\varphi_p(0) < \infty$, then $C_{\alpha,\Phi_{p,\varphi}}\big(\{0\}; B(0,1)\big) > 0$, and $I_\alpha f$ is continuous when $f \in L^{\Phi_{p,\varphi}}(\mathbf{R}^n)$ has compact support, in view of Mizuta – Shimomura [19].

If $\alpha p < n$, then we see from (φ4) below that

$$\varphi_p(r) \sim [r^{n-\alpha p}\varphi(r^{-1})]^{-1/(p-1)}$$

when $0 < r < \frac{1}{2}$, where we write $f \sim g$ if there exists a constant $A > 0$ so that $A^{-1}g \leq f \leq Ag$.

Theorem 1.1. *If $\alpha p < n$, then*

$$C_{\alpha,\Phi_{p,\varphi}}\big(B(0,r); B(0,1)\big) \sim r^{n-\alpha p}\varphi(r^{-1})$$

whenever $0 < r < \frac{1}{2}$.

Example 1.2. For $n = \alpha p$, consider the function

$$\varphi(t) = (\log(\mathrm{e}+t))^\beta.$$

If $\beta < p-1$, then

$$\varphi_p(r) \sim (\log(\mathrm{e}+1/r))^{-\beta/(p-1)+1}$$

for $0 < r < 1$. In this case

$$C_{\alpha,\Phi_{p,\varphi}}\big(B(0,r);B(0,1)\big) \sim (\log(\mathrm{e}+1/r))^{\beta-p+1}$$

whenever $0 < r < \frac{1}{2}$.

If $\beta = p-1$, then

$$\varphi_p(r) \sim \log\Big(\mathrm{e} + \big(\log(\mathrm{e}+1/r)\big)\Big)$$

for $0 < r < 1$. In this case

$$C_{\alpha,\Phi_{p,\varphi}}\big(B(0,r);B(0,1)\big) \sim \bigg(\log\Big(\mathrm{e} + \big(\log(\mathrm{e}+1/r)\big)\Big)\bigg)^{-p+1}$$

whenever $0 < r < \frac{1}{2}$.

Remark 1.3. We here introduce another capacity. Define

$$\|g\|_{L^{\Phi_{p,\varphi}}(\mathbf{R}^n)} = \inf\{\lambda > 0 : \Phi_{p,\varphi}(g/\lambda) \le 1\}.$$

This is a quasi-norm in $L^{\Phi_{p,\varphi}}(\mathbf{R}^n)$. For a set $E \subset G$, we define

$$N_{\alpha,\Phi_{p,\varphi}}(E;G) = \inf_f \|f\|^p_{L^{\Phi_{p,\varphi}}(\mathbf{R}^n)},$$

where the infimum is taken over all nonnegative measurable functions f on $\mathbf{R}^n$ such that $f = 0$ outside G and $I_\alpha f(x) \ge 1$ for all $x \in E$. With the aid of [8, 16], one can find a constant $A > 1$ such that

$$A^{-1}\varphi_p(r)^{-(p-1)} \le N_{\alpha,\Phi_{p,\varphi}}\big(B(0,r);B(0,1)\big) \le A\varphi_p(r)^{-(p-1)}$$

for $0 < r < \frac{1}{2}$. This implies that two quantities $C_{\alpha,\Phi_{p,\varphi}}\big(B(0,r);B(0,1)\big)$ and $N_{\alpha,\Phi_{p,\varphi}}\big(B(0,r);B(0,1)\big)$ are comparable. We do not know whether two capacities $C_{\alpha,\Phi_{p,\varphi}}\big(\cdot\,;B(0,1)\big)$ and $N_{\alpha,\Phi_{p,\varphi}}\big(\cdot\,;B(0,1)\big)$ are comparable or not.

Next consider the Morrey-Orlicz space $\mathcal{M}^{\nu,\Phi_{p,\varphi}}(\mathbf{R}^n)$ of all functions g such that

$$\sup_{x\in\mathbf{R}^n,r>0} r^{\nu-n}\int_{B(x,r)} \Phi_{p,\varphi}(|g(y)|)\,\mathrm{d}y < \infty$$

with the quasi-norm

$$\|g\|_{\mathcal{M}^{\nu,\Phi_{p,\varphi}}(\mathbf{R}^n)} = \inf\bigg\{\lambda > 0 : \sup_{x\in\mathbf{R}^n,r>0} r^{\nu-n}\int_{B(x,r)} \Phi_{p,\varphi}(|g(y)|/\lambda)\,\mathrm{d}y \le 1\bigg\}.$$

For fundamental properties of Morrey – Orlicz space, we refer the reader to the papers by Adams – Xiao [4], Mizuta – Nakai – Ohno – Shimomura [15] and Mizuta – Shimomura [20, 21].

Our second aim in the present note is to give an estimate of Morrey – Orlicz capacity of balls defined by

$$N_{\alpha,\mathcal{M}^{\nu,\Phi_{p,\varphi}}}(E;G) = \inf_f \|f\|^p_{\mathcal{M}^{\nu,\Phi_{p,\varphi}}(\mathbf{R}^n)},$$

where the infimum is taken over all functions f such that $f = 0$ outside G and

$$I_\alpha f(x) \ge 1 \quad \text{for all } x \in E.$$

For fixed α, ν, consider

$$\tilde{\varphi}_p(r) = \int_r^1 [t^{\nu-\alpha p}\varphi(t^{-1})]^{-1/p}\,\mathrm{d}t/t$$

when $0 < r < 1$.

Theorem B. *Suppose $\alpha p \le \nu < n$. There exists a constant $A_2 > 0$ such that*

$$A_2^{-1}\tilde{\varphi}_p(r)^{-p} \le N_{\alpha,\mathcal{M}^{\nu,\Phi_{p,\varphi}}}\big(B(0,r);B(0,1)\big) \le A_2\tilde{\varphi}_p(r)^{-p}$$

whenever $0 < r < \frac{1}{2}$.

This is an extension of Adams – Xiao [4, Theorem 5.3].

The case $\nu = n$ was treated by Theorem A. If $\tilde{\varphi}_p(0) < \infty$, then

$$N_{\alpha,\mathcal{M}^{\nu,\Phi_{p,\varphi}}}\big(\{0\};B(0,1)\big) > 0,$$

and $I_\alpha f$ is continuous on $\mathbf{R}^n$ when $f \in \mathcal{M}^{\nu,\Phi_{p,\varphi}}(\mathbf{R}^n)$ has compact support, in view of Mizuta – Shimomura [20, 21].

Theorem 1.4. *If $\alpha p < \nu$, then*

$$N_{\alpha,\mathcal{M}^{\nu,\Phi_{p,\varphi}}}\big(B(0,r);B(0,1)\big) \sim r^{\nu-\alpha p}\varphi(r^{-1})$$

whenever $0 < r < \frac{1}{2}$.

Example 1.5. Let $\varphi(t) = \big(\log(\mathrm{e}+t)\big)^\beta$. If $\alpha p < \nu$, then

$$\tilde{\varphi}_p(r) \sim [r^{\nu-\alpha p}\big(\log(\mathrm{e}+1/r)\big)^\beta]^{-1/p}$$

for $0 < r < \frac{1}{2}$, so that

$$N_{\alpha,\mathcal{M}^{\nu,\Phi_{p,\varphi}}}\big(B(0,r);B(0,1)\big) \sim r^{\nu-\alpha p}\big(\log(\mathrm{e}+1/r)\big)^\beta$$

whenever $0 < r < \frac{1}{2}$.

If $\alpha p = \nu$ and $\beta < p$, then

$$\tilde{\varphi}_p(r) \sim \big(\log(\mathrm{e}+1/r)\big)^{1-\beta/p}$$

for $0 < r < \frac{1}{2}$, so that

$$N_{\alpha,\mathcal{M}^{\nu,\Phi_{p,\varphi}}}\big(B(0,r);B(0,1)\big) \sim \big(\log(\mathrm{e}+1/r)\big)^{\beta-p}$$

whenever $0 < r < 1/2$.

For further related results, see Aïssaoui – Benkirane [5], Adams – Hurri-Syrjänen [2], Edmunds – Evans [7] and Mizuta – Shimomura [17–19].

Throughout this note, let A denote various constants independent of the variables in question and $A(a,b,\dots)$ be a constant that depends on $a,b,\dots$.

2. Proof of Theorem A

First we collect properties which follow from condition (φ1) (see [14; 15, Section 7; 17, Lemma 2.3]).

(φ2) φ satisfies the doubling condition, that is, there exists $c_2 > 1$ such that

$$c_2^{-1}\varphi(r) \le \varphi(2r) \le c_2\varphi(r) \quad \text{whenever } r > 0.$$

(φ3) For each $\gamma > 0$, there exists $c_3 = c_3(\gamma) \ge 1$ such that

$$c_3^{-1}\varphi(r) \le \varphi(r^\gamma) \le c_3\varphi(r) \quad \text{whenever } r > 0.$$

(φ4) For each $\gamma > 0$, there exists $c_4 = c_4(\gamma) \ge 1$ such that

$$s^\gamma\varphi(s) \le c_4 t^\gamma\varphi(t) \quad \text{whenever } 0 < s < t.$$

(φ5) For each $\gamma > 0$, there exists $c_5 = c_5(\gamma) \ge 1$ such that

$$t^{-\gamma}\varphi(t) \le c_5 s^{-\gamma}\varphi(s) \quad \text{whenever } 0 < s < t.$$

($\varphi 6$) If φ and φ_1 are positive monotone functions on $[0,\infty)$ satisfying ($\varphi 1$), then for each $\gamma > 0$ there exists a constant $c_6 = c_6(\gamma) \geq 1$ such that

$$c_6{}^{-1}\varphi(r) \leq \varphi(r^\gamma \varphi_1(r)) \leq c_6\varphi(r) \quad \text{whenever } r > 0.$$

Consider the function

$$\bar{\Phi}_{p,\varphi}(r) = \int_0^r \sup_{0<s<t} s^{-1}\Phi_{p,\varphi}(s)\, dt.$$

Then $\bar{\Phi}_{p,\varphi}$ is nondecreasing and

$$\bar{\Phi}_{p,\varphi}(r) \sim \Phi_{p,\varphi}(r)$$

by ($\varphi 4$). Thus $\|\cdot\|_{L^{\bar{\Phi}_{p,\varphi}}(\mathbf{R}^n)}$ defines an equivalent norm to $\|\cdot\|_{L^{\Phi_{p,\varphi}}(\mathbf{R}^n)}$, and

$$C_{\alpha,\Phi_{p,\varphi}}\big(E, B(0,1)\big) \sim C_{\alpha,\bar{\Phi}_{p,\varphi}}\big(E, B(0,1)\big)$$

for $E \subset B(0,\frac{1}{2})$.

Lemma 2.1. *There exists a constant $A > 0$ such that*

$$C_{\alpha,\Phi_{p,\varphi}}\big(B(0,r), B(0,1)\big) \leq A\varphi_p(r)^{-(p-1)}$$

for $0 < r < \frac{1}{2}$.

PROOF. Set $K = \varphi_p(r)$. For $0 < r < r_0$, consider the function

$$f_r(y) = K^{-1}|y|^{-\alpha}\big[|y|^{n-\alpha p}\varphi\big(K^{-1/\alpha}|y|^{-1}\big)\big]^{-1/(p-1)}$$

for $r < |y| < K^{-1/\alpha}$ and $f_r = 0$ elsewhere. We treat only the case $\alpha p = n$, since the case $\alpha p < n$ can be obtained similarly. If $r > 0$ is small, say $0 < r < r_0$ $(< \frac{1}{2})$, then we see from ($\varphi 4$) that

$$rK^{1/\alpha} < \sqrt{r}.$$

If $y \in B(0,1) \setminus B(0,r)$ and $x \in B(0,r)$, then $|x-y| < 2|y|$. Hence we have

$$\begin{aligned} I_\alpha f_r(x) &\geq 2^{\alpha-n}\int_{B(0,1)\setminus B(0,r)} |y|^{\alpha-n} f_r(y)\, dy \\ &\geq 2^{\alpha-n}\omega_{n-1}K^{-1}\int_{rK^{1/\alpha}}^1 [\varphi(t^{-1})]^{-1/(p-1)}\, dt/t \\ &\geq AK^{-1}\int_{\sqrt{r}}^1 [\varphi(t^{-1})]^{-1/(p-1)}\, dt/t \\ &\geq A, \end{aligned}$$

since $\varphi_p(r)$ is of log-type, where ω_{n-1} denotes the area of the unit sphere. Hence we obtain

$$C_{\alpha,\Phi_{p,\varphi}}\big(B(0,r), B(0,1)\big) \leq A\Phi_{p,\varphi}(f_r).$$

Since $\varphi(f_r) \leq A\varphi(K^{-1/\alpha}|y|^{-1})$ by ($\varphi 6$), we obtain

$$\begin{aligned} &\int_{B(0,1)} \Phi_{p,\varphi}(f_r(y))\, dy \\ &\quad \leq AK^{-p}\int_{B(0,K^{-1/\alpha})\setminus B(0,r)} |y|^{-\alpha p}\varphi(K^{-1/\alpha}|y|^{-1})^{-p/(p-1)+1}\, dy \leq AK^{-p+1}. \end{aligned}$$

Thus it follows that

$$C_{\alpha,\Phi_{p,\varphi}}\big(B(0,r), B(0,1)\big) \leq AK^{-p+1}$$

for $0 < r < r_0$.

If $r_0 \le r < \frac{1}{2}$, then

$$C_{\alpha,\Phi_{p,\varphi}}\big(B(0,r),B(0,1)\big) \le C_{\alpha,\Phi_{p,\varphi}}\big(B(0,\tfrac{1}{2}),B(0,1)\big) = A \le AK^{-p+1},$$

which proves the lemma. □

Lemma 2.2. *There exists a constant $A > 0$ such that*

$$C_{\alpha,\Phi_{p,\varphi}}\big(B(0,r),B(0,1)\big) \ge A\varphi_p(r)^{-(p-1)}.$$

for $0 < r < \frac{1}{2}$.

Proof. We prove this only when $\alpha p = n$, since the case $\alpha p < n$ is proved similarly. For $0 < r < \frac{1}{2}$, take a nonnegative measurable function f on $B(0,1)$ such that

$$I_\alpha f(x) \ge 1 \quad \text{for } x \in B(0,r).$$

Then we have by Fubini's theorem

$$\begin{aligned}\int_{B(0,r)} \mathrm{d}x \le \int_{B(0,r)} I_\alpha f(x)\,\mathrm{d}x &\le \int_{B(0,1)} \left(\int_{B(0,r)} |x-y|^{\alpha-n}\,\mathrm{d}x\right) f(y)\,\mathrm{d}y \\ &\le Ar^n \int_{B(0,1)} (r+|y|)^{\alpha-n} f(y)\,\mathrm{d}y,\end{aligned}$$

so that

$$1 \le A\int_{B(0,1)} (r+|y|)^{\alpha-n} f(y)\,\mathrm{d}y.$$

We show that

$$1 \le A[\varphi_p(r)]^{p-1}\Phi_{p,\varphi}(f). \tag{2.1}$$

Let $0 < \varepsilon < 1$ and $K = \varphi_p(r)$. Considering

$$k(y) = \varepsilon K^{-1}(r+|y|)^{-\alpha}[\varphi(K^{-1/\alpha}(r+|y|)^{-1})]^{-1/(p-1)},$$

we obtain

$$\begin{aligned}&\int_{B(0,1)} (r+|y|)^{\alpha-n} f(y)\,\mathrm{d}y \\ &\quad \le \int_{B(0,1)} (r+|y|)^{\alpha-n} k(y)\,\mathrm{d}y + A\int_{B(0,1)} (r+|y|)^{\alpha-n} f(y)\left(\frac{f(y)}{k(y)}\right)^{p-1} \frac{\varphi(f(y))}{\varphi(k(y))}\,\mathrm{d}y \\ &\quad \le A\varepsilon K^{-1}\int_0^1 [\varphi(K^{-1/\alpha}(r+t)^{-1})]^{-1/(p-1)}(r+t)^{-1}\,\mathrm{d}t \\ &\qquad\qquad + A(\varepsilon)K^{p-1}\int_{B(0,1)} f(y)^p\varphi(f(y))\,\mathrm{d}y \\ &\quad \le A\varepsilon + A(\varepsilon)K^{p-1}\Phi_{p,\varphi}(f),\end{aligned}$$

since

$$\begin{aligned}\varphi(k(y)) &\ge A(\varepsilon)\varphi((K^{-1/\alpha}(r+|y|)^{-1})^\alpha[\varphi(K^{-1/\alpha}(r+|y|)^{-1})]^{-1/(p-1)}) \\ &\ge A(\varepsilon)\varphi(K^{-1/\alpha}(r+|y|)^{-1})\end{aligned}$$

by ($\varphi 2$) and ($\varphi 6$), and

$$\int_r^1 [\varphi(K^{-1/\alpha}t^{-1})]^{-1/(p-1)}\,dt/t \le \int_{rK^{1/\alpha}}^{K^{1/\alpha}} [\varphi(s^{-1})]^{-1/(p-1)}\,ds/s$$
$$\le AK + \int_1^{K^{1/\alpha}} [\varphi(s^{-1})]^{-1/(p-1)}\,ds/s$$
$$\le AK + A\max\{1, [\varphi(K^{1/\alpha})]^{-1/(p-1)}\}\log K$$
$$\le AK$$

by ($\varphi 6$). Now it follows that

$$1 \le A\varepsilon + A(\varepsilon)K^{p-1}\Phi_{p,\varphi}(f).$$

By taking $A\varepsilon = \frac{1}{2}$, we obtain (2.1), and hence

$$1 \le A\varphi_p(r)^{p-1}C_{\alpha,\Phi_{p,\varphi}}\big(B(0,r); B(0,1)\big),$$

which proves the conclusion. □

Now we establish Theorem A by use of Lemmas 2.1 and 2.2.

3. Proof of Theorem B

To prove Theorem B, we prepare some lemmas.

Lemma 3.1. *If $\alpha p \le \nu < n$, then there exists a constant $A > 0$ such that*

$$N_{\alpha,\mathcal{M}^{\nu,\Phi_{p,\varphi}}}\big(B(0,r); B(0,1)\big) \le A\tilde{\varphi}_p(r)^{-p}$$

whenever $0 < r \le \frac{1}{2}$.

Proof. For $0 < r < \frac{1}{2}$ and $K = \tilde{\varphi}_p(r)$, consider the function

$$f_r(y) = K^{-1}|y|^{-\alpha}[|y|^{\nu-\alpha p}\varphi(|y|^{-1})]^{-1/p}$$

for $r < |y| < 1$ and $f_r = 0$ elsewhere. We show only the case $\alpha p = \nu$, because the case $\alpha p < \nu$ is obtainded similarly. If $x \in B(0,r)$, then, as in the proof of Lemma 2.1, we have

$$I_\alpha f_r(x) \ge A,$$

so that

$$N_{\alpha,\mathcal{M}^{\nu,\Phi_{p,\varphi}}}\big(B(0,r); B(0,1)\big) \le A\|f_r\|^p_{\mathcal{M}^{\nu,\Phi_{p,\varphi}}(\mathbf{R}^n)}.$$

Setting

$$g_r(y) = |y|^{-\alpha}[\varphi(|y|^{-1})]^{-1/p},$$

we have

$$\int_{B(x,t)} \Phi_{p,\varphi}(g_r(y))\,dy \le A\int_{B(0,t)} |y|^{-\alpha p}\,dy \le At^{N-\alpha p},$$

which implies that $\|g_r\|_{\mathcal{M}^{\nu,\Phi_{p,\varphi}}(\mathbf{R}^n)} \le A$. Hence it follows that

$$\|f_r\|_{\mathcal{M}^{\nu,\Phi_{p,\varphi}}(\mathbf{R}^n)} \le AK^{-1}\|g_r\|_{\mathcal{M}^{\nu,\Phi_{p,\varphi}}(\mathbf{R}^n)} \le AK^{-1},$$

which proves the lemma. □

Lemma 3.2. *If $\alpha p \le \nu < n$, then there exists a constant $A > 0$ such that*

$$N_{\alpha,\mathcal{M}^{\nu,\Phi_{p,\varphi}}}\big(B(0,r); B(0,1)\big) \ge A\tilde{\varphi}_p(r)^{-p}$$

whenever $0 < r \le \frac{1}{2}$.

PROOF. As before, we show this lemma only when $\alpha p = \nu$. For $0 < r < \frac{1}{2}$, take a nonnegative measurable function f on $B(0,1)$ such that

$$I_\alpha f(x) \geq 1 \quad \text{for } x \in B(0,r).$$

Then, as in the proof of Lemma 2.2, we have

$$1 \leq A \int_{B(0,1)} (r + |y|)^{\alpha-n} f(y)\, \mathrm{d}y.$$

We show that

$$1 \leq A\tilde{\varphi}_p(r)^p \|f\|^p_{\mathcal{M}^{\nu,\Phi_{p,\varphi}}(\mathbf{R}^n)}. \tag{3.1}$$

To show this, we may assume that $\|f\|_{\mathcal{M}^{\nu,\Phi_{p,\varphi}}(\mathbf{R}^n)} \leq 1$. Considering

$$k(y) = |y|^{-\alpha}[\varphi(|y|^{-1})]^{-1/p},$$

we find

$$\begin{aligned}
&\int_{B(0,t)} f(y)\, \mathrm{d}y \\
&\quad \leq \int_{B(0,t)} k(y)\, \mathrm{d}y + A \int_{B(0,t)} f(y) \left(\frac{f(y)}{k(y)}\right)^{p-1} \frac{\varphi(f(y))}{\varphi(k(y))}\, \mathrm{d}y \\
&\quad \leq At^{n-\alpha}[\varphi(t^{-1})]^{-1/p} + At^{n+\alpha(p-1)}[\varphi(t^{-1})]^{-1/p} \int_{B(0,t)} f(y)^p \varphi(f(y))\, \mathrm{d}y \\
&\quad \leq At^{n-\alpha}[\varphi(t^{-1})]^{-1/p},
\end{aligned}$$

so that

$$\begin{aligned}
\int_{B(0,1)} (r + |y|)^{\alpha-n} f(y)\, \mathrm{d}y &\leq A \int_0^1 \left(\int_{B(0,t)} f(y)\, \mathrm{d}y \right) (r+t)^{\alpha-n-1}\, \mathrm{d}t \\
&\leq A \int_0^1 (r+t)^{-1} [\varphi((r+t)^{-1})]^{-1/p}\, \mathrm{d}t \leq A\tilde{\varphi}(r),
\end{aligned}$$

which gives (3.1). Thus we obtain

$$1 \leq A\tilde{\varphi}_p(r)^p N_{\alpha, \mathcal{M}^{\nu,\Phi_{p,\varphi}}}\big(B(0,r); B(0,1)\big),$$

which proves the conclusion. □

Now we establish Theorem B by use of Lemmas 3.1 and 3.2.

References

1. D. R. Adams and L. I. Hedberg, *Function spaces and potential theory*, Grundlehren Math. Wiss., vol. 314, Springer, Berlin, 1996.
2. D. R. Adams and R. Hurri-Syrjänen, *Vanishing exponential integrability for functions whose gradients belong to $L^n(\log(\mathrm{e}+L))^\alpha$*, J. Funct. Anal. **197** (2003), no. 1, 162–178.
3. ———, *Capacity estimates*, Proc. Amer. Math. Soc. **131** (2003), no. 4, 1159–1167 (electronic).
4. D. R. Adams and J. Xiao, *Nonlinear potential analysis on Morrey spaces and their capacities*, Indiana Univ. Math. J. **53** (2004), no. 6, 1629–1663.
5. N. Aïssaoui and A. Benkirane, *Capacités dans les espaces d'Orlicz*, Ann. Sci. Math. Québec **18** (1994), no. 1, 1–23.
6. C. Bennett and R. Sharpley, *Interpolation of operators*, Pure Appl. Math, vol. 129, Academic Press, Boston, MA, 1988.
7. D. E. Edmunds and W. D. Evans, *Hardy operators, function spaces and embeddings*, Springer Monogr. Math., Springer, Berlin, 2004.

8. T. Futamura, Y. Mizuta, T. Ohno, and T. Shimomura, *Orlicz – Sobolev capacity of balls*, Illinois J. Math., to appear.
9. J. Joensuu, *On null sets of Sobolev – Orlicz capacities*, Illinois J. Math. **52** (2008), no. 4, 1195–1211.
10. ———, *Orlicz – Sobolev capacities and their null sets*, Rev. Mat. Complut. **23** (2010), no. 1, 217–232.
11. ———, *Estimates for certain Orlicz – Sobolev capacities of an Euclidean ball*, Nonlinear Anal. **72** (2010), no. 11, 4316–4330.
12. N. G. Meyers, *A theory of capacities for potentials of functions in Lebesgue classes*, Math. Scand. **26** (1970), 255–292 (1971).
13. Y. Mizuta, *Continuity properties of potentials and Beppo – Levi – Deny functions*, Hiroshima Math. J. **23** (1993), no. 1, 79–153.
14. ———, *Potential theory in Euclidean spaces*, GAKUTO Internat. Ser. Math. Sci. Appl., vol. 6, Gakkōtosho, Tokyo, 1996.
15. Y. Mizuta, E. Nakai, T. Ohno, and T. Shimomura, *Boundedness of fractional integral operators on Morrey spaces and Sobolev embeddings for generalized Riesz potentials*, J. Math. Soc. Japan **62** (2010), no. 3, 707–744.
16. Y. Mizuta, T. Ohno, and T. Shimomura, *Weighted Orlicz – Riesz capacity of balls*, Proc. Amer. Math. Soc. **138** (2010), no. 12, 4291–4302.
17. Y. Mizuta and T. Shimomura, *Differentiability and Hölder continuity of Riesz potentials of Orlicz functions*, Analysis (Munich) **20** (2000), no. 3, 201–223.
18. ———, *Vanishing exponential integrability for Riesz potentials of functions in Orlicz classes*, Illinois J. Math. **51** (2007), no. 4, 1039–1060.
19. ———, *Continuity properties of Riesz potentials of Orlicz functions*, Tohoku Math. J. (2) **61** (2009), no. 2, 225–240.
20. ———, *Continuity properties for Riesz potentials of functions in Morrey spaces of variable exponent*, Math. Inequal. Appl. **13** (2010), no. 1, 99–122.
21. ———, *Sobolev's inequality for Riesz potentials of functions in Morrey spaces of integral form*, Math. Nachr. **283** (2010), no. 9, 1336–1352.
22. W. P. Ziemer, *Weakly differentiable functions: Sobolev spaces and functions of bounded variation*, Grad. Texts in Math., vol. 120, Springer, New York, 1989.

Department of Mechanical Systems Engineering, Hiroshima Institute of Technology, 2-1-1 Miyake, Saeki-ku, Hiroshima 731-5193, Japan
E-mail address: yoshihiromizuta3@gmail.com

Faculty of Education and Welfare Science, Oita University, Dannoharu Oita-city 870-1192, Japan
E-mail address: t-ohno@oita-u.ac.jp

Centre de Recherches Mathématiques
CRM Proceedings and Lecture Notes
Volume **55**, 2012

Potential Analysis on Nonsmooth Domains — Martin Boundary and Boundary Harnack Principle

Hiroaki Aikawa

ABSTRACT. Martin introduced an ideal boundary (referred to as the Martin boundary) which represents every positive harmonic function in a domain with Green function. The identification of the Martin boundary for specific domains is an interesting question. We explain various potential theoretical notions giving information on the Martin boundary. Boundary Harnack principles are typical important examples. We introduce several classes of nonsmooth domains which enjoy such potential theoretical properties.

1. Martin boundary and boundary Harnack principle

This is an expository paper on potential analysis on nonsmooth domains, mainly concerning the Martin boundary and related potential theoretical properties.

The *Martin boundary* Δ describes the family of all positive harmonic functions in a domain $D \subset \mathbb{R}^n$ with Green function $G(x,y)$ ([30]). Roughly speaking, Δ is the totality of all ideal boundary points ξ for which the limit $\lim_{y\to\xi} G(x,y)/G(x_0,y) = K(x,\xi)$ exists. Here $x_0 \in D$ is a fixed reference point. We say that a positive harmonic function h is minimal if every positive harmonic function majorized by h must coincide with a constant multiple of h. We say that ξ is a minimal Martin boundary point if $K(\cdot,\xi)$ is a minimal positive harmonic function in D. The set of minimal Martin boundary points is called the minimal Martin boundary and is denoted by Δ_1. The remaining boundary $\Delta_0 = \Delta \setminus \Delta_1$ is called the non-minimal Martin boundary. Marin proved that every positive harmonic function h on D is represented as the integral $\int_{\Delta_1} K(\cdot,\xi)\,\mathrm{d}\mu_h(\xi)$ with unique measure μ_h on Δ_1.

We naturally have the following questions:

(i) What is the Martin boundary of a specific domain?
(ii) How do we identify it?

2010 *Mathematics Subject Classification.* 31B05, 31B25, 31C35.
Key words and phrases. Martin boundary, boundary Harnack principle, nonsmooth domain, Carleson estimate, capacitary width, box argument, Domar's argument.
This work was supported in part by Kakenhi No. 20244007.
This is the final form of the paper.

(iii) Do the (minimal) Martin boundary and the topological boundary coincide, i.e., $\Delta = \Delta_1 = \partial D$?

If D is sufficiently smooth, then the above questions have satisfactory solutions:

(i) The Green function $G(x_0, y)$ with pole at x_0 is comparable to the distant function $\delta_D(y) = \mathrm{dist}(y, \partial D)$ outside a small ball with center at x_0, i.e., $G(x_0, y) \approx \delta_D(y)$.

(ii) Let $P(x, y) = -\partial G(x, y)/\partial n_y$ be the Poisson kernel, where n_y stands for the outward normal at $y \in \partial D$. Observe that the Martin kernel $K(x, y)$ is equal to the Poisson kernel $P(x, y)$ multiplied by a positive function of $y \in \partial D$. Hence the Martin representation essentially coincides with the Poisson integral representation.

(iii) $\Delta = \Delta_1 = \partial D$.

If D is general, then the above problems are very difficult and have been attracting attentions of a number of mathematicians. In this note we show several families of domains between smooth domains and general domains, for which some properties of the Martin boundary are obtained. Of course, the more general D, the coarser information.

We use the following notation. We write $B(x, r), C(x, r)$ and $S(x, r)$ for the open ball, closed ball and sphere of center at x and radius r, respectively. By the symbol C we denote a positive constant whose value is unimportant and may change from one occurrence to the next. If necessary, we use $C_0, C_1, \ldots$, to specify them. If two positive quantities f and g satisfies $C^{-1} \le f/g \le C$ with some constant $C \ge 1$, then we say f and g are comparable and write $f \approx g$. The constant C is referred to as the constant of comparison. We must be careful for the dependency of the constant of comparison.

Let us begin with the classical the Harnack principle.

Theorem 1.1 (Harnack principle (1886) [22]). *Let K be a compact subset of D. Then there exists a constant $C > 1$ depending only on K and D such that*

$$\frac{u(x)/v(x)}{u(y)/v(y)} \le C \quad \textit{for } x, y \in K,$$

whenever u and v are positive harmonic functions in D.

Let us remove the compactness of K and suppose K touches ∂D. What happens? The hypothesis u and v be positive and harmonic in D is insufficient for the above comparison; the boundary values of u and v may affect. So, in order to have some information between u and v, we have to assume something for boundary values of u and v, say $u = v = 0$ on $K \cap \partial D$.

Definition 1.2 (Boundary Harnack principle). Let K be a compact set in $\mathbb{R}^n$ intersecting D and ∂D, and let V be an open set containing K. We say that the *boundary Harnack principle* holds if there exists a constant $C > 1$ depending only on K, V and D such that

$$\frac{u(x)/v(x)}{u(y)/v(y)} \le C \quad \text{for } x, y \in K \cap D,$$

whenever u and v are positive harmonic functions in D with $u = v = 0$ on $\partial D \cap V$.

The geometry of D is crucial for the validity of the boundary Harnack principle. First, we have to clarify the meaning: $u = v = 0$ on ∂D, since D may be irregular for the Dirichlet problem. We cannot assume $u = v = 0$ on ∂D continuously. Instead, we impose a milder assumption: u and v are bounded and $u = v = 0$ q.e. (quasi everywhere), i.e., the equalities hold outside a polar set. Here, a polar set is a set on which some superharmonic function diverges to $+\infty$. A polar set is very small; for instance, it is known that the Hausdorff dimension of a polar set is at most $n-2$ (see, e.g., [10, Theorem 5.9.6]).

Next, we vary the couple (V, K) to focus on the geometry of D. Since we are interested in the smoothness of D, it is relevant to study properties valid for every pair (V, K) such that

$$K \subset V, \quad K \cap D \neq \varnothing, \quad K \cap \partial D \neq \varnothing. \tag{1.1}$$

Definition 1.3 (Global boundary Harnack principle). We say that D satisfies the *global boundary Harnack principle* if for every (V, K) with (1.1) we find a constant $C_1 > 1$ depending only on D and (V, K) such that

$$\frac{u(x)/u(y)}{v(x)/v(y)} \leq C_1 \quad \text{for } x, y \in K \cap D,$$

whenever u and v are positive superharmonic functions in D such that

(i) u and v are bounded and harmonic in $V \cap D$;
(ii) $u = v = 0$ q.e. on $V \cap \partial D$,

Let us recall the Poisson kernel $P(x, y) = -\partial G(x, y)/\partial n_y$ for a smooth domain. It is easy to see that $P(\cdot, y)$ is a positive harmonic function in D such that $P(\cdot, y) = 0$ on $\partial D \setminus \{y\}$ and $P(\cdot, y)$ is unbounded near y. This observation yields the following definition.

Definition 1.4 (Kernel function). Let D be a bounded domain and let $x_0 \in D$. Let $\xi \in \partial D$. We say that h is a *kernel function at ξ normalized at x_0* and write $h \in \mathcal{H}_\xi$ if

(i) h is a positive harmonic function in D and $h = 0$ q.e. on ∂D;
(ii) h is bounded on $D \setminus B(\xi, r)$ for every $r > 0$;
(iii) $h(x_0) = 1$.

The relationship between Martin kernels and kernel functions is subtle. Let D be a bounded domain. Let $\{y_j\}$ be a sequence in D which has no accumulation point in D. Then there exists a boundary point $\xi \in \partial D$ to which a subsequence of $\{y_j\}$ converges. Taking a subsequence, we may assume that $\{y_j\}$ itself converges to ξ and furthermore $G(\cdot, y_j)/G(x_0, y_j)$ converges to a Martin kernel. We say that this Martin kernel lies over ξ and denote by Δ_ξ the set of all Martin kernels lie over ξ. Note that $\Delta_\xi \neq \varnothing$. In many cases, a Martin kernel over ξ is a kernel function of ξ in the sense of Definition 1.4. The global boundary Harnack principle ensures that this is the case.

Theorem 1.5. *Let D be a bounded domain and let $\xi, \xi' \in \partial D$. Suppose D satisfies the global boundary Harnack principle. Then*

(i) $\Delta_\xi \subset \mathcal{H}_\xi$.
(ii) *If $\xi \neq \xi'$, then $\mathcal{H}_\xi \cap \mathcal{H}_{\xi'} = \varnothing$.*
(iii) *If $\xi \neq \xi'$, then $\Delta_\xi \cap \Delta_{\xi'} = \varnothing$.*
(iv) *$\partial D \subset \Delta$, i.e., there is an injection from ∂D to Δ.*

PROOF. Let us begin with (i). Let $h \in \Delta_\xi$. By definition, we find a sequence of points $y_j \in D$ such that $y_j \to \xi$ and $G(\cdot, y_j)/G(x_0, y_j) \to h$. Let $r > 0$ be small. Fix $y_r \in D \cap B(\xi, r/2)$. Suppose j is sufficiently large and $|y_j - \xi| < r/2$. Apply the global boundary Harnack principle to $u = G(\cdot, y_j)/G(x_0, y_j)$, $v = G(\cdot, y_r)/G(x_0, y_r)$, $V = D \setminus \overline{B(\xi, r/2)}$, and $K = \overline{D} \setminus B(\xi, r)$. Then we have

$$\frac{G(x, y_j)}{G(x_0, y_j)} \approx \frac{G(x, y_r)}{G(x_0, y_r)} \le C_r \quad \text{for } x \in D \setminus B(\xi, r).$$

Taking the limit as $j \to \infty$, we obtain $h \le C_r$ on $D \setminus B(\xi, r)$. Moreover, using the harmonic measure, we obtain

$$\frac{G(\cdot, y_j)}{G(x_0, y_j)} \le C_r \omega\big(D \cap S(\xi, r), D \setminus C(\xi, r)\big) \quad \text{on } D \setminus C(\xi, r)$$

uniformly for j. Passing to the limit as $j \to \infty$, we obtain

$$h \le C_r \omega\big(D \cap S(\xi, r), D \setminus C(\xi, r)\big) \quad \text{on } D \setminus C(\xi, r).$$

This implies that $h = 0$ q.e. on $\partial D \setminus C(\xi, r)$, and so $h = 0$ q.e. on ∂D by the arbitrariness of $r > 0$. Since $h(x_0) = 1$, it follows that $h \in \mathcal{H}_\xi$.

Now let us prove (ii), (iii), and (iv). Let $\xi \ne \xi'$. If $h' \in \mathcal{H}_{\xi'}$, then $h' \notin \mathcal{H}_\xi$ by boundedness. Hence (ii) follows. This, together with (i), shows (iii). Define a mapping Φ from ∂D to Δ by $\Phi(\xi) = h_\xi$ with $h_\xi \in \Delta_\xi$. Then Φ is injective by (iii). □

Definition 1.6 (Local boundary Harnack principle). We say that D satisfies the *local boundary Harnack principle* if there exist constants $C_2 > 1$, $C_3 > 1$ and $r_0 > 0$ depending only on D such that

$$\frac{u(x)/u(y)}{v(x)/v(y)} \le C_2 \quad \text{for } x, y \in D \cap B(\xi, r),$$

whenever

(i) $\xi \in \partial D$ and $0 < r < r_0$;
(ii) u and v are positive harmonic functions on $D \cap B(\xi, C_3 r)$;
(iii) $u = v = 0$ q.e. on $\partial D \cap B(\xi, C_3 r)$.

The local boundary Harnack principle is a property stronger than the global boundary Harnack principle. The local boundary Harnack principle ensures that the Martin boundary is homeomorphic to the topological boundary.

Theorem 1.7 (Kemper [28]). *Let D be a bounded domain. If D satisfies the local boundary Harnack principle, then $\Delta = \Delta_1 = \partial D$, i.e., the Martin boundary is homeomorphic to the topological boundary; the Martin boundary consists of minimal Martin boundary points.*

PROOF. In view of Theorem 1.5, it is sufficient to show that $\mathcal{H}_\xi$ is a singleton of a minimal harmonic function. The important step is to show that

$$\frac{u}{v} \le C \quad \text{on } D, \tag{1.2}$$

whenever $u, v \in \mathcal{H}_\xi$. We shall postpone the proof of this claim and complete the proof of the theorem.

Let

$$c = \sup_{\substack{u,v\in\mathcal{H}_\xi \\ x\in D}} \frac{u(x)}{v(x)}.$$

Then $1 \le c < \infty$ by (1.2). Let us prove $c = 1$ by contradiction. Suppose $c > 1$. By definition we have

$$u \le cv \quad \text{on } D \text{ for every } u, v \in \mathcal{H}_\xi. \tag{1.3}$$

Take u, $v \in \mathcal{H}_\xi$. Then $cv - u$ is a nonnegative harmonic function in D, so that $cv - u$ is a positive harmonic function in D by the minimum principle as $cv(x_0) - u(x_0) = c - 1 > 0$. Normalization gives $v_1 = (cv - u)/(c - 1) \in \mathcal{H}_\xi$. Hence $u \le cv_1 = c(cv - u)/(c - 1)$ by (1.3) with u and v_1. By a simple calculation we have $(2c - 1)u \le c^2 v$, so that

$$c = \sup_{\substack{u,v\in\mathcal{H}_\xi \\ x\in D}} \frac{u(x)}{v(x)} \le \frac{c^2}{2c-1} < c.$$

This contradiction shows that $c = 1$, which immediately implies that $\mathcal{H}_\xi$ is a singleton.

Next, we show that $u \in \mathcal{H}_\xi$ is minimal. Suppose v is a positive harmonic function majorized by u. Then $v/v(x_0) \in \mathcal{H}_\xi$, so $u = v/v(x_0)$ by the previous part. Hence v is a constant multiple of u, so that u is minimal.

Now, let us prove (1.2). For small $r > 0$ let u_r be the regularized reduced function $\widehat{R}_v^{D\setminus B(\xi,r)}$, i.e., u_r is the lower regularization of the infimum of all nonnegative superharmonic functions w in D such that $w \ge u$ on $D \setminus B(\xi, r)$. Observe that

(i) $u_r \ge 0$ is bounded and superharmonic in D;
(ii) $u_r = 0$ q.e. on ∂D;
(iii) u_r is a Green potential $G\mu_r$ of a measure μ_r concentrated on $D \cap S(\xi, r)$.

Take another $v \in \mathcal{H}_\xi$ and let $v_r = G\nu_r$ with a measure ν_r concentrated on $D \cap S(\xi, r)$. Observe that

$$\int_{D\cap S(\xi,r)} G(x_0, y)\,\mathrm{d}\mu_r(y) = u_r(x_0) = 1 = v_r(x_0) = \int_{D\cap S(\xi,r)} G(x_0, y)\,\mathrm{d}\nu_r(y).$$

Let $x \in D \setminus \overline{B(\xi, C_3 r)}$. Apply the local boundary Harnack principle to $u = G(x, \cdot)$ and $v = G(x_0, \cdot)$. Then

$$\frac{G(x,y)}{G(x_0,y)} \bigg/ \frac{G(x,y')}{G(x_0,y')} \le C_2 \quad \text{for } y, y' \in D \cap B(\xi, r).$$

so that

$$G(x,y) \le C_2 \frac{G(x,y')}{G(x_0,y')} G(x_0, y).$$

Integrate the both sides with respect to $\mathrm{d}\mu_r(y)$. Then we obtain

$$u_r(x) = \int G(x,y)\,\mathrm{d}\mu_r(y) \le C_2 \frac{G(x,y')}{G(x_0,y')} u_r(x_0) = C_2 \frac{G(x,y')}{G(x_0,y')}.$$

Hence $u_r(x) G(x_0, y') \le C_2 G(x, y')$. Integrate with respect to $\mathrm{d}\nu_r(y')$. Then

$$u_r(x) = u_r(x) \int G(x_0, y')\,\mathrm{d}\nu_r(y') \le C_2 \int G(x, y')\,\mathrm{d}\nu_r(y') = C_2 v_r(x),$$

so that $u = u_r \le C_2 v_r = Cv$ on $D \setminus \overline{B(\xi, C_3 r)}$. Letting $r \to 0$ we obtain $u \le C_2 v$ on D. □

Remark 1.8. Kemper [28] formulated the boundary Harnack principle for a Lipschitz domain, and then identified the Martin boundary with the aid of the boundary Harnack principle as above. Unfortunately, his proof of the boundary Harnack principle had a gap. This gap was filled by Dahlberg, Ancona, and Wu independently. See the next section.

2. From smooth domains to NTA domains

2.1. Smooth domain. In PDE theory, a smooth domain usually means $C^{2,\alpha}$-domain ([21]). In potential theory, more precise smoothness is considered.

Definition 2.1. We say that a domain D satisfies the *interior ball condition* if there exists $r_0 > 0$ such that every boundary point $\xi \in \partial D$ has an open ball of radius r_0 in D whose boundary intersects ∂D exactly at ξ. We say that D satisfies the *exterior ball condition* if there exists $r_0 > 0$ such that every boundary point $\xi \in \partial D$ has an open ball of radius r_0 in the complement of D whose boundary intersects ∂D exactly at ξ.

We say that D satisfies the *ball condition* if it satisfies both the interior and exterior ball conditions.

Remark 2.2. It turns out that D satisfies the ball condition if and only if D is a $C^{1,1}$-domain ([6]).

Remark 2.3. Suppose D satisfies the ball condition. If u is a positive harmonic function in D such that $u = 0$ on a portion E of ∂D, then $u(x) \approx \delta_D(x)$ near E. Using this estimate, we can easily prove the local boundary Harnack principle. Similar estimates hold for a $C^{1,\alpha}$-domain (Widman [35]), and the local boundary Harnack principle holds.

2.2. Lipschitz domain. We say that a bounded domain D is a Lipschitz domain if ∂D is given locally as the graph of a Lipschitz continuous function. Let us illustrate the importance of Lipschitz domains by observing that Lipschitz domains are relevant for the boundary behavior of a harmonic function even in the upper half space $\mathbb{R}^n_+$ in $\mathbb{R}^n$. Let $E \subset \partial\mathbb{R}^n_+$. We say that a function h in $\mathbb{R}^n_+$ is nontangentially bounded at E if each $\xi \in E$ has a nontangential truncated cone $\Gamma(\xi)$ of vertex at ξ on which h is bounded.

Theorem 2.4 (Local Fatou theorem (Carleson [16])). *Let h be harmonic in $\mathbb{R}^n_+$ and suppose h is nontangentially bounded on $E \subset \partial\mathbb{R}^n_+$. Then h has nontangential boundary values for a.e. $\xi \in E$.*

In the proof of the theorem we may assume that nontangential cones $\Gamma(\xi)$ have the same height and aperture uniformly for $\xi \in E$ by a point of density argument ([32, p. 202]). (See [25, Section 6] for precise arguments in the context of NTA domains.) Then the union of nontangential cones $\cup_{\xi \in E}\Gamma(\xi)$ forms a Lipschitz domain. Hence Fatou's theorem reduces to the existence of nontangential boundary values of positive harmonic functions in a Lipschitz domain.

Note that $u(x) \not\approx \delta_D(x)$ for a Lipschitz domain (cf. Remark 2.3). The following simple example illustrates the situation (cf. Figure 1). Identify $\mathbb{R}^2$ and $\mathbb{C}$. Then the positive harmonic function $h(z) = xy$ in the first quadrant decays at the speed

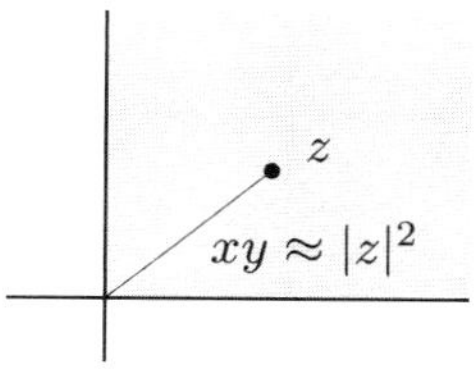

FIGURE 1

of $|z|^2$ on the diagonal line $\{x = y\}$ as $|z| \to 0$, whereas it decays at the rate of $\delta_D(z)$ as z approaches positive real and imaginary axes apart from the origin.

Nevertheless, Carleson [16] proved the local Fatou theorem by the *Carleson estimate*. Let D be a domain and let $\xi \in \partial D$. For $r > 0$ small, we say $\xi_r \in S(\xi, r) \cap D$ is a *nontangential point* if $\delta_D(\xi_r) \approx r$. It is easy to see that every boundary point of a Lipschitz domain has a nontangential point for every $r > 0$ small. If a domain has an outward cusp, then the tip of the cusp has no nontangential point.

Definition 2.5 (Carleson estimate). Suppose every boundary point ξ of D has a nontangential point ξ_r for every $r > 0$ small. We say that D satisfies the *Carleson estimate* if $\xi \in \partial D$ and $r > 0$ is small, and if $u > 0$ is bounded and harmonic in D and $u = 0$ on $\partial D \cap B(\xi, Cr)$, then

$$u(x) \leq Cu(\xi_r) \quad \text{for } x \in D \cap B(\xi, r),$$

where $C > 1$ is independent of u, ξ and r.

Theorem 2.6 ([16]). *Every Lipschitz domain satisfies the Carleson estimate.*

The above Carleson estimate will be extended to a general domain as the local Carleson estimate. It turns out that the local boundary Harnack principle and the local Carleson estimate are equivalent. See Definition 5.1 and Theorem 5.2 below.

Remark 2.7. Carleson's argument depends on the interior and exterior cone condition: each $\xi \in \partial D$ has interior and exterior cones of vertex at ξ with fixed height and aperture. Hunt – Wheeden [23, 24] applied Carleson's method to a Lipschitz domain and proved $\partial D = \Delta = \Delta_1$.

Remark 2.8. Ancona [8] and Wu [36] gave the local boundary Harnack principle for a Lipschitz domain independently, while Dahlberg [17] gave the global boundary Harnack principle.

2.3. NTA domain. By a *Harnack chain* of length N (Figure 2) connecting x and y we mean a sequence $\{B(x_j, \frac{1}{2}\delta_D(x_j))\}_{j=1}^{N}$ of consecutive balls in D such that

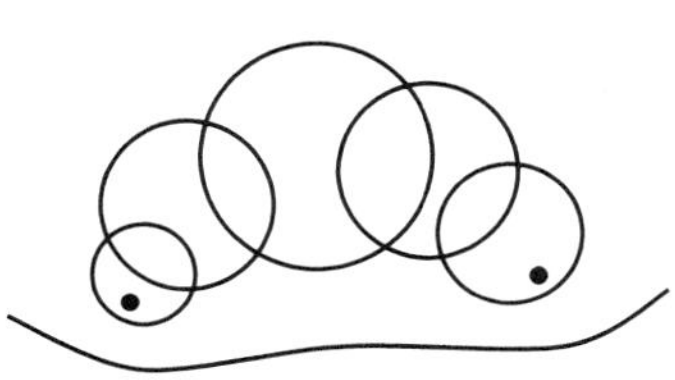

FIGURE 2

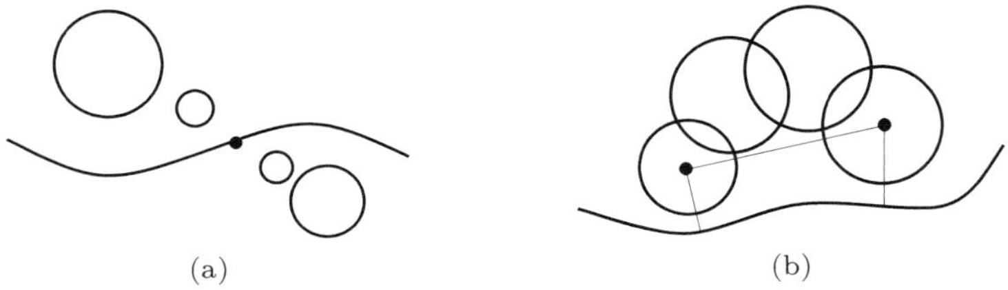

FIGURE 3

(i) $B\left(x_j, \frac{1}{2}\delta_D(x_j)\right) \cap B\left(x_{j+1}, \frac{1}{2}\delta_D(x_{j+1})\right) \neq \varnothing$,
(ii) $x \in B\left(x_1, \frac{1}{2}\delta_D(x_1)\right)$, $y \in B\left(x_N, \frac{1}{2}\delta_D(x_N)\right)$.

The repeated application of the Harnack inequality yields a constant $C > 1$ depending only on n such that if $h > 0$ is harmonic in D and if x and y is connected by a Harnack chain of length N, then $h(x)/h(y) \leq C^N$.

Jerison – Kenig [25] generalized a Lipschitz domain to an *NTA domain* (Non-Tangentially Accessible domain). Namely, a bounded domain D is an NTA domain if there exist constants $C > 1$ and $r_0 > 0$ with the following three conditions:

(i) *Corkscrew condition*: If $\xi \in \partial D$ and $0 < r < r_0$, then $D \cap B(\xi, r)$ contains a ball of radius r/C (Figure 3(a)).
(ii) *Exterior corkscrew condition*: If $\xi \in \partial D$ and $0 < r < r_0$, then $B(\xi, r) \setminus D$ contains a ball of radius r/C.
(iii) *Harnack chain condition*: If $x, y \in D$ and $|x - y| \approx \delta_D(x) \approx \delta_D(y)$, then x and y is connected by a Harnack chain of bounded length (Figure 3(b)).

Remark 2.9. NTA domains are much nastier than Lipschitz domains; the Hausdorff dimension of the boundary may be bigger than $n - 1$. A typical example is a snow flake domain.

In spite of the nonsmooth nature, Carleson's method is applicable to NTA domains in principle. Jerison – Kenig showed the local boundary Harnack principle, $\Delta = \Delta_1 = \partial D$, the local Fatou theorem, and the doubling property of harmonic measure. However, the harmonic measure need not be absolutely continuous with respect to the surface measure. Harmonic analysis on NTA domains is developed, based on harmonic measure.

Jerison – Kenig [25] is very influential; numbers of studies have been made on NTA domains. However, this is not the whole story. There are domains more irregular than NTA domains for which we can develop fruitful potential analysis such as the Martin boundary, the boundary Harnack principle and the Carleson estimate as we shall see in the succeeding sections.

3. Various nonsmooth domains

3.1. Hölder domain. By an *α-Hölder domain* we mean a bounded domain whose boundary is given locally as the graph of an α-Hölder function ($0 < \alpha \leq 1$). If α is not important, we call it a Hölder domain.

If $\alpha = 1$, then a 1-Hölder domain is a Lipschitz domain. If $0 < \alpha < 1$, then an α-Hölder domain may be irregular for $n \geq 3$. There is no uniform barrier and the method initiated by Carleson does not work. Nevertheless, the global boundary Harnack principle holds for a Hölder domain.

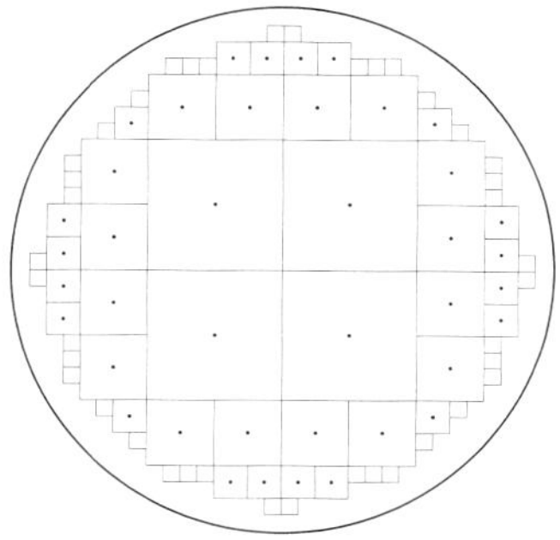

FIGURE 4. A uniform domain: an open ball minus the centers of Whitney cubes.

Theorem 3.1 (Bañuelos – Bass – Burdzy [13,14]**).** *A Hölder domain enjoys the global boundary Harnack principle.*

Bañuelos – Bass – Burdzy's argument is purely probabilistic. It requires no exterior conditions such as a uniform barrier. It turns out that interior conditions are crucial.

3.2. Uniform domain. Uniform domains are introduced in different contexts such as BMO-extension, Sobolev-extension, and quasiconformal mappings ([19,20, 26,27,33]).

Definition 3.2 (Uniform domain). We say that D is a *uniform domain* if every pair of points $x, y \in D$ can be connected by a curve $\gamma \subset D$ such that

(i) $\ell(\gamma) \le C|x-y|$ (*Bounded Turning*);

(ii) $\min\{\ell(\gamma(x,z)), \ell(\gamma(z,y))\} \le C\delta_D(z)$ for every $z \in \gamma$ (*Cigar*).

Here, $\ell(\gamma)$ stands for the length of the curve γ, and $\gamma(x,z)$ for the subcurve of γ connecting x and z.

An example of uniform domains is given by an open ball minus the centers of Whitney cubes. See Figure 4. Because of the lack of exterior conditions, uniform domains need not be regular. Nevertheless, uniform domains are nice domains:

(i) A bounded domain is a uniform domain if and only if the corkscrew condition and the Harnack chain condition for NTA domains hold.

(ii) A uniform domain enjoys the local boundary Harnack principle and $\Delta = \Delta_1 = \partial D$. The ratio u/v of two positive harmonic functions u, v vanishing on $E \subset \partial D$ is Hölder continuous up to E, although u and v need not continuously vanish on E. See [1].

(iii) Under the capacity density condition stated below, the validity of the local boundary Harnack principle characterizes uniform domains. See [2].

3.3. John domain. In the definition of uniform domains, by fixing $y = x_0$ and moving alone x, we obtain the *carrot condition*, which characterizes John domains.

Definition 3.3 (John domain). We say that D is a *John domain* with *John constant* c_J if every point $x \in D$ can be connected to x_0 by a curve $\gamma \subset D$ such that $\delta_D(z) \ge c_J \ell(\gamma(x,z))$ for every $z \in \gamma$. If x_0 is replaced by a fixed compact set, then D is called a *generalized John domain*.

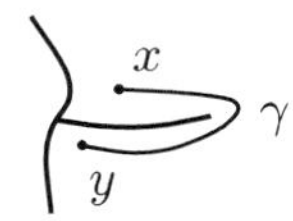

FIGURE 5

John domains are nice domains:

(i) The John constant c_J is not greater than 1. The closer c_J to 1, the smoother D.
(ii) If a bounded domain satisfies the cigar condition of a uniform domain (without the bounded turning condition), then it is a John domain.
(iii) A John domain enjoys the global boundary Harnack principle ([14]).
(iv) A John domain enjoys the *weak boundary Harnack principle.* The number of minimal Martin boundary points over a topological boundary point is estimated ([5]).
(v) Under the capacity density condition (see Definition 3.14 below), John domains are characterized by estimates of harmonic measure ([2])

3.4. Inner uniform domain. Define the *inner length distance* $\lambda_D(x,y)$ and *internal metric* $\rho_D(x,y)$ in D by

$$\lambda_D(x,y) = \inf\{\ell(\gamma)\},$$
$$\rho_D(x,y) = \inf\{\operatorname{diam}(\gamma)\},$$

where the infimum is taken over all paths γ in D with end points x and y (cf. Figure 5). We observe that $|x-y| \le \rho_D(x,y) \le \lambda_D(x,y)$ by definition, and that $\rho_D(x,y) \approx \lambda_D(x,y)$ for John domains (Väisälä (1998) [34]).

Definition 3.4 (Inner uniform domain). We say that D is an inner uniform domain if every pair of points $x, y \in D$ can be connected by $\gamma \subset D$ with $\ell(\gamma) \le C\rho_D(x,y)$ such that γ satisfies the cigar condition. See Balogh – Volberg [11, 12], Bonk – Heinonen – Koskela [15].

Inner uniform domains are nice domains:

(i) Inner uniform domains are John domains.
(ii) An inner uniform domain enjoys the local boundary Harnack principle with respect to ρ_D.
(iii) The Martin boundary is identified: $\Delta = \Delta_1 = \partial^* D$, the ideal boundary with respect to ρ_D.

3.5. Semi-uniform domain. Let us recall the definition of uniform domains. Every pair of points of D is connected by a cigar curve with bounded turning condition. This connecting property extends to points of $D \cup \partial D$. If we restrict one point on the boundary ∂D, then we obtain a semi-uniform domain. A *Denjoy domain* is a typical example.

Definition 3.5 (Semi-uniform domain). We say that D is a *semi-uniform domain* if every $x \in D$ and $y \in \partial D$ can be connected by a cigar curve γ such that $\gamma \setminus \{y\} \subset D$ and $\ell(\gamma) \le C|x-y|$.

Definition 3.6 (Inner semi-uniform domain). We say that D is an *inner semi-uniform domain* if every $x \in D$ and $y \in \partial D$ can be connected by a cigar curve γ such that $\gamma \setminus \{y\} \subset D$ and $\ell(\gamma) \le C\rho_D(x,y)$.

Semi-uniform domains and inner semi-uniform domains are nice domains. Let D be a John domain with the capacity density condition:

(i) D is a semi-uniform domain if and only if harmonic measure is strongly doubling.
(ii) D is an inner semi-uniform domain if and only if harmonic measure is strongly doubling with respect to inner metric.

See [4].

3.6. Quasihyperbolic boundary condition (QHB).

Definition 3.7. Let D be a general domain such that $\partial D \neq \varnothing$. The *quasihyperbolic metric* $k_D(x,y)$ is defined by

$$k_D(x,y) = \inf_\gamma \int_\gamma \frac{ds}{\delta_D\big(\gamma(s)\big)},$$

where the infimum is taken over all curves γ connecting x and y in D. Here, the parametrization of the curve γ with respect to arc length is denoted by the same symbol $\gamma(s)$, $0 \le s \le \ell(\gamma)$.

Observe that $k_D(x,y)$ is comparable to the shortest length of Harnack chain connecting x and y in D.

Theorem 3.8. *Let D be a general domain such that $\partial D \neq \varnothing$. Then*

$$k_D(x,y) \ge \log\Big(\frac{|x-y|}{\min\{\delta_D(x),\delta_D(y)\}} + 1\Big) \quad \textit{for } x,y \in D.$$

PROOF. Let γ be a curve connecting x and y in D of length L. Then $|x-y| \le L$. Moreover, $\delta_D\big(\gamma(s)\big) \le \delta_D(x) + s$ for $0 \le s \le L$ with arc length parameter s. Hence

$$\int_\gamma \frac{ds}{\delta_D\big(\gamma(s)\big)} \ge \int_0^L \frac{ds}{\delta_D(x)+s} = \big[\log(\delta_D(x)+s)\big]_0^L \ge \log\Big(1 + \frac{|x-y|}{\delta_D(x)}\Big).$$

Replacing x and y proves the theorem. □

The reverse inequality (up to a multiplicative constant) characterizes uniform domains.

Theorem 3.9 ([20]). *Let D be a general domain such that $\partial D \neq \varnothing$. Then D is a uniform domain if and only if*

$$k_D(x,y) \le C\log\Big(\frac{|x-y|}{\min\{\delta_D(x),\delta_D(y)\}} + 1\Big) + C' \quad \textit{for } x,y \in D.$$

Remark 3.10. Appealing to Theorem 3.9, we have shown in [2] that uniform domains are characterized by the validity of the local boundary Harnack principle.

Definition 3.11. We say that D satisfies the *quasihyperbolic boundary* (*QHB*) *condition* if

$$k_D(x,x_0) \le C\log\frac{\delta_D(x_0)}{\delta_D(x)} + C' \quad \text{for } x \in D,$$

where $x_0 \in D$ is a fixed reference point. We simply call D is a *QHB domain* if it satisfies the quasihyperbolic boundary (QHB) condition.

Theorem 3.12. *John domains are QHB domains.*

PROOF. Let γ be a John curve connecting x and x_0 in D of length L. Then

$$\begin{aligned}\int_\gamma \frac{\mathrm{d}s}{\delta_D\big(\gamma(s)\big)} &= \int_0^{\delta_D(x)/4} + \int_{\delta_D(x)/4}^{L} \frac{\mathrm{d}s}{\delta_D\big(\gamma(s)\big)} \\ &\le \frac{\delta_D(x)/4}{3\delta_D(x)/4} + \int_{\delta_D(x)/4}^{L} \frac{\mathrm{d}s}{c_J s} \le \frac{1}{c_J}\log\frac{\delta_D(x_0)}{\delta_D(x)} + C. \qquad \square\end{aligned}$$

Remark 3.13. There exists a QHB domain which is not a John domain. Smith – Stegenga [31] called a QHB domain a *Hölder domain*. This is different from a Hölder domain, a straightforward generalization of a Lipschitz domain, as illustrated in the beginning of this section. In order to avoid confusion, we propose the terminology *QHB domain.*

QHB domains are nice domains:

(i) The quasihyperbolic metric $k_D(x, y)$ is exponentially integrable, i.e., there exists an $\varepsilon > 0$ such that
$$\int_D \exp\big(\varepsilon k_D(x, x_0)\big)\, \mathrm{d}x < \infty.$$
See [31].

(ii) A QHB domain enjoys the global boundary Harnack principle ([3]).

3.7. Summary of interior conditions. So far, we have introduced various nonsmooth domains. From the view point of interior conditions, they are summarized as follows:

$$C^{2,\alpha} \subsetneqq C^{1,1} = \text{Ball condition} \subsetneqq \text{Lipschitz} \subsetneqq \text{NTA} \subsetneqq \text{Uniform} \subsetneqq \text{John} \subsetneqq \text{QHB}.$$

We have finer classifications between Uniform and John:

$$\begin{array}{ccccc} & & \text{Inner uniform} & & \\ & \subsetneqq & & \subsetneqq & \\ \text{Uniform} & & & & \text{Inner semi-uniform} \subsetneqq \text{John}. \\ & \subsetneqq & & \subsetneqq & \\ & & \text{Semi-uniform} & & \end{array}$$

3.8. Capacity density condition. By $\mathrm{Cap}_U(E)$ we denote the Green capacity of E relative to an open set U with Green function. More precisely, let $G_U(x, y)$ be the Green function for U. If K is a compact subset of U, then $\widehat{R}_1^K$ is a Green potential $G_U\mu_K$ of a Radon measure μ_K on K and let $\mathrm{Cap}_U(K)$ be the total mass of μ_K. If V is an open subset of U, then $\mathrm{Cap}_U(V)$ is defined to be the supremum of $\mathrm{Cap}_U(K)$ for compact subsets K of V. If E is a general subset of U, then $\mathrm{Cap}_U(E)$ is defined to be the infimum of $\mathrm{Cap}_U(V)$ for open subsets V containing E. This definition is consistent.

Definition 3.14 (Capacity density condition). We say that a domain D satisfies the *capacity density condition* (*CDC*), if there exist positive constants λ and r_0 such that

$$\frac{\mathrm{Cap}_{B(\xi,2r)}\big(C(\xi, r) \setminus D\big)}{\mathrm{Cap}_{B(\xi,2r)}\big(C(\xi, r)\big)} \ge \lambda,$$

whenever $\xi \in \partial D$ and $0 < r < r_0$. See Figure 6.

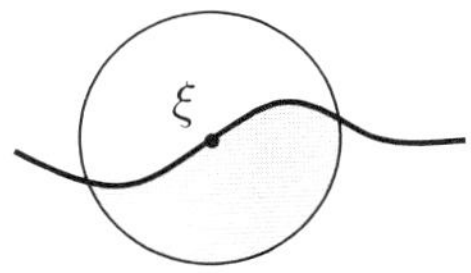

FIGURE 6

Remark 3.15. In the above definition, we use the closed ball $C(\xi, r)$, which can be replaced by the open ball $B(\xi, r)$. Replacing the Green capacity by the logarithmic capacity ($n = 2$) and the Newtonian capacity ($n \geq 3$), we can reformulate the capacity density condition by

$$\operatorname{Cap}(C(\xi, r) \setminus D) \geq \begin{cases} Cr & \text{if } n = 2, \\ Cr^{n-2} & \text{if } n \geq 3. \end{cases}$$

Definition 3.16 (Volume density condition). We say that a domain D satisfies the *volume density condition*, if there exist positive constants C and r_0 such that

$$\frac{|B(\xi, r) \setminus D|}{|B(\xi, r)|} \geq C$$

whenever $\xi \in \partial D$ and $0 < r < r_0$.

Remark 3.17. Ancona [9] introduced the notion of capacity density condition in connection with strong barriers and the Hardy inequality. By a *strong barrier* we mean a positive superharmonic function s on D such that $\Delta s(x) + (\varepsilon/\delta_D(x)^2)s(x) \leq 0$ for some $\varepsilon > 0$. A domain D has a strong barrier if and only if the Hardy inequality

$$\int_D \left| \frac{\psi(x)}{\delta_D(x)} \right|^2 dx \leq C \int_D |\nabla \psi(x)|^2 \, dx \quad \text{for } \psi \in W_0^{1,2}(D)$$

holds. The capacity density condition implies the existence of a strong barrier. See Ancona [9] for these accounts.

Lewis [29] gave the notion of *uniform fatness*, an extension in the context of Riesz capacity and the L^p-Hardy inequalities.

3.9. Summary of exterior conditions. Exterior conditions are summarized as follows:

$$\begin{aligned} \text{Exterior Ball} &\subsetneqq \text{Exterior cone} \subsetneqq \text{Exterior corkscrew} \\ &\subsetneqq \text{Volume density} \subsetneqq \text{CDC} \subsetneqq \text{Strong barrier.} \end{aligned}$$

4. The box argument — heart of the proof

Let us illustrate the heart of the proof of the boundary Harnack principle. While the argument founded by Carleson makes use of the exterior cone condition and uniform barriers, these are not available for uniform domains and John domains satisfying only interior conditions. Bass – Burdzy [14] found the box argument which requires no exterior conditions. This is a significant difference between NTA domains and John domains. The following *capacitary width* explains why no exterior condition is needed.

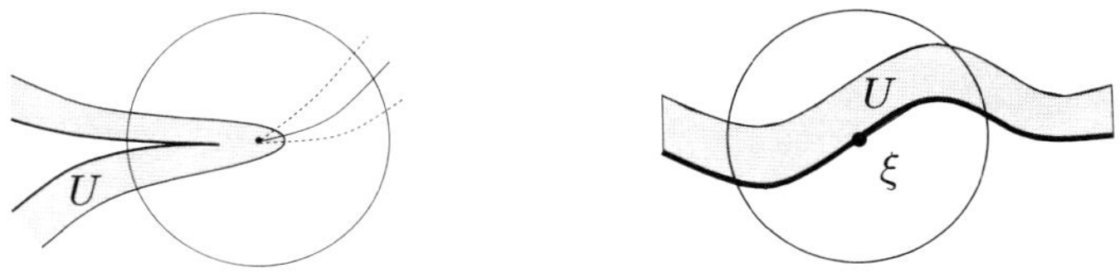

FIGURE 7. John domains and CDC domains share the common property.

Definition 4.1 (Capacitary width). Let $0 < \eta < 1$. Define the *capacitary width* for an open set U by

$$w_\eta(U) = \inf\left\{R > 0 : \frac{\mathrm{Cap}_{B(x,2R)}(C(x,R) \setminus U)}{\mathrm{Cap}_{B(x,2R)}(C(x,R))} \geq \eta \ (x \in U)\right\}.$$

Here $\eta > 0$ has no significance; different η gives a different but comparable capacitary width. A domain with the capacity density condition and a John domain share the following property.

Lemma 4.2. *Let D be a John domain or let D satisfy the capacity density condition. Then*

$$w_\eta(\{x \in D : \delta_D(x) < r\}) \leq Cr. \tag{4.1}$$

See Figure 7.

PROOF. Let $U = \{x \in D : \delta_D(x) < r\}$. Suppose D is a John domain. Then $x \in U$ can be connected to x_0 by a John curve γ. If $z \in \gamma$ and $|x - z| \geq 3r/c_J$, then $\delta_D(z) \geq c_J \ell(\gamma(x,z)) \geq c_J|x - z| \geq 3r$. This means $C(z,r) \cap U = \varnothing$. Hence (4.1) holds with $C = 3/c_J + 1$.

Suppose D satisfies the capacity density condition. Then (4.1) holds with $C = 2$ by definition. □

Capacity and harmonic measure have a close relationship.

Lemma 4.3. *Let $r > 0$ and suppose E is a closed subset of $B(0,r)\setminus\{0\}$. Then*

$$\omega\big(0; S(0,2r), B(0,2r) \setminus E\big) \leq 1 - C\frac{\mathrm{Cap}_{B(0,2r)}(E)}{\mathrm{Cap}_{B(0,2r)}\big(C(0,r)\big)},$$

where C is a positive constant depending only on the dimension.

PROOF. By dilation we have

$$\mathrm{Cap}_{B(0,2r)}\big(C(0,r)\big) = r^{n-2}\,\mathrm{Cap}_{B(0,2)}\big(C(0,1)\big) = Cr^{n-2}. \tag{4.2}$$

Observe that the regularized reduced function $\widehat{R}_1^E$ relative to $B(0,2r)$ is a Green potential $G_{B(0,2r)}\mu_E$ of a measure μ_E concentrated on E such that $\mathrm{Cap}_{B(0,2r)}(E) = \|\mu_E\|$, and that

$$\omega\big(S(0,2r), B(0,2r) \setminus E\big) = 1 - \widehat{R}_1^E \quad \text{on } B(0,2r) \setminus E.$$

Hence

$$\begin{aligned}\omega\big(0; S(0,2r), B(0,2r) \setminus E\big) &= 1 - \widehat{R}_1^E(0) = 1 - \int_E |y|^{2-n}\, d\mu_E(y)\\ &\leq 1 - r^{2-n}\,\|\mu_E\| = 1 - r^{2-n}\,\mathrm{Cap}_{B(0,2r)}(E)\end{aligned}$$

for $n \geq 3$. We have a similar estimate for $n = 2$ with $\log(2r/|y|)$ and $\log 2$ in place of $|y|^{2-n}$ and r^{2-n}, respectively. These estimates, together with (4.2), yield the required inequality. □

Capacitary width gives an estimate of harmonic measure, which is crucial in the so called box argument. See, e.g.,[1, Lemma 1].

Lemma 4.4. *There exists a constant $C_4 > 1$ depending only on n and η such that for an open set $U \neq \varnothing$, $x \in U$ and $R > 0$*

$$\omega\big(x; U \cap S(x,R), U \cap B(x,R)\big) \leq \exp\left(2 - C_4 \frac{R}{w_\eta(U)}\right). \tag{4.3}$$

Proof. Write $\Omega = \omega\big(U \cap S(x,R), U \cap B(x,R)\big)$ for simplicity. Without loss of generality, we may assume that $R/w_\eta(U) > 2$. Let k be the least integer such that $2kw_\eta(U) < R \leq 2(k+1)w_\eta(U)$. Find $r > w_\eta(U)$ such that $2kr < R$ and

$$\frac{\mathrm{Cap}_{B(y,2r)}(E)}{\mathrm{Cap}_{B(y,2r)}\big(C(y,r)\big)} \geq \eta \quad \text{for } y \in U,$$

where $E = C(y,r) \setminus U$. Hence Lemma 4.3, translation and the maximum principle yield

$$\omega\big(y; U \cap S(y,2r), U \cap B(y,2r)\big) \leq \omega\big(y; S(y,2r), B(y,2r) \setminus E\big) \leq 1 - C\eta$$

for $y \in U$. By the maximum principle we have

$$\omega\big(y; U \cap S(y,2r), U \cap B(y,2r)\big) \leq 1 - C\eta \quad \text{for } y \in U, \tag{4.4}$$

where C depends only on the dimension.

We claim that

$$\sup_{U \cap B(x, R-2jr)} \Omega \leq (1 - C\eta)^j \tag{4.5}$$

for $j = 0, 1, \ldots, k$. Once we obtain (4.5), then we get immediately (4.3) as $k \approx R/w_\eta(U)$.

Now let us prove the claim by induction on j. Obviously, we have (4.5) for $j = 0$. Suppose (4.5) holds for $j - 1$ with $j \geq 1$. In view of the maximum principle, it is sufficient to show

$$\sup_{U \cap S(x, R-2jr)} \Omega \leq (1 - C\eta)^j. \tag{4.6}$$

Let $y \in U \cap S(x, R - 2jr)$. Since $B(y,2r) \subset B(x, R - 2(j-1)r)$, it follows from (4.5) for $j - 1$ that

$$\Omega \leq (1 - C\eta)^{j-1} \quad \text{on } U \cap B(y,2r).$$

Since $\Omega = 0$ q.e. on $B(y,2r) \cap \partial U$, we have from the maximum principle

$$\Omega \leq (1 - C\eta)^{j-1} \omega\big(U \cap S(y,2r), U \cap B(y,2r)\big) \quad \text{on } U \cap B(y,2r).$$

Observe from (4.4) that

$$\omega\big(y; U \cap S(y,2r), U \cap B(y,2r)\big) \leq 1 - C\eta.$$

Substitute this inequality. Then we have $\Omega(y) \leq (1 - C\eta)^j$. Since y is an arbitrary point in $U \cap S(x, R - 2jr)$, we obtain (4.6), and hence (4.5). See Figure 8 for $j = 2$. □

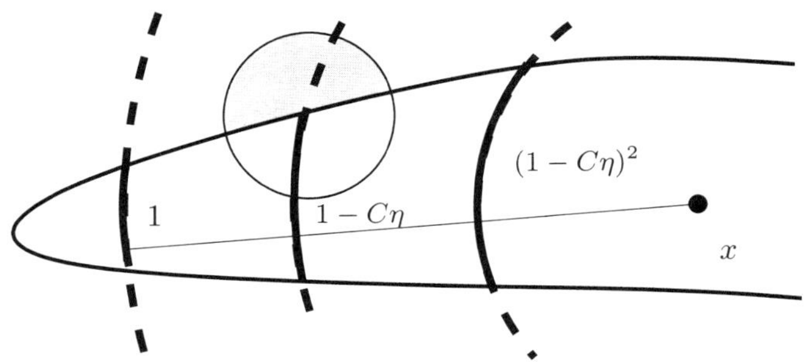

FIGURE 8. Estimate of Ω for $j = 2$.

Remark 4.5. See [1] for the derivation of the boundary Harnack principle from the above estimate.

5. Carleson estimate and Domar's argument

Besides the box argument there is another approach to the boundary Harnack principle, i.e., Domar's argument and the equivalence between the boundary Harnack principle and the Carleson estimate.

Definition 5.1 (Local Carleson estimate). We say that a domain D enjoys the *local Carleson estimate* if there is a constant $C_5 > 1$ with the following property: If $\xi \in \partial D$, $0 < r < r_0$ and

(i) u is bounded, positive and harmonic in $B(\xi, C_5 r) \cap D$,
(ii) u vanishes q.e. on $B(\xi, C_5 r) \cap \partial D$,

then $u(x) \leq C_\varepsilon u(y)$ for $x \in B(\xi, r) \cap D$, whenever $y \in B(\xi, r) \cap D$ and $\delta_D(y) \geq \varepsilon r$ with $0 < \varepsilon < 1$. Here C_ε depends only on D and ε.

The boundary Harnack principle and the Carleson estimate are closely related. Actually, the Carleson estimate in this note was sometimes referred to as the boundary Harnack principle by some authors. In [3] it is proved that they are equivalent.

Theorem 5.2. *Assume that a domain has positive constants ε and r_0 such that if $\xi \in \partial D$ and $0 < r < r_0$, then $D \cap B(\xi, r)$ contains a ball of radius εr. Then D enjoys the local boundary Harnack principle if and only if D enjoys the local Carleson estimate.*

The proof of the theorem is not so difficult and is carried out by using the relationship between the harmonic measure and the Green function (Brelot technique). See [3] for details.

While the local boundary Harnack principle involves two positive harmonic functions, the local Carleson estimate involves one positive harmonic function. The proof of the local Carleson estimate is somewhat easier. In this section we prove the local Carleson estimate for a uniform domain D by using Domar's argument, based on the submean value property of subharmonic functions. Domar's argument is widely applicable not only to subharmonic functions but also to p-subharmonic functions satisfying weak submean value property. Let us illustrate Domar's argument for subharmonic functions. From a positive harmonic function vanishing on a portion of the boundary, we can construct a subharmonic function beyond the portion. Then the nasty boundary disappears for an extended subharmonic function.

Let D be a domain and V be an open set intersecting ∂D. Let u be a positive harmonic function in D vanishing q.e. on $\partial D \cap V$ and bounded in $D \cap V$. Then the regularization of the extension $u = 0$ in $V \setminus D$ is subharmonic in $D \cup V$. Note that nasty boundary disappears for the extended subharmonic function.

For $u \geq 0$ we write

$$\log^+ u = \begin{cases} \log u & \text{if } u \geq 1, \\ 0 & \text{otherwise.} \end{cases}$$

Lemma A (Domar [5, 18]). *Let $u \geq 0$ be subharmonic in $\Omega \subset \mathbb{R}^n$. If there exists $\varepsilon > 0$ such that*

$$I = \int_\Omega (\log^+ u)^{n-1+\varepsilon}\, dx < \infty,$$

then $u(x) \leq C \exp(CI^{1/\varepsilon} \delta_\Omega(x)^{-n/\varepsilon})$, where $C = C(\varepsilon, n)$.

Domar's argument yields the Carleson estimate, which in turn to be equivalent to the boundary Harnack principle.

Theorem 5.3. *Every uniform domain enjoys the local Carleson estimate and hence the local boundary Harnack principle.*

PROOF. Observe

$$k_D(x, \xi_r) \leq C \log \frac{Cr}{\delta_D(x)} + C' \qquad \text{for } x \in D \cap B(\xi, Cr), \tag{5.1}$$

$$\int_{D \cap B(\xi, Cr)} \left[\frac{Cr}{\delta_D(x)}\right]^\tau dx \leq Cr^n \qquad \text{for some } \tau > 0. \tag{5.2}$$

In fact, we can show (5.1) in the same way as in the proof of Theorem 3.12. The integral condition (5.2) holds even for a John domain ([5, Lemma 4.1]). Let u be a bounded positive harmonic function in $D \cap B(\xi, Cr)$ such that $u = 0$ q.e. on $\partial D \cap B(\xi, Cr)$. Extend u to $B(\xi, Cr) \setminus D$ by 0 and make it subharmonic on $B(\xi, Cr)$ as explained. The Harnack inequality, together with (5.1), gives

$$\frac{u(x)}{u(\xi_r)} \leq C \exp\big(Ck_D(x, \xi_r)\big) \quad \text{for } x \in D \cap B(\xi, Cr). \tag{5.3}$$

Let $\varepsilon > 0$. By (5.1), (5.3) and (5.2), we have

$$\begin{aligned} I &:= \int_{B(\xi, Cr)} \left[\log^+ \frac{u(x)}{u(\xi_r)}\right]^{n-1+\varepsilon} dx \leq C \int_{D \cap B(\xi, Cr)} [k_D(x, \xi_r)]^{n-1+\varepsilon}\, dx \\ &\leq C \int_{D \cap B(\xi, Cr)} \left[\log \frac{Cr}{\delta_D(x)}\right]^{n-1+\varepsilon} dx \leq C \int_{D \cap B(\xi, Cr)} \left[\frac{Cr}{\delta_D(x)}\right]^\tau dx \leq Cr^n, \end{aligned}$$

where we have used an elementary inequality $(\log t)^{n-1+\varepsilon} \leq Ct^\tau$ for $t > 1$ in the third inequality. Hence Domar's argument gives

$$\begin{aligned} \frac{u(x)}{u(\xi_r)} &\leq C \exp(CI^{1/\varepsilon} \delta_{B(\xi, Cr)}(x)^{-n/\varepsilon}) \\ &\leq C \exp(Cr^{n/\varepsilon} r^{-n/\varepsilon}) \leq C \quad \text{for } x \in D \cap B(\xi, r). \end{aligned}$$

This shows the Carleson estimate

$$u(x) \le Cu(\xi_r) \quad \text{for } x \in D \cap B(\xi, r),$$

and hence the local boundary Harnack principle by the equivalence between the boundary Harnack principle and the Carleson estimate ([3, Theorem 2]). □

Remark 5.4. We can extend the Carleson estimate to p-harmonic functions in a *metric measure space* ([7]). However, the boundary Harnack principle for p-harmonic functions does not follow because of the lack of the relationship between p-harmonic measure and p-Green function.

Acknowledgments. The author would like to thank the referee for many valuable comments.

References

1. H. Aikawa, *Boundary Harnack principle and Martin boundary for a uniform domain*, J. Math. Soc. Japan **53** (2001), no. 1, 119–145.
2. ———, *Potential-theoretic characterizations of nonsmooth domains*, Bull. London Math. Soc. **36** (2004), no. 4, 469–482.
3. ———, *Equivalence between the boundary Harnack principle and the Carleson estimate*, Math. Scand. **103** (2008), no. 1, 61–76.
4. H. Aikawa and K. Hirata, *Doubling conditions for harmonic measure in John domains*, Ann. Inst. Fourier (Grenoble) **58** (2008), no. 2, 429–445.
5. H. Aikawa, K. Hirata, and T. Lundh, *Martin boundary points of a John domain and unions of convex sets*, J. Math. Soc. Japan **58** (2006), no. 1, 247–274.
6. H. Aikawa, T. Kilpeläinen, N. Shanmugalingam, and X. Zhong, *Boundary Harnack principle for p-harmonic functions in smooth Euclidean domains*, Potential Anal. **26** (2007), no. 3, 281–301.
7. H. Aikawa and N. Shanmugalingam, *Carleson-type estimates for p-harmonic functions and the conformal Martin boundary of John domains in metric measure spaces*, Michigan Math. J. **53** (2005), no. 1, 165–188.
8. A. Ancona, *Principe de Harnack à la frontière et théorème de Fatou pour un opérateur elliptique dans un domaine lipschitzien*, Ann. Inst. Fourier (Grenoble) **28** (1978), no. 4, 169–213, x.
9. ———, *On strong barriers and an inequality of Hardy for domains in* $\mathbb{R}^n$, J. London Math. Soc. (2) **34** (1986), no. 2, 274–290.
10. D. H. Armitage and S. J. Gardiner, *Classical potential theory*, Springer Monogr. Math., Springer, London, 2001.
11. Z. Balogh and A. Volberg, *Boundary Harnack principle for separated semihyperbolic repellers, harmonic measure applications*, Rev. Mat. Iberoamericana **12** (1996), no. 2, 299–336.
12. ———, *Geometric localization, uniformly John property and separated semihyperbolic dynamics*, Ark. Mat. **34** (1996), no. 1, 21–49.
13. R. Bañuelos, R. F. Bass, and K. Burdzy, *Hölder domains and the boundary Harnack principle*, Duke Math. J. **64** (1991), no. 1, 195–200.
14. R. F. Bass and K. Burdzy, *A boundary Harnack principle in twisted Hölder domains*, Ann. of Math. (2) **134** (1991), no. 2, 253–276.
15. M. Bonk, J. Heinonen, and P. Koskela, *Uniformizing Gromov hyperbolic spaces*, Astérisque **270** (2001).
16. L. Carleson, *On the existence of boundary values for harmonic functions in several variables*, Ark. Mat. **4** (1962), 393–399 (1962).
17. B. E. J. Dahlberg, *Estimates of harmonic measure*, Arch. Rational Mech. Anal. **65** (1977), no. 3, 275–288.
18. Y. Domar, *On the existence of a largest subharmonic minorant of a given function*, Ark. Mat. **3** (1957), 429–440.
19. F. W. Gehring and O. Martio, *Lipschitz classes and quasiconformal mappings*, Ann. Acad. Sci. Fenn. Ser. A I Math. **10** (1985), 203–219.
20. F. W. Gehring and B. G. Osgood, *Uniform domains and the quasihyperbolic metric*, J. Analyse Math. **36** (1979), 50–74 (1980).

21. D. Gilbarg and N. S. Trudinger, *Elliptic partial differential equations of second order*, Classics Math., Springer, Berlin, 2001. Reprint of the 1998 edition.
22. A. Harnack, *Existenzbeweise zur Theorie des Potentiales in der Ebene und im Raume*, Ber. Verh. K. Sächs. Ges. Wiss. Leipz., Math.-Phys. Kl. **38** (1886), 144–169; Math. Ann. **35** (1889), no. 1-2, 19–40.
23. R. A. Hunt and R. L. Wheeden, *On the boundary values of harmonic functions*, Trans. Amer. Math. Soc. **132** (1968), 307–322.
24. ———, *Positive harmonic functions on Lipschitz domains*, Trans. Amer. Math. Soc. **147** (1970), 507–527.
25. D. S. Jerison and C. E. Kenig, *Boundary behavior of harmonic functions in nontangentially accessible domains*, Adv. in Math. **46** (1982), no. 1, 80–147.
26. P. W. Jones, *Extension theorems for BMO*, Indiana Univ. Math. J. **29** (1980), no. 1, 41–66.
27. ———, *Quasiconformal mappings and extendability of functions in Sobolev spaces*, Acta Math. **147** (1981), no. 1-2, 71–88.
28. J. T. Kemper, *A boundary Harnack principle for Lipschitz domains and the principle of positive singularities*, Comm. Pure Appl. Math. **25** (1972), 247–255.
29. J. L. Lewis, *Uniformly fat sets*, Trans. Amer. Math. Soc. **308** (1988), no. 1, 177–196.
30. R. S. Martin, *Minimal positive harmonic functions*, Trans. Amer. Math. Soc. **49** (1941), 137–172.
31. W. Smith and D. A. Stegenga, *Exponential integrability of the quasi-hyperbolic metric on Hölder domains*, Ann. Acad. Sci. Fenn. Ser. A I Math. **16** (1991), no. 2, 345–360.
32. E. M. Stein, *Singular integrals and differentiability properties of functions*, Princeton Math. Ser., vol. 30, Princeton Univ. Press, Princeton, NJ, 1970.
33. J. Väisälä, *Uniform domains*, Tohoku Math. J. (2) **40** (1988), no. 1, 101–118.
34. ———, *Relatively and inner uniform domains*, Conform. Geom. Dyn. **2** (1998), 56–88 (electronic).
35. K.-O. Widman, *Inequalities for the Green function and boundary continuity of the gradient of solutions of elliptic differential equations*, Math. Scand. **21** (1967), 17–37 (1968).
36. J. M. G. Wu, *Comparisons of kernel functions, boundary Harnack principle and relative Fatou theorem on Lipschitz domains*, Ann. Inst. Fourier (Grenoble) **28** (1978), no. 4, 147–167, vi.

Department of Mathematics, Hokkaido University, Kita 10, Nishi 8, Kita-Ku, Sapporo, Hokkaido, 060-0810, Japan

E-mail address: `aik@math.sci.hokudai.ac.jp`

Centre de Recherches Mathématiques
CRM Proceedings and Lecture Notes
Volume **55**, 2012

Potential Theory on Trees and Multiplication Operators

David Singman

ABSTRACT. The article surveys a number of potential theory results in the discrete setting of trees and in an application to complex analysis. On trees for which the associated random walk is recurrent, we discuss Riesz decomposition, flux, a type of potential called H-potential, and present a new result dealing with the boundary behavior of H-potentials on a specific recurrent homogeneous tree. On general trees we discuss Brelot structures and their classification. On transient homogeneous trees we discuss clamped and simply-supported biharmonic Green functions.

We also describe an application of potential theory, namely a certain minimum principle for multiply superharmonic functions, that is used to prove a result concerning the norm of a class of multiplication operators from $H^{\infty}(D)$ to the Bloch space on D, where D is a bounded symmetric domain. The proof of the minimum principle involves the use of the Cartan – Brelot topology.

1. Introduction

In this article we survey some of our potential theory results in discrete and classical settings. Most of the paper considers potential theory in the discrete setting of trees. In one section we consider it in a Brelot space, and in the final section we consider an application of potential theory to a problem involving complex and functional analysis.

On trees, functions are viewed as real-valued, with domain the set of vertices. Once transition probabilities are prescribed on each of the directed edges, harmonic functions are defined by the condition that the value at each vertex equals the average of the values at the neighboring vertices. Much of classical potential theory can then be developed in this setting. The type of theorems considered depends on whether or not the associated random walk is transient or recurrent. From the probabilistic point of view, the walk is transient if and only if, for each vertex v, the probability that the random walk visits v infinitely often is 0. Otherwise it is recurrent. From the analytic point of view, transience is equivalent to the existence of positive potentials.

2010 *Mathematics Subject Classification.* Primary 31C20; Secondary 31A30, 31C05, 31D05.
Key words and phrases. tree, flux, standard, H-potential, Brelot space, biharmonic Green function, Cartan – Brelot topology, multiplication operator, multiply superharmonic function, symmetric domain.
This is the final form of the paper.

The transient theory was largely developed in [24] where Cartier considers basic properties of harmonic and superharmonic functions, integral representation of positive harmonic functions, and limit theorems of positive superharmonic functions at the boundary of the tree along random paths. Limit theorems along deterministic paths are considered in [26, 34, 35].

After a preliminary section on the basic facts concerning trees, we describe in §3 the results of [28], which deal with general recurrent trees. The potential theory in this setting is analogous to the classical potential theory in the complex plane. We were motivated by the work of Anandam in [7] and subsequent articles, where he developed the theory of Brelot spaces without positive potentials. In the recurrent setting, there are no positive nonconstant superharmonic functions, so the focus is on admissible superharmonic functions, that is, superharmonic functions having a harmonic minorant outside a compact set. Anandam also introduced several notions of "potential." A key idea is to develop a theory of flux for admissible superharmonic functions, and further, for any function superharmonic outside a compact set. In the Brelot space setting, Anandam first defined flux for functions harmonic outside of a compact set, then for globally defined superharmonic functions, and finally for functions superharmonic outside a compact set. It is true that if one linearly extends along the edges functions that are harmonic on a tree, then one gets a Brelot space, and so the results of Anandam translate directly to results on the tree. However, our approach is different from his. We have a condensed definition of flux, where we directly define it for functions superharmonic outside finitely many vertices. We also introduce a new notion of potential, called H-potential, which seems quite natural in that it is defined by means of a "greatest harmonic minorant zero" condition, and it can be represented in terms of a kernel, called the H-Green function, which itself is represented by an explicit formula. We use the H-potentials to give a Riesz decomposition theorem for the set of admissible superharmonic functions.

In §4, we develop the idea of producing Brelot spaces by starting with a discrete harmonic structure and extending in ways more general than linearly on the edges. After giving a brief background on Brelot spaces, we describe the work in [18], where we first characterize, up to Brelot space isomorphism, all the Brelot space structures on intervals, and then discuss the problem of putting these structures together to produce a Brelot space structure on the tree viewed as a one-dimensional simplicial complex.

In §5, we describe the work in [17] where we show that our theories of flux and H-potential can be formulated in a general Brelot space without positive potentials, and so we obtain a Riesz decomposition theorem for admissible superharmonic functions in this general setting. We show that the theories described here and in §3 agree in the special case that we extend linearly along the edges of the tree.

In §6, we look at a particular example of the trees in §3, namely the $(\frac{1}{4}, \frac{1}{4}, \frac{1}{2})$-tree. This is the tree in which all vertices have exactly three neighbors, and aside from the root of the tree in which the outward transition probabilities are $\frac{1}{3}$, all forward probabilities are $\frac{1}{4}$, and all backward probabilities are $\frac{1}{2}$. On this tree, we write down explicit formulas for the various quantities introduced in §3, and in particular for the H-potentials. We then depart from our survey in order to prove a new result concerning the boundary behavior of the H-potentials in the spirit of the boundary results of [34] for ordinary potentials on the same tree with a transient structure.

In §7, we describe the results in [29] concerning clamped and simply-supported biharmonic Green functions. These functions arise classically on domains (say on $\mathbb{D}$, the unit disk in the complex plane) in connection with various physical problems, such as the deflection of a thin plate. For applications in complex analysis, see [33]. The biharmonic Green functions can be used to construct functions on $\mathbb{D}$ having a prescribed bilaplacian, zero boundary values, and either vanishing normal derivative at the boundary (for the clamped one) or vanishing Laplacian at the boundary (for the simply-supported one). We consider the problem of defining and calculating these functions on a homogeneous tree.

In §8, we leave the discrete topics and consider instead a problem of complex and functional analysis. The results are given in [31]. Let D be a bounded homogeneous domain in $\mathbb{C}^N$, and consider the Banach spaces $X = H^\infty(D)$ of bounded holomorphic functions on D and $Y = \mathcal{B}$ of Bloch functions on D. Let $\psi\colon D \to \mathbb{C}$ have the property that for all $f \in X$, ψf is an element of Y. Then the mapping $M_\psi\colon X \to Y$, $f \mapsto \psi f$ is called a multiplication operator. In this section we describe a theorem which gives the norm of this operator. Our focus is on the following minimum principle concerning multiply superharmonic functions v defined on a product $\omega_1 \times \omega_2$ of relatively compact domains in Euclidean space. The minimum principle says that if v is lower bounded and has a lim inf greater or equal to 0 on the distinguished boundary $\partial\omega_1 \times \partial\omega_2$, then $v \geq 0$ on $\omega_1 \times \omega_2$. Though the result seems natural, the proof doesn't seem to be completely trivial. The proof we have of it makes use of a topology on the set of differences of positive superharmonic functions, called the Cartan-Brelot topology, which we describe in §8.

For our other collaborations which deal with potential theory on trees but which we do not survey here, see [16, 20, 26, 27, 35].

2. Preliminaries on trees

A **tree** is a connected graph with no loops. We fix one vertex **e** and refer to it as the **root** of the tree. We write $v \sim w$ if there is an edge joining vertices v and w, and say that v and w are **neighbors**. The number of neighbors of a vertex is referred to as its **degree**. In this paper we assume that T is a tree with infinitely many vertices, each vertex has finite degree, and there are no **terminal vertices**, that is, no vertices with only one neighbor. For each vertex v, let $[e, v] = [e, v_1, v_2, \ldots, v_n = v]$ be the unique geodesic path from e to v. Denote by $|v| = n$ the number of edges in $[e, v]$ and call it the **modulus** of v. More generally, for two vertices v and w, $|v - w|$ is the number of edges in the geodesic path from v to w. We refer to this as the **hyperbolic distance** from v to w, since by analogy, the hyperbolic distance between pairs of points on the unit disk in the complex plane is unbounded. By a **function** on T we mean a real-valued function on the vertices of T.

If v, w are vertices, we write $v \leq w$ if v is a vertex of $[e, w]$. We say that v is an **ancestor** of w and w is a **descendant** of v. The **children** of a vertex v are the vertices w such that $v \sim w$ and $|w| = |v| + 1$. If v is a vertex, v_k will always refer to the vertex of modulus k in the geodesic $[e, v]$. The **wedge product** $v \wedge w$ is the unique ancestor of both v and w having largest modulus. The **sector** $S(v)$ is the set consisting of v and the descendants of v. For a fixed vertex $w \neq e$ we define the half-sectors $\overline{S(w_k)} := S(w_k) \setminus S(w_{k+1})$ for $k = 0, \ldots, |w| - 1$. Note that the sets $\overline{S(w_k)}$, $k = 0, \ldots, |w| - 1$ together with $S(w)$ give a partition of T. Denote this partition by $\mathcal{P}(w)$:

$$\mathcal{P}(w) := \{\overline{S(w_0)}, \dots, \overline{S(w_{|w|-1})}, S(w)\}.$$

We say that a function f on T is of **finite type** if it is constant on each of the elements of $\mathcal{P}(w)$ for some vertex w.

If $a > 1$ and $u \neq e$, then a simple calculation shows that, for a homogeneous tree of degree 3,

$$\sum_{v \in S(u)} 2^{-a|v|} = \frac{2^{-a|u|}}{1 - 2^{1-a}} \tag{2.1}$$

The boundary ∂T of T is the set of infinite geodesic paths beginning at e. If $\omega = [e, \omega_1, \omega_2, \dots)$ is a boundary point of T and v a vertex, we write $v \leq \omega$ if $v = \omega_k$ for some k. The **wedge product** $v \wedge \omega$ is the vertex of largest modulus common to $[e, v]$ and ω. The **interval I(v)** is the set of boundary points ω such that $v \leq \omega$, or equivalently, $v \wedge \omega = v$. If $v, w \in T \cup \partial T$, we define the **Euclidean distance** from v to w to be 0 if $v = w$ and otherwise $2^{-|v \wedge w|}$. This is by way of analogy with the Euclidean distance between points on the unit disk in the complex plane, since it is bounded. This makes $T \cup \partial T$ a metric space which is a compactification of T, and the sets $\{I(\omega_j), j = 1, \dots\}$ form a base of neighborhoods for the boundary point $\omega = [e, \omega_1, \dots)$.

In order to define harmonic and superharmonic functions, we introduce **transition probabilities** on the directed edges: $p\colon T \times T \to [0, 1]$ such that (i) $p(u, v) > 0$ if and only if $u \sim v$, and, fixing $u \in T$, (ii) $\sum_{v \sim u} p(u, v) = 1$. The **Laplacian** Δf of a function f is given by $\Delta f(u) = \sum_{v \sim u} p(u, v) f(v) - f(u)$. We say that f is **harmonic** (respectively, **superharmonic**) at u if $\Delta f(u) = 0$ (respectively, ≤ 0). A function f is **subharmonic** at u if $-f$ is superharmonic at u.

The transition probabilities determine a random walk on the vertices of T that is either **transient** (for all u, v, the walk starting at u visits v finitely many times a.s.) or **recurrent** (for all u, v, the walk starting at u visits v infinitely many times a.s.). For simplicity, we shall simply refer to the tree as being transient or recurrent. A good reference for potential theory on transient trees is [24].

It can be shown that a tree is recurrent if and only if there exists a positive function that is superharmonic but not harmonic on T ([24, §2.3] or [28, pp. 3–4]).

If $[u, v]$ is a directed edge with $|u| = |v| - 1$, we call $[u, v]$ an **outward** edge and $p(u, v)$ an **outward** probability. We similarly define inward edges and inward probabilities.

3. Riesz decomposition on recurrent trees

In this section we consider a general recurrent tree, and describe the work in [28].

A **potential** is a positive superharmonic function whose greatest harmonic minorant is 0. Potentials exist precisely on trees whose associated random walk is transient. In this case there is a **Green function**, $G\colon T \times T \to (0, \infty)$, where $G(u, v)$ is the expected number of times the random walk which begins at vertex u visits vertex v. The potentials are characterized as the functions of the form $Gf(u) := \sum_{v \in T} G(u, v) f(v)$, for f a nonnegative function on T satisfying $\sum_{v \in T} G(e, v) f(v) < \infty$. The **Martin kernel** $K_\omega(w) := G(w, w \wedge \omega)/G(e, w \wedge \omega)$

generates all positive harmonic functions h on T: for each such function, there exists a Borel measure μ_h on ∂T such that $h(\cdot) = \int K_\omega(\cdot)\, \mathrm{d}\mu_h(\omega)$.

Example 3.1 (**$\frac{1}{3}$-tree**). If every vertex of T has degree 3, we say T is **homogeneous** of degree 3. For such a tree, if all the transition probabilities are $\frac{1}{3}$, we refer to T as the $\frac{1}{3}$-tree. The Green function is given by $G(u, v) = 2 \times 2^{-|u-v|}$ ([24, p. 264]) from which it follows that the Martin kernel is $K_\omega(w) = 2^{2|w\wedge\omega|-|w|}$. Note the similarity with the formula for the Poisson kernel on the unit disk in the complex plane: $P_{\mathrm{e}^{\mathrm{i}\theta}}(z) = (1-|z|^2)/(2\pi|\mathrm{e}^{\mathrm{i}\theta} - z|^2)$. The latter behaves roughly like the distance of z to the boundary of the disk divided by the square of the distance from z to the given boundary point and the former is exactly the hyperbolic distance of w to the boundary divided by the square of the hyperbolic distance from w to the given boundary point ω.

Example 3.2 (**$(\frac{1}{4}, \frac{1}{4}, \frac{1}{2})$-tree**). Again we take T to be homogeneous of degree 3, but instead we define the transition probabilities by

$$p(u, v) = \begin{cases} \frac{1}{3}, & u = e,\ v \sim e, \\ \frac{1}{2}, & [u, v] \text{ a backward edge,} \\ \frac{1}{4}, & [u, v] \text{ a forward edge with } u \neq e. \end{cases}$$

This is an example of a recurrent tree.

It follows easily from the definition that on a transient tree, every positive superharmonic function s on T can be written uniquely as $h_T + Gf$, where h_T is harmonic on T; h_T is just the greatest harmonic minorant of s and it is immediate that $s - h_T$ has greatest harmonic minorant 0, so is a potential.

The above result is known as the Riesz decomposition theorem. More generally we refer to **a Riesz decomposition theorem** as one which gives a unique decomposition of each member of some class of superharmonic functions as the sum of a global harmonic function and a member of some special class of "potentials." "Potentials" should be defined by a sentence that includes all of the words "greatest," "harmonic," "minorant" and "zero," and ideally they can be characterized in terms of a "Green function."

Our aim in this section is to describe a Riesz Decomposition Theorem on recurrent trees. In the recurrent setting there are no nonconstant positive superharmonic functions, and so all superharmonic functions with a global harmonic minorant are themselves harmonic. However, there are lots of nonconstant globally superharmonic functions that have a minorant on T which is harmonic except for at most finitely many vertices. Such superharmonic functions are referred to as **admissible**.

We state the main result of this section in Theorem 3.8. It is proved in [28]. Before we state the result, we will define and discuss several terms. However, looking at parts (ii) and (iii) of Theorem 3.8 , it is clear that it includes a Riesz Decomposition Theorem in the sense we have given above.

We begin with the term **standard**. Its definition is motivated by the following theorem.

Theorem 3.3 ([14, Theorem 9.7]). *Let Ω be an unbounded subset of $\mathbb{R}^2$ and K a compact subset of Ω. If u is harmonic on $\Omega \backslash K$, then u has a unique decomposition of the form $u = u_H + w$, where u_H is harmonic on Ω and w is a harmonic function on $\mathbb{R}^2 \setminus K$ satisfying* $\lim_{x\to\infty} w(x) - \alpha \log|x| = 0$ *for some constant α.*

Thus u can be written in the form $u(x) = u_H(x) + b + \alpha \log|x|$, where b is bounded outside a compact set. The number α is called the **flux** of u, and we call the function $\chi_{|x|\geq 1}(x) \ln|x|$ a **standard** for Ω. This motivates the following definition on T.

Definition 3.4. Let T be a recurrent tree. A function H on T is called a **standard** if (i) $H \geq 0$, (ii) H is harmonic outside the root, and (iii) given any function h on T which is harmonic off an arbitrary finite set of vertices, there exists a function h_T harmonic on T, a unique real number α, and a bounded function b such that

$$h = \alpha H + h_T + b.$$

One of the advantages of working in the tree setting is that we can make explicit calculations. In [28] we calculated a unique function $H \geq 0$ satisfying the following properties:

(1) $H(e) = 0$,
(2) $\Delta H(u) = \delta_e(u)$,
(3) H is constant on children.

The function H is given explicitly by

$$H(v) = \begin{cases} 0, & \text{if } v = e, \\ \sum_{k=0}^{n-1} \epsilon_0(v)\epsilon_1(v)\cdots\epsilon_k(v), & \text{if } |v| = n \geq 1, \end{cases}$$

where

$$\epsilon_0(v) = 1, \qquad \epsilon_k(v) = \frac{p(v_k.v_{k-1})}{1 - p(v_k, v_{k-1})} \quad \text{for } 1 \leq k \leq n-1.$$

To show the above H is a standard, we introduce and explicitly calculate a family of functions $\{H_v \colon v \in T\}$ uniquely determined by the following properties:

$$\text{(i)}\quad H_v(v) = 0, \qquad \text{(ii)}\quad \Delta H_v(u) = \delta_v(u),$$
$$\text{(iii)}\quad H_v \geq 0, \qquad \text{(iv)}\quad H_v \text{ constant on children.}$$

The family $\{-H_v, v \in T\}$ has some of the properties that a Green function would have if there were a Green function on T. Since H will turn out to be a standard and H_v is harmonic outside a finite set (namely $\{v\}$), we anticipate being able to write H_v as

$$H_v(u) = \alpha_v H(u) + h_T(u) - b^v(u),$$

where α_v is a unique constant, h_T is globally harmonic, and b^v is bounded. Direct calculation gives that $h_T \equiv 0$ and α_v is the ratio of the product of the forward and backward probabilities:

$$\alpha_v = \frac{p[e,v]}{p[v,e]} := \frac{p(e,v_1)p(v_1,v_2)\cdots p(v_{n-1},v)}{p(v_1,e)p(v_2,v_1)\cdots p(v,v_{n-1})}.$$

The function b^v turns out to be constant on each of the elements of the partition $\mathcal{P}(v)$ (and so is of finite type), the constant values given by $b^v(v) = \alpha_v H(v)$ and

$$b^v(v_k) = b^v(v) - \frac{1}{p(v,v_{n-1})}\left[1 + \sum_{m=k+1}^{n-1} \prod_{j=m}^{n-1} \frac{p(v_j,v_{j+1})}{p(v_j,v_{j-1})}\right].$$

The above calculations allow us to prove that H is indeed a standard. Observe that H is harmonic outside the root, it is subharmonic at the root, and it is necessarily unbounded. Indeed if H were bounded above by a constant M, then

$M - H$ would be nonnegative superharmonic, and so constant, which is clearly not the case.

We now discuss the concept of **flux** in the tree setting. The concept was introduced in [7] in a Brelot space having no potentials. There the flux was first defined for functions harmonic outside a compact set, then for admissible superharmonic functions, then finally for functions superharmonic outside a compact set. We give instead the following single definition of flux.

Definition 3.5. Let s be superharmonic except possibly for finitely many vertices. Then

$$\text{flux}(s) := \sup \{c \in \mathbb{R} : s - cH \text{ is bounded below by a function harmonic on } T\}.$$

Take for example the case that s is nonconstant and globally superharmonic, but not harmonic. Then the numbers c on the list defining flux(s) are all negative (since positive nonharmonic superharmonic functions cannot have a global harmonic minorant). Since H is unbounded, the idea is that we can hope to add a large enough positive multiple of H to s so that it has a global harmonic minorant and that the negative of the smallest such positive multiple should be the flux of s. The superharmonic functions on T thus have flux$(s) \leq 0$, and the admissible superharmonic functions are precisely the ones having finite flux.

We summarize in the next result some of the elementary properties of flux.

Theorem 3.6 ([28, Theorem 2.5]). (a) *If h is harmonic outside a finite set, then the flux of h is the unique constant α in the definition of standard such that $h = \alpha H + h_T + b$ with h_T harmonic on T and b bounded.*

(b) *If h is bounded and harmonic outside a finite set or harmonic everywhere, then the* flux$(s) = 0$.

(c) *If s is superharmonic on T, then* flux$(s) \leq 0$, flux$(s) = 0$ *if and only if s is harmonic, and* flux$(s) > -\infty$ *if and only if s is admissible.*

(d) *If s is superharmonic outside a finite set K of vertices, then the flux of s is equal to the flux of the greatest harmonic minorant of s outside K.*

(e) *If s is superharmonic outside a finite set and the set A_s in Definition* 3.5 *whose supremum defines the flux of s satisfies $A_s \neq \varnothing$, then $A_s = (-\infty, \text{flux}(s)]$.*

By part (a) of the theorem, knowing the flux of a function harmonic outside a finite set of vertices allows us to associate to it, in a natural way, a global harmonic function.

We now turn to the definitions of H-potentials and the H-Green function.

Definition 3.7. For f superharmonic outside the root, let $D(f)$ be the **greatest harmonic minorant** of f outside the root. Let in particular $f = s - \text{flux}(s)H$, where s is admissible. We say that s is an ***H*-potential** if $D(s - \text{flux}(s)H) = 0$.

The H-Green functions G^H is defined by the formula

$$G^H(w, v) := H_v(e) - H_v(w).$$

The following theorem gives the main results of this section and of [28].

Theorem 3.8 ([28]). *Let T be a recurrent tree, H a standard on T and G^H the H-Green function.*

(i) *For s superharmonic except possibly for finitely many vertices, the flux of s is given by the formulas*

$$\operatorname{flux}(s) = \sum_{v\in T} \Delta s(v)\alpha_v = \lim_{n\to\infty} \sum_{|v|=n} (s(v) - s(v^-))\alpha_v p(v, v^-).$$

(ii) (Riesz decomposition) *Let s be an admissible superharmonic function on T. Then s can be written uniquely as the sum of a global harmonic function and an H-potential.*

(iii) *The H-potentials are precisely the functions of the form*

$$G^H f(u) := \sum_{v\sim u} G^H(u, v)f(v),$$

where $f \geq 0$ and $f \in L^1_\alpha(T)$, i.e., $\sum_{v\in T} \alpha_v f(v) < \infty$.

A few comments concerning the function $v \mapsto \alpha_v = p[e, v]/p[v, e]$ are in order. We introduced the functions H_v in order to be able to construct functions with prescribed Laplacian on a given finite set, and to in particular be able to prove that H is a standard. It does, however, appear in the literature in other ways. A simple calculation shows that if $v \sim w$, then $\alpha_v p(v, w) = \alpha_w p(w, v)$. This allows one to view the tree as an electrical network, where the edge function $c(\{v, w\}) := \alpha_v p(v, w)$ is known as the **conductance** of the edge. It also allows us to view the tree as a reversible Markov chain. Up to a constant multiple, the function α_v appears in the text [40] in the chapter on recurrent Markov chains. If we view the tree as a Markov chain with transition matrix P and states given by the vertices, the authors define a measure α on the state space having the property $\alpha P = \alpha$ and refer to it as a **regular measure**. Its existence is guaranteed by the ergodic theorem. If $\sum_v \alpha_v < \infty$, the chain is called **ergodic**, and if $\sum_v \alpha_v = \infty$, the chain is called **null**. In the ergodic case, α can be normalized to give the equilibrium distribution of the various states, that is the fraction of time a random walk spends in each of the states.

The second formula of part (i) of Theorem 3.8 relates flux to the physics notion of flux of a vector field; the sum can be viewed as an integral of the normal derivative of s over a sphere, where the surface measure is given by the conductance. The first formula in (i) gives a criterion for admissibility of a superharmonic function, namely $\Delta s \in L^1(\alpha)$, where we view α_v as a measure on the tree.

4. Classification of Brelot structures on trees

In this section we describe the work in [17] where trees are viewed as Brelot spaces, and the task is to classify the possible Brelot structures one can have.

Brelot spaces were developed to provide an axiomatization of the properties satisfied by solutions of various elliptic partial differential equations on open subsets of $\mathbb{R}^n$ ([21]). Specifically, a **Brelot space** is a connected, locally connected, locally compact but not compact Hausdorff space Ω such that on each nonempty open set there is a vector space of continuous real-valued functions, called harmonic functions, satisfying Axioms 1, 2 and 3, given as follows.

Axiom 1 (Sheaf property). A function harmonic on an open set $U \subset \Omega$ is harmonic on any open subset of U; If a function on U is harmonic on a neighborhood of each of the points of U, then it is harmonic on U.

A relatively compact domain $U \subset \Omega$ is called **regular** if every function f continuous real-valued on the boundary of U has a unique continuous harmonic extension H_f^U to U, and that extension is nonnegative if f is nonnegative.

Axiom 2 (Regularity property). The regular domains form a base for the topology on Ω.

Axiom 3 (Harnack property). Any sequence of harmonic function that increases pointwise on a connected open subset of Ω has a pointwise limit that is either harmonic or identically infinity.

By Axiom 2, if U is regular, then for each $x \in U$, there exists a measure ρ_x^U on ∂U such that $H_f^U(x) = \int f d\rho_x^U$ for each f continuous on ∂U. A function $g\colon \Omega \to (-\infty, \infty]$ is called **superharmonic** if it is lower semicontinuous, not identically ∞, and for every regular domain U and $x \in U$, $\int g\, \mathrm{d}\rho_x^U \leq g(x)$. A nonnegative superharmonic function is called a **potential** if it doesn't majorize any positive harmonic function.

Two Brelot spaces Ω, Ω' are called **Brelot isomorphic** if there is an isomorphism $f\colon \Omega \to \Omega'$ which carries one harmonic structure onto the other.

Of course the set of harmonic functions on an open set in $\mathbb{R}^n$ is an example of a Brelot space. Another class of examples consists of the set of harmonic functions on a tree in which they are extended linearly along the edges. This is considered in [15].

Actually one can create other Brelot structures on trees by extending along the edges in various ways. This point of view was adopted in [18] where the problem of considering what, up to Brelot isomorphism, are the Brelot structures on trees, and what is their relation to discrete harmonic structures on the vertices. A classification of Brelot structures on one-dimensional manifolds was considered in [42], based on an earlier work on one-dimensional harmonic spaces [43]. The classification described here is different from those in that it has more equivalence classes of Brelot spaces. We describe next the main results in [18].

We first describe the possible Brelot structures on an interval I. A Brelot structure on I is called **quasi-linear** if constants are harmonic, **quasi-hyperbolic** if it is not quasi-linear but there does exist a positive harmonic function on I, and **quasi-trigonometric** if there are no positive harmonic functions on I. On each interval I the set of harmonic functions is two-dimensional, and we refer to a basis $\{f, g\}$ as a **Brelot basis**. The next theorem characterizes the quasi-linear and quasi-hyperbolic Brelot structures on $I = (0, 1)$.

Theorem 4.1. (a) *Every continuous* 1*-to-*1 *function f on* $[0, 1]$ *gives rise to a quasi-linear Brelot structure on* $(0, 1)$ *such that* $\{1, f\}$ *is a Brelot basis and, conversely, every quasi-linear structure arises in this way. Consequently a quasi-linear Brelot structure on an interval is Brelot isomorphic to linear functions on some interval, and hence there are three distinct classes of quasi-linear structures on intervals represented by the linear structures on* $(-\infty, \infty)$, $(0, \infty)$ *and* $(0, 1)$.

(b) *Let $\mathcal{H}$ be a quasi-hyperbolic structure on* $(0, 1)$ *and let k be a positive harmonic function on* $(0, 1)$. *Then there exists a harmonic function h on* $(0, 1)$ *such that* $\{h, k\}$ *is a Brelot basis. For any such k, h/k is* 1*-to-*1. *Conversely, let k be any positive continuous function on* $(0, 1)$ *and let ϕ be any homeomorphism from* $(0, 1)$ *to some interval. Then the pair* $\{k, k\phi\}$ *is a Brelot basis for a Brelot structure* *which is either quasi-linear or quasi-hyperbolic. It is quasi-linear if and only if* k *is constant or* ϕ *is a linear combination of* $1/k$ *and* 1.

We explain next why a quasi-trigonometric structure is closely related to a structure generated by $\sin x$ and $\cos x$ on some interval. Let $-\infty \leq a < b \leq \infty$ with $b - a > \pi$. A **trigonometric Brelot structure** on (a, b) is a structure generated by two continuous functions $S(x)$ and $C(x)$ such that for all $x \in (a, b)$ the sign of $S(x)$ equals the sign of $\sin x$, the sign of $C(x)$ equals the sign of $\cos x$, and $S(x)/C(x) = \tan x$. Then the quasi-trigonometric Brelot structures are described by the following theorem.

Theorem 4.2. (a) *Any quasi-trigonometric Brelot structure on* $(0, 1)$ *is Brelot isomorphic to a trigonometric structure on some interval* (a, b).

(b) *Let* f, g *be continuous functions on* $(0, 1)$ *whose corresponding zero sets are disjoint, discrete, alternating, and assume that* f *has at least two zeros. Suppose that the restriction of* f/g *(respectively,* g/f*) to any interval containing no zero of* g *(respectively, no zero of* f*) is 1-to-1. Then* $\{f, g\}$ *is a Brelot basis for a quasi-trigonometric Brelot structure on* $(0, 1)$. *Conversely, every quasi-trigonometric Brelot structure has a Brelot basis of this form.*

Having described the Brelot structures on intervals, the next step is to understand them on trees. A discrete harmonic structure on a tree T is determined by a function $p\colon T \times T \to [0, \infty)$ such that $p(u, v) > 0$ if and only if $u \sim v$. In this section we don't assume that $\sum_{u \sim v} p(u, v) = 1$ for each u, since we don't wish to assume that constant functions are harmonic. We refer to the discrete harmonic structure as $\mathcal{H}_p$.

Let $\widetilde{T}$ be the tree viewed as a one-dimensional simplicial complex, that is, for all $u \sim v$, consider the set $[u, v] = \{(1 - t)u + tv : -1 \leq t \leq 1\}$. Then $\widetilde{T} = \bigcup_{u \sim v}[u, v]$. Before we give the main results, we give several definitions.

Definition 4.3. (a) A Brelot structure on $(0, 1)$ is called **extendible** if it is the restriction of some Brelot structure on $(-1, 1)$. This is equivalent to having every harmonic function f on $(0, 1)$ satisfy $f(0) := \lim_{t \to 0^+} f(t)$ exist and be nonzero.

(b) A relatively compact domain U is called a **Dirichlet domain** if, for any boundary function f, there exists a unique solution h_f^U to the corresponding Dirichlet problem.

(c) A Dirichlet domain is called a **positive Dirichlet domain** if the solution to the Dirichlet problem with nonnegative boundary values is nonnegative in some neighborhood of the boundary.

(d) A Dirichlet domain U is called **weakly regular with respect to** $\boldsymbol{x \in U}$ if for any function f defined on the boundary of U which is nonnegative and not identically zero, $h_f^U(x) > 0$.

(e) The **weak ball regularity axiom** says that the unit ball centered at any vertex is weakly regular with respect to its center.

We remark that examples given in [18] show that a Dirichlet domain can be weakly regular with respect to some points but not to others, a positive Dirichlet domain need not be regular, Dirichlet domains need not be positive Dirichlet domains, and the weak ball regularity axiom can hold without the unit ball being regular.

The next theorem describes how to generate examples of Brelot structures on trees.

Theorem 4.4. *Let T be a tree with an extendible harmonic structure on each edge of $\widetilde{T}$. For each directed edge τ, let $\imath(\tau) = u$ be the initial vertex of τ and pick a point $u_\tau \in \tau$ different from u such that (u, u_τ) is regular and let $p(u, u_\tau)$ be an arbitrary positive number. Let $U \subset \widetilde{T}$ be a connected open set. For each directed edge τ with $\imath(\tau) = u \in U$, let $u'_\tau \in (u, u_\tau) \cap U$. Let $\mathcal{H}(U)$ be the set of functions f such that for each edge $[u, v]$ intersecting U, the restriction of f to $(u, v) \cap U$ is harmonic and for each vertex $u \in U$ we have $f(u) = \sum_{\imath(\tau)=u} p(u, u'_\tau) f(u'_\tau)$. Then $\mathcal{H}$ yields a Brelot structure on $\widetilde{T}$.*

And finally, we give a theorem which describes the relation between discrete harmonic structures on T and Brelot structures on $\widetilde{T}$.

Theorem 4.5. (a) *Let T be a tree with a discrete harmonic structure. Assume that each edge of $\widetilde{T}$ has a Brelot structure for which the whole edge is a positive Dirichlet domain. Then there is a unique Brelot structure on $\widetilde{T}$ which induces the given discrete structure and whose restriction to each edge is the given Brelot structure. Furthermore, it satisfies the weak ball regularity axiom.*

(b) *A Brelot structure on $\widetilde{T}$ satisfying the weak ball regularity axiom induces the discrete harmonic structure with transition matrix p as follows: if $u \sim v$, $p(u, v)$ is the value at u of the solution to the Dirichlet problem on the unit ball $B_1(u)$ with boundary values equal to the characteristic function of v.*

5. Extension of results of Section 3 to Brelot spaces

In this section we describe the work in [17]. We let Ω be a Brelot space in which constants are harmonic and there are no potentials. This means that every superharmonic function with a harmonic minorant is necessarily constant. In this setting, admissible superharmonic functions were introduced in [7]. A superharmonic function s is called **admissible** if outside some compact set K there is a harmonic function h such that $h(x) \leq s(x)$ for all $x \in \Omega \setminus K$.

Some of the discrete ideas discussed above in Section 3 can be formulated in this setting. We describe here some of the results in [17] where standard and H-potential are defined, and a Riesz decomposition theorem for admissible superharmonic functions is proved.

Definition 5.1 ([17, Definition 2.1]). Let $K \subset \Omega$ be a nonempty compact set that is not polar (that is, there does not exist a superharmonic function on Ω which is identically ∞ on K). A function H harmonic off K is called a **standard** for Ω associated with K if the following is true: given any function h which is harmonic off a compact set, there exist a unique function h_Ω harmonic on Ω and a unique real number α such that $b = h - \alpha H - h_\Omega$ is bounded off a compact set and $\liminf_{x \to \infty} b(x) = 0$, where the lim inf is taken with respect to the Alexandrov one-point compactification of Ω.

That a standard exists was proved in [8]. We next define flux. As we mentioned earlier, Anandam defined flux in this setting. However, the condensed definition we gave in Section 3 works well here.

Definition 5.2 ([17, Definition 2.3]). Let s be a function on Ω superharmonic outside a compact set. Define the flux of s by $\text{flux}(s) = \sup A_s$, where $A_s = \{\alpha \in \mathbb{R} : \text{there exists } h_\Omega \text{ harmonic on } \Omega \text{ such that } s - \alpha H \geq h_\Omega\}$.

The H-potentials were introduced in [28] in order to give a Riesz decomposition theorem for admissible superharmonic functions on recurrent trees. They were later introduced in [17] in the Brelot space setting for a similar purpose. Before we give the definition, we describe other potential type functions given in [9, 10].

Definition 5.3. Let s be superharmonic on $\Omega \setminus K_0$, where K_0 is a compact outer regular set. Let $\mathcal{E} = \{U_n\}$ be an increasing exhaustion of Ω consisting of relatively compact regular sets containing K_0. Let $h_n = h_s^{U_n}$, the solution of the Dirichlet problem on U_n with boundary values s on the boundary of U_n. Define $D_{\mathcal{E}}s(x) = \lim_{n\to\infty} h_n(x)$ if this limit exists locally uniformly.

Then the potentials introduced in [9] and [10] are defined as follows.

Definition 5.4. An admissible superharmonic function s is said to be in the class $\mathcal{P}$ if there exists an exhaustion $\mathcal{E}$ such that $D_{\mathcal{E}}(s-\alpha H)$ exists and is constant, where α is the flux of s. If, furthermore, that constant is 0 for some exhaustion $\mathcal{E}$, then s is called a **BS potential**. Define the class $\mathcal{Q}$ as the collection of all admissible superharmonic functions s satisfying the property: there exists $s' \in \mathcal{P}$ such that the difference of the greatest harmonic minorants of s and s' outside a compact set is bounded.

The following partial Riesz decomposition holds.

Theorem 5.5 ([11]). *Any admissible superharmonic function s on a BS space is a sum of a function in the class $\mathcal{Q}$ and a harmonic function. This decomposition is unique up to an additive constant. If s is harmonic outside a compact set, then the element of $\mathcal{Q}$ can be chosen uniquely to be a BS potential.*

Besides the fact that this theorem doesn't give a unique representation of admissible superharmonic functions, it might prove difficult to show whether or not a given function is in class $\mathcal{P}$, class $\mathcal{Q}$ or is a BS potential. For this reason the H-potentials were introduced in [17]. We now define them.

Definition 5.6. Let K_0 be an outer regular compact set and let s be an admissible superharmonic function with flux α. Define $D(s-\alpha H)$ to be the greatest harmonic minorant of $s - \alpha H$ on $\Omega \setminus K_0$. Then s is called an H-potential if

$$\liminf_{x\to\infty} D(s - \alpha H)(x) = 0,$$

where the lim inf is taken with respect to the Alexandrov one-point compactification of Ω.

Note that in the above definition, $D(s-\alpha H)$ is calculated by taking any regular exhaustion $\{U_n\}$ of ω, solving the Dirichlet problem on $U_n \setminus K_0$ with $s - \alpha H$ on the boundary, and taking the pointwise limit as n goes to ∞.

With these definitions in place, we state the following Riesz decomposition theorem.

Theorem 5.7 ([17, Theorem 3.1]). *Every admissible superharmonic function can be written uniquely as the sum of an H-potential and a harmonic function.*

We now make a few comments concerning the relation between the discrete results of Section 3 and the Brelot space results of this section. As we noted in the previous section, a recurrent tree can be viewed as a Brelot space without potentials. Thus one could use the work of Anandam to obtain some of the discrete results discussed here. One example is the proof of the existence of a standard H in the discrete setting. However, one doesn't need such sophisticated ideas to understand standards on a tree, and it turned out be very useful to have an explicit formula, something which is not available in a Brelot space. Another example is our definition of flux, which we think simplifies its study.

Finally we note the apparent difference in the definition of H-potential in the discrete and the Brelot space setting: in the discrete setting we require that the greatest harmonic minorant outside the root of $s - \alpha H$ is 0, whereas in the Brelot space we require the apparently weaker condition that the lim inf at the point at ∞ of the greatest harmonic minorant outside the compact set K_0 of $s - \alpha H$ is 0.

Consider the tree setting again. Let s be admissible with flux α, and let $h = D(s - \alpha H)$ be the greatest harmonic minorant outside the root of $s - \alpha H$. Assume that $\liminf_{|v| \to \infty} h(v) = 0$. We show that s is an H-potential. What makes things simpler here than in a Brelot space is that the only place where h may fail to be harmonic is at the root. Since the Laplacian $\Delta h(e)$ is either positive, zero, or negative, it follows that h is either globally subharmonic, harmonic, or superharmonic. But from Theorem 3.6(d) it follows that $\text{flux}(h) = \text{flux}(s - \alpha H) = \text{flux}(s) - \alpha\, \text{flux}(H) = \alpha - \alpha = 0$, so from Theorem 3.6(c), h must be harmonic on T. The lim inf condition then implies that h is lower bounded, and so since the tree is recurrent, h must be constant. Since the liminf is zero at ∞, h must be zero, proving that s is an H-potential.

6. Boundary behavior of H-potentials in the $(\frac{1}{4}, \frac{1}{4}, \frac{1}{2})$-tree

In this section we return to the theory discussed in Section 3 and apply it in particular to prove a new result on the $(\frac{1}{4}, \frac{1}{4}, \frac{1}{2})$-tree We write explicitly what are the H-potentials, and study their boundary behavior. This is perhaps the simplest nontrivial example of a recurrent tree, and studying the behavior of H-potentials is analogous to the classical problem of studying the behavior at ∞ of logarithmic potentials on $\mathbb{R}^2$.

Let T be the homogeneous tree of degree 3, that is, each vertex has exactly three neighbors. If the transition probabilities are all taken to be $\frac{1}{3}$, we obtain the $\frac{1}{3}$-tree. The $\frac{1}{3}$-tree is transient, and the potentials are all of the form $Gf(w) = \sum_{v \in T} G(w, v) f(v)$, where the Green function $G(w, v)$ is given by $G(w, v) = 2 \times 2^{-|w-v|}$, and f is any nonnegative function on T such that $\sum_{v \in T} 2^{-|v|} f(v) < \infty$ (see [24, Section 4.5]).

In [34] we considered the boundary behavior of these potentials. In this section we prove some results concerning the boundary behavior of H-potentials on the same homogeneous tree but with the transition probabilities of the $(\frac{1}{4}, \frac{1}{4}, \frac{1}{2})$-tree.

We first describe the results in [34] beginning with the exceptional sets that arise. For $E \subset \partial T$ and $0 < \beta \leq 1$ we define

$$H_\beta(E) = \sup_{\delta > 0} \inf \Big\{ \sum_i 2^{-\beta |v_i|} : E \subset \bigcup_i I(v_i), |v_i| > \log_i (1/\delta) \Big\}$$

and

$$C_\beta(E) = \inf\Big\{\sum_i 2^{-\beta|v_i|} : E \subset \bigcup_i I(v_i)\Big\}.$$

We call $H_\beta(E)$ the **β-dimensional Hausdorff measure** of E and $C_\beta(E)$ the **β-dimensional content** of E. The null sets are the same for H_β and C_β.

In case $\beta = 1$, $H_1(I_v) = 2^{-|v|}$. On the $\frac{1}{3}$-tree, up to a constant factor H_1 is the representing measure on the Martin boundary (which is just ∂T) of the constant harmonic function 1. We thus refer to H_1 as the **Lebesgue measure** on ∂T.

The approach regions to a boundary point $\omega = [e, \omega_1, \omega_2, \dots)$ which we consider are defined as follows: For $\tau \geq 1$, $a \geq 0$, let

$$\Omega_{\tau,a}(\omega) := \bigcup_j \{w \in T : \omega \wedge w = \omega_j, |w - \omega_j| \leq (\tau - 1)j + a\}.$$

We view it as an approach region to ω which is a **radial** approach region if $a = 0$ and $\tau = 1$, a **nontangential** approach region with aperture a if $\tau = 1$ and $a > 0$, and a **tangential** approach region of tangency τ if $\tau > 1$. We say that a function has τ-limit L at a boundary point ω provided that for every $a \geq 0$, $f(w)$ converges to L as $w \to \omega$ within $\Omega_{\tau,a}(\omega)$.

Theorem 6.1 ([34]). *Let T be homogeneous with degree 3 with the transition probabilities of the $\frac{1}{3}$-tree. Let $0 < \gamma < 1$ for $p > 1$ and $0 < \gamma \leq 1$ for $p = 1$. Let further $1 \leq \tau \leq 1/\gamma$. Let f be be a nonnegative function defined on T such that Gf is finite and $\sum_{v \in T} f^p(v) 2^{-\gamma|v|} < \infty$. Then the limit of $Gf(v)$ as v tends to $\omega \in \partial T$ with v in the approach region $\Omega_{\tau,a}(\omega)$ is 0 for all $\omega \in \partial T$ except possibly for a set $E \subset \Omega$ having $\tau\gamma$-dimensional Hausdorff measure 0.*

In [34] it was also shown that the approach regions and the exceptional sets in the theorem were in some sense the best possible.

Consider now T with the structure of the $(\frac{1}{4}, \frac{1}{4}, \frac{1}{2})$-tree. Thus, except for the three forward edges starting at the root, all forward probabilities are $\frac{1}{4}$ and all backward probabilities are $\frac{1}{2}$. Straightforward calculations and results from the previous section give the following formulas:

$$\begin{aligned}
\alpha_v &= \begin{cases} 1, & \text{if } v = e, \\ \frac{4}{3} \cdot 2^{-|v|}, & \text{if } v \neq e, \end{cases} \\
H(v) &= |v| \\
b^v(v_k) &= 4\Big[-1 + \frac{|v|}{3} 2^{-|v|} + 2^{k-|v|}\Big], \qquad k = 0, \dots, |v| \\
H_v(w) &= 4 + \tfrac{4}{3}\big[|w| - |v| - 3 \times 2^{|w \wedge v|}\big] 2^{-|v|} \\
G^H(w, v) &= 4 \times 2^{-|v|}\big(2^{|w \wedge v|} - 1\big) - \alpha_v H(w).
\end{aligned}$$

Define the nonnegative function B_v on T by

$$B_v(w) := 4 \times 2^{-|v|}\big(2^{|w \wedge v|} - 1\big),$$

and for $f \geq 0$ on T

$$Bf(w) := \sum_{v \in T} B_v(w) f(v).$$

By Theorem 3.8(iii), the H-potentials are precisely the functions

$$G^H f(w) = \sum_{v\in T} G^H(w,v)f(v) = \sum_{v\in T} B_v(w)f(v) - H(w)\sum_{v\in T}\alpha_v f(v),$$

where f is a nonnegative function on T for which $\sum_{v\in T}\alpha_v f(v) < \infty$. Since $\Delta G^H f = -f$, it follows from Theorem 3.8(i) that

$$G^H f = Bf + \text{flux}(Gf)H. \tag{6.1}$$

What is a natural boundary result to consider for the H-potentials? We motivate this with an informal discussion. Consider the potential theory of the usual Laplacian on the entire real line (which is a recurrent potential theory) and on the right half-line $\{x > 0\}$ (which is a transient potential theory). The Green function on the right half-line is given by

$$G(x,y) = \begin{cases} x, & \text{if } x < y, \\ y, & \text{if } x \geq y, \end{cases}$$

Thus

$$G_y(x)/G_y(1) \to \begin{cases} 1, & \text{if } y \to 0, \\ x, & \text{if } y \to \infty, \end{cases}$$

and so the Martin boundary consists of 0 and ∞ with the Martin kernel given by

$$P_y(x) = \begin{cases} 1, & \text{if } y = 0, \\ x, & \text{if } y = \infty. \end{cases}$$

For the entire real line, the "standard" is given by $H(x) = |x|$, and on the right half-line $|x|$ is the minimal harmonic function corresponding to the Martin boundary point ∞.

Returning to the tree, by analogy we fix a vertex $e' \sim e$, and consider the sector $S(e')$, which, if we make the root an absorbing state, is a transient tree T' having boundary $\{e\} \cup I(e')$. We anticipate from the above example that the standard $H(w) = |w|$ is, when restricted to T', the positive harmonic function with representing measure given by a multiple of the restriction of Lebesgue measure to $I(e')$. Let $G^H f$ be an H-potential on T and let α be the flux of $G^H f$. Then by definition of H-potential, $G^H f - \alpha H$ is a potential on T', so by the Fatou–Naïm–Doob theorem, we would expect to be able to prove that $(G^H f - \alpha H)/H$ has a limit of 0 (in some sense) at Lebesgue a.e. point of $I(e')$. Thus we should consider the limiting behavior of $G^H f/H = Bf/H + \alpha$, and expect to be able to prove it has a radial limit of α at Lebesgue-a.e. point of $I(e')$.

One can view T as a Brelot space if harmonic functions on T are extended linearly along the edges ([15]). Thus one can hope to prove the above radial limit theorem as a consequence of the general Fatou–Naïm–Doob theorem on Brelot spaces. However, we intend to prove the following more general limit theorem, and it does not have an analogue in a general Brelot space setting.

Theorem 6.2. *Let $G^H f$ be an H-potential on the $(\frac{1}{4},\frac{1}{4},\frac{1}{2})$-tree, where f satisfies the growth condition $\sum_{v\in T} f^p(v)2^{-\gamma|v|} < \infty$. Here, $0 < \gamma < 1$ for $p > 1$ and $0 < \gamma \leq 1$ for $p = 1$. Let $1 \leq \tau \leq 1/\gamma$. Then the limit of $G^H f(v)/H(v)$ as $v \to \omega \in \partial T$ with $v \in \Omega_{\tau,a}(\omega)$ is the flux of $G^H f$ for all ω except possibly for a set $E \subset \partial T$ of $\tau\gamma$–Hausdorff measure 0.*

Note the similarity between Theorem 6.1 and Theorem 6.2. Theorem 6.1 asserts that Gf has $\tau-$limit 0 at $\tau\gamma$-a.e. boundary point, and, by (6.1), Theorem 6.2 asserts that Bf/H has $\tau-$limit 0 at $\tau\gamma$-a.e. boundary point. For the first theorem, this amounts to estimating $w \mapsto \sum_{v\in T} 2^{-|v-w|}f(v)$ and the second theorem to estimating $w \mapsto \sum_{v\in T} 2^{-|v|}(2^{|w\wedge v|-1})/|w|f(v)$, and so it is worth asking if the second theorem follows immediately from the first. Estimating the ratio of the kernel defining Bf/H and the kernel defining Gf we get about $(2^{-|v|}2^{|w\wedge v|}/|w|)/2^{-|v-w|} = 2^{|w-w\wedge v|}/|w|$ and this is not upper bounded. For example, if $|v \wedge w| = |w|/2$, the above quantity is $2^{|w|/2}/|w|$, and this is not upper bounded. For another example, if $w \in \Omega_{\tau,a}(\omega)$ and $|v\wedge\omega| > |w\wedge\omega|$, then $w\wedge v = w\wedge\omega$ and so $2^{|w-w\wedge v|} = 2^{|w-w\wedge\omega|}$; this can be as large as $2^{(\tau-1)|w\wedge\omega|+a}$, making $2^{|w-w\wedge v|}/|w|$ unbounded in the interesting case where $\tau > 1$. Thus Theorem 6.2 does not follow immediately from Theorem 6.1.

Nevertheless, the idea of the proof of Theorem 6.1 can be modified to give a proof of Theorem 6.2, and this is what we do here. The main tools needed, as given in [34], are as follows. For a nonnegative function h on T, we define, for $w \in T$,

$$h^*(w) := \sum_{v\in S(w)} h(v)$$

and for $\beta > 0, \omega \in \partial T$ we define

$$M_\beta h(\omega) := \sup_{i\geq 1} 2^{i\beta} h^*(\omega_i).$$

It is shown in [34] that for every $\lambda > 0$,

$$C_\beta\{\omega \in \partial T : M_\beta h(\omega) > \lambda\} \leq \frac{\|h\|_1}{\lambda}, \tag{6.2}$$

where $\|h\|_1 = \sum_{v\in T} h(v)$. If f is as in the statement of Theorem 6.2, let

$$A := \left\{\omega \in \partial T : \limsup_{i\to\infty} 2^i \sum_{v\in S(\omega_i)} f(v)2^{-|v|} > 0\right\} \tag{6.3}$$

and

$$B := \left\{\omega \in \partial T : \limsup_{i\to\infty} 2^{i\tau\gamma} \sum_{v\in S(\omega_i)} f^p(v)2^{-\gamma|v|} > 0\right\}. \tag{6.4}$$

In the proof of Theorem 6.1 as done in [34], it is shown that $H_{\tau\gamma}(A \cup B) = 0$.

For $\omega \in \partial T$ and $w \in \Omega_{\tau,a}(\omega)$, let the vertices of the geodesic from $w \wedge \omega$ to w be denoted by $w^0, w^1, \dots, w^{|w|-|w\wedge\omega|}$ and let $n = |w| - |w \wedge \omega| - 1$. In order to estimate $Bf(w)/H(w)$ we write

$$\sum_{v\in T} \frac{2^{|v\wedge w|}-1}{|w|} 2^{-|v|} f(v) = \sum_{i=1}^{4} \sum_{v\in T_i} \frac{2^{|v\wedge w|}-1}{|w|} 2^{-|v|} f(v),$$

where

$$T_1 := \bigcup_{j=0}^{|w\wedge\omega|-1} \overline{S(\omega_j)}, \qquad T_2 := \bigcup_{i=1}^{n} \overline{S(w^j)}, \tag{6.5}$$

$$T_3 := S(w), \qquad T_4 := S(w \wedge \omega) \setminus (T_2 \cup T_3).$$

Define

$$\sum_i := \sum_{v\in T_i} \frac{2^{|v\wedge w|}-1}{|w|}2^{-|v|}f(v), \quad i=1,\dots,4. \tag{6.6}$$

We now write the proof of Theorem 6.2.

PROOF. In the proof, c is used to denote a number which may be different with each occurrence, but which does not depend on any of the parameters or functions of interest.

We first prove the theorem in case $p=1$. Let $\epsilon,\delta>0$. Let $h(v)=f(v)2^{-\gamma|v|}$. By assumption $\|h\|_1<\infty$. Note that if f had bounded support, then for some fixed N,

$$0\le Bf(w)=4\sum_{|v|\le N}(2^{|v\wedge w|}-1)2^{-|v|}f(v)\le 4(2^N-1)\sum_{|v|\le N} f(v)$$

is bounded and so $Bf(w)/|w|\to 0$ as $|w|\to\infty$. Thus we may assume that h has no support on any given bounded subset of T, and so it is enough to prove the result in case $\|h\|_1<\epsilon\delta$. It thus suffices to prove that

$$C_{\tau\gamma}\left\{\omega\in\partial T:\ \limsup_{\Omega_\tau,a(\omega)\ni w\to\omega}\frac{Bf(w)}{|w|}>\delta\right\}<c\epsilon.$$

By (6.2), this will be achieved if, for each i from 1 to 4, $\omega\in\partial T$, and $w\in\Omega_{\tau\gamma}(\omega)$, we can show that

$$\sum_i \le cM_{\tau\gamma}h(\omega),$$

where $\sum_i$ is defined in (6.6). We have

$$\begin{aligned}
\sum_1 &\le \sum_{j=0}^{|w\wedge\omega|-1}\sum_{v\in\overline{S(\omega_j)}}\frac{2^j-1}{|w|}2^{-(1-\gamma)|v|}h(v)\\
&\le \sum_{j=0}^{|w\wedge\omega|-1}\sum_{v\in\overline{S(\omega_j)}}2^{-\gamma\tau j}\frac{2^j-1}{|w|}2^{-(1-\gamma)j}h(v)2^{\gamma\tau j}\\
&\le \sum_{j=0}^{|w\wedge\omega|-1}\sum_{v\in\overline{S(\omega_j)}}\frac{2^{\gamma j(1-\tau)}}{|w|}h(v)2^{\gamma\tau j}\\
&\le \sum_{j=0}^{|w\wedge\omega|-1}\frac{h^*(\omega_j)}{|w|}2^{\gamma\tau j}\le \sum_{j=0}^{|w\wedge\omega|-1}\frac{M_{\tau\gamma}h(\omega)}{|w|}\le M_{\tau\gamma}h(\omega);
\end{aligned}$$

$$\begin{aligned}
\sum_2 &\le \sum_{j=1}^{n} \sum_{v\in \overline{S(w^j)}} \frac{2^j-1}{|w|} 2^{-(1-\gamma)|v|} h(v) \\
&\le \sum_{j=1}^{n} \sum_{v\in \overline{S(w^j)}} \frac{2^j-1}{|w|} 2^{-(1-\gamma)[|w\wedge\omega|+j]} h(v) \\
&\le \sum_{j=1}^{n} \frac{2^{\gamma[|w\wedge\omega|+j]}}{|w|} h^*(w\wedge\omega) \\
&\le \sum_{j=1}^{n} \frac{2^{\gamma[\tau|w\wedge\omega|+a]}}{|w|} h^*(w\wedge\omega) \le 2^{\gamma a} \sum_{j=1}^{n} \frac{M_{\gamma\tau}h(\omega)}{|w|} \le 2^{\gamma a} M_{\gamma\tau}h(\omega); \\
\sum_3 &\le \sum_{v\in S(w)} \frac{2^{|w|}-1}{|w|} 2^{-(1-\gamma)|v|} h(v) \\
&\le \sum_{v\in S(w)} \frac{2^{|w|}-1}{|w|} 2^{-(1-\gamma)|w|} h(v) \\
&\le \frac{2^{\gamma|w|}}{|w|} h^*(w) \le \frac{2^{\gamma[\tau|w\wedge\omega|+a]}}{|w|} h^*(w\wedge\omega) \le 2^a M_{\gamma\tau}h(\omega); \\
\sum_4 &\le \sum_{v\in \overline{S(w\wedge\omega)}} \frac{2^{|w\wedge\omega|}-1}{|w|} 2^{-(1-\gamma)|v|} h(v) \\
&\le \sum_{v\in \overline{S(w\wedge\omega)}} \frac{2^{|w\wedge\omega|}-1}{|w|} 2^{-(1-\gamma)|w\wedge\omega|} h(v) \\
&\le \frac{2^{\gamma|w\wedge\omega|}}{|w|} h^*(w\wedge\omega) \le 2^{\gamma\tau|w\wedge\omega|} h^*(w\wedge\omega) \le M_{\gamma\tau}h(\omega),
\end{aligned}$$

completing the proof in case $p=1$.

We now consider the case of $p>1$. Thus $0<\gamma<1$. Fix $\omega \in \partial T \setminus A \cup B$, defined above in (6.3) and (6.4). Let $\epsilon > 0$. Choose j_0 such that

$$C_{j_0} := \sup_{j\ge j_0} 2^j \sum_{v\in S(\omega_j)} f(v) 2^{-|v|} < \epsilon;$$

we can do this because $\omega \notin A$. Now consider $w \in \Omega_{\tau,a}(\omega)$ with $|w| > j_0$. As before, we consider $\sum_i := \sum_{v\in T_i} \frac{2^{|v\wedge w|}-1}{|w|} 2^{-|v|} f(v)$ separately for $i=1,\ldots,4$. We have

$$\sum_1 = \sum_{j=0}^{j_0-1} \sum_{v\in \overline{S(\omega_j)}} \frac{2^{|v\wedge w|}-1}{|w|} 2^{-|v|} f(v) + \sum_{j=j_0}^{|w\wedge\omega|-1} \sum_{v\in \overline{S(\omega_j)}} \frac{2^{|v\wedge w|}-1}{|w|} 2^{-|v|} f(v).$$

The first term is bounded by $\left(j_0 2^{j_0} \sum_T 2^{-|v|} f(v)\right)/|w|$, and this goes to 0 as $|w| \to \infty$. The second term is bounded above by

$$c \sum_{j=j_0}^{|w\wedge\omega|-1} \sum_{v\in \overline{S(\omega_j)}} \frac{2^j 2^{-|v|} f(v)}{|w|} \le c C_{j_0} \sum_{j=j_0}^{|w\wedge\omega|-1} 1/|w| < c\epsilon,$$

and so we are done with $\sum_1$.

To deal with $\sum_2, \sum_3$ and $\sum_4$ we will apply Hölder's inequality. Let $p' := p/(p-1)$ be the exponent conjugate to p. We write

$$\frac{2^{|v\wedge w|}}{|w|}2^{-|v|}f(v) = \left(\frac{2^{|v\wedge w|}}{|w|}2^{-(1-\gamma/p)|v|}\right)\left(2^{-(\gamma/p)|v|}f(v)\right). \tag{6.7}$$

Before we apply Hölder's inequality we first estimate $\sum_{v\in T_i} 2^{|v\wedge w|p'}2^{-(1-\gamma/p)p'|v|}$ for $i = 2, 3, 4$. We claim that each is bounded above by a multiple of $2^{(p'/p)\gamma\tau|w\wedge\omega|}$. To prove it, we make use of (2.1) and the fact that $(1-\gamma/p)p'$ is greater than 1. Recalling that $n = |w| - |w\wedge\omega| - 1$, we have

$$\begin{aligned}
\sum_{v\in T_2} 2^{|v\wedge w|p'}2^{-(1-\gamma/p)p'|v|} &\le c\sum_{j=1}^{n} 2^{(j+|w\wedge\omega|)p'}2^{-(1-\gamma/p)p'(j+|w\wedge\omega|)}\\
&\le c\sum_{j=1}^{n} 2^{(j+|w\wedge\omega|)p'\gamma/p}\\
&\le c2^{(n+|w\wedge\omega|)p'\gamma/p}\\
&\le c2^{|w\wedge\omega|\tau\gamma p'/p},
\end{aligned}$$

since $n + |w\wedge\omega| < |w| \le \tau|w\wedge\omega| + a$. This proves the claim for the sum over T_2. We omit the proofs for the other two cases (actually they are easier to prove) and the claim is established.

By (6.7), we thus have for $i = 2, 3, 4$, that

$$\begin{aligned}
\sum_i &\le \left(\sum_{v\in T_i} 2^{|v\wedge w|p'}2^{-(1-\gamma/p)|v|p'}\right)^{1/p'}\left(\sum_{S(w\wedge\omega)} 2^{-\gamma|v|}f^p(v)\right)^{1/p}\\
&\le c2^{(\gamma\tau/p)|w\wedge\omega|}\left(\sum_{S(w\wedge\omega)} 2^{-\gamma|v|}f^p(v)\right)^{1/p}\\
&= c\left(2^{\gamma\tau|w\wedge\omega|}\sum_{v\in S(w\wedge\omega)} 2^{-\gamma|v|}f^p(v)\right)^{1/p},
\end{aligned}$$

and, since ω is not in the set B, this goes to 0 as $|w|$ goes to ∞. This completes the proof. □

Note that in the above proof for $p > 1$, the factor of $1/|w|$ was needed in dealing with $\sum_1$, but not in dealing with $\sum_2, \sum_3$ or $\sum_4$.

7. Biharmonic Green functions on homogeneous trees

Our aim in this section is to explain how one might define various biharmonic Green functions on the $\frac{1}{3}$-tree and to see what one can say about them. The details appear in [29]. One of the interesting features of that paper is the calculations we were able to do, in particular explicit calculations of the Euclidean and hyperbolic simply-supported biharmonic Green functions as well as a function we called the biharmonic Martin kernel. On the surface these calculations appear to be hard, so we made use of a computer algebra system in order to do some of them.

We first describe what is meant by a biharmonic Green function Γ on the unit disk in the complex plane $\mathbb{D} = \{z \in \mathbb{C} : |z| < 1\}$. If we think of $\mathbb{D}$ as being a thin metal plate, then for points $P, Q \in \mathbb{D}$, $\Gamma(P, Q)$ represents the deflection at

P due to a unit load placed at Q. Outside of Q, the function $P \mapsto \Gamma(P, Q)$ is not harmonic, but rather is biharmonic, that is, it is annihilated by the Euclidean bilaplacian, Δ_E^2. In order to uniquely determine Γ, one must impose boundary conditions. In addition to the condition that the deflection on the boundary be zero, one must prescribe one additional condition, namely a **clamped condition** or a **simply-supported condition**. The clamped condition is that the plate is clamped horizontally at the boundary, which means that the normal derivative is 0 at the boundary; the simply-supported condition is that the boundary rests on a support, which means that the Euclidean Laplacian is 0 at the boundary. We thus get two biharmonic Green functions Γ_C and Γ_S uniquely determined as follows:

Clamped Euclidean biharmonic Green function.

$$\begin{cases} (\Delta_E)_P^2 \Gamma_C(P, Q) = \delta_Q(P) & \text{for } P, Q \in \mathbb{D}, \\ \Gamma_C(p, Q) = 0 & \text{for } (p, Q) \in \partial\mathbb{D} \times \mathbb{D}, \\ (\mathrm{d}/\,\mathrm{d}n)_P \Gamma_C(p, Q) = 0 & \text{for } (p, Q) \in \partial\mathbb{D} \times \mathbb{D}. \end{cases}$$

Simply-supported Euclidean biharmonic Green function.

$$\begin{cases} (\Delta_E)_P^2 \Gamma_S(P, Q) = \delta_Q(P) & \text{for } P, Q \in \mathbb{D}, \\ \Gamma_S(p, Q) = 0 & \text{for } (p, Q) \in \partial\mathbb{D} \times \mathbb{D}, \\ \lim_{P \to p} (\Delta_E)_P \Gamma_S(P, Q) = 0 & \text{for } (p, Q) \in \partial\mathbb{D} \times \mathbb{D}. \end{cases}$$

Biharmonic Green functions can be used to construct functions having prescribed bilaplacian and various prescribed boundary conditions. For a discussion of the clamped Euclidean biharmonic Green function on $\mathbb{D}$, see [33]. For a detailed discussion of both biharmonic Green functions on a Riemannian manifold, see [50].

Each Riemannian metric on $\mathbb{D}$ defines a distance between points and a Laplace–Beltrami operator. The Euclidean metric defines the usual distance on D, and so is bounded, whereas the hyperbolic metric defines a distance for which the diameter of $\mathbb{D}$ is infinite. The Euclidean metric $\mathrm{d}s^2 = \mathrm{d}x^2 + \mathrm{d}y^2$ associates the Euclidean Laplacian $\Delta_E f = (\partial^2/\partial x^2 + \partial^2/\partial y^2) f$ and the hyperbolic metric $\mathrm{d}s^2 = (1 - x^2 - y^2)^{-2} \times (\mathrm{d}x^2 + \mathrm{d}y^2)$ associates the hyperbolic Laplacian

$$\Delta_H f = (1 - x^2 - y^2)^2 (\partial^2/\partial x^2 + \partial^2/\partial y^2) f.$$

A function is called **Euclidean (respectively, hyperbolic-biharmonic)** if it is annihilated by the bilaplace operator Δ_E^2 (respectively, Δ_H^2). The Euclidean and hyperbolic harmonic functions are the same, but the Euclidean and hyperbolic biharmonic functions are quite different due to the extra factor of $(1 - x^2 - y^2)^2$ in Δ_H.

In Section 2 we defined, for vertices v, w, the hyperbolic and Euclidean distances between them, namely $|v - w|$ and $2^{-|v \wedge w|}$, respectively. Since the Laplacian we have so far discussed in this paper is defined by averaging a given function over nearest neighbors, i.e., over vertices a hyperbolic distance away of 1, we refer to this Laplacian as the **hyperbolic Laplacian**: $\Delta_H f(v) = \sum_{w \sim v} p(v, w) f(w) - f(w)$.

Motivated by the fact that on the unit disk $\mathbb{D}$ we have

$$\Delta_E f(x, y) = (1 - x^2 - y^2)^{-2} \Delta_H f(x, y),$$

and the factor $1 - x^2 - y^2$ behaves roughly like the Euclidean distance of $x + iy$ to the boundary of the disk, we define the **Euclidean Laplacian** on the $\frac{1}{3}$-tree to be

$$\Delta_E f(u) = 2^{2|u|}\Delta_H f(u),$$

since the factor $2^{2|u|}$ is the Euclidean distance of u to the boundary of T raised to the power -2. We can thus define the **Euclidean simply-supported biharmonic Green function** Γ_S on T to be a function satisfying

$$\begin{cases} (\Delta_E)^2_u \Gamma_S(u,v) = \delta_v(u) & \text{for all } u, v \in T \\ \lim_{n\to\infty} \Gamma_S(\omega_n, v) = 0 & \text{for all } \omega \in \partial T,\ v \in T \\ \lim_{n\to\infty} (\Delta_E)_u \Gamma_S(\omega_n, v) = 0 & \text{for all } \omega \in \partial T,\ v \in T. \end{cases}$$

In order to define what is the clamped Euclidean biharmonic Green function, we must first define what is the normal derivative of a function f at boundary point ω. Assuming that $f(\omega) = \lim_{m\to\infty} f(\omega_m)$ exists, we define it as a limit of difference quotients by

$$\frac{\mathrm{d}f}{\mathrm{d}n}(\omega) = \lim_{m\to\infty} \frac{f(\omega_m) - f(\omega)}{2^{-m}},$$

since the Euclidean distance of ω_m to ω is 2^{-m}. Thus we define the **clamped biharmonic Green function** Γ_C to satisfy

$$\begin{cases} (\Delta_E)^2_u \Gamma_C(u,v) = \delta_v(u) & \text{for all } u, v \in T \\ \lim_{n\to\infty} \Gamma_C(\omega_n, v) = 0 & \text{for all } \omega \in \partial T, v \in T \\ (\mathrm{d}/\,\mathrm{d}n)_\omega \Gamma_C(\omega, v) = 0 & \text{for all } \omega \in \partial T,\ v \in T. \end{cases}$$

The main result in [29] is the following.

Theorem 7.1. (a) *Γ_S and Γ_C both exist and are unique.*
(b) *Γ_S exists and is given by the two formulas*

$$\Gamma_S(u,v) = 4 \cdot 2^{-2|v|} \sum_{w\in T} 2^{-2|w|} 2^{-|u-w|} 2^{-|v-w|} \tag{7.1}$$

$$= \frac{x_S(u\wedge v, v)}{2^{|u-u\wedge v|}} + \frac{y_S(u \wedge v, v)}{2^{3|u-u\wedge v|}},$$

where

$$x_S(u\wedge v, v) = \frac{2\big(2^{-3|v|-|u\wedge v|}\big[25 + 15|u\wedge v|\big] - 4\cdot 2^{-5|v|+|u\wedge v|}\big)}{7} \tag{7.2}$$

$$y_s(u\wedge v, v) = -\tfrac{8}{7}\, 2^{-3|v|-|u\wedge v|}. \tag{7.3}$$

Let's write a few comments about the proof. The uniqueness of Γ_S is proved using a simple maximum principle argument. The first formula in (7.1) is proved by just simply checking that the sum has the required properties of Γ_S, and so the existence of Γ_S in (a) is established. In order to understand where the second equality in (7.1) comes from, let's consider what are the radial Euclidean biharmonic functions on $T \setminus \{e\}$, that is the functions f such that $f(u)$ is constant on each n sphere $|u| = n$ for $n \geq 1$. Denote by x_n the value $f(u)$ for $|u| = n$. Then for $|u| = n$, $\Delta_E f(u) = 2^{2n}[\frac{2}{3}x_{n+1} + \frac{1}{3}x_{n-1} - x_n]$ and

$$\Delta_E^2 f(u) = 2^{2n}\big[\tfrac{2}{3}2^{2n+2}\big(\tfrac{2}{3}x_{n+2} + \tfrac{1}{3}x_n - x_{n+1}\big) + \tfrac{1}{3}2^{2n-2}\big(\tfrac{2}{3}x_n + \tfrac{1}{3}x_{n-2} - x_{n-1}\big) - 2^{2n}\big(\tfrac{2}{3}x_{n+1} + \tfrac{1}{3}x_{n-1} - x_n\big)\big].$$

Setting this to zero and simplifying gives the fourth order homogeneous difference equation

$$64x_{n+2} - 120x_{n+1} + 70x_n - 15x_{n-1} + x_{n-2} = 0.$$

The corresponding characteristic polynomial is $64r^4 - 120r^3 + 70r^2 - 15r + 1$ and this factors as $(r-1)(2r-1)(4r-1)(8r-1)$ and so the roots are $1, 1/2, 1/2^2, 1/2^3$. Thus the radial biharmonic function f can be written in the form

$$f(u) = c_0 + c_1 2^{-|u|} + c_2 2^{-2|u|} + c_3 2^{-3|u|}. \tag{7.4}$$

Now consider each of these four terms with respect to their boundary behavior and the boundary behavior of their Euclidean Laplacian. The first term is the only one that does not go to zero at infinity. Outside the root, the Euclidean Laplacian of the second term is 0, the Euclidean Laplacian of the third term is constant, and the Euclidean Laplacian of the fourth term is a multiple of $2^{-|u|}$. Thus of the four terms, only the second and fourth terms both go to 0 at infinity and have Euclidean Laplacian which also goes to 0 at infinity.

Now fix $v \in T$ and consider the corresponding partition $\mathcal{P}(v)$ of T. A typical member of it is a set $\{u \in T : |u \wedge v| = k\}$ for $k = 0, \ldots, |v|$. On each such set, it follows by symmetry that $u \mapsto \Gamma_S(u,v)$ is radial, and so by the result of the previous paragraph, it can be written as a linear combination of $2^{-|u-u\wedge v|}$ and $2^{-3|u-u\wedge v|}$. The coefficients in this linear combination are denoted by $x_S(u \wedge v, v)$ and $y_S(u \wedge v, v)$ as in (7.2) and (7.3) above. It turns out that the infinite sum in (7.1) can be explicitly found in case u is either $u \wedge v$ or a forward neighbor of $u \wedge v$. This then gives two linear equation satisfied by $x_S(u \wedge v, v)$ and $y_S(u \wedge v, v)$, and so can be solved, resulting in the values of x_S and y_S in (7.2) and (7.3).

The results concerning Γ_C are more demanding, both for existence and uniqueness. We only comment briefly on the existence proof. It makes use of the explicit formulas for Γ_S as well as a new function, $B : T \times \partial T \to \mathbb{R}^+$, which we call the **biharmonic Martin kernel**, used to produce Euclidean biharmonic functions. It is defined by

$$B(u,\omega) = 2 \sum_{w \in T} \frac{2^{2|w\wedge\omega| - |w|}}{2^{|u-w|}\, 2^{2|w|}},$$

chosen this way in order to have the property that for each $\omega \in \partial T$, $\Delta_H B(u,\omega) = -K_H(u,\omega) = -2^{2|u\wedge\omega|-|u|}$, (recall the Martin kernel discussed in Example 3.1, which we now call the hyperbolic Martin kernel, K_H). By integrating out ω with respect to a measure μ one obtains a Euclidean biharmonic function $B\mu(u) = \int B(u,\omega)d\mu(\omega)$ such that $\Delta_E(B\mu) = -K_H\mu$. In order to use it to prove the existence theorem, one needs to sum it in closed form. This can be done, and the result is

$$B(u,\omega) = \frac{2^{-|u|}}{7}\big[25 + 15|u\wedge\omega| - 2^{2(1-|u-u\wedge\omega|)}\big].$$

From here one can calculate the normal derivative at ∂T, and thus construct examples of Euclidean biharmonic functions with prescribed normal derivative. Then $\Gamma_C(u,v)$ is obtained from $\Gamma_S(u,v)$ by subtracting off an appropriate function $\Phi(u,v)$ which for each fixed v is of the form $(B\mu)(u)$, with μ chosen so that $(B\mu)(u)$ has the appropriate normal derivative at ∂T.

8. Multiplication operators and a minimum principle

In this section we begin by discussing the results in [31]. We will then focus attention on a certain minimum principle which is needed to prove the main result, and which can be formulated in a fairly general Brelot space.

Let X, Y be Banach spaces of holomorphic function on a domain $D \subset \mathbb{C}^N$. Let $\psi\colon D \to \mathbb{C}$ be holomorphic such that $\psi f \in Y$ whenever $f \in X$. Then the **multiplication operator with symbol ψ** is the function $M_\psi\colon X \to Y$, $f \mapsto \psi f$. The aim is to characterize the symbol ψ in order that M_ψ is of a certain special type, such as for example bounded, or compact, or an isometry. It is also of interest to calculate the norm of M_ψ.

There are many possible choices for D (for example the unit disk in $\mathbb{C}$, the unit ball in $\mathbb{C}^N$, the unit polydisk in $\mathbb{C}^N$, a bounded homogeneous domain, a bounded symmetric domain) and for the spaces X and Y (Hardy spaces, Bergman spaces, Dirichlet spaces, Bloch spaces, weighted versions of these spaces). There are numerous examples in the literature of this kind of research, for example [1–6, 13, 22, 39, 45–49, 52, 53, 57]. The one which we focus on in this section, [31], takes D to be a bounded symmetric domain, X to be $H^\infty(D)$ and Y to be the Bloch space on U. We explain these terms below.

The **Bloch space** $\mathcal{B}$ on $\mathbb{D} := \{z \in \mathbb{C} : |z| < 1\}$ is the set of holomorphic functions on $\mathbb{D}$ such that $\beta_f := \sup_{|z|<1}(1 - |z|^2)|f'(z)|$ is finite. Then $\|f\|_{\mathcal{B}} := |f(0)| + \beta_f$ is a norm, and it makes $\mathcal{B}$ a Banach space. A geometric condition equivalent to the Bloch condition is that $f \in \mathcal{B}$ if and only if the radii of schlicht disks in $f(\mathbb{D})$ are upper bounded. One has the inclusions that H^∞ and BMOA are properly contained in $\mathcal{B}$, none of the Hardy spaces H^p are contained in $\mathcal{B}$ if $p < \infty$, and $\mathcal{B}$ is not a subset of the Nevanlinna class.

The Bloch space can be defined on bounded homogeneous domains in $\mathbb{C}^N$. The domain D is called a **homogeneous domain** if for all $z, w \in D$ there exists $\phi \in \operatorname{Aut}(D)$ (i.e., a holomorphic self map that is a bijection) such that $\phi(z) = w$. In [36, 54, 55] the definition of Bloch function was extended to homogeneous domains by the condition

$$Q_f := \sup_{z \in D} \sup_{u \in \mathbb{C}^n \setminus \{0\}} \frac{|\sum_{j=1}^N (\partial f/\partial z_j)(z) u_j|}{\sqrt{H_z(u, \bar{u})}} < \infty,$$

where for each $z \in D$, H_z is the Bergman metric on D at z.

A domain D in $\mathbb{C}^N$ is called **symmetric** if for every $z \in D$ there exists $\phi \in \operatorname{Aut}(D)$ such that $\phi \circ \phi = \mathrm{id}$, and z is an isolated fixed point of ϕ. Such domains are actually homogeneous. Cartan in [23] showed that any bounded symmetric domain is biholomorphically equivalent to a unique finite product of irreducible bounded symmetric domains. He classified the irreducible domains into the four classical domains $R_{\mathrm{I}}, R_{\mathrm{II}}, R_{\mathrm{III}}, R_{\mathrm{IV}}$ (see [41]), and two exceptional domains $R_{\mathrm{V}}, R_{\mathrm{VI}}$ (see [32]).

A bounded symmetric domain D is said to be in **standard form** if it can be written as a finite product of Cartan domains. To each domain D one can associate the **Bloch constant** c_D defined by

$$c_D = \sup\{Q_f : f \in H^\infty(D), \|f\|_\infty \le 1\}.$$

The values of the Bloch constants are known for each of the Cartan domains (see [25, 56]), $c_D \le 1$ with strict inequality unless $D = \mathbb{D}$, and if $D = D_1 \times \cdots \times D_k$ is a product of Cartan domains, then $c_D = \max_{1\le j\le k} c_{D_j}$ ([25]). Thus for any bounded symmetric domain, $c_D \le 1$ with equality if and only if D has the unit disk as a factor.

We now state the main result of [31].

Theorem 8.1. *Let D be a bounded symmetric domain in $\mathbb{C}^N$ and $M_\psi\colon H^\infty(D) \to \mathcal{B}(D)$.*

(1) *M_ψ is bounded if and only if $\psi \in H^\infty(D)$.*
(2) *If $\psi(0) = 0$ then $\|M_\psi\| = c_D\|\psi\|_\infty$.*
(3) *Suppose D has $\mathbb{D}$ as a factor. Then*

$$\|M_\psi\| = |\psi(0)| + c_D\|\psi\|_\infty.$$

(4) *If D is a bounded symmetric domain without exceptional factors, then there exist no isometric multiplication operators from $H^\infty(D)$ to $\mathcal{B}(D)$.*

The theorem suggests the following two **conjectures**:

(1) The norm equality in (3) above holds even if $\mathbb{D}$ is not a factor of D;
(2) There exist no isometries among the bounded multiplication operators from $H^\infty(D)$ to the Bloch space of any bounded symmetric domain D.

The proof of Theorem 8.1 makes use of a minimum principle for multiply superharmonic functions.

Definition 8.2. Let $\omega_1 \times \omega_2$ be a product of relatively compact open subsets of $\mathbb{R}^m, \mathbb{R}^n$, respectively. We say that $v\colon \omega_1 \times \omega_2 \to \mathbb{R} \cup \{\infty\}$ is **multiply superharmonic** if

(1) v is not identically ∞,
(2) $v(x) > -\infty$ for all x,
(3) v is lower semicontinuous,
(4) for each fixed $x_1 \in \omega_1$ and $x_2 \in \omega_2$, $v(x_1, \cdot)$ is hyperharmonic (i.e., $\equiv \infty$ or superharmonic) on ω_2 and $v(\cdot, x_2)$ is hyperharmonic on ω_1.

Then the minimum principle which is used to prove Theorem 8.1 is the following.

Theorem 8.3 (Minimum principle). *Let v be multiply superharmonic on $\omega_1 \times \omega_2$ and bounded below. If*

$$\liminf_{(z,z')\to(x,y)} v(z, z') \ge 0,$$

for all $(x, y) \in \partial\omega_1 \times \partial\omega_2$, then $v \ge 0$ on $\omega_1 \times \omega_2$.

The proof of this which we have in mind makes use of the **Cartan-Brelot topology**.

Definition 8.4 (Cartan – Brelot topology). Let U be a fixed bounded open subset of $\mathbb{R}^m$. Let $S = S^+(U) \times S^+(U)/\sim$, where $\sim$ is the equivalence relation defined by $(u_1, v_1) \sim (u_2, v_2)$ if $u_1 + v_2 = u_2 + v_1$ and $S^+(U)$ denotes the set of positive superharmonic functions on U. Let $\mathcal{O}$ be the set of balls of rational radii and X any countable dense subset of $\mathbb{R}^m$. For $\omega \in \mathcal{O}$ and $x \in \omega \cap X$, let ρ_x^ω denote the harmonic measure. Define the seminorm $\Pi_{\omega,x}$ on S by

$$\Pi_{\omega,x}[(u, v)] = \left|\int u \,\mathrm{d}\rho_x^\omega - \int v\, \rho_x^\omega\right|.$$

The Cartan – Brelot topology is the topology generated by this countable family of seminorms.

A related topology is considered in [38]. The following basic properties of the Cartan – Brelot topology hold:

Lemma 8.5. (1) *The Cartan – Brelot topology is Hausdorff and* $S^+(U)$ *is closed.*
(2) *The mapping* $f\colon S^+(U)\times U \to \mathbb{R}\cup\{\infty\}$ *defined by* $f(v,x)=v(x)$ *is lower semicontinuous.*
(3) *Every uniformly locally bounded sequence in* $S^+(U)$ *has a subsequence converging in the Cartan – Brelot topology.*

In proving Theorem 8.1 above in [31], all the details are given except for showing the proofs of the above basic properties.

Actually one can formulate the Cartan-Brelot topology in any Brelot space having a countable base of regular domains. The lemma and the minimum principle can then be proved in such a Brelot space if in addition it satisfies the following axiom.

Axiom of Domination. For U any relatively compact set and g a locally bounded positive superharmonic function on Ω that is harmonic on U, any positive superharmonic function which majorizes g on $\Omega \setminus U$ also majorizes it on U.

The proof of the lemma and the minimum principle in the Brelot space setting can be found in [51]. Actually the proof of Theorem 8.3 above was adapted from the one in [51].

References

1. A. Aleman, P. Duren, M. J. Martín, and D. Vukotić, *Multiplicative isometries and isometric zero-divisors*, Canad. J. Math. **62** (2010), no. 5, 961–974.
2. R. F. Allen and F. Colonna, *On the isometric composition operators on the Bloch space in* $\mathbb{C}^n$, J. Math. Anal. Appl. **355** (2009), no. 2, 675–688.
3. ———, *Isometries and spectra of multiplication operators on the Bloch space*, Bull. Aust. Math. Soc. **79** (2009), no. 1, 147–160.
4. ———, *Multiplication operators on the Bloch space of bounded homogeneous domains*, Comput. Methods Funct. Theory **9** (2009), no. 2, 679–693.
5. ———, *Weighted composition operators on the Bloch space of a bounded homogeneous domain*, Topics in Operator Theory. Volume 1: Operators, Matrices and Analytic Functions (Williamsburg, VA, 2008) (J. A. Ball, V. Bolotniko, J. W. Helton, L. Rodman, and I. M. Spitkovsky, eds.), Oper. Theory Adv. Appl., vol. 202, Birkhäuser, Basel, 2010, pp. 11–37.
6. ———, *Weighted composition operators from* H^∞ *to the Bloch space of a bounded homogeneous domain*, Integral Equations Operator Theory **66** (2010), no. 1, 21–40.
7. V. Anandam, *Espaces harmoniques sans potentiel positif*, Ann. Inst. Fourier (Grenoble) **22** (1972), no. 4, 97–160.
8. ———, *Harmonic spaces with positive potentials and nonconstant harmonic functions*, Rend. Circ. Mat. Palermo (2) **21** (1972), 149–167.
9. ———, *Admissible superharmonic functions and associated measures*, J. London Math. Soc. (2) **19** (1979), no. 1, 65–78.
10. ———, *Subharmonic functions outside a compact set in* $\mathbf{R}^n$, Proc. Amer. Math. Soc. **84** (1982), no. 1, 52–54.
11. ———, *Pseudo-potentiels dans un espace harmonique sans potentiel positif*, Bull. Sci. Math. (2) **100** (1976), no. 4, 369–376 (French).
12. J. M. Anderson, J. Clunie, and Ch. Pommerenke, *On Bloch functions and normal functions*, J. Reine Angew. Math. **270** (1974), 12–37.

13. J. Arazy, *Multipliers of Bloch functions*, University of Haifa Mathematics Publications, vol. 54, 1982.
14. S. Axler, P. Bourdon, and W. Ramey, *Harmonic function theory*, 2nd ed., Grad. Texts in Math., vol. 137, Springer, New York, 2001.
15. I. Bajunaid, J. M. Cohen, F. Colonna, and D. Singman, *Trees as Brelot spaces*, Adv. in Appl. Math. **30** (2003), no. 4, 706–745.
16. ______, *Function series, Catalan numbers, and random walks on trees*, Amer. Math. Monthly **112** (2005), no. 9, 765–785.
17. ______, *A Riesz decomposition theorem on harmonic spaces without positive potentials*, Hiroshima Math. J. **38** (2008), no. 1, 37–50.
18. ______, *Classification of harmonic structures on graphs*, Adv. in Appl. Math. **43** (2009), no. 2, 113–136.
19. ______, *Corrigendum to "Trees as Brelot spaces" [Adv. in Appl. Math.* 30 (4) (2003) 706–745], Adv. in Appl. Math. **47** (2011), no. 2, 401–402.
20. ______, *Biharmonic extensions on trees without positive potentials*, J. Math. Anal. Appl. **378** (2011), no. 2, 710–722.
21. M. Brelot, *Axiomatique des fonctions harmoniques et surharmoniques dans un espace localement compact*, Séminaire Brelot-Choquet-Deny. Théorie du potentiel, vol. 2, Secrétariat mathématique, Paris, 1958, pp. 1–40.
22. L. Brown and A. L. Shields, *Multipliers and cyclic vectors in the Bloch space*, Michigan Math. J. **38** (1991), no. 1, 141–146.
23. É. Cartan, *Sur les domains bornés de l'espace de n variable complexes*, Abh. Math. Sem. Univ. Hamburg **11** (1935), 116–162.
24. P. Cartier, *Fonctions harmoniques sur un arbre*, Symposia Mathematica, Vol. IX (Rome, 1971), Academic Press, London, 1972, pp. 203–270.
25. J. M. Cohen and F. Colonna, *Bounded holomorphic functions on bounded symmetric domains*, Trans. Amer. Math. Soc. **343** (1994), no. 1, 135–156.
26. J. M. Cohen, F. Colonna, K. Gowrisankaran, and D. Singman, *Polyharmonic functions on trees*, Amer. J. Math. **124** (2002), no. 5, 999–1043.
27. J. M. Cohen, F. Colonna, and D. Singman, *Distributions and measures on the boundary of a tree*, J. Math. Anal. Appl. **293** (2004), no. 1, 89–107.
28. ______, *A global Riesz decomposition theorem on trees without positive potentials*, J. Lond. Math. Soc. (2) **75** (2007), no. 1, 1–17.
29. ______, *Biharmonic Green functions on homogeneous trees*, Mediterr. J. Math. **6** (2009), no. 3, 249–271.
30. ______, *A global Riesz decomposition theorem on trees without positive potentials*, J. Lond. Math. Soc. (2) **83** (2011), no. 3, 810.
31. F. Colonna, G. R. Easley, and D. Singman, *Norm of the multiplication operators from H^∞ to the Bloch space of a bounded symmetric domain*, J. Math. Anal. Appl. **382** (2011), no. 2, 621–630.
32. D. Drucker, *Exceptional Lie algebras and the structure of Hermitian symmetric spaces*, Mem. Amer. Math. Soc. **16** (1978), no. 208.
33. P. Duren and A. Schuster, *Bergman spaces*, Math. Surveys Monogr., vol. 100, Amer. Math. Soc., Providence, RI, 2004.
34. K. Gowrisankaran and D. Singman, *Tangential limits of potentials on homogeneous trees*, Potential Anal. **18** (2003), no. 1, 79–96.
35. ______, *Minimal fine limits on trees*, Illinois J. Math. **48** (2004), no. 2, 359–389.
36. K. T. Hahn, *Holomorphic mappings of the hyperbolic space into the complex Euclidean space and the Bloch theorem*, Canad. J. Math. **27** (1975), 446–458.
37. S. Helgason, *Differential geometry and symmetric spaces*, Pure Appl. Math., vol. 12, Academic Press, New York, 1962.
38. R.-M. Hervé, *Recherches axiomatiques sur la théorie des fonctions surharmoniques et du potentiel*, Ann. Inst. Fourier (Grenoble) **12** (1962), 415–571.
39. T. Hosokawa, K. Izuchi, and S. Ohno, *Topological structure of the space of weighted composition operators on H^∞*, Integral Equations Operator Theory **53** (2005), no. 4, 509–526.
40. J. G. Kemeny, J. L. Snell, and A. W. Knapp, *Denumerable Markov chains*, 2nd ed., with a chapter on Markov random fields by D. Griffeath, Grad. Texts in Math., vol. 40, Springer, New York, 1976.

41. S. Kobayashi, *Hyperbolic manifolds and holomorphic mappings: An introduction*, 2nd ed., World Sci. Publ., Hackensack, NJ, 2005.
42. J. Král and J. Lukeš, *Brelot spaces on one-dimensional manifolds*, Časopis Pěst. Mat. **99** (1974), 179–185 (Czech).
43. J. Král, J. Lukeš, and I. Netuka, *Elliptic points in one-dimensional harmonic spaces*, Comment. Math. Univ. Carolinae **12** (1971), 453–483.
44. S. G. Krantz and D. Ma, *Bloch functions on strongly pseudoconvex domains*, Indiana Univ. Math. J. **37** (1988), no. 1, 145–163.
45. S. Li and S. Stević, *Weighted composition operators from H^∞ to the Bloch space on the polydisc*, Abstr. Appl. Anal. (2007), Art. ID 48478.
46. ———, *Weighted composition operators from α-Bloch space to H^∞ on the polydisc*, Numer. Funct. Anal. Optim. **28** (2007), no. 7-8, 911–925.
47. ———, *Weighted composition operators between H^∞ and α-Bloch spaces in the unit ball*, Taiwanese J. Math. **12** (2008), no. 7, 1625–1639.
48. S. Ohno, *Weighted composition operators between H^∞ and the Bloch space*, Taiwanese J. Math. **5** (2001), no. 3, 555–563.
49. S. Ohno and R. Zhao, *Weighted composition operators on the Bloch space*, Bull. Austral. Math. Soc. **63** (2001), no. 2, 177–185.
50. L. Sario, M. Nakai, C. Wang, and L. O. Chung, *Classification theory of Riemannian manifolds: Harmonic, quasiharmonic and biharmonic functions*, Lecture Notes in Math., vol. 605, Springer, Berlin, 1977.
51. D. Singman, *Exceptional sets in a product of harmonic spaces and applications*, Doctoral dissertation, McGill University, Montréal, 1980.
52. S. Stević, *Norm of weighted composition operators from Bloch space to H^∞_μ on the unit ball*, Ars Combin. **88** (2008), 125–127.
53. S. Stevich, P. Chen, and Z. Chzhou, *Weighted composition operators acting from a Bloch space in the polydisk to another Bloch space*, Mat. Sb. **201** (2010), no. 2, 131–160 (Russian); English transl., Sb. Math. **201** (2010), no. 1-2, 289–319.
54. R. M. Timoney, *Bloch functions in several complex variables*. I, Bull. London Math. Soc. **12** (1980), no. 4, 241–267.
55. ———, *Bloch functions in several complex variables*. II, J. Reine Angew. Math. **319** (1980), 1–22.
56. G. Zhang, *Bloch constants of bounded symmetric domains*, Trans. Amer. Math. Soc. **349** (1997), no. 7, 2941–2949.
57. M. Z. Zhang and H. H. Chen, *Weighted composition operators of H^∞ into α-Bloch spaces on the unit ball*, Acta Math. Sin. (Engl. Ser.) **25** (2009), no. 2, 265–278.

DEPARTMENT OF MATHEMATICAL SCIENCES, GEORGE MASON UNIVERSITY, FAIRFAX, VA 22030, USA

E-mail address: dsingman@gmu.edu

Centre de Recherches Mathématiques
CRM Proceedings and Lecture Notes
Volume **55**, 2012

Recent Progress on Fine Differentiability and Fine Harmonicity

Stephen J. Gardiner

ABSTRACT. This paper describes recent results concerning the notions of differentiability and harmonicity with respect to the fine topology of classical potential theory.

1. Introduction

The fine topology of classical potential theory is the coarsest topology on $\mathbb{R}^n$ $(n \geq 2)$ that makes all subharmonic functions continuous. It is closely related to the notion of thinness, as characterized by Wiener's criterion. More precisely, a set U in $\mathbb{R}^n$ that contains a point x is a fine neighborhood of x if and only if $\mathbb{R}^n \setminus U$ is thin at x. Thus, if V is a (Euclidean) open set in $\mathbb{R}^n$ and E is the set of boundary points that are irregular for the Dirichlet problem on V, then $V \cup E$ would be a finely open set. We refer to Chapter 7 of the book [2] for an introduction to this topology. We identify $\mathbb{R}^2$ with $\mathbb{C}$ in the usual way.

Since about 1970, Fuglede and others have developed elegant theories of holomorphic and harmonic functions with respect to the fine topology. The purpose of this paper is to describe some recent work on the notions of real differentiability and harmonicity with respect to the fine topology.

We begin by recalling some definitions. Let U be a finely open subset of the complex plane $\mathbb{C}$ and let $z_0 \in U$. A function $f\colon U \to \mathbb{C}$ is called *finely differentiable* (in the complex sense) at z_0 if the usual difference quotient

$$\frac{f(z) - f(z_0)}{z - z_0}$$

has a fine limit at z_0. We then call this limit the *fine derivative of f at z_0* and denote it by $f'(z_0)$. Thus, for any $\varepsilon > 0$, there is a fine neighborhood V of z_0 such that

$$\left| \frac{f(z) - f(z_0)}{z - z_0} - f'(z_0) \right| < \varepsilon \quad (z \in V).$$

2010 *Mathematics Subject Classification.* Primary 31C40.
Key words and phrases. fine differentiability, finely harmonic functions.
The author was supported in part by Science Foundation Ireland under Grant 09/RFP/MTH2149.
This is the final form of the paper.

It is readily seen that fine differentiability implies fine continuity at any given point. (The target set $\mathbb{C}$ is endowed with the usual Euclidean topology.) We call f *finely holomorphic* on U if f is finely differentiable (in the complex sense) at each point of U and the function f' is finely continuous on U. The theorem below lists some key properties of finely holomorphic functions. Further results and original references may be found in Fuglede [7]. By a *fine domain* we mean a nonempty finely connected finely open set.

Theorem A. *Let U be a nonempty finely open subset of $\mathbb{C}$.*

(a) *A function $f\colon U \to \mathbb{C}$ is finely holomorphic if and only if every point of U has a fine neighborhood $V \subset U$ on which f coincides with a (Euclidean) C^1-function $\tilde{f}\colon \mathbb{C} \to \mathbb{C}$ that satisfies the Cauchy–Riemann equation $\bar{\partial}\tilde{f} = 0$ on V.*
(b) *A finely holomorphic function $f\colon U \to \mathbb{C}$ has fine derivatives of all orders, which are themselves finely holomorphic on U.*
(c) *The composition of two finely holomorphic functions is finely holomorphic, as is the inverse of an injective finely holomorphic function.*
(d) *A finely holomorphic function on a fine domain is uniquely determined by its fine derivatives at a given point.*
(e) *A finely holomorphic function on a fine domain has at most countably many zeros (unless it is identically zero).*

2. Finely continuously differentiable functions

Now let us turn to the notion of real differentiability with respect to the fine topology on $\mathbb{R}^n$. Let $f\colon U \to \mathbb{R}$, where U is a finely open set in $\mathbb{R}^n$. A linear mapping $L\colon \mathbb{R}^n \to \mathbb{R}$ is called the *fine differential* of f at a point $x^0 \in U$ if the quotient

$$\frac{f(x) - f(x^0) - L(x - x^0)}{|x - x^0|} \tag{2.1}$$

has fine limit 0 at x^0. It is easy to see that there is at most one linear mapping with this property, since any fine neighborhood of x^0 will intersect most rays from x^0 arbitrarily closely to x^0. We will denote this linear mapping by $\mathrm{d_f} f(x^0)$ when it exists. Again, it is easily seen that the existence of $\mathrm{d_f} f(x^0)$ implies the fine continuity of f at x^0. If $\mathrm{d_f} f$ exists and is finely continuous on U, then f is said to belong to the space fine-$\mathcal{C}^1(U)$ of *finely continuously differentiable functions* on U. An example of a finely continuously differentiable function on $\mathbb{R}^n$ which is discontinuous at 0 is given by

$$f(x) = \begin{cases} (1 - |x - x^k|^2/r_k^2)^2 & (x \in B_k; k \geq 1) \\ 0 & (x \notin \bigcup_k B_k), \end{cases}$$

where (x^k) is a sequence of distinct points in $\mathbb{R}^n \setminus \{0\}$ that converges to 0 and (r_k) is a sequence of radii such that the balls $B_k = \{x : |x - x^k| \leq r_k\}$ are pairwise disjoint and $\bigcup_k B_k$ is thin at 0.

Clearly a sufficient condition for f to belong to fine-$\mathcal{C}^1(U)$ is that, for each $x \in U$, there is a fine neighborhood $V \subset U$ of x and a function $\tilde{f} \in \mathcal{C}^1(\mathbb{R}^n)$ such that $\tilde{f} = f$ on V. Functions satisfying this condition are said to belong to $\mathcal{C}^1_{\text{f-loc}}(U)$. Thus $\mathcal{C}^1_{\text{f-loc}}(U) \subseteq$ fine-$\mathcal{C}^1(U)$. In view of Theorem A(a) it is natural to ask if the reverse inclusion also holds. When $n = 2$ this was verified by Lávička [13], who adapted arguments from the holomorphic case. However, these proofs are all based on a special property of thin sets in the plane, namely that a fine neighborhood of a point x^0 contains arbitrarily small circles centred at x^0. The higher dimensional case proved to be more difficult, and it was even an open question whether a function which has zero fine differential at every point of a fine domain must be constant. These questions were recently resolved in [10] by the following results.

Theorem 2.1. *Let $U \subseteq \mathbb{R}^n$ be finely open. Then* fine-$\mathcal{C}^1(U) = \mathcal{C}^1_{\text{f-loc}}(U)$.

Corollary 2.2. *Let $f: U \to \mathbb{R}$, where $U \subseteq \mathbb{R}^n$ is a fine domain. If f has zero fine differential at every point of U, then f is constant.*

The corollary can be deduced as follows. Let $f: U \to \mathbb{R}$, where $U \subseteq \mathbb{R}^n$ is a fine domain and f has zero fine differential throughout U. Further, let $x^0 \in U$ and $E = \{x \in U : f(x) = f(x^0)\}$. Since f is finely continuous, the set E is finely closed relative to U. Let $x^1 \in E$. By Theorem 2.1 there is a fine neighborhood V of x^1 and a function $\tilde{f} \in \mathcal{C}^1(\mathbb{R}^n)$ such that $V \subset U$ and $\tilde{f} = f$ on V. We know from a result of Lyons [14] that V is polygonally connected. Since $\mathrm{d}\tilde{f} = \mathrm{d}_{\mathrm{f}} f = 0$ on V, we see that $V \subseteq E$. Hence E is also finely open, and so $E = U$ by the fine connectedness of U.

The proof of Theorem 2.1 combines ideas from work of Fuglede, Lyons and Lávička with some new ingredients. In particular, it relies on the following results, where $[x, y]$ denotes the closed line segment with endpoints $x, y \in \mathbb{R}^n$ and $P_{[x,y]}$ denotes the hyperplane $\{z \in \mathbb{R}^n : |z - x| = |z - y|\}$.

Lemma A. *Let V be a finely open set and $x^0 \in V$. Then there is a (Euclidean) compact fine neighborhood $K \subset V$ of x^0 with the following property: for any two distinct points $x, y \in K$, there is a set $A \subset P_{[x,y]}$, of positive $(n-1)$-dimensional measure, such that*

$$|z - x| < \frac{|x - y|}{\sqrt{3}} \quad \text{and} \quad [x, z] \cup [z, y] \subset V \qquad (z \in A).$$

Lemma B. *Let $U \subseteq \mathbb{R}^n$ be finely open and $f: U \to \mathbb{R}$ be finely continuous. Then, for each $x^0 \in U$, there is a compact fine neighborhood $K \subset U$ of x^0 such that $f|_K$ is continuous (with respect to the Euclidean topology).*

Theorem B. *Let $K \subset \mathbb{R}^n$ be compact, and let both $f: K \to \mathbb{R}$ and $d: K \to \mathbb{R}^n$ be continuous. Suppose that, for any $\varepsilon > 0$, there exists $\delta > 0$ such that*

$$|f(y) - f(x) - \langle d(x), y - x \rangle| \leq \varepsilon |y - x| \quad \text{whenever } x, y \in K \text{ and } |x - y| < \delta.$$

Then there exists $\tilde{f} \in \mathcal{C}^1(\mathbb{R}^n)$ such that $\tilde{f} = f$ and $\nabla \tilde{f} = d$ on K.

Lemma A is an enhanced version of a result of Lyons (see Theorem 4.4 of [14] or Lemma 3 of Fuglede [6]), which follows from the same proof. Lemma B is due to Fuglede (see Section 1 of [5]), and Theorem B is due to Whitney (see [16], or Section 6.5 of Evans and Gariepy [3]).

Outline proof of Theorem 2.1. Let $U \subseteq \mathbb{R}^n$ be finely open. We need to show that fine-$\mathcal{C}^1(U) \subseteq \mathcal{C}^1_{\text{f-loc}}(U)$ when $n \geq 3$. Let $f \in$ fine-$\mathcal{C}^1(U)$ and $x^0 \in U$. Then f and $\mathrm{d}_{\mathrm{f}} f$ are finely continuous on U, so by Lemma B we can find a compact fine neighborhood L of x^0 such that both f and $\mathrm{d_f}f$ are continuous on L. Let $\varepsilon > 0$. By uniform continuity we can choose $\delta > 0$ such that

$$\big|\big(\mathrm{d_f}f(x^1) - \mathrm{d_f}f(x^2)\big)(h)\big| < \frac{\varepsilon}{4}|h| \quad (h \in \mathbb{R}^n) \tag{2.2}$$

whenever $x^1, x^2 \in L$ and $|x^1 - x^2| < \delta$. Let V denote the fine interior of L, whence $x^0 \in V$, and choose K as in the statement of Lemma A. Next, let $x, y \in K$, where $0 < |x - y| < \delta$, and let $A \subset P_{[x,y]}$ be as in Lemma A. We may choose A to be compact. By a result of Ancona [1] we can also arrange that A is nonthin at each of its points. By a suitable choice of coordinate system we can arrange that $x = (-t, 0, \dots, 0)$ and $y = (t, 0, \dots, 0)$, where $t = |x - y|/2$. Let $E = \bigcup_{z \in A}([x, z] \cup [z, y])$ and

$$a = \sup\{z_1 : z \in E \text{ and } |f(z) - f(x) - \big(\mathrm{d_f}f(x)\big)(z - x)| \le \varepsilon|z - x|\},$$

where z_1 denotes the first coordinate of the point z. Clearly $E \subset V$ and $a \ge -t$.

Now suppose, for the sake of contradiction, that $a < 0$. By the continuity of f on the compact set E, there is a point $z^1 \in E$ such that $z_1^1 = a$ and

$$\big|f(z^1) - f(x) - \big(\mathrm{d_f}f(x)\big)(z^1 - x)\big| \le \varepsilon|z^1 - x|. \tag{2.3}$$

Consider the conical set $C = \{w \in \mathbb{R}^n : w_1 - z_1^1 \ge (\sqrt{3}/2)|w - z^1|\}$. A projection argument, when combined with the nonthinness property of A mentioned above (see [10] for details), shows that the set $E \cap C$ is nonthin at z^1. Since $f \in$ fine-$\mathcal{C}^1(U)$, there exists $z^2 \in E \cap C$ such that $z_1^2 < 0$ and

$$\big|f(z^2) - f(z^1) - \big(\mathrm{d_f}f(z^1)\big)(z^2 - z^1)\big| \le \frac{\varepsilon}{4}|z^2 - z^1|. \tag{2.4}$$

Hence

$$\begin{aligned}
&\big|f(z^2) - f(x) - \big(\mathrm{d_f}f(x)\big)(z^2 - x)\big| \\
&\quad\le \big|f(z^2) - f(z^1) - \big(\mathrm{d_f}f(x)\big)(z^2 - z^1)\big| + \big|f(z^1) - f(x) - \big(\mathrm{d_f}f(x)\big)(z^1 - x)\big| \\
&\quad\le \big|f(z^2) - f(z^1) - \big(\mathrm{d_f}f(z^1)\big)(z^2 - z^1)\big| + \big|\big(\mathrm{d_f}f(z^1) - \mathrm{d_f}f(x)\big)(z^2 - z^1)\big| + \varepsilon|z^1 - x| \\
&\quad\le \varepsilon\left(|z^1 - x| + \frac{|z^2 - z^1|}{2}\right) \\
&\quad\le \varepsilon|z^2 - x|,
\end{aligned}$$

using (2.3), (2.2) and (2.4). (The final step relies on the observation that the angle between $[x, z^1]$ and $[z^1, z^2]$ does not exceed $\pi/3$.) However, since $z_1^1 < z_1^2$, this contradicts the definition of a.

We conclude that $a \ge 0$. A similar argument shows that it is also impossible to have $0 \le a < t$. Hence $a = t$ and so

$$\big|f(y) - f(x) - \big(\mathrm{d_f}f(x)\big)(y - x)\big| \le \varepsilon|y - x|.$$

We can now appeal to Theorem B to complete the argument. □

Let $e^1, \dots, e^n$ be the standard basis vectors for $\mathbb{R}^n$. Given a function f in fine-$\mathcal{C}^1(U)$, we can define the first order *fine partial derivatives* of f at $x \in U$ by $(D_{\mathrm{f}}^{e^i} f)(x) = (d_{\mathrm{f}}f(x))(e^i)$. The *fine gradient* $\nabla_{\mathrm{f}} f(x)$ is then the associated vector of fine partial derivatives. If all these fine partial derivatives themselves also belong to fine-$\mathcal{C}^1(U)$, we write $f \in$ fine-$\mathcal{C}^2(U)$. The space fine-$\mathcal{C}^k(U)$ is defined inductively for all $k \in \mathbb{N}$ in the obvious way, and we put fine-$\mathcal{C}^\infty(U) = \cap_k$fine-$\mathcal{C}^k(U)$. Finally, for any $k \in \mathbb{N} \cup \{\infty\}$, we write $f \in \mathcal{C}^k_{\text{f-loc}}(U)$ if, for each $x \in U$, there is a fine neighborhood $V \subset U$ of x and a function $\tilde{f} \in \mathcal{C}^k(\mathbb{R}^n)$ such that $\tilde{f} = f$ on V. A proof of the following generalization of Theorem 2.1 may be found in [10].

Theorem 2.3. *Let $U \subseteq \mathbb{R}^n$ be finely open and $k \in \mathbb{N} \cup \{\infty\}$. Then* fine-$\mathcal{C}^k(U) = \mathcal{C}^k_{\text{f-loc}}(U)$.

Let $f\colon U \to \mathbb{R}$ and $x^0 \in U$, where U is finely open. We will say that f has a *fine local minimum* at x^0 if there is a fine neighborhood V of x^0 such that $f(x^0) \le f(x)$ for all $x \in U \cap V$. The notion of a fine local maximum is similarly defined.

From Theorem 2.3 we easily deduce:

Corollary 2.4. *Let $U \subseteq \mathbb{R}^n$ be finely open and $x^0 \in U$.*

(a) *If $f \in$ fine-$\mathcal{C}^1(U)$ and f has a fine local minimum (or maximum) at x^0, then $\nabla_{\mathrm{f}} f(x^0) = 0$.*

(b) *If $f \in$ fine-$\mathcal{C}^2(U)$ and $\nabla_{\mathrm{f}} f(x^0) = 0$, and if the Hessian matrix of second order fine partial derivatives at x^0 is positive definite, then f has a fine local minimum at x^0.*

3. The fine Laplacian

Harmonic functions on (Euclidean) open sets can be equivalently defined either as C^2 functions which satisfy Laplace's equation or as continuous functions which have the mean value property. The latter approach is better suited for defining finely harmonic functions on finely open sets U. We will explain below its connection with the fine Laplacian Δ_{f} defined by

$$\Delta_{\mathrm{f}} f = D_{\mathrm{f}}^{e^1}(D_{\mathrm{f}}^{e^1} f) + \cdots + D_{\mathrm{f}}^{e^n}(D_{\mathrm{f}}^{e^n} f) \quad (f \in \text{fine-}\mathcal{C}^2(U)).$$

If $E \subset \mathbb{R}^n$ $(n \ge 3)$ and v is a positive superharmonic function on $\mathbb{R}^n$, we define

$$R_v^E = \inf\{u : u \text{ is positive and superharmonic on } \mathbb{R}^n \text{ and } u \ge v \text{ on } E\},$$

and denote by $\widehat{R}_v^E$ its lower semicontinuous regularization. If $n \ge 3$ and $v(x) = |x-y|^{2-n}$ for some $y \in \mathbb{R}^n$, then $\widehat{R}_v^E$ is the potential of a measure ν_y^E on $\partial E \cup \{y\}$. This "swept" measure is a generalization of the notion of harmonic measure. (When $n = 2$ we define swept measures in terms of the Green function of a Greenian open set.) We will write E^{c} for $\mathbb{R}^n \setminus E$.

Let U be a finely open set in $\mathbb{R}^n$ and let $B(U)$ denote the collection of bounded finely open sets V such that $\overline{V} \subset U$. A function $h\colon U \to \mathbb{R}$ is called *finely harmonic* on U if h is finely continuous and

$$h(x) = \int h \, \mathrm{d}\nu_x^{V^{\mathrm{c}}} \quad (x \in V; V \in B(U)).$$

Similarly, a function $u\colon U \to [-\infty, \infty)$ is called *finely subharmonic* on U if u is finely upper semicontinuous,

$$u(x) \le \int u \, \mathrm{d}\nu_x^{V^{\mathrm{c}}} < \infty \quad (x \in V; V \in B(U)),$$

and u is not identically valued $-\infty$ on any fine component of U. (We require that the above integrals exist.) The main properties of such functions are described in the book of Fuglede [4].

For ordinary C^2 functions, subharmonicity is equivalent to nonnegativity of the Laplacian. We can now give a fine topology analogue of this fact.

Theorem 3.1. *Let $f \in \text{fine-}\mathcal{C}^2(U)$, where $U \subseteq \mathbb{R}^n$ is finely open. Then:*

(a) *f is finely subharmonic on U if and only if $\Delta_{\mathrm{f}} f \geq 0$ on U;*
(b) *f is finely harmonic on U if and only if $\Delta_{\mathrm{f}} f = 0$ on U.*

SAMPLE PROOF OF THE "IF" PART OF THEOREM 3.1(a). Let $U_x(y) = c_n|x - y|^{2-n}$, where c_n is chosen so that the distributional Laplacian of $-U_x$ is the unit mass concentrated at x. (If $n = 2$, we work instead with $c_2 \log(1/|x - y|)$.) Let $x^0 \in U$. By Theorem 2.3 we can choose a compact fine neighborhood $K \subset U$ of x^0 and $\tilde{f} \in C^2(\mathbb{R}^n)$ such that $\tilde{f} = f$ on K, and we can further arrange that $\tilde{f}$ has compact support. Then

$$f(x) = \tilde{f}(x) = -\int U_x(y)(\Delta\tilde{f})(y)\,\mathrm{d}y \quad (x \in K).$$

Let V be a finely open subset of K. For $z \in V$ we have

$$f(z) - \int f\,\mathrm{d}\nu_z^{V^c} = \int_V \left(U_z(y) - \int U_x(y)\,\mathrm{d}\nu_z^{V^c}(x)\right)(-\Delta\tilde{f})(y)\,\mathrm{d}y.$$

Since the first factor in this last integrand is nonnegative, and $\Delta\tilde{f} = \Delta_{\mathrm{f}} f \geq 0$ on V, we conclude that f is finely subharmonic on U, as claimed. □

Related reasoning leads to the following theorem, which, when $n > 2$, generalizes a result of Stein and Weiss [15].

Theorem 3.2. *Let $f \in \text{fine-}\mathcal{C}^2(U)$, where $U \subseteq \mathbb{R}^n$ is a fine domain, and suppose that $\Delta_{\mathrm{f}} f = 0$ on U. Then the function*

$$\log|\nabla_{\mathrm{f}} f| \quad (n = 2), \qquad |\nabla_{\mathrm{f}} f|^{(n-2)/(n-1)} \quad (n \geq 3)$$

is finely subharmonic on U (provided f is nonconstant when $n = 2$).

Let $h \in \text{fine-}C^2(U)$, where U is finely open and $\Delta_{\mathrm{f}} h = 0$. If $n = 2$, then $h \in \text{fine-}C^\infty(U)$. This follows from Theorem 2.3 and parts (a) and (b) of Theorem A, since $\bar{\partial}(\partial u/\partial x - \mathrm{i}\partial u/\partial y) = 0$ when $\Delta u = 0$ for any $u \in C^2(\mathbb{R}^2)$. However, the corresponding implication fails in higher dimensions, as the following example shows.

Example 3.3. Let $1 < \gamma < 1 + 1/n$ and $E = \bigcup_k B_k$, where $n \geq 3$,

$$B_k = \{x : |x - x^k| \leq r_k\}, \quad x^k = (2^{-k}, 0, \ldots, 0) \quad \text{and} \quad r_k = 2^{-k\gamma}.$$

Further, let u_* be a C^∞ superharmonic function on $\mathbb{R}^n$ such that $u_*(x) = |x|^{2-n}$ when $|x| \geq 1$, let $n < \beta < (n+1)/\gamma$ and let

$$u = \sum m_k u_k, \quad \text{where } u_k(x) = r_k^{2-n} u_*\left(\frac{x - x^k}{r_k}\right) \text{ and } m_k = 2^{-k\beta\gamma}.$$

Since any partial derivative v of u_k up to order 2 satisfies an inequality of the form $|v| \leq C(n, u_*)2^{kn\gamma}$, and since $\sum m_k 2^{kn\gamma} = \sum 2^{k(n-\beta)\gamma} < \infty$, we see that $u \in C^2(\mathbb{R}^n)$. Further, $\Delta u = 0$ on $\mathbb{R}^n \setminus E$, which is finely open by Wiener's criterion, since

$$\sum 2^{k(n-2)} r_k^{n-2} = \sum 2^{(1-\gamma)(n-2)k} < \infty.$$

However, if $y \in \mathbb{R}^{n-1}$ and $|y|^2 \le \frac{1}{2}$, then

$$\frac{\partial^3 u}{\partial x_1^3}(-t, ty) \ge C(n) \sum m_k (t + 2^{-k})^{-n-1} \to C(n) \sum 2^{(n+1-\beta\gamma)k} = \infty \qquad (t \to 0+),$$

so $u \notin$ fine-$C^3(\mathbb{R}^n \setminus E)$.

4. Coincidence of harmonic and finely harmonic functions

We will now turn our attention to another question concerning the relationship between harmonicity and fine harmonicity. Any harmonic function on a (Euclidean) open set is certainly finely harmonic there. The converse is true when $n = 2$ but not in higher dimensions. The most that one can say about a finely harmonic function u on an open set U is that there is a dense open subset V of U on which u is harmonic (see [5, 9]). What about finely harmonic functions on finely open sets without (Euclidean) interior points? An old question of Fuglede (see [4, p. 75]) is as follows:

Problem 4.1. Given a finely open set U without interior points, does there exist a finely harmonic function u on U such that no restriction of u to a nonempty finely open subset of U admits a harmonic extension?

The next theorem, which is taken from [11], shows that the answer to this problem is, in general, negative. It goes further and characterizes the finely open sets U which admit such finely harmonic functions u.

Theorem 4.2. *Let U be a nonempty finely open set in $\mathbb{R}^n$. The following statements are equivalent*:

(a) *for any finely harmonic function u on U there is a nonempty finely open set $V \subseteq U$ and a harmonic function w on an open set $W \supseteq V$ such that $w = u$ on V;*
(b) *for any finely harmonic function u on U there is a harmonic function w on a nonempty open set W such that $W \setminus U$ is polar and $w = u$ on $U \cap W$;*
(c) *U is of second Baire category (with respect to the Euclidean topology).*

The removal of a polar set from any Euclidean domain results in a fine domain. A counterexample to Fuglede's question is thus $U = B \setminus Z$, where B is an open ball and Z is a countable dense subset of B, since U is then a fine domain without interior points that is of second Baire category. On the other hand, fine domains of the form $U = B \setminus Y$, where Y is a G_δ polar dense subset of B, would admit functions with the nowhere-extendable property referred to in the problem.

Since (b) clearly implies (a) in Theorem 4.2, the crucial implication for the counterexample to Fuglede's question is that (c) implies (b). We will outline the proof of this below. The remaining implication is easy when $n = 2$, but more involved in higher dimensions. As with the earlier question about finely continuously differentiable functions, this difference is again due to the special property of thinness in the plane.

PROOF THAT (c) IMPLIES (b) IN THEOREM 4.2. We will give the argument for higher dimensions. It is easily modified for the plane. Let U be a nonempty finely open set in $\mathbb{R}^n$ and u be a finely harmonic function on U. For each $m \in \mathbb{Z}$ we define S_m to be the set $\mathbb{R}^n \setminus \{x \in U : |u(x) - m| < 1\}$. Then, for $k \in \mathbb{N}$, we put

$$A_{m,k} = \{x \in \mathbb{R}^n : \widehat{R}_1^{S_m \cap B(x,k^{-1})}(x) \leq \tfrac{1}{2}\}.$$

If $x_0 \notin A_{m,k}$, then we can choose $\varepsilon \in (0, k^{-1})$ such that

$$\widehat{R}_1^{S_m \cap B(x_0,k^{-1}-\varepsilon)}(x) > \tfrac{1}{2} \quad \text{and so} \quad \widehat{R}_1^{S_m \cap B(x,k^{-1})}(x) > \tfrac{1}{2} \qquad (x \in B(x_0, \varepsilon)).$$

Thus $A_{m,k}$ is closed. If $x_1 \in U$, we can choose $m_1 \in \mathbb{Z}$ such that $|u(x_1) - m_1| < 1$ and, since $|u - m_1|$ is finely continuous on U, we see that S_{m_1} is thin at x_1. Thus $\widehat{R}_1^{S_{m_1} \cap B(x_1,k^{-1})}(x_1) \to 0$ as $k \to \infty$ (see [2, Theorem 7.3.1]), and so $x_1 \in A_{m_1,k}$ when k is sufficiently large. This leads to the conclusion that $U \subseteq \bigcup_{m,k} A_{m,k}$. If (c) holds, then there must exist m_0, k_0 such that A_{m_0,k_0} has nonempty Euclidean interior W. By the definition of $A_{m,k}$ the set S_m is thin at each point of $A_{m,k}$, so S_{m_0} is thin at each point of W. Hence $W \cap S_{m_0}$ is polar. Since $\mathbb{R}^n \setminus U \subseteq S_{m_0}$, the set $W \setminus U$ is also polar. Further, $|u - m_0| \leq 1$ on $W \cap U$. Thus the restriction $u|_{W \cap U}$ admits a harmonic extension to W, since polar sets are removable for bounded finely harmonic functions (see [4, Theorem 9.15]) and bounded finely harmonic functions on Euclidean open sets are actually harmonic (see [4, Theorem 9.8]). □

5. An open question in fine potential theory

We conclude with another recent result concerning fine potential theory and highlight an associated open question. Let u be a nonnegative superharmonic function on an open Ω that possesses a Green function. The Riesz decomposition theorem then tells us that the following conditions are equivalent:

(i) the only nonnegative harmonic minorant of u is 0;
(ii) the only nonnegative subharmonic minorant of u is 0;
(iii) there is a Borel measure μ on Ω such that $u = \int G_\Omega(\cdot, y)\, d\mu(y)$.

The fine topology counterparts of (ii) and (iii) were shown to be equivalent by Fuglede [8], and so either can be used as the definition of a *fine potential*. However, the following question, which was first posed in [4], remained open for a long time.

Problem 5.1. Suppose that u is a nonnegative finely superharmonic function on a fine domain, and that the only nonnegative finely harmonic minorant of u is 0. Does it follow that u is a fine potential?

The higher dimensional case of this problem was finally resolved negatively in [12], but the planar case is still open. This contrasts with our earlier discussion, where the planar case was consistently the easier one to treat.

References

1. A. Ancona, *Sur une conjecture concernant la capacité et l'effilement*, Théorie du potentiel (Orsay, 1983) (G. Mokobodzki and D. Pinchon, eds.), Lecture Notes in Math., vol. 1096, Springer, Berlin, 1984, pp. 34–68.
2. D. H. Armitage and S. J. Gardiner, *Classical potential theory*, Springer Monogr. Math., Springer, London, 2001.
3. L. C. Evans and R. F. Gariepy, *Measure theory and fine properties of functions*, Stud. Adv. Math., CRC Press, Boca Raton, FL, 1992.
4. B. Fuglede, *Finely harmonic functions*, Lecture Notes in Math., vol. 289, Springer, Berlin, 1972.

5. ———, *Finely harmonic mappings and finely holomorphic functions*, Ann. Acad. Sci. Fenn. Ser. A I Math. **2** (1976), 113–127.
6. ———, *Asymptotic paths for subharmonic functions and polygonal connectedness of fine domains*, Séminaire de Théorie du Potentiel, Paris, No. 5 (F. Hirsch and G. Mokobodzki, eds.), Lecture Notes in Math., vol. 814, Springer, Berlin, 1980, pp. 97–116.
7. ———, *Fine topology and finely holomorphic functions*, 18th Scandinavian Congress of Mathematicians (Aarhus, 1980) (E. Balslev, ed.), Progr. Math., vol. 11, Birkhäuser, Boston, MA, 1981, pp. 22–38.
8. ———, *Représentation intégrale des potentiels fins*, C. R. Acad. Sci. Paris Sér. I Math. **300** (1985), no. 5, 129–132.
9. S. J. Gardiner, *Sets of harmonicity for finely harmonic functions*, Potential Anal. **21** (2004), no. 1, 1–6.
10. ———, *Finely continuously differentiable functions*, Math. Z. **266** (2010), no. 4, 851–861.
11. ———, *Coincidence of harmonic and finely harmonic functions*, Potential Anal. **34** (2011), no. 1, 81–88.
12. S. J. Gardiner and W. Hansen, *The Riesz decomposition of finely superharmonic functions*, Adv. Math. **214** (2007), no. 1, 417–436.
13. R. Lávička, *A remark on fine differentiability*, Adv. Appl. Clifford Algebr. **17** (2007), no. 3, 549–554.
14. T. J. Lyons, *Finely holomorphic functions*, J. Funct. Anal. **37** (1980), no. 1, 1–18.
15. E. M. Stein and G. Weiss, *On the theory of harmonic functions of several variables.* I: *The theory of H^p-spaces*, Acta Math. **103** (1960), 25–62.
16. H. Whitney, *Analytic extensions of differentiable functions defined in closed sets*, Trans. Amer. Math. Soc. **36** (1934), no. 1, 63–89.

School of Mathematical Sciences, University College Dublin, Belfield, Dublin 4, Ireland

E-mail address: stephen.gardiner@ucd.ie

Centre de Recherches Mathématiques
CRM Proceedings and Lecture Notes
Volume **55**, 2012

Reversibility Questions in Groups Arising in Analysis

Anthony G. O'Farrell

ABSTRACT. An element g of a group is called *reversible* if it is conjugate in the group to its inverse. In this paper we review some results about the structure of groups involving the reversible elements and we pose some questions about groups associated to a Banach algebra.

1. Introduction

Definition 1.1. An element g of a group is called *reversible* if it is conjugate in the group to its inverse, i.e., there exists some map h, belonging to the group, such that the conjugate $g^h = h^{-1}gh$ equals g^{-1}. We say that h *reverses* g, in this case.

This concept had its origins in the study of dynamical systems [13]. Classical conservative systems such as the harmonic oscillator and a system of n bodies moving under their mutual (Newtonian) gravitational attraction, and billiards (on a table without pockets or corners!) have what is called a time-reversal symmetry: a bijection of the phase space which conjugates the dynamical system to its inverse.

I became interested in reversibility because I encountered reversible dynamical systems when studying a number of different problems related to approximation:

- Approximation by functions of the form $F(x, y) = f(x)+g(y)$ on compact sets in $\mathbb{R}^2$ [59, 60].
- Approximation by polynomials of the form $p(z^2, \bar{z}^2 + \bar{z}^3)$ on a disk in the complex plane [66].

I was not the first to find that an apparently undynamical problem had an essential connection to some discrete dynamical system. In fact, this phenomenon has been named *hidden dynamics* by V. I. Arnol′d [78]. Other problems of this type are:

- Biholomorphic classification of a pair of tangent real-analytic arcs in the plane [1, 49, 62, 78, 79].
- The polynomial hull of a disk having an isolated complex tangent [61, 71].

2010 *Mathematics Subject Classification.* Primary 20E34, 37-02, 46H05, 46L05.
Key words and phrases. Reversible map, Banach algebra, group, dynamical system.
The author is grateful for the support of the CRM, and the hospitality of Paul Gauthier and K. N. GowriSankaran.
This is the final form of the paper.

Each of these problems involves a pair of non-commuting involutions, and so relates to a reversible element in some group of maps. Moreover, it turned out that this connection to dynamics provided the key to resolving the problem.

With this in mind, I decided to come at the phenomenon from the other end, and to try to understand the phenomenon of reversibility in general. The natural context is the theory of groups, and I have been calculating examples, assembling what is known from the literature, and encouraging people to examine reversibility in their favourite groups. In fact, it turns out that there is a gigantic literature on reversibility in various different groups. Much of this falls into islands (or continents) of work, where the workers are unaware of the connection between this aspect of their field and of other fields. The work cuts right across the various fields of mainstream mathematics. See [4, 7–9, 15, 18, 22, 23, 35–37, 39, 40, 42, 48, 54–56, 68–70, 72, 73, 80–82], and further references therein and below.

In this short paper, I review some results about reversibility and particularly factorization into reversible factors, and pose some questions about groups related to Banach algebras, specifically.

I have drawn upon material in the draft of a book on aspects of reversibility that is in preparation with Ian Short, and wish to acknowledge his help with this.

2. Notation

Let G be a group. We use the following notation:

$I = I(G) := \{f \in G : f^2 = \mathrm{id}\}$ — the set of *involutions* of G (including the identity, id).

$R_f = R_f(G) := \{h \in G : f^h = f^{-1}\}$ (where $f^h = h^{-1}fh$) — the set of *reversers of f*.

$R = R(G) := \{f \in G : R_f \neq \varnothing\}$ — the set of *reversible elements*.

For $A \subset G$, $A^n := \{f_1 \cdots f_n : f_j \in A\}$, and $A^\infty := \bigcup_{n=1}^\infty A_n$.

Elements of I^2 are called *strongly-reversible*. They are reversed by an involution.

Membership in I^n or R^n is a conjugacy invariant, and $I^2 \subset R$.

I^∞ and R^∞ are normal subgroups of G.

3. Example: $\mathbf{GL}(n, \mathbb{C})$

Classification of linear reversible maps on $\mathbb{C}^n$ is simple. Suppose $F \in \mathrm{GL}(n, \mathbb{C})$ (the general linear group over the field $\mathbb{C}$ of complex numbers) is reversible. Since the Jordan normal form of F^{-1} consists of blocks of the same size as F with reciprocal eigenvalues, the eigenvalues of F that are not ± 1 must split into groups of pairs $\lambda, 1/\lambda$. Furthermore, we must have the same number of Jordan blocks of each size for λ as for $1/\lambda$. Vice versa, if the eigenvalues of F are either ± 1 or split into groups of pairs $\lambda, 1/\lambda$ with the same number of Jordan blocks of each size, then both F and F^{-1} have the same Jordan normal form and are therefore conjugate to each other.

Incidentally, each matrix $A \in \mathrm{GL}(n, \mathbb{C})$ is mapped to a conjugate of its inverse by some outer automorphism of the group. In fact $A \mapsto (A^t)^{-1}$ is an automorphism, and A^t is conjugate to A. There is a wider context concerning "outer reversibility" in a group.

4. The basic questions

In each group, G, we ask:

- Which f are reversible in G?
- Which h reverse a given f?
- Describe I^∞.
- Describe R^∞.
- Is $I^n = I^\infty$ for some n?
- Is $R^n = R^\infty$ for some n?
- Does every nonempty R_g have an element of finite order? If so, what orders occur? Is $\min\{o(h) : h \in R_g\}$ bounded, for $g \in R$?

If g is reversible by some element of finite order, then it is the product of two elements of that (even) order. Thus results about R^n, combined with results about the order of reverses, also give information about factorizing elements of G as a product of elements of at most a given order. Some find this interesting [43].

We wish to encourage people to investigate the questions above in their favourite groups. Many groups have been analyzed, and in the next section we survey some of these results. But many groups remain to be investigated. In the final section, we shall draw particular attention to groups associated to Banach algebras.

5. Survey of known results

We now give a summary of some answers, in examples of various categories of groups, with a sampling of relevant sources (by no means complete). Here, G always denotes the group under consideration. We give no detail about the derivation of these results, just references to the related literature. Some of the proofs are quite deep, and they draw on diverse branches of mathematics.

5.1. If G is abelian, then obviously $R = I = I^\infty$.
If G is free, then it is not hard to see that $R = I = \{1\}$.

5.2. Finite groups [21, 52, 75–77].

Dihedral $\mathbf{D}_n$: $I^2 = G = R$.
Symmetric $\mathbf{S}_n$: $I^2 = G = R$.
Finite Coxeter: $I^2 = G = R$.
Alternating $\mathbf{A}_n$: $I^2 = R$.
$R \neq G$, except when $n \in \{1, 2, 5, 6, 10, 14\}$.
Quaternion 8-group: $I^2 \neq R = G$.
Finite, simple G: $R = \{1\}$, if $|G|$ is odd.
$G = R^2$, if $|G|$ is even, except for $\mathrm{PSU}(3, 3^2)$.
In general, $G \neq I^2$; when it happens is known.

5.3. Classical groups [7, 10, 11, 19, 24–30, 38, 44, 50, 51, 57, 58, 63].

General linear $\mathbf{GL}(n, F)$ $(n > 1)$: $I^2 = R$.
$I^4 = I^\infty = R^2$.
Special linear $\mathbf{SL}(n, \mathbb{C})$: $I^2 = R$, unless $n = 2 \pmod 4$.
$R^2 = G$.
Orthogonal $\mathbf{O}(n, \mathbb{R})$ (≈ spherical isometries): $I^2 = G$.
Special orthogonal $\mathbf{SO}(n, \mathbb{R})$: $I^2 = R$.
$I^3 = G$, if $n \geq 3$.

$I^2 = G$, if $n \neq 2 \pmod 4$.
Unitary $\mathbf{U}(n, \mathbb{C})$: $I^2 = R$
$I^4 = I^\infty$.
Special unitary $\mathbf{SU}(n, \mathbb{C})$: $I^2 \neq R$.
$I^3 \neq I^6 = G = R^2$.
Unitary quaternionic $\mathbf{Sp}(n, \mathbb{C}) = \mathbf{Symp}(2n, \mathbb{C}) \cap \mathbf{U}(2n, \mathbb{C})$:
$I^2 \neq R = G = I^6$.
Spinor $\mathbf{Spin}(n, \mathbb{C})$: $I^2 = G$, if $n = 0, 1, 7 \bmod 8$.
$R = G$, unless $n = 2 \bmod 4$.
$I^4 = G$, if $n \geq 5$.

5.4. Discrete matrix groups [7, 9, 45, 53].

$\mathbf{GL}(n, \mathbb{Z})$: $I^{3n+9} = G$. $I^{41} = G$ for $n \geq 84$.
$I^2 \neq R \subset I^4$ when $n = 2$.
Modular $\mathbf{PSL}(2, \mathbb{Z})$: $I^2 = R$.

5.5. Finite-dimensional isometry groups [74].

Euclidean $\mathbf{Isom}(\mathbb{R}^n)$: $I^2 = G$.
Orientation-preserving $\mathbf{Isom}^+(\mathbb{R}^n)$: $I^3 = G$, if $n \geq 3$.
$I^2 = G$, if $n = 0$ or $3 \pmod 4$.
Hyperbolic $\mathbf{Isom}(\mathbf{H}^n)$: $I^3 = G$, if $n \geq 2$.

5.6. Homeomorphism groups [6, 20, 31–33, 46, 47, 64, 83].

$\mathbf{Homeo}(\mathbb{R})$: $I^2 \neq R$.
$I^3 \neq I^4 = G = R^2$.
$\mathbf{Homeo}^+(\mathbb{R})$: $R^4 = G$.
$\mathbf{Homeo}(\mathbb{S}^1)$: $I^2 \neq R$.
$I^3 = R^2 = G$.
$\mathbf{Homeo}^+(\mathbb{S}^1)$: $I^2 \neq R$.
$I^3 = R^2 = G$.
$\mathbf{Homeo}(\mathbb{S}^n)$: $G = I^6$ when $n = 2$ or 3. (Open for $n > 3$).
Compact surface MCG: $I^n \neq G = I^\infty$, $\forall n \in \mathbb{N}$, if genus > 2.

5.7. Maps with extra structure [2, 3, 12, 14, 34, 67].

Diffeomorphism $\mathbf{Diffeo}(\mathbb{R})$: $I^2 \neq R$.
$I^3 \neq I^4 = G = R^2$.
$\mathbf{Diffeo}^+(\mathbb{R})$: $R^4 = G$.
Formal germs on $(\mathbb{C}^n, \mathbf{0})$ [2, 65]: $I^4 = I^\infty$ when $n = 1$.
$R^2 = R^\infty$ when $n = 1$.
$R^k = R^\infty$ with $k = 3 + 2 \cdot \text{ceiling}(\log_2 n)$ when $n \geq 2$.
$I^{15} = R^\infty$ when $n = 2$.
Piecewise-linear $\mathbf{PL}(\mathbb{R})$ [16, 17]: $I^2 \neq R$.
$I^3 \neq I^4 = G = R^2$.
$\mathbf{PL}^+(\mathbb{R})$: $I = \{1\}$.
$R^4 = G$.
PL with finitely many nodes $\mathbf{PLF}(\mathbb{R})$: $I^2 = R$.
$\mathbf{PLF}^+(\mathbb{R})$: $R^4 = G$.

6. Banach algebras

Let A be a Banach algebra. We may associate two collections of groups to A. In each case, we ask the usual questions.

6.1. A^{-1} and its subgroups. Suppose A has identity (or adjoin one, if not) and $\|1\| = 1$.

Reversibility in A^{-1} is not interesting unless A is noncommutative. Also central reversibles are just central involutions, so the real problems are about the quotient

$$\frac{A^{-1}}{Z(A^{-1})} \equiv \mathrm{Inn}(A).$$

One interesting subgroup is

$$\mathrm{Iso}(A) = \{x \in A : \|x\| = \|x^{-1}\| = 1\}.$$

This coincides with the subgroup (often denoted $\mathrm{U}(A)$ [5]) of unitary elements, in case A is a C^* algebra.

One may also focus on the connected component of 1 in either group, G, and on the intersection $G \cap E^c$ with the commutator of any subset $E \subset A$.

One also has the normal subgroup $\{x \in A : \|a - 1\| < 1\}^\infty$, which lies in the group $(\exp A)^\infty$.

For instance, Gustafson, Halmos, and Radjavi [41] showed that for finite-dimensional (real or complex) Hilbert spaces H, the group $G = \mathrm{GL}(H)$ has $I^4 = I^\infty$, and they noted that for infinite-dimensional H, we have $I^4 \neq I^7 = \mathrm{GL}(H)$. I don't know whether 7 is the best possible value in that statement. What happens with other C^* algebras?

It is known that for finite-dimensional $\mathrm{GL}(H)$, we have $I^2 = R$ in G, and also in the unitary subgroup. What about other C^* algebras?

6.2. Aut(A) and its subgroups. The main interesting subgroup (apart from the inner automorphism group, already mentioned) is the group of isometric isomorphisms. This is often the same as $\mathrm{Aut}(A)$.

As an example, when X is a locally-compact Hausdorff space and $A = C_0(X, \mathbb{C})$, then $\mathrm{Aut}(A)$ is isomorphic to $\mathrm{Homeo}(X)$, so we have seen answers in case $X = \mathbb{R}^1$ and $X = \mathbb{S}^1$. For the disk algebra, the automorphism group is isomorphic to $\mathrm{PSL}(2, \mathbb{R})$, so all elements are reversible. Similarly, for the polydisk algebra, the automorphism group is isomorphic to the group of conformal automorphisms of the polydisk, which is isomorphic to the wreath product $\mathrm{S}_n \wr \mathrm{PSL}(2, \mathbb{R})$, and again all elements are reversible.

We remark that the algebra of all formal power series in n indeterminates, with complex coefficients, has a natural Fréchet algebra structure, and embeds in some Banach algebras [5]. Its automorphism group is isomorphic to the group of formally-invertible formal germs mentioned already.

References

1. P. Ahern and X. Gong, *A complete classification for pairs of real analytic curves in the complex plane with tangential intersection*, J. Dyn. Control Syst. **11** (2005), no. 1, 1–71.
2. P. Ahern and A. G. O'Farrell, *Reversible biholomorphic germs*, Comput. Methods Funct. Theory **9** (2009), no. 2, 473–484.
3. P. Ahern and J.-P. Rosay, *Entire functions, in the classification of differentiable germs tangent to the identity, in one or two variables*, Trans. Amer. Math. Soc. **347** (1995), no. 2, 543–572.

4. P. Ahern and W. Rudin, *Periodic automorphisms of* $\mathbf{C}^n$, Indiana Univ. Math. J. **44** (1995), no. 1, 287–303.
5. G. R. Allan, *Introduction to Banach spaces and algebras*, Oxf. Grad. Texts Math., vol. 20, Oxford Univ. Press, Oxford, 2011.
6. R. D. Anderson, *On homeomorphisms as products of conjugates of a given homeomorphism and its inverse*, Topology of 3-Manifolds and Related Topics (Atherns, GA, 1961), Prentice-Hall, Englewood Cliffs, NJ, 1962, pp. 231–234.
7. M. Baake and J. A. G. Roberts, *Reversing symmetry group of* $\mathrm{Gl}(2, \mathbf{Z})$ *and* $\mathrm{PGl}(2, \mathbf{Z})$ *matrices with connections to cat maps and trace maps*, J. Phys. A **30** (1997), no. 5, 1549–1573.
8. ______, *Symmetries and reversing symmetries of toral automorphisms*, Nonlinearity **14** (2001), no. 4, R1–R24.
9. ______, *The structure of reversing symmetry groups*, Bull. Austral. Math. Soc. **73** (2006), no. 3, 445–459.
10. A. Baker, *Matrix groups: An introduction to Lie group theory*, Springer Undergrad. Math. Ser., Springer, London, 2002.
11. C. S. Ballantine, *Products of involutory matrices*. I, Linear and Multilinear Algebra **5** (1977/78), no. 1, 53–62.
12. T. Banakh and T. Yagasaki, *The diffeomorphism groups of the real line are pairwise bihomeomorphic*, Topology **48** (2009), no. 2-4, 119–129.
13. G. D. Birkhoff, *The restricted problem of three bodies*, Rend. Circ. Mat. Palermo **39** (1915), 265–334.
14. ______, *Déformations analytiques et fonctions auto-équivalentes*, Ann. Inst. H. Poincaré **9** (1939), 51–122.
15. M. G. Brin, *The chameleon groups of Richard J. Thompson: automorphisms and dynamics*, Inst. Hautes Études Sci. Publ. Math. **84** (1996), 5–33 (1997).
16. M. G. Brin and C. C. Squier, *Groups of piecewise linear homeomorphisms of the real line*, Invent. Math. **79** (1985), no. 3, 485–498.
17. ______, *Presentations, conjugacy, roots, and centralizers in groups of piecewise linear homeomorphisms of the real line*, Comm. Algebra **29** (2001), no. 10, 4557–4596.
18. S. Bullett, *Dynamics of quadratic correspondences*, Nonlinearity **1** (1988), no. 1, 27–50.
19. F. Bünger, F. Knüppel, and K. Nielsen, *Products of symmetries in unitary groups*, Linear Algebra Appl. **260** (1997), 9–42.
20. A. B. Calica, *Reversible homeomorphisms of the real line*, Pacific J. Math. **39** (1971), 79–87.
21. H. S. M. Coxeter, *The product of three reflections*, Quart. Appl. Math. **5** (1947), 217–222.
22. R. L. Devaney, *Reversible diffeomorphisms and flows*, Trans. Amer. Math. Soc. **218** (1976), 89–113.
23. D. Ž. Djoković, *Product of two involutions*, Arch. Math. (Basel) **18** (1967), 582–584.
24. ______, *Pairs of involutions in the general linear group*, J. Algebra **100** (1986), no. 1, 214–223.
25. D. Ž. Djoković and J. G. Malzan, *Products of reflections in* $\mathrm{U}(p, q)$, Mem. Amer. Math. Soc. **37** (1982), no. 259.
26. E. W. Ellers, *Bireflectionality in classical groups*, Canad. J. Math. **29** (1977), no. 6, 1157–1162.
27. ______, *The reflection length of a transformation in the unitary group over a finite field*, Linear and Multilinear Algebra **35** (1993), no. 1, 11–35.
28. E. W. Ellers and J. Malzan, *Products of reflections in the kernel of the spinorial norm*, Geom. Dedicata **36** (1990), no. 2-3, 279–285.
29. E. W. Ellers and W. Nolte, *Bireflectionality of orthogonal and symplectic groups*, Arch. Math. (Basel) **39** (1982), no. 2, 113–118.
30. E. W. Ellers and O. Villa, *The special orthogonal group is trireflectional*, Arch. Math. (Basel) **82** (2004), no. 2, 122–127.
31. W. Feit and G. J. Zuckerman, *Reality properties of conjugacy classes in spin groups and symplectic groups*, Algebraists' Homage: Papers in Ring Theory and Related Topics (New Haven, CT, 1981) (S. A. Amitsur, D. J. Saltman, and G. B. Seligman, eds.), Contemp. Math., vol. 13, Amer. Math. Soc., Providence, R.I., 1982, pp. 239–253.
32. N. J. Fine and G. E. Schweigert, *On the group of homeomorphisms of an arc*, Ann. of Math. (2) **62** (1955), 237–253.
33. N. Gill, A. G. O'Farrell, and I. Short, *Reversibility in the group of homeomorphisms of the circle*, Bull. Lond. Math. Soc. **41** (2009), no. 5, 885–897.

34. N. Gill and I. Short, *Reversible maps and composites of involutions in groups of piecewise linear homeomorphisms of the real line*, Aequationes Math. **79** (2010), no. 1-2, 23–37.
35. A. Gómez and J. D. Meiss, *Reversible polynomial automorphisms of the plane: the involutory case*, Phys. Lett. A **312** (2003), no. 1-2, 49–58.
36. ———, *Reversors and symmetries for polynomial automorphisms of the complex plane*, Nonlinearity **17** (2004), no. 3, 975–1000.
37. X. Gong, *Fixed points of elliptic reversible transformations with integrals*, Ergodic Theory Dynam. Systems **16** (1996), no. 4, 683–702.
38. G. R. Goodson, *The inverse-similarity problem for real orthogonal matrices*, Amer. Math. Monthly **104** (1997), no. 3, 223–230.
39. ———, *Inverse conjugacies and reversing symmetry groups*, Amer. Math. Monthly **106** (1999), no. 1, 19–26.
40. R. Gow, *Real-valued characters of solvable groups*, Bull. London Math. Soc. **7** (1975), 132.
41. W. H. Gustafson, P. R. Halmos, and H. Radjavi, *Products of involutions*, Linear Algebra and Appl. **13** (1976), no. 1/2, 157–162.
42. P. R. Halmos and S. Kakutani, *Products of symmetries*, Bull. Amer. Math. Soc. **64** (1958), 77–78.
43. M. Hladnik, M. Omladič, and H. Radjavi, *Products of roots of the identity*, Proc. Amer. Math. Soc. **129** (2001), no. 2, 459–465.
44. F. Hoffman and E. C. Paige, *Products of two involutions in the general linear group*, Indiana Univ. Math. J. **20** (1970/1971), 1017–1020.
45. H. Ishibashi, *Involutary expressions for elements in* $\mathrm{GL}_n(\mathbf{Z})$ *and* $\mathrm{SL}_n(\mathbf{Z})$, Linear Algebra Appl. **219** (1995), 165–177.
46. W. Jarczyk, *Reversibility of interval homeomorphisms without fixed points*, Aequationes Math. **63** (2002), no. 1-2, 66–75.
47. ———, *Reversible interval homeomorphisms*, J. Math. Anal. Appl. **272** (2002), no. 2, 473–479.
48. R. Kane, *Reflection groups and invariant theory*, CMS Books Math./Ouvrages Math. SMC, vol. 5, Springer, New York, 2001.
49. E. Kasner, *Conformal classification of analytic arcs or elements: Poincaré's local problem of conformal geometry*, Trans. Amer. Math. Soc. **16** (1915), no. 3, 333–349.
50. F. Knüppel, *Products of involutions in orthogonal groups*, Combinatorics '86 (Trento, 1986) (A. Barlotti, M. Marchi, and G. Tallini, eds.), Ann. Discrete Math., vol. 37, North-Holland, Amsterdam, 1988, pp. 231–247.
51. F. Knüppel and K. Nielsen, *On products of two involutions in the orthogonal group of a vector space*, Linear Algebra Appl. **94** (1987), 209–216.
52. S. G. Kolesnikov and Ja. N. Nuzhin, *On strong reality of finite simple groups*, Acta Appl. Math. **85** (2005), no. 1-3, 195–203.
53. T. J. Laffey, *Lectures on integer matrices*, unpublished.
54. J. S. W. Lamb, *Reversing symmetries in dynamical systems*, J. Phys. A **25** (1992), no. 4, 925–937.
55. J. S. W. Lamb and J. A. G. Roberts, *Time-reversal symmetry in dynamical systems: a survey*, Phys. D **112** (1998), no. 1-2, 1–39.
56. J. S. W. Lamb, J. A. G. Roberts, and H. W. Capel, *Conditions for local (reversing) symmetries in dynamical systems*, Phys. A **197** (1993), no. 3, 379–422.
57. R. Lávička, A. G. O'Farrell, and I. Short, *Reversible maps in the group of quaternionic Möbius transformations*, Math. Proc. Cambridge Philos. Soc. **143** (2007), no. 1, 57–69.
58. K. M. Liu, *Decomposition of matrices into three involutions*, Linear Algebra Appl. **111** (1988), 1–24.
59. D. E. Marshall and A. G. O'Farrell, *Uniform approximation by real functions*, Fund. Math. **104** (1979), no. 3, 203–211.
60. ———, *Approximation by a sum of two algebras. The lightning bolt principle*, J. Funct. Anal. **52** (1983), no. 3, 353–368.
61. J. K. Moser and S. M. Webster, *Normal forms for real surfaces in* $\mathbf{C}^2$ *near complex tangents and hyperbolic surface transformations*, Acta Math. **150** (1983), no. 3-4, 255–296.
62. I. Nakai, *The classification of curvilinear angles in the complex plane and the groups of* $\pm$ *holomorphic diffeomorphisms*, Ann. Fac. Sci. Toulouse Math. (6) **7** (1998), no. 2, 313–334.

63. K. Nielsen, *On bireflectionality and trireflectionality of orthogonal groups*, Linear Algebra Appl. **94** (1987), 197–208.
64. A. G. O'Farrell, *Conjugacy, involutions, and reversibility for real homeomorphisms*, Irish Math. Soc. Bull. **54** (2004), 41–52.
65. ______, *Composition of involutive power series, and reversible series*, Comput. Methods Funct. Theory **8** (2008), no. 1-2, 173–193.
66. A. G. O'Farrell and M. A. Sanabria-García, *De Paepe's disc has nontrivial polynomial hull*, Bull. London Math. Soc. **34** (2002), no. 4, 490–494.
67. A. G. O'Farrell and I. Short, *Reversibility in the diffeomorphism group of the real line*, Publ. Mat. **53** (2009), no. 2, 401–415.
68. G. R. W. Quispel and H. W. Capel, *Local reversibility in dynamical systems*, Phys. Lett. A **142** (1989), no. 2-3, 112–116.
69. H. Radjavi, *The group generated by involutions*, Proc. Roy. Irish Acad. Sect. A **81** (1981), no. 1, 9–12.
70. J. A. G. Roberts and M. Baake, *Symmetries and reversing symmetries of area-preserving polynomial mappings in generalised standard form*, Phys. A **317** (2003), no. 1-2, 95–112.
71. A. Sanabria-García, *Polynomial hulls of smooth discs: a survey*, Irish Math. Soc. Bull. **45** (2000), 135–153.
72. P. Sarnak, *Reciprocal geodesics*, Analytic Number Theory (Göttingen, 2005) (W. Duke and Yu. Tschinkel, eds.), Clay Math. Proc., vol. 7, Amer. Math. Soc., Providence, RI, 2007, pp. 217–237.
73. M. B. Sevryuk, *Reversible systems*, Lecture Notes in Math., vol. 1211, Springer, Berlin, 1986.
74. I. Short, *Reversible maps in isometry groups of spherical, Euclidean and hyperbolic space*, Math. Proc. R. Ir. Acad. **108** (2008), no. 1, 33–46.
75. A. Singh and M. Thakur, *Reality properties of conjugacy classes in algebraic groups*, Israel J. Math. **165** (2008), 1–27.
76. A. Stein, $1\frac{1}{2}$*-generation of finite simple groups*, Beiträge Algebra Geom. **39** (1998), no. 2, 349–358.
77. P. H. Tiep and A. E. Zalesski, *Real conjugacy classes in algebraic groups and finite groups of Lie type*, J. Group Theory **8** (2005), no. 3, 291–315.
78. S. M. Voronin, *Analytic classification of germs of conformal mappings* $(\mathbf{C},0) \to (\mathbf{C},0)$, Funktsional. Anal. i Prilozhen. **15** (1981), no. 1, 1–17, 96 (Russian); English transl., Functional Anal. Appl. **15** (1981), no. 1, 1–13.
79. ______, *Analytic classification of pairs of involutions and its applications*, Funktsional. Anal. i Prilozhen. **16** (1982), no. 2, 21–29, 96 (Russian); English transl., Functional Anal. Appl. **16** (1982), no. 2, 94–100.
80. S. M. Webster, *Double valued reflection in the complex plane*, Enseign. Math. (2) **42** (1996), no. 1-2, 25–48.
81. ______, *Real ellipsoids and double valued reflection in complex space*, Amer. J. Math. **120** (1998), no. 4, 757–809.
82. M. J. Wonenburger, *Transformations which are products of two involutions*, J. Math. Mech. **16** (1966), 327–338.
83. S. W. Young, *The representation of homeomorphisms on the interval as finite compositions of involutions*, Proc. Amer. Math. Soc. **121** (1994), no. 2, 605–610.

DEPARTMENT OF MATHEMATICS & STATISTICS, NUI MAYNOOTH, MAYNOOTH, CO. KILDARE, IRELAND

E-mail address: admin@maths.nuim.ie

Centre de Recherches Mathématiques
CRM Proceedings and Lecture Notes
Volume **55**, 2012

Subordinate Harmonic Structures in an Infinite Network

Victor Anandam

ABSTRACT. In an infinite network X, the Laplace equation and the Schrödinger equation have been studied, using different techniques from random walks and electric networks. In an abstract sense, it is as though one studies two related discrete harmonic structures on X. This note emphasizes this point of view, by using potential-theoretic methods.

1. Preliminaries

An infinite network X is an infinite graph consisting of a countably infinite vertices and countably infinite edges, each edge joining a pair of vertices. If the vertices x and y are joined by an edge, we say that x and y are neighbors in X and write $x \sim y$. X is assumed to be connected (that is, any two vertices can be joined by a path), locally finite (that is, any vertex has only a finite number of neighbors) and without self-loops (that is, there is no edge connecting a vertex to itself).

To any pair of vertices x and y is associated a conductance $t(x,y) \geq 0$ such that $t(x,y) > 0$ if and only if x and y are neighbors; $t(x,y)$ need not be the same as $t(y,x)$. If $t(x,y) = t(y,x)$ for every pair of vertices, then we say that X has the symmetric conductance $\{t(x,y)\}$. Write $t(x) = \sum_{y \in X} t(x,y) = \sum_{y \sim x} t(x,y)$ for every x in X and note that $t(x) > 0$. Let $p(x,y) = t(x,y)/t(x)$ for every pair of vertices x and y. Given a network $\{X, t(x,y)\}$, we often work with $\{X, p(x,y)\}$ because of the advantage $\sum_y p(x,y) = 1$ for every x. P symbolizes the set of conductance $\{p(x,y)\}$.

Let E be an arbitrary set in X. A vertex x in E is said to be an interior point of E if all its neighbors are also in E. Denote by $\mathring{E}$ the set of all interior points of E and $\partial E = E \setminus \mathring{E}$. A real-valued function u on E is said to be P-superharmonic on E if, for every $x \in \mathring{E}$, $u(x) \geq \sum_{y \sim x} p(x,y)u(y)$. When there is only one P-structure on X, we leave out the prefix P-from the term P-superharmonic. A function v is said to be subharmonic on E if $-v$ is superharmonic on E and a function h is said to be harmonic on E if h and $-h$ are superharmonic on E. Write $(-\Delta)u(x) = u(x) - \sum_{y \sim x} p(x,y)u(y)$. A nonnegative superharmonic function p on

---

2010 *Mathematics Subject Classification.* Primary 31C20; Secondary 31B30.
Key words and phrases. discrete Laplace, Schrödinger operators; subordinate harmonic structures; parahyperbolic networks.
This is the final form of the paper.

X is called a potential if v is a subharmonic function on X such that $v \leq p$, then $v \leq 0$. A subset F of X is said to be circled if any $y \in \partial F$ has at least one neighbor in $\mathring{F}$. For a subset E, let $V(E)$ denote the union of E and all the neighbors of each vertex of E. Then $V(E)$ is circled. Moreover, if E is connected, then $V(E)$ is connected and circled.

An infinite tree T is an infinite network in which there is no closed path and consequently in T, if a and b are two vertices then there exists a unique path from a to b. In the case of a tree T, suppose $P = \{p(x,y)\}$ is a set of conductance. Fix a vertex e and for any vertex x, if $\{e = x_0, x_1, \ldots, x_n = x\}$ is the unique path from e to x, then set $\varphi(x) = \big(p(x_0,x_1)\cdots p(x_{n-1},x_n)\big)/\big(p(x_n,x_{n-1})\cdots p(x_1,x_0)\big)$ to remark that $\varphi(x)p(x,y) = \varphi(y)p(y,x)$ for any pair of vertices x and y in T. Hence, if we write $t(x,y) = \varphi(x)p(x,y)$, then the infinite tree with conductance $\{t(x,y)\}$ can be considered as an infinite network with symmetric conductance. Potential theory in T associated with a set of conductance $P = \{p(x,y)\}$ is studied in Cartier [4], using the Green kernel. Yamasaki presents in [10, 11] a theory of potentials in the context of an infinite network X with symmetric conductance $\{t(x,y)\}$, using the Dirichlet norm. A part of [1] considers potential theory in an infinite network without symmetric conductance.

The following theorem [1, Theorem 3.1.7] dealing with a generalized form of the Dirichlet problem is of great value in the study of potentials in an infinite network. Some important results like balayage, equilibrium principle, condenser principle, solutions to mixed boundary value problems, etc. can be obtained as solutions to appropriately posed Dirichlet problems.

Theorem 1.1. *Let F be an arbitrary subset of an infinite network X, $E \subset \mathring{F}$, and f a real-valued function defined on $F \backslash E$. Suppose u and v are real-valued functions on F such that $v \leq u$ on F, $v \leq f \leq u$ on $F \backslash E$, $\Delta v(x) \geq 0$ and $\Delta u(x) \leq 0$ at each vertex x in E. Then, there exists a function h on F such that $v \leq h \leq u$ on F, $h = f$ on $F \backslash E$, and $\Delta h(x) = 0$ at each $x \in E$.*

Remark. The solution h in the above theorem is unique in the following cases (where the presence of u and v can be dispensed with).

(i) E is a finite subset.

(ii) f is a bounded function and X is a parabolic network (that is, there is no positive potential on X); and h is bounded also on F.

(iii) There is a positive potential p defined on X (in which case, X is called a hyperbolic network) such that $|f| \leq p$ on $F \setminus E$; and $|h|$ is dominated by a potential in X.

In an infinite network $\{X, t(x,y)\}$, apart from the Laplace operator Δ, another extensively studied operator is the Schrödinger operator given by the equation $\Delta_q u(x) = \sum_{y \sim x} t(x,y)[u(y) - u(x)] - q(x)u(x)$ where $q(x) \geq 0$ for all $x \in X$ and at least for one $x_0 \in X$, $q(x_0) > 0$. Remark that

$$(-\Delta_q)u(x) = [t(x) + q(x)]\left[u(x) - \sum_{y \sim x} \frac{t(x,y)}{t(x) + q(x)} u(y)\right].$$

Let us write $p'(x,y) = t(x,y)/\big(t(x) + q(x)\big)$. Then $p'(x,y) \geq 0$, $p'(x,y) > 0$ if and only if $x \sim y$, $p'(x,y) \leq t(x,y)/t(x) = p(x,y)$ for any pair of vertices x and y and $p'(x_0,y) < p(x_0,y)$ if $y \sim x_0$. To understand the Schrödinger operator theory in relation to the Laplace operator theory in the infinite network, it is convenient to define a real-valued function u on a subset E as a q-superharmonic function if $u(x) \geq \sum_{y \sim x} p'(x,y)u(y)$ for every $x \in \mathring{E}$.

The inspiration for this article comes from Royden's paper [8] on the equation $\Delta u = Pu$, and the classification of open Riemann surfaces.

2. Green kernel and potentials in an infinite network

In an infinite network X, the Green function $G_y(x)$ with pole at $\{y\}$, or the Green kernel $G(x,y)$, and associated potentials in X are variously defined:

(i) If T is an infinite tree with the nearest neighbor probability $p(x,y)$, for a path $\gamma = \{v_0, v_1, \ldots, v_n\}$ in X, set $p(\gamma) = \prod_{j=1}^{n} p(v_{j-1}, v_j)$. For x, y in X, let $\Gamma_{x,y}$ denote the set of all finite paths from x to y. Define $G(x,y) = \sum_{\gamma \in \Gamma_{x,y}} p(\gamma)$. If $G(x,y)$ is finite for some vertices x, y, then it is finite for all pairs in T. When finite, $G(x,y)$ is called the Green kernel of T, and a real-valued function p on X is called a potential if p is of the form $p(x) = \sum_y f(y)G(x,y)$ for some $f \geq 0$, see Cartier [4, p. 237] and Bajunaid et al. [2, p. 709; 3].

(ii) Probabilistically, if $P = \{p(x,y)\}$ is the transition matrix and if $p^n(x,y)$ denotes the entries of the matrix P^n, then $G(x,y) = \sum_{n=0}^{\infty} p^n(x,y)$ represents the expected number of visits to y when the random walk starts at x, see also Soardi [9, p. 17].

(iii) In an infinite network X with symmetric conductance, if $D(x)$ is the Hilbert space of the Dirichlet finite functions, then let $D_0(x)$ be the closure in $D(x)$ of the set of functions with finite support. If $D_0(x) \neq D(x)$, Yamasaki [11, pp. 36–37] constructs the (unique) Green function $G_y(x) \in D_0^+(x)$ such that $(-\Delta)G_y(x) = \delta_y(x)$ and shows that $G_y(x) = G_x(y)$ for any pair of vertices x, y in X. A real-valued function p is called a Green potential if and only if $p(x) = \sum_y f(y)G_y(x)$ for some $f \geq 0$ and [11, Theorem 5.1] shows that a nonnegative superharmonic function s is a potential if and only if the greatest harmonic minorant of s is 0.

(iv) In this article, X is an infinite network in which the conductance may not be symmetric. Here we change the order of appearance of the Green function and potentials. A nonnegative superharmonic function s on X is said to be a potential if for any subharmonic function u such that $u \leq s$ on X, we have $u \leq 0$. Suppose there are positive potentials on X. Let y be a fixed vertex in X. Let $p > 0$ be a potential on X. Let $\Im$ be the family of all positive superharmonic functions s on X such that $s(y) \geq p(y)$. Let $q(x) = \inf_{s \in \Im} s(x)$. Then (as in the proof of Theorem 2.4.11) we show that q is a superharmonic function on X such that $0 \leq q \leq p$ on X, $q(y) = p(y)$ and $\Delta q(z) = 0$ if $z \neq y$. If q were harmonic on X, then p being a potential, q should be 0; that is not correct since $q(y) = p(y) > 0$. Hence $\Delta q(y) < 0$. Define $G_y(x) = q(x)/\big((-\Delta)q(y)\big)$. Then $G_y(x)$ is a positive potential on X such that $(-\Delta)G_y(x) = \delta_y(x)$. These two properties of $G_y(x)$ define it uniquely. For, if $v > 0$ is a potential satisfying $(-\Delta)v(x) = \delta_y(x)$ then $G_y(x) - v(x)$ is a harmonic function h on X. Remark that $h = 0$ by the uniqueness of Riesz decomposition. Write now $G_y(x) = G(x,y)$ for any pair of vertices x, y in X and call $G(x,y)$ the Green kernel of X. Note that $G(x,y)$ and $G(y,x)$ may be different. We remark that [1, Theorem 3.2.6] gives a different method to construct the Green kernel $G(x,y)$ in this case of an infinite network without symmetric conductance.

Moreover the relation between the Green kernel and potentials is given in [1, Theorem 3.31]: A nonnegative superharmonic function p on X is a potential if and only if $p(x) = \sum_y (-\Delta)p(y)G_y(x)$ for all $x \in X$. Remark that $G_y(x) \leq G_y(y) < \infty$ for all x in X (Domination principle [1, Theorem 3.3.6]). Hence for any $f \geq 0$ with finite support A in X, $q(x) = \sum_y f(y)G_y(x)$ is a bounded potential in X such that $\Delta q(x) = 0$ if $x \notin A$. Remark also that $G_y(y) > 1$ for each y in X. For $1 = (-\Delta)G_y(y) = G_y(y) - \sum_z p(y,z)G_y(z) < G_y(y)$.

3. Subordinate structures

Let $\{X, p(x,y)\}$ be an infinite network, $p(x,y) \geq 0, p(x,y) > 0$ if and only if $x \sim y$ and $p(x) = \sum_y p(x,y) = 1$ for every $x \in X$. Suppose there is another set of conductance $\{p'(x,y)\}$ defined on the infinite graph X as given below in Definition 3.1. Then $\{X, p'(x,y)\}$ introduces a new harmonic structure P', called the P'-harmonic structure. In this section, we relate the P-harmonic potential theory with the P'-harmonic potential theory.

Definition 3.1. Let $\{p'(x,y)\}$ be a set of conductance on X such that $p(x,y) \geq p'(x,y) \geq 0$ for any pair of vertices x and y, $p'(x,y) > 0$ if and only if $x \sim y$, and $p'(x,y) < p(x,y)$ for at least one pair x and y. Then we say that $P' = \{p'(x,y)\}$ defines a harmonic structure on X that is subordinate to the harmonic structure defined by $P = \{p(x,y)\}$; or simply that P' is subordinate to P on X.

Remark. As noted earlier, a Schrödinger operator Δ_q defines a subordinate structure on X. But for a subordinate structure $\{X, P'\}$ to define a Schrödinger operator on X, it is necessary that for every $x \in X$, the quantity $p(x,y)/p'(x,y)$ is independent of y, for any $y \sim x$.

A real-valued function u on a subset E of X is said to be P'-superharmonic on E if $u(x) \geq \sum_y p'(x,y)u(y)$ for every x in $\mathring{E}$. If P' is subordinate to P, then for any $u \geq 0$ on a subset E of $X, x \in \mathring{E}$, $\sum p(x,y)u(y) \geq \sum p'(x,y)u(y)$. Hence if $u \geq 0$ is P-superharmonic on E, then u is P'-superharmonic on X. In particular, the constant 1 is P'-superharmonic that is not P'-harmonic on X. Hence if P' is any subordinate structure in (X, P), then (X, P') is a hyperbolic network.

In the sequel, we shall always assume that P' is a subordinate harmonic structure in (X, P). For any real valued function u on X, write

$$(-\Delta)u(x) = u(x) - \sum_y p(x,y)u(y)$$

and

$$(-\Delta')u(x) = u(x) - \sum_y p'(x,y)u(y).$$

Proposition 3.2. *If q is a P-potential, then it is a P'-potential.*

PROOF. First note that q is P'-superharmonic in X. Let $u \geq 0$ be a P'-subharmonic function such that $u \leq q$ on X. Note that u is P-subharmonic on X and hence $u = 0$ since q is a P-potential. Consequently q is a P'-potential on X. □

Lemma 3.3. *Let $s \geq 0$ be a P'-superharmonic function and p be a P'-potential on X. If $(-\Delta')s \geq (-\Delta')p$, then $s \geq p$ on X.*

PROOF. By hypothesis, $s = p + u$ where u is P'-superharmonic on X. Since $s \geq 0$, $-u \leq p$ on X. Hence $-u \leq 0$ so that $s \geq p$ on X. □

Since (X, P') is a hyperbolic network, for any vertex $e \in X$, there exists a P'-potential $G'_e(x)$ such that $(-\Delta')G'_e(x) = \delta_e(x)$.

Proposition 3.4. *Let (X, P) be a hyperbolic network. Let $G_e(x)$ be the P-harmonic Green function on X with harmonic support $\{e\}$. Then $G'_e(x) \leq G_e(x)$ for every $x \in X$.*

PROOF. $G_e(x)$ is a P'-superharmonic function on X. If $x \neq e$, $(-\Delta')G_e(x) \geq 0$ and $(-\Delta')G'_e(x) = 0$. When $x = e$,

$$\begin{aligned}(-\Delta')G_e(e) &= G_e(e) - \sum_y p'(e,y)G_e(y) \\ &\geq G_e(e) - \sum_y p(e,y)G_e(y) \\ &= (-\Delta)G_e(e) \\ &= 1 \\ &= (-\Delta')G'_e(e).\end{aligned}$$

Hence for all $x \in X, (-\Delta')G_e(x) \geq (-\Delta')G'_e(x)$. Then, by Lemma 3.3, $G_e(x) \geq G'_e(x)$ for any $x \in X$. □

Lemma 3.5. *There exists a nonnegative P'-harmonic function h on X such that, for any $x \in X$, $1 = \sum_y [1 - p'(y)]G'_y(x) + h(x)$ where $p'(y) = \sum_z p'(y,z)$.*

PROOF. First note that for any positive P'-superharmonic function s on X, $s(x) = \sum_y (-\Delta')s(y)G'_y(x) + v(x)$ where $v(x)$ is a nonnegative P'-harmonic function on X. For, if $s(x)$ is a P'-harmonic function X, there is nothing to prove; otherwise, if $u(x) = s(x) - \sum_{y \in E}(-\Delta')s(y)G'_y(x)$ where E is a finite set in X, then $(-\Delta')u(x) = 0$ if $x \in E$ and $(-\Delta')u(x) \geq 0$ if $x \notin E$. Hence u is P'-superharmonic on X such that $-u$ is majorized by the P'-potential $\sum_{y \in E}(-\Delta')s(y)G'_y(x)$. Hence $-u \leq 0$ which implies that $\sum_{y \in E}(-\Delta')s(y)G'_y(x) \leq s(x)$ for all $x \in X$. Consequently, when E is enlarged to cover X, we obtain $\sum_{y \in X}(-\Delta')s(y)G'_y(x) \leq s(x)$ for $x \in X$. Write $v(x) = s(x) - \sum_{y \in X}(-\Delta')s(y)G'_y(x)$. Since $(-\Delta')v(x) = 0$ for any $x \in X$, v is a nonnegative P'-harmonic function on X.

Now, take $s \equiv 1$. Then $(-\Delta')s(x) = s(x) - \sum_y p'(x,y)s(y) = 1 - p'(x)$. Consequently, $1 = \sum_y [1 - p'(y)]G'_y(x) + h(x)$ for any $x \in X$, where $h \geq 0$ is a P'-harmonic function on X. □

Since 1 is P'-superharmonic that is not P'-harmonic on X, 1 is a P'-potential on X or there exists a bounded P'-harmonic function on X. In the former case, though X has P'-potentials, the P'-superharmonic functions have many properties reminiscent of the properties of superharmonic functions in a parabolic Riemann surface. (That is, a hyperbolic network behaving like a parabolic network.) Since this parabolic–hyperbolic structure on the network X has some distinguishing features, we introduce the following definition.

Definition 3.6. (X, P') is said to be a parahyperbolic network, if 1 is a P'-potential on X.

Theorem 3.7. *(X, P') is a parahyperbolic if and only if $\sum_y [1-p'(y)]G'_y(x) = 1$ for some (and hence any) x in X.*

Proof. Since $1 = \sum_y [1 - p'(y)]G'_y(x) + h(x)$, if $h(x) = 0$ at some vertex, then $h(x) = 0$ for all x in X. If $1 = \sum_y [1 - p'(y)]G'_y(x)$, then 1 is a P'-potential since $\sum_y [1 - p'(y)]G'_y(x)$ is a P'-potential. Conversely, if 1 is a P'-potential, then by the uniqueness of Riesz representation $1 = \sum_y [1 - p'(y)]G'_y(x)$. □

Corollary 3.8. *There exists a bounded positive P'-harmonic function on X if and only if $\sum_y [1 - p'(y)]G'_y(x) < 1$ for some (and hence any) x in X.*

Proposition 3.9. *Suppose for some $\delta > 0$, $p'(x) \le 1 - \delta$ for any x in X. Then for any bounded function f on X, the Poisson equation $(-\Delta')u = f$ has a solution.*

Proof. From Lemma 3.5, $1 \ge \sum_y [1 - p'(y)]G'_y(x) \ge \delta \sum_y G'_y(x)$. Hence if f is bounded, then $u_1(x) = \sum_y f^+(y)G'_y(x)$ and $u_2(x) = \sum_y f^-(y)G'_y(x)$ are P'-potentials on X; if $u = u_1 - u_2$, then $(-\Delta')u(x) = f(x)$. □

Lemma 3.10. *Let u be a P'-superharmonic function defined outside a finite set in X. Then $u = p - q$ outside a finite set where p is a P'-superharmonic function on X and q is a P'-potential on X with finite P'-harmonic support and hence bounded.*

Proof. Suppose $(-\Delta')u(x) \ge 0$ if $x \in X \setminus \mathring{A}$ where A is a finite set. Let v be the Dirichlet solution in A with boundary values u and then extended by u outside A.

Let $p_1(x) = v(x) - \sum_{y \in \partial A}(-\Delta')v(y)G'_y(x)$ for x in X. Then $(-\Delta')p_1(x) \ge 0$ if $x \notin A$ and $(-\Delta')p_1(x) = 0$ if $x \in A$. Hence p_1 is P'-superharmonic on X and $v(x) = p(x) - q(x)$, where $q(x) = \sum_{y \in \partial A}[(-\Delta')v(y)]^- G'_y(x)$ is a P'-potential with finite P'-harmonic support and hence bounded in X and $p(x) = p_1(x) + \sum_{y \in \partial A}[(-\Delta')v(y)]^+ G'_y(x)$. Since $v = u$ in $X \setminus A$, the lemma is proved. □

Theorem 3.11. *The following are equivalent:*

(i) *Any bounded P'-superharmonic function u defined outside a finite set is of the form $u = p - q$ where p and q are bounded P'-potentials in X.*
(ii) *Any bounded P'-superharmonic function in X is a P'-potential.*
(iii) *0 is the only bounded P'-harmonic function in X.*
(iv) *(X, P') is parahyperbolic.*

Proof. (i) ⇒ (ii) Let v be a bounded P'-superharmonic function in X. Then by (i), $v = p - q$ outside a finite set A. Hence $|v| \le p + q$ on $X \setminus A$. Since A is a finite set, $|v| \le M$ on A, and we choose a large constant $\alpha > 1$ such that $M \le \alpha(p+q)$ on A. Consequently, $|v| \le \alpha(p+q)$ on X. Since $-v \le \alpha(p+q)$, we conclude $-v \le 0$; then $0 \le v \le \alpha(p+q)$ so that v is a P'-potential on X.

(ii) ⇒ (iii) If h is a bounded P'-harmonic function on X, then by (ii), it is a P'-potential. Hence $h = 0$.

(iii) ⇒ (iv) Since 1 is P'-superharmonic on X, by (iii) the greatest P'-harmonic minorant of 1 is 0. Hence 1 is a P'-potential, that is (X, P') is parahyperbolic.

(iv) ⇒ (i) As in Lemma 3.10, write $u = p - q$ outside a finite set in X. Since u is bounded by hypothesis and q is bounded, we find that p is bounded on X. Since 1 is a P'-potential by (iv) the bounded P'-superharmonic function p is a P'-potential. □

Corollary 3.12. *If (X, P) is parabolic, then (X, P') is parahyperbolic.*

PROOF. Let h be a bounded P'-harmonic function on X. Then $|h|$ is P'-subharmonic on X and hence bounded P-subharmonic on X. Since there are no positive P-potentials on X, $|h|$ should be a constant α. If $\alpha \neq 0$, in the equation $|h| = \alpha$, the left side is P'-subharmonic and the right side is a P'-superharmonic function that is not P'-harmonic, a contradiction. Hence $\alpha = 0$, that is $h = 0$. Thus 0 is the only bounded P'-harmonic function in X. Hence X is parahyperbolic by the above theorem. □

Proposition 3.13. *If (X, P') is parahyperbolic, then any lower bounded P'-superharmonic function is nonnegative. Conversely, if any lower bounded P'-harmonic function is nonnegative, then (X, P') is parahyperbolic.*

PROOF. Let (X, P') be parahyperbolic. Suppose s is a P'-superharmonic function on X such that $s \geq -M$ for some $M > 0$. Since M is a P'-potential by assumption, $-s \leq M$ implies that $s \geq 0$.

Conversely, suppose any lower bounded P'-harmonic function on X is nonnegative. If (X, P') is not parahyperbolic, then there exists a P'-harmonic function h on $X, 0 < h < 1$. Since $-h$ is lower bounded, $-h \geq 0$ a contradiction. □

Corollary 3.14. *Suppose h is a P'-harmonic function bounded on one side in X. If h takes both positive and negative values in X, then there exists a bounded P'-harmonic function H, $0 < H < 1$, on X.*

PROOF. The existence of such a function h implies that X is not parahyperbolic, hence the existence of a bounded positive harmonic function in X. □

Theorem 3.15 (Maximum principle). *The following are equivalent:*

(i) *(X, P') is parahyperbolic.*
(ii) *In an arbitrary subset F of X, if u is an upper bounded subharmonic function such that $u \leq 0$ on ∂F, then $u \leq 0$ on F.*

PROOF. (i) $\Rightarrow$ (ii). Let v be the function on X, equal to u^+ on F extended by 0 outside F. Then v is P'-subharmonic on X and for some $M > 0, v \leq M$ on X by hypothesis. Since M is a P'-potential on X by hypothesis (i), then $v \leq 0$ on X. In particular, $u \leq 0$ on F.

(ii) $\Rightarrow$ (i). By Theorem 3.11, it is enough to show that 0 is the only bounded P'-harmonic function on X. Let h be a bounded harmonic function on X. Let F be an infinite subset of X such that $A = X \setminus \mathring{F}$ is finite. Let $\max_{x\in A}|h(x)| = M$. Then for some x_0 in A, $|h(x_0)| = M$ and $u(x) = |h(x)| - M$ is a bounded subharmonic function on X such that $u \leq 0$ on ∂F. Hence by using the assumption (ii), $u \leq 0$ on X and consequently the P'-subharmonic function $|h(x)|$ attains its nonnegative maximum value M at $x = x_0$. Hence $|h(x)| = M$ for all $x \in X$. This is possible only if $M = 0$ (for if $M > 0$, the right side is a P'-superharmonic function that is not P'-harmonic while the left side is a P'-subharmonic function). □

***P'*-harmonic measure of the vertex at infinity.** Let A be a nonempty finite set in X. Let $\{E_n\}$ be an exhaustion of X by finite subsets E_n such that $A \subset \mathring{E}_1$, $E_n \subset \mathring{E}_{n+1}$ for $n \geq 1$ and $X = \bigcup_n E_n$. Let h_n be the function on X (Condenser Principle [1, Theorem 4.1.10]) such that $h_n = 0$ on A, $h_n = 1$ on $X \setminus E_n$ and $\Delta' h_n = 0$ at each vertex in $E_n \setminus A$. Then h_n is a decreasing sequence in X.

Let $h = \lim_{n\to\infty} h_n$. Then $0 \le h \le 1$ on X, $\Delta' h = 0$ at each vertex in $X \setminus A$. There are two possibilities now, either $h = 0$ on X or the P'-subharmonic function h on X is majorized by the P'-superharmonic function 1 on X so that there is a P'-harmonic function u on X such that $0 < u < 1$. In the first case, we say that the P'-harmonic measure of the vertex at infinity is zero and in the second case, the P'-harmonic measure at infinity is nonzero. It can be seen that this P'-harmonic measure at infinity is 0 if and only if X is parahyperbolic.

Let us adopt the above procedure in the context of the original network (X, P) and take A as the set consisting of a single vertex $\{a\}$. Let $u_n = 1 - h_n$ and $u = 1 - h$. The transmission probability of the system $S_n = \{X, p(x, y), a, X \setminus E_n\}$ is given by $\tau(S_n) = (-\Delta)u_n(a)$. Now the well-known Nash-Williams theorem [7] shows that (X, P) is recurrent if and only if $\lim_{n\to\infty} \tau(S_n) = 0$, see T. Lyons [5] and McGuinness [6, p. 90]. That is (X, P) is recurrent if and only if the P-harmonic measure of the vertex at infinity is 0.

In the context of an electrical network (X, P), $a \in X$, the effective resistance of the network between a and the vertex at infinity is proportional to $1/\big((-\Delta)u(a)\big) = \lim_{n\to\infty} 1/\big((-\Delta)u_n(a)\big)$. This effective resistance is finite if and only if (X, P) is transient (see Soardi [9, p. 46]). That is, the effective resistance between a and ∞ is finite if and only if the P-harmonic measure of the point at infinity is nonzero.

4. Relation between nonnegative P'- and P-harmonic functions

To what extent does the existence of bounded or nonnegative P-harmonic functions in X influence the existence of similar P'-harmonic functions in X? We know that [1, Theorem 3.2.10] in a hyperbolic network (X, P), if e is a fixed vertex, if $H_0^+(e)$ is the set of nonnegative functions h on X such that $h(e) = 0$ and $\Delta h(x) = 0$ if $x \ne e$, and if H^+ is the set of nonnegative harmonic functions on X, then there is a bijection $\lambda \colon H_0^+(e) \to H^+$ such that if $h_1, h_2 \in H_0^+(e)$ and c_1, c_2 are nonnegative constants, then $\lambda(c_1 h_1 + c_2 h_2) = c_1 \lambda(h_1) + c_2 \lambda(h_2)$; $h \in H_0^+(e)$ is bounded if and only if $\lambda(h)$ is bounded.

Since X is always P'-hyperbolic, there exists at least one positive P'-harmonic function on X. For, if $\{E_n\}$ is an exhaustion of X by circled connected sets, such that $e \in \mathring{E}_1$, $E_n \subset \mathring{E}_{n+1}$, $X = \bigcup E_n$, and every $x \in \partial E_n$ has at least one neighbor in $\mathring{E}_n$ (that is, E_n is circled), let s_n be the P'-Dirichlet solution in E_n with boundary value 1 and extended by 1 outside E_n. Let $h_n(x) = s_n(x)/s_n(e)$. Then using the Harnack property [1, p. 47], we can extract a convergent subsequence $\{u_n\}$ of $\{h_n\}$ such that $u(x) = \lim u_n(x)$ is a positive P'-harmonic function on X with $u(e) = 1$.

Corollary 3.8 gives a necessary and sufficient condition for the existence of bounded positive P'-harmonic functions on X. If the network (X, P) is parabolic, then Corollary 3.12 implies that 0 is the only bounded P'-harmonic function on X. When (X, P) is hyperbolic, by using Theorem 3.7, we can prove the following proposition on the existence of bounded P'-harmonic function on X. This however needs a restriction that $\sup_{x\in X} G_x(x)$ is bounded. This condition is satisfied for example in a standard homogenous tree where each vertex has exactly $q \ge 3$ neighbors with $p(x, y) = 1/q$ if $x \sim y$, (Cartier [4, pp. 262–264]).

Proposition 4.1. *Let (X, P) be hyperbolic with its Green potential $G_y(x)$ satisfying the condition $\sup_{x\in X} G_x(x) \le M$. If $\sum_x [1 - p'(x)] < \infty$, then X has bounded positive P'-harmonic functions on X.*

PROOF. If 0 is the only bounded positive P'-harmonic functions on X, the constant 1 is a P'-potential in X and

$$\begin{aligned} 1 &= \sum_y [1 - p'(y)] G'_y(x) && \text{for } x \in X \\ &\leq \sum_y [1 - p'(y)] G_y(x) && \text{(Proposition 3.4)} \\ &\leq \sum_y [1 - p'(y)] G_y(y) \\ &\leq M \sum_y [1 - p'(y)] \\ &< \infty. \end{aligned}$$

Hence $u(x) = \sum_y [1 - p'(y)] G_y(x)$ should be a P-potential. But this is not possible since $u(x)$ majorizes the P-harmonic function 1. □

Theorem 4.2. *Let B (respectively B') be the set of all bounded nonnegative P- (respectively P'-)harmonic functions in X. Then there is an injective map $S\colon B' \to B$ such that $S(\alpha_1 h_1 + \alpha_2 h_2) = \alpha_1 S(h_1) + \alpha_2 S(h_2)$ where α_1, α_2 are nonnegative constants and h_1, h_2 are in B'.*

PROOF. Let $h \in B'$. Then h is a bounded P-subharmonic function. Let $S(h)$ be the least P-harmonic majorant of h. Then $S(\alpha_1 h_1 + \alpha_2 h_2) = \alpha_1 S(h_1) + \alpha_2 S(h_2)$. Suppose $S(h_1) = S(h_2)$. Note that for $h \in B'$, $S(h) - h$ is a P-potential and hence a P'-potential. Consequently, if $S(h_1) = S(h_2)$, then $|h_1 - h_2| = |[S(h_1) - h_1] - [S(h_2) - h_2]| \leq p_1 + p_2$ where p_1 and p_2 are P'-potential on X. Since $|h_1 - h_2|$ is P'-subharmonic on X, $h_1 = h_2$. □

Corollary 4.3. *If there are nonproportional bounded positive P'-harmonic functions in X, then there is at at least one nonconstant bounded P-harmonic function in X.*

PROOF. If h_1 and h_2 are nonproportional in B', then $S(h_1)$ and $S(h_2)$ are nonproportional bounded P-harmonic functions in X. Hence at least one of them is nonconstant. □

Lemma 4.4. *Let h be a P'-harmonic function in X, such that $|h| \leq s$ where s is P'-superharmonic on X. Then $h = h_1 - h_2$ where h_1 and h_2 are nonnegative P'-harmonic functions such that $h_1 - h^+$ and $h_2 - h^-$ are P'-potentials. This decomposition is unique.*

PROOF. Let h_1 be the least P'-harmonic majorant of h^+ and h_2 be the least P'-harmonic majorant of h^-. Then $p_1 = h_1 - h^+$ and $p_2 = h_2 - h^-$ are P'-potentials on X. Hence $h = h^+ - h^- = (h_1 - h_2) - (p_1 - p_2)$. Then by the uniqueness of Riesz decomposition, $h = h_1 - h_2$ on X.

Suppose $h = u_1 - u_2$ is another such decomposition. Since $u_1 - h^+$ and $h_1 - h^+$ are P'-potentials, as before we conclude that $u_1 = h_1$ and then $u_2 = h_2$. □

Proposition 4.5. *If there exists a bounded P'-harmonic function on X that takes both positive and negative values, then there is at least one bounded nonconstant P-harmonic function X.*

PROOF. Let h be a bounded P'-harmonic function, write $h = h_1 - h_2$ as in Lemma 4.4; since h takes both positive and negative values by the assumption, h_1 and h_2 are positive. Suppose $h_1 = \lambda h_2$. Then $h = (\lambda - 1)h_2$, contradicting the assumption that h takes both positive and negative values on X. Since h_1, h_2 are nonproportional, by Corollary 4.3, there is at least one nonconstant bounded P-harmonic function on X. □

Operator T. Suppose $u \in B$. Then u is a nonnegative P'-superharmonic function. Let Tu be the greatest P'-harmonic minorant of u. Then $u - Tu$ is a P'-potential. Thus $T\colon B \to B'$ is a map such that $T(\alpha_1 u_1 + \alpha_2 u_2) = \alpha_1 T(u_1) + \alpha_2 T(u_2)$ where α_1, α_2 are nonnegative constants. We remark that $T(Sh) = h$ for $h \in B'$. For $Sh \in B$. Hence $Sh - T(Sh)$ is a P'-potential. Now $Sh - h$ is a P-potential and hence a P'-potential. Consequently, $T(Sh) - h = p_1 - p_2$ where p_1 and p_2 are P'-potentials. Since the P'-subharmonic function $|T(Sh) - h|$ is majorized by the P'-potential $p_1 + p_2$, we conclude $T(Sh) = h$.

Suppose X is P'-parahyperbolic, that is 1 is a P'-potential on X. Then $T1 = 0$. Hence, we do not always have $ST = I$. The following proposition gives a sufficient condition for ST to be the identity.

Proposition 4.6. *Suppose every P'-potential on X is dominated by a P-potential on X. Then $ST\varphi = \varphi$ for every $\varphi \in B$.*

PROOF. φ is bounded P'-superharmonic and $T\varphi$ is its greatest P'-harmonic minorant in X. Hence $\varphi - T\varphi = p'$ is a P'-potential. $T\varphi$ is a bounded P-subharmonic function and $ST\varphi$ is the least P-harmonic majorant of $T\varphi$. Hence $ST\varphi - T\varphi = p$ is a P-potential. Note that φ is P-harmonic and $\varphi \geq T\varphi$ on X, so that $\varphi \geq ST\varphi$. Hence $0 \leq \varphi - ST\varphi = p' - p \leq p'$. By the assumption p' is dominated by a P-potential p_0. Thus the nonnegative P-harmonic function $\varphi - ST\varphi$ is dominated by the P-potential p_0, so that $\varphi - ST\varphi = 0$. □

5. Biharmonic functions in a network

In [1, Chapter 5] some properties of polyharmonic functions in an infinite tree are given. In this section, first we review some of these results in the context of an infinite network (X, P). Later, we relate the theory of P-biharmonic functions in X with the theory of P'-biharmonic functions.

(i) A real-valued function u on E, a subset in (X, P), is said to be *P-biharmonic* on E if there exists a P-harmonic function h on E such that $\Delta u(x) = h(x)$ for every $x \in \mathring{E}$.

(ii) A P-potential q in (X, P) is said to be a *P-bipotential* if and only if $(-\Delta)q = p$ where p is a P-potential in X. We say that X is a P-bipotential network if there exists a positive P-bipotential on X.

(iii) For a fixed vertex z in X, a P-potential $Q_z(x)$ in (X, P) is said to be the *P-biharmonic Green function* with P-biharmonic support $\{z\}$ if and only if $(-\Delta)Q_z(x) = G_z(x)$, where $G_z(x)$ is the P-harmonic Green function with P-harmonic support z.

Proposition 5.1. *The P-biharmonic Green function exists on X if and only if there is a positive P-bipotential on X.*

PROOF. Clearly the P-biharmonic Green function is a P-bipotential. Conversely, let q be a positive P-bipotential, $(-\Delta)q = p$. Then $q(x) = \sum_y G(x, y)p(y)$.

Let z be a fixed vertex. Then for some $\lambda > 0$, $G_z(y) \leq \lambda p(y)$ for any $y \in X$ (Domination Principle [1, Theorem 3.3.6]). Hence $Q_z(x) = \sum_y G(x,y)G_z(y)$ is a well-defined P-potential such that $(-\Delta)Q_z(x) = G_z(x)$. $\square$

Theorem 5.2. *Let X be a P-bipotential network, $Q_y(x)$ being the P-biharmonic Green potential. If $\sum_y f(y)Q_y(x)$ is finite at some vertex x_0, for some $f \geq 0$ then $q(x) = \sum_y f(y)Q_y(x)$ is a bipotential on X. Conversely any P-bipotential $q(x)$ can be represented as $q(x) = \sum_y f(y)Q_y(x)$, where $f(x) = (-\Delta)^2 q(x)$.*

PROOF. Let $(-\Delta)q = p$ on X. For a finite set E in X, write $s(x) = q(x) - \sum_{y\in E}(-\Delta)p(y)Q_y(x)$. Then

$$(-\Delta)s(x) = p(x) - \sum_{y\in E}(-\Delta)p(y)G_y(x) = \sum_{y\notin E}(-\Delta)p(y)G_y(x) \geq 0.$$

Hence s is P-superharmonic on X and since $-s(x) \leq \sum_{y\in E}(-\Delta)p(y)Q_y(x)$ we conclude $-s \leq 0$. Hence $q(x) \geq \sum_{y\in E}(-\Delta)p(y)Q_y(x)$. Allow E to grow into X, to conclude $q(x) \geq \sum_{y\in X}(-\Delta)p(y)Q_y(x)$. Write

$$t(x) = q(x) - \sum_{y\in X}(-\Delta)p(y)Q_y(x).$$

Then

$$(-\Delta)t(x) = p(x) - \sum_{y\in X}(-\Delta)p(y)G_y(x) = 0.$$

Hence $t(x)$ is a nonnegative P-harmonic function, majorized by the P-potential $q(x)$. Hence $t = 0$, so that $q(x) = \sum_{y\in X}(-\Delta)p(y)Q_y(x)$, for any $x \in X$.

Conversely, suppose $q(x) = \sum_y f(y)Q_y(x)$ which is a convergent sum of P-potentials, if $q(x_0)$ is finite at some vertex x_0. Then $q(x)$ is a P-potential in X and

$$(-\Delta)q(x) = \sum_y f(y)(-\Delta)Q_y(x) = \sum_y f(y)G_y(x)$$

being finite at each x, defines a P-potential $p(x)$. Thus $(-\Delta)q(x) = p(x)$. $\square$

Proposition 5.3. *Let X be a P-bipotential network. For $z \in X$, if $Q_z(x)$ and $G_z(x)$ are the P-biharmonic and P-harmonic Green potentials, then $Q_z(x) > G_z(x)$ for any $x \in X$.*

PROOF. Since $(-\Delta)Q_z(x) = G_z(x)$, $(-\Delta)G_z(x) = \delta_z(x)$ and since $G_z(z) \geq G_z(x)$ for all x (Domination Principle), we have

$$(-\Delta)\left[\frac{Q_z(x)}{G_z(z)}\right] = \frac{G_z(x)}{G_z(z)} \geq \delta_z(x) = (-\Delta)G_z(x).$$

Hence $u(x) = Q_z(x)/G_z(z) - G_z(x)$ is a P-superharmonic function such that $-u(x) \leq G_z(x)$; hence $-u \leq 0$ on X. Consequently, $Q_z(x) \geq G_z(z)G_z(x) > G_z(x)$, since $G_z(z) > 1$ (see Section 2). $\square$

Proposition 5.4. *Let q be a P-bipotential in X, $(-\Delta)q = p$. Suppose $0 \leq f \leq p$. Then there exists a P-potential v, $v \leq q$, such that $(-\Delta)v = f$ on X.*

PROOF.

$$q(x) = \sum_y G(x,y)p(y) \geq \sum_y G(x,y)f(y) = v(x).$$

Then $v(x)$ is a P-potential, $v \leq q$ and $(-\Delta)v(x) = f(x)$ for all $x \in X$. □

Corollary 5.5. *Let X be a P-bipotential network. If u is a P-potential with finite P-harmonic support in X, then there is a P-bipotential v on X such that $(-\Delta)v = u$ on X.*

PROOF. By hypothesis, there are positive P-potentials p and q such that $(-\Delta)q = p$ on X. Since u has finite P-harmonic support, $u \leq \lambda p$ on X for some $\lambda > 0$ (Domination Principle [1, Theorem 3.3.6].) Hence by Proposition 5.4, there is a P-potential $v \leq \lambda q$ such that $(-\Delta)v = u$ on X. □

Lemma 5.6. *Let F be a finite subset in any infinite network (X, P). Let $E \subset \mathring{F}$ and $f \geq 0$ be a real-valued function on E. Then there exists a P-potential u on F such that $(-\Delta)u(x) = f(x)$ for any $x \in E$.*

PROOF. Assume f is defined on F by giving values 0 to f in $F \setminus E$. Let $G^F_y(x)$ be the P-Green function in F [1, Theorem 2.2.12] with point P-harmonic support $y \in \mathring{F}$ such that $G^F_y(x) = 0$ if $x \in \partial F$. Let $u(x) = \sum_{y \in F} f(y) G^F_y(x)$. Then u is a P-potential in F such that $(-\Delta)u(x) = f(x)$ if $x \in E$. □

Theorem 5.7 (Riquier problem). *Let F be a finite subset of an arbitrary infinite network (X, P), and $E \subset \mathring{F}$. Let f and g be two real-valued functions on $F \backslash E$. Then there exists a unique pair of functions (b, h) on F such that if $z \in F \backslash E$, then $b(z) = f(z)$ and $h(z) = g(z)$; and if $x \in E$, then $\Delta b(x) = h(x)$ and $\Delta h(x) = 0$.*

PROOF. There exists (Theorem 1.1) a bounded function h on F such that $h = g$ on $F \backslash E$ and $\Delta h(x) = 0$ if $x \in E$. Then, by using the above Lemma 5.6, we can choose a function u on F such that $\Delta u(x) = h(x)$ if $x \in E$. Again by Theorem 1.1 there exists a function v on F such that $v = f - u$ on $F \backslash E$ and $\Delta v(x) = 0$ if $x \in E$. Take now $b = u + v$ on F. Then $b(z) = f(z)$ if $z \in F \backslash E$ and $\Delta b(x) = h(x)$ if $x \in E$.

As for the uniqueness, suppose (b_1, h_1) is another such pair in F. Now by the uniqueness of the Dirichlet solution in a finite set, we remark immediately that $h = h_1$ on F. Let then $\varphi = b - b_1$ on F. Remark that $\Delta\varphi = 0$ on E and $\varphi(z) = 0$ if $z \in F \backslash E$. Consequently, $\varphi = 0$ on X [1, Corollary 2.2.2]. □

Proposition 5.8. *If X is P'-parahyperbolic, then 0 is the only bounded P'-biharmonic function on X.*

PROOF. If u is a bounded function on X, $|u| \leq M$, then $|(-\Delta')u(x)| = |\sum_y p'(x,y)[u(x) - u(y)]| \leq 2M \sum_y p'(x,y) \leq 2M$. Hence if b is a bounded P'-biharmonic function on X, then $(-\Delta')b = h$ is a bounded P'-harmonic function. Since X is P'-parahyperbolic, $h = 0$ and hence b is bounded P'-harmonic on X, so that $b = 0$. □

Proposition 5.9. *If X is a P-bipotential network, then X is a P'-bipotential network. If $Q_y(x)$ and $Q^{'}_y(x)$ are the P- and P'-biharmonic Green functions, then $Q_y(x) \geq Q^{'}_y(x)$.*

PROOF. Let $(-\Delta)q = p$, where p and q are P-potentials in X. Then

$$\begin{aligned} q(x) &= \sum_y G_y(x)p(y) \\ &\geq \sum_y G_y{}'(x)p(y) \qquad \text{(Proposition 3.4)}. \end{aligned}$$

Hence $v(x) = \sum_y G_y{}'(x)p(y)$ is a P'-potential such that $(-\Delta')v(x) = p(x)$. Since $p(x)$ is also a P'-potential, we conclude that (X, P') is a P'-bipotential network.

Let $Q_y(x)$ and $Q_y^{'}(x)$ be the P- and P'-biharmonic Green functions in X with support $\{y\}$. Then

$$\begin{aligned} (-\Delta')Q_y(x) &= Q_y(x) - \sum_z p'(x,z)Q_y(z) \\ &\geq Q_y(x) - \sum_z p(x,z)Q_y(z) \\ &= (-\Delta)Q_y(x) \\ &= G_y(x) \\ &\geq G_y^{'}(x) \\ &= (-\Delta')Q_y^{'}(x). \end{aligned}$$

Hence for a P'-superharmonic function s on X, $Q_y(x) - Q_y^{'}(x) = s(x)$. Since $-s(x) \leq Q_y^{'}(x)$, we conclude $-s(x) \leq 0$ and hence $Q_y(x) \geq Q_y^{'}(x)$. □

Example. Let P' be a subordinate structure in (X, P) such that for some $\delta > 0$, $1 - p'(x) \geq \delta$ for every $x \in X$. Then

(1) X is a P'-bipotential network: if p is any bounded P'-potential (in particular if p has finite P'-harmonic support), then there is a bounded P'-bipotential q such that $(-\Delta')q = p$ on X. For, if $G'_y(x)$ is the P'-harmonic Green potential in X, then (Lemma 3.5)

$$\delta \sum_{y \in X} G'_y(x) \leq \sum_{y\in X} [1 - p'(y)]G'_y(x) \leq 1.$$

Hence if p is a bounded P'-potential, $p \leq M$, then

$$q(x) = \sum_{y \in X} p(y)G'_y(x) \leq \frac{M}{\delta}$$

so that q is a P'-potential such that $(-\Delta')q = p$ on X. Hence, q is a bounded P'-bipotential on X.

(2) If $z \in X$, then the P'-biharmonic Green potential $Q'_z(x)$ is a bounded function of x. In fact $Q'_z(x) \leq \frac{1}{\delta}G'_z(z)$. For,

$$Q'_z(x) = \sum_{y\in X} G'_z(y)G'_y(x) \leq G'_z(z) \sum_{y \in X} G'_y(x) \leq \frac{1}{\delta}G'_z(z).$$

(3) If there is a bounded P'-harmonic function h in X, $0 < h < 1$, then there is a bounded positive P'-biharmonic function b in X, that is not P'-harmonic. In fact, $b(x) = \sum_{y \in X} h(y)G'_y(x) \leq 1/\delta$.

(4) If $f \geq 0$ is a bounded function on X, then there exists a bounded P'-potential φ on X such that $(-\Delta')\varphi = f$.

Acknowledgement. I thank the referee for the valuable comments.

References

1. V. Anandam, *Harmonic functions and potentials on finite or infinite networks*, Lect. Notes Unione Mat. Ital., vol. 12, Springer, Heidelberg, 2011.
2. I. Bajunaid, J. M. Cohen, F. Colonna, and D. Singman, *Trees as Brelot spaces*, Adv. in Appl. Math. **30** (2003), no. 4, 706–745.
3. ———, *Corrigendum to "Trees as Brelot spaces" [Adv. in Appl. Math. 30 (4) (2003) 706–745]*, Adv. in Appl. Math. **47** (2011), no. 2, 401–402.
4. P. Cartier, *Fonctions harmoniques sur un arbre*, Convegno di Calcolo delle Probabilità (Rome, 1971), Symposia Mathematica, Vol. IX, Academic Press, London, 1972, pp. 203–270.
5. T. Lyons, *A simple criterion for transience of a reversible Markov chain*, Ann. Probab. **11** (1983), no. 2, 393–402.
6. S. McGuinness, *Recurrent networks and a theorem of Nash-Williams*, J. Theoret. Probab. **4** (1991), no. 1, 87–100.
7. C. St. J. A. Nash-Williams, *Random walk and electric currents in networks*, Proc. Cambridge Philos. Soc. **55** (1959), 181–194.
8. H. L. Royden, *The equation $\Delta u = Pu$, and the classification of open Riemann sufaces*, Ann. Acad. Sci. Fenn. Ser. A I **271** (1959), 27.
9. P. M. Soardi, *Potential theory on infinite networks*, Lecture Notes in Math., vol. 1590, Springer, Berlin, 1994.
10. M. Yamasaki, *Parabolic and hyperbolic infinite networks*, Hiroshima Math. J. **7** (1977), no. 1, 135–146.
11. ———, *Discrete potentials on an infinite network*, Mem. Fac. Sci. Shimane Univ. **13** (1979), 31–44.

The Institute of Mathematical Sciences, Taramani, Chennai 600 113, India
E-mail address: vanandam@hotmail.com

Centre de Recherches Mathématiques
CRM Proceedings and Lecture Notes
Volume **55**, 2012

The Generalized Binomial Theorem

Murali Rao

ABSTRACT. In this note, we give a generalization of the classical binomial theorem and then discuss its application in probability theory.

1. Introduction

For $\lambda \geq 0$ and non-negative integer n, let

$$p_\lambda(n) = \frac{\lambda^n}{n!}\,\mathrm{e}^{-\lambda}.$$

Then $p_\lambda(n) \geq 0$ and $\sum_{n=0}^{\infty} p_\lambda(n) = 1$. In other words, the numbers $p_\lambda(n)$ define a probability distribution on the set of non negative integers. This is the well-known *Poisson distribution.* The Gaussian and the Poisson distributions are among the most important distributions in Probability Theory.

Consul and Jain studied a generalization of the Poisson Distribution given by

$$a(n) = \frac{\theta(\theta + n\lambda)^{n-1}}{n!}\,\mathrm{e}^{-\theta-n\lambda}, \tag{1.1}$$

where $\lambda \geq 0$, $\theta \geq 0$ and n is an integer [2]. This gives a two parameter family of distributions on the set of non negative integers. When $\lambda = 0$ this reduces to the Poisson distribution. That is why they called this family the *Generalized Poisson distribution.* The proof that $a(n)$ is a distribution is not that straightforward. Consul and Jain refer to a paper of Jensen [4] for a proof. Jensen uses Lagrange's expansion formula which expresses w as a series in z if we know that they are satisfying the functional equation $f(w) = z$. A celebrated example for such an expansion is *Lambert's W-function* which fulfills the equation $w\mathrm{e}^w = z$ and then we deduce that

$$w = \sum_{n=1}^{\infty} \frac{(-n)^{n-1}}{n!}\,z^n \quad (|z| < \mathrm{e}^{-1}).$$

Consul seems to suggest that (1.1) is valid for all $\theta \geq 0$ and $\lambda \geq 0$ [1]. However, as we shall see, this is only true if $0 \leq \lambda \leq 1$. This was proved by a different method in [5]. In this paper, we reprove Euler's difference of powers formula (see

---

2010 *Mathematics Subject Classification.* Primary: 05A10, 11B65, Secondary: 60E05.
Key words and phrases. Binomial identity, Poisson distribution, Gaussian distribution, W-function.
This is the final form of the paper.

Theorem 3.4) and use this to prove the identity (1.2) which then leads to the proof of (1.1).

Interestingly in 2000, H. J. H. Tuenter [6] uses Euler's difference formula to prove the identity

$$\sum_{n=0}^{\infty} \frac{(\theta + n\lambda)^n}{n!} \mathrm{e}^{-\theta - n\lambda} = (1-\lambda)^{-1}, \tag{1.2}$$

where $\lambda_0 < \lambda < 1$ and λ_0 is the solution to $\lambda \mathrm{e}^{\lambda} = 1$. But, as said before, this identity already appeared in [5].

2. The main result

The classical binomial formula says

$$\sum_{n=0}^{N} \binom{N}{n} a^n b^{N-n} = (a+b)^N.$$

We now give a generalization of this result.

Theorem 2.1 (Generalized Binomial Theorem (GBT)). *For all complex numbers a, b, c and for all integers N, we have*

$$\sum_{n=0}^{N} \binom{N}{n} a(a+nc)^{n-1} b(b+(N-n)c)^{N-n-1} = (a+b)(a+b+Nc)^{N-1}. \tag{2.1}$$

PROOF. Let

$$A(t) = \sum_{n=0}^{N} \binom{N}{n} ta(ta+nc)^{n-1} b(b+(N-n)c)^{N-n-1} \quad (t \in \mathbb{R}). \tag{2.2}$$

Then a little rearranging after differentiation gives

$$A'(t) = Na \sum_{n=0}^{N-1} \binom{N-1}{n} (ta+c)(ta+c+nc)^{n-1} b(b+(N-1-n)c)^{N-1-n-1}.$$

Clearly the claim is true when $N = 1$. Using induction (a replaced by $ta + c$) we get

$$A'(t) = a(ta+c+b+(N-1)c)^{N-2}$$

and the latter equals

$$\frac{\mathrm{d}}{\mathrm{d}t}[(ta+b)(ta+b+Nc)^{N-1}].$$

Integration finishes the proof. □

The above theorem and its proof hold verbatim if a, b and c are commuting matrices.

3. Applications

3.1. Lambert's W-function. Lambert's W-function has many applications. See [3].

Theorem 3.1. *Let*

$$w = \sum_{n=1}^{\infty} \frac{n^{n-1}}{n!} z^n \quad (|z| < e^{-1}). \tag{3.1}$$

Then w satisfies the equation

$$we^{-w} = z. \tag{3.2}$$

PROOF. Let

$$g(a,z) = \sum_{n=0}^{\infty} \frac{a(a+n)^{n-1}}{n!} z^n \quad (|z| < e^{-1}). \tag{3.3}$$

We have

$$|(a+n)^{n-1}| \le n^{n-1}(1+|a|/n)^{n-1} \le n^{n-1}(1+|a|/n)^n \le n^{n-1}e^{|a|}.$$

Using this and Stirling's formula, we find that the series in (3.3) converges for all complex parameters a (indeed, uniformly on compacts) if $|z| < e^{-1}$. Using the GBT, we find $g(a,z)g(b,z) = g(a+b,z)$. Thus, we must have

$$g(a,z) = e^{af(z)} \tag{3.4}$$

for some function f. Subtract 1 from each side of (3.3), divide by a and let a tend to zero to get

$$f(z) = \sum_{n=1}^{\infty} \frac{n^{n-1}}{n!} z^n. \tag{3.5}$$

From (3.1) and (3.5) we deduce $w = f(z)$. Writing $n = m+1$ in (3.5) one finds

$$w = f(z) = \sum_{m=0}^{\infty} \frac{(m+1)^m}{(m+1)!} z^{m+1} = \sum_{m=0}^{\infty} \frac{(m+1)^{m-1}}{m!} z^{m+1}$$

By (3.1), the last series is $zg(1,z)$. Thus, by (3.4), we get $w = zg(1,z) = ze^{f(z)} = ze^w$. □

Incidentally, the above proof also shows that

$$e^{aw} = \sum_{n=0}^{\infty} \frac{a(a+n)^{n-1}}{n!} z^n \quad (|z| < e^{-1}).$$

The above theorem leads to a simple Corollary.

Corollary 3.2. *Let $\theta \ge 0$. Then*

$$\sum_{n=1}^{\infty} \frac{n^{n-1}}{n!} \left(\theta e^{-\theta}\right)^n = \theta_1, \tag{3.6}$$

where θ_1 is the unique number in the interval $[0,1]$ such that $\theta e^{-\theta} = \theta_1 e^{-\theta_1}$.

Proof. If $\theta \geq 0$, then $\theta \mathrm{e}^{-\theta} \leq \mathrm{e}^{-1}$. Taking $z = \theta \mathrm{e}^{-\theta}$ in (3.2), we get $\theta \mathrm{e}^{-\theta} = \theta_1 \mathrm{e}^{-\theta_1}$. Now, the function $x\mathrm{e}^{-x}$ is strictly increasing in the interval $[0,1]$ and strictly decreasing in the interval $[1,+\infty]$, and maps each of these intervals bijectively to $[0, \mathrm{e}^{-1}]$. Thus, for the equation $x\mathrm{e}^{-x} = y\mathrm{e}^{-y}$ we have only two possibilities: both x and y are in $[0,1]$ in which case $x = y$ or x is in $[0,1]$ and y is in $[1,+\infty]$. Clearly, θ_1, as a function of θ, is continuous and as observed above $\theta \mathrm{e}^{-\theta} = \theta_1 \mathrm{e}^{-\theta_1}$. Near $\theta = 0$, θ_1 is also near zero. Hence $\theta_1 = \theta$ if θ is near zero. In particular, $\theta_1 \leq 1$ for θ near zero. By continuity,

$$\theta_1 \leq 1 \quad \text{for all } \theta \leq 1. \tag{3.7}$$

If $\theta \geq 1$, there is a unique x in $[0,1]$ such that $x\mathrm{e}^{-x} = \theta \mathrm{e}^{-\theta}$. By (3.6),

$$\theta_1 = \sum_{n=1}^{\infty} \frac{n^{n-1}}{n!} \left(\theta \mathrm{e}^{-\theta}\right)^n = \sum_{n=1}^{\infty} \frac{n^{n-1}}{n!} \left(x\mathrm{e}^{x}\right)^n.$$

From (3.7), the last series is precisely x, i.e., $\theta_1 = x$. In particular, $\theta_1 \leq 1$ for all $\theta \geq 0$. Thus we have shown

$$\theta_1 \leq 1 \quad \text{for all } \theta \geq 0 \qquad \text{and} \qquad \theta_1 = \theta \quad \text{iff} \quad \theta \leq 1. \qquad \square$$

3.2. Generalized Poisson distribution (GPD). The following result shows that the distribution (1.1) is a probability distribution if $\theta \leq 1$. If $\theta \geq 1$, a normalizing factor is needed.

Theorem 3.3. *For complex number a and $\theta \geq 0$, let*

$$P(a,\theta) = \sum_{n=0}^{\infty} \frac{a(a+n\theta)^{n-1}}{n!} \mathrm{e}^{-a-n\theta}.$$

Then

$$P(a,\theta) = \mathrm{e}^{-a(1-\theta_1/\theta)},$$

where $0 \leq \theta_1 \leq 1$ satisfies $\theta \mathrm{e}^{-\theta} = \theta_1 \mathrm{e}^{-\theta_1}$.

Proof. Clearly the above result holds if $\theta = 0$. Note that $P(a\theta,\theta)\mathrm{e}^{a\theta}$ is just $g(a,z)$ defined in (3.3) with z replaced by $\theta \mathrm{e}^{-\theta}$. And from (3.4), (3.5) and (3.6)

$$P(a\theta,\theta)\mathrm{e}^{a\theta} = g(a, \theta \mathrm{e}^{-\theta}) = \exp\left(af(\theta \mathrm{e}^{-\theta})\right) = \mathrm{e}^{a\theta_1},$$

where, by Corollary 3.2, θ_1 is the unique number in $[0,1]$ satisfying $\theta \mathrm{e}^{-\theta} = \theta_1 \mathrm{e}^{-\theta_1}$. $\square$

3.3. Euler's formula for difference of powers.

Theorem 3.4. *For any positive integers N and $p \leq N$ and complex numbers a and c we have*

$$\sum_{0}^{N} \binom{N}{n} (-1)^{N-n} (a+nc)^p = N!(p-N+1)^+ c^n. \tag{3.8}$$

Remark 1. Note that the right side of (3.8) is 0 if $p \leq (N-1)$ and is $N!\,c^N$ if $p = N$.

PROOF. In (2.1) let $a+b+Nc=0$, i.e., $b+Nc=-a$ to get after some simplification

$$\sum_0^N \binom{N}{n}(-1)^{N-n}(a+nc)^{N-2}=0. \tag{3.9}$$

Write

$$F(a,c)=\sum_0^N \binom{N}{n}(-1)^{N-n}(a+nc)^{N-1}. \tag{3.10}$$

It is seen from (3.9) that the derivative of $F(a,c)$ with respect to a equals zero. It follows that $F(a,c)$ is independent of a namely $F(a,c)=F(b,c)$ for all a,b. In particular $F(-Nc,c)=F(0,c)$. But from (3.9) it is easily seen that $F(-Nc,c)=-F(0,c)$ implying $F(0,c)=-F(0,c)$, i.e.,

$$F(0,c)=0 \quad \text{and hence also} \quad F(a,c)=0. \tag{3.11}$$

Finally write

$$G(N,a,c)=\sum_0^N \binom{N}{n}(-1)^{N-n}(a+nc)^{N}.$$

The derivative of $G(N,a,c)$ with respect to a is $NF(a,c)$ which is zero by (3.11). So

$$G(N,a,c) \text{ is independent of } a. \tag{3.12}$$

We compute

$$\begin{aligned} G(N,a,c) &= \sum_0^N \binom{N}{n}(-1)^{N-n}(a+nc)(a+nc)^{N-1} \\ &= aF(a,c)+c\sum_0^N \binom{N}{n}(-1)^{N-n}n(a+nc)^{N-1} \\ &= c\sum_1^N \binom{N}{n}(-1)^{N-n}n(a+nc)^{N-1}, \end{aligned}$$

Since $F(a,c)=0$ from (3.11),

$$\begin{aligned} G(N,a,c) &= \sum_1^N N\binom{N-1}{n-1}(-1)^{N-1-(n-1)}(a+nc)^{N-1} \\ &= Nc\sum_0^{N-1} \binom{N-1}{n}(-1)^{N-1-n}(a+c+nc)^{N-1} \\ &= Nc\,G(N-1,a+c,c)=Nc\,G(N-1,a,c), \end{aligned} \tag{3.13}$$

because $G(N,a,c)=G(N,b,c)$ by (3.12). Finally, (3.13) implies $G(N,a,c)=N!\,c^N$. This concludes the proof. □

3.4. A novel distribution. Manfred Denker of Penstate told us in a talk at the University of Florida that he and his student discovered this distribution and that it is a new distribution. He did not give a proof. Below we show how this follows from GBT.

Let $0 < \lambda < 1$ and N a positive integer. For $1 \leq n \leq N$, let

$$p(n, N, \lambda) = \frac{N(1-\lambda)}{N-(N-1)\lambda} n^{n-2} \binom{N-1}{n-1} N^{1-n} (\lambda)^{n-1} \left(1 - \frac{n\lambda}{N}\right)^{N-n-1}.$$

Theorem 3.5. *$p(n, N, \lambda)$ is a distribution. Namely*

$$\sum_{n=1}^{N} p(n, N, \lambda) = 1. \tag{3.14}$$

PROOF. To get this as a consequence of GBT rewrite $p(n, N, \lambda)$ as

$$p(n, N, \lambda) = \frac{1-\lambda}{N-(N-1)\lambda} N^{2-N} \binom{N}{n} (\lambda n)^{n-1} (N - \lambda n)^{N-n-1}.$$

From (2.1) and (2.2)

$$A(t) = (ta + b)(ta + b + Nc)^{N-1}. \tag{3.15}$$

Differentiating both sides of (3.15) relative to t, with a little simplification, taking $t = 0$ and $a = 1$ or $a = I$ if a is a matrix we get

$$\sum_{0}^{N} \binom{N}{n} (nc)^{(n-1)} b(b + (N-n)c)^{N-n-1} = N(b+c)(b+Nc)^{N-2}.$$

This identity is valid for all complex b and c. Take $c = \lambda$ and $b = N(1-\lambda)$ gives

$$\sum_{0}^{N} \binom{N}{n} (n\lambda)^{n-1} N(1-\lambda) \big(N(1-\lambda) + (N-n)\lambda)\big)^{N-n-1} = (N-(N-1)\lambda)N^{N-1}$$

or equivalently

$$\sum_{0}^{N} \binom{N}{n} (n\lambda)^{n-1} N(1-\lambda)(N-n\lambda)^{N-n-1} = (N-(N-1)\lambda)N^{N-1}.$$

But rewriting the last equation as

$$\sum_{0}^{N} \binom{N}{n} \frac{1-\lambda}{N-(N-1)\lambda} N^{2-N} (n\lambda)^{n-1} (N-n\lambda)^{N-n-1} = 1,$$

we obtain (3.14). □

References

1. P. C. Consul, *Generalized Poisson distributions: Properties and applications*, Statist. Textbooks Monogr., vol. 99, Dekker, New York, 1989.
2. P. C. Consul and G. C. Jain, *A generalization of the Poisson distribution*, Technometrics **15** (1973), 791–799.
3. R. M. Corless, G. H. Gonnet, D. E. G. Hare, D. J. Jeffrey, and D. E. Knuth, *On the Lambert W function*, Adv. Comput. Math. **5** (1996), no. 4, 329–359.
4. J. L. W. V. Jensen, *Sur une identité d'Abel et sur d'autres formules analogues*, Acta Math. **26** (1902), no. 1, 307–318 (French).
5. B. Lerner, A. Lone, and M. Rao, *On generalized Poisson distributions*, Probab. Math. Statist. **17** (1997), no. 2, Acta Univ. Wratislav. No. 2029, 377–385.

6. H. J. H. Tuenter, *On the generalized Poisson distribution*, Statist. Neerlandica **54** (2000), no. 3, 374–376.

Department of Mathematics, University of Florida, 358 Little Hall, PO Box 118105 Gainesville, FL 32611-8105, USA
E-mail address: mrao@ufl.edu

Centre de Recherches Mathématiques
CRM Proceedings and Lecture Notes
Volume **55**, 2012

Uniform and C^m-Approximation by Polyanalytic Polynomials

Konstantin Yu. Fedorovskiy

Abstract. We are dealing with problems of uniform and C^m-approximability of functions by polyanalytic polynomials on compact subset of the complex plane. One new sufficient approximability condition in the respective uniform approximation problem is obtained.

1. Introduction

Let U be an open set in the complex plane $\mathbb{C}$ and $n \geqslant 1$ be an integer. We recall, that a function f is said to be *polyanalytic of order* n (or, shorter, n-*analytic*) in U if it is of the form

$$f(z) = \bar{z}^{n-1} f_{n-1}(z) + \cdots + \bar{z} f_1(z) + f_0(z), \tag{1.1}$$

where $f_0, f_1, \dots, f_{n-1}$ are holomorphic functions in U. Let us denote by $\mathcal{O}_n(U)$ the set of all n-analytic functions in U. Notice that $\mathcal{O}_n(U)$ consists of all continuous functions f on U such that $\bar{\partial}^n f = 0$ in U in the distributional sense, where $\bar{\partial} = \partial/\partial\bar{z}$ is the standard Cauchy–Riemann operator in $\mathbb{C}$. By n-*analytic polynomials* and n-*analytic rational functions* we mean the functions of the form (1.1), where $f_0, \dots, f_{n-1}$ are polynomials and rational functions in the complex variable z respectively.

Polyanalytic functions have appeared, as far as we know, at the beginning of 20th century in connection with investigations of problems of plane elasticity theory. A more or less exhaustive survey of the properties of polyanalytic functions as well as of the history of their investigation may be found in the book [1]. Among known interesting and important problems concerning polyanalytic functions there are problems of uniform and C^m-approximation (for integer $m \geqslant 1$) of functions by polyanalytic polynomials, which are considered in the present paper.

2010 *Mathematics Subject Classification.* Primary 30E10; Secondary 30G30.

Key words and phrases. Polyanalytic functions and polynomials, uniform and C^m-approximation.

The author's research is partially supported by grants 10-01-00837-a and 12-01-00434-a of the *Russian Foundation for Basic Research* and by the grant NSh-3476.2010.1 of the *President Program of Support for Leading Scientific Schools of Russia.*

This is the final form of the paper.

2. Uniform approximation by polyanalytic polynomials and concepts of Nevanlinna and locally Nevanlinna domains

Let X be a compact set in $\mathbb{C}$ and let $C(X)$ be the space of all continuous complex-valued functions on X endowed with the uniform norm. Let, moreover, $n \geqslant 2$ be an integer. We are interested in the following uniform approximation problem for polyanalytic functions ($\overset{\circ}{X}$ stands, as usual, for the interior of X):

Problem 2.1. What conditions on X are necessary and sufficient in order that

$$A_n(X) = P_n(X), \tag{2.1}$$

where $A_n(X) = C(X) \cap \mathcal{O}_n(\overset{\circ}{X})$ and $P_n(X)$ is the $C(X)$-closure of the subspace $\{p_{|X} : p \text{ is } n\text{-analytic polynomial}\}$?

In other words one needs to find necessary and sufficient conditions on X in order that each function which is continuous on X and n-analytic on its interior can be uniformly on X approximated by n-analytic polynomials. We exclude the case $n = 1$ from the consideration, because Mergelyan's classical theorem (see [14]) says that $A_1(X) = P_1(X)$ if and only if the set $\mathbb{C} \setminus X$ is connected, and so the respective problem is solved completely (in terms of topological properties of X).

The investigation of Problem 2.1 was started in 1980th years. For instance, as it was proved in [4], if the set $\mathbb{C} \setminus X$ is connected, then $A_n(X) = P_n(X)$ for any integer $n \geqslant 2$. However, even in the simplest case of compact sets with disconnected complement, namely in the case when X is a closed rectifiable Jordan curve in $\mathbb{C}$, the solution of Problem 2.1 is formulated in terms of special *analytic* characteristic of X (see [10, Theorem 1]). For instance, if X is any circle in $\mathbb{C}$, then the equality (2.1) fails, but this equality holds for any closed polygonal line or ellipse (which is not a circle) X in $\mathbb{C}$.

In the case of *Carathéodory compact sets* (we recall that a compact set X is called a Carathéodory compact set if $\partial X = \partial \widehat{X}$, where $\widehat{X}$ denotes the union of X and all bounded connected components of $\mathbb{C} \setminus X$) Problem 2.1 was solved in [7, Theorem 2.2]. In order to formulate and discuss this result one needs to recall the concept of a Nevanlinna domain which was introduced in [7, Definition 2.1; 10, Definition 3] and which has turned out crucial to the problem under consideration. Let $\mathbb{D}$ be the unit disk and $\mathbb{T}$ be the unit circle. For an open set U denote by $H^\infty(U)$ the set of all bounded holomorphic functions in U. It is well known, that for any $f \in H^\infty = H^\infty(\mathbb{D})$ the angular boundary values $f(\xi)$ are well defined for almost all points $\xi \in \mathbb{T}$.

Definition 2.2. A bounded simply connected domain Ω in $\mathbb{C}$ is said to be a *Nevanlinna domain* if there exist two functions $u, v \in H^\infty(\Omega)$, $v \not\equiv 0$, such that the equality

$$\overline{\zeta} = u(\zeta)/v(\zeta) \tag{2.2}$$

holds on $\partial\Omega$ almost everywhere in the sense of conformal mappings. It means that the equality $\overline{\varphi(\xi)} = (u \circ \varphi)(\xi)/(v \circ \varphi)(\xi)$ of angular boundary values holds for almost all $\xi \in \mathbb{T}$, where φ is some conformal mapping from the unit disk $\mathbb{D}$ onto Ω.

This is a consistent definition not depending on one's choice of φ. Moreover, in view of the Luzin–Privalov boundary uniqueness theorem, the quotient u/v is uniquely determined in Ω (for a Nevanlinna domain Ω). It should be noted, that if a Nevanlinna domain Ω is a Jordan domain with analytic boundary, then the corresponding quotient u/v coincides with the Schwarz function of $\partial\Omega$. Let us briefly recall this notion, which is studied in details in the book [8]. Let G be a Jordan domain with analytic boundary Γ. Then there exists holomorphic in some domain $W \supset \Gamma$ function S such that $\Gamma = \{z \in W : \bar{z} = S(z)\}$; the function S is called the Schwarz function of Γ.

It is easy to find out examples of Nevanlinna (any disk) and not Nevanlinna domains (any domain bounded by an ellipse which is not a circle, or by a polygonal line) among the domains with piecewise analytic boundaries. Several interesting examples and useful properties of Nevanlinna domains were obtained in [3, 5, 7, 13]. However, the construction of Nevanlinna domains with non analytic boundaries as well as with other prescribed analytical and geometrical properties is fairly difficult problem which requires some special and delicate analysis. This problem was studied in [2, 11] where, in particular, for each $\alpha \in (0, 1)$ several examples of Nevanlinna domains with C^1 but not $C^{1,\alpha}$ boundaries were constructed. Moreover, in [2] several examples of Nevanlinna domains with much irregular boundaries were obtained and it was shown that Nevanlinna domains may have "almost" unrectifiable boundaries. However the question about existence of Nevanlinna domains with unrectifiable boundaries (see [7, Problem 3.2]) remains open.

Let us formulate now the result of [7, Theorem 2.2]:

Theorem. *For a Carathéodory compact set X and for an integer $n \geqslant 2$ the equality* (2.1) *holds if and only if each bounded connected component of the set $\mathbb{C} \setminus X$ is not a Nevanlinna domain.*

For instance, if Ω is a Jordan domain, then $C(\partial\Omega) = P_2(\partial\Omega)$ if and only if $C(\partial\Omega) = P_n(\partial\Omega)$ for some integer $n \geqslant 2$ and if and only if Ω is not a Nevanlinna domain. So that the solution of Problem 2.1 in the case of Carathéodory compact sets is formulated in terms of analytic properties of the compact set where approximation is considered and does not depend on the order of polyanalyticity.

At the same time for non Carathéodory compact sets Problem 2.1 is more difficult and remains open in the general case. Several necessary and sufficient approximability conditions in this case were recently obtained in [3, 5–7].

As it was proved in [6, Theorem 1] for each $n \geqslant 1$ there exists a compact set Y_n such that $P_{2n}(Y_n) = C(Y_n) \neq P_n(Y_n)$ (in fact Y_n is the nest of n confocal ellipses). It means that the final solution of Problem 2.1 (which is still not found) should depend on the order of polyanalyticity. One ought to notice that the nature of this dependence is not clear up to now and its study is one of the most important problems in the themes under consideration.

For compact sets X and $Y \supseteq X$ let $R_n(X, Y)$ be the $C(X)$-*closure* of the subspace $\{f_{|X} : f$ is n-analytic rational function such that all (rational) functions $f_0, \dots, f_{n-1}$ from (1.1) have their poles outside $Y\}$.

Let us recall the following approximability criterion in Problem 2.1 which was proved in [3, Theorem 3]:

Theorem. *Let X be a compact set in $\mathbb{C}$ and let $\{\Omega_j\}_{j\in\mathcal{J}}$ be the* (*connected*) *components of $\widehat{X}^\circ$ that are not contained in X. Then the equality* (2.1) *holds if and only if the equalities $A_n(X \cap \overline{\Omega}_j) = R_n(X \cap \overline{\Omega}_j, \overline{\Omega}_j)$ hold for all $j \in \mathcal{J}$.*

In order to apply this criterion we need to be able to answer the following question: Let Ω be a Carathéodory domain (it means that Ω is a bounded simply connected domain such that $\partial\Omega = \partial\Omega_\infty$, where Ω_∞ stands for the unbounded component of the set $\overline{\mathbb{C}}\setminus\overline{\Omega})$ and let X be a compact subset of $\overline{\Omega}$ such that $A_n(X) = R_n(X, \overline{\Omega})$ for some $n \geqslant 2$; under what additional conditions does the equation $A_n(X \cup \partial\Omega) = R_n(X \cup \partial\Omega, \overline{\Omega})$ hold?

This question was studied in [5, Theorem 3; 7, Theorem 4.3] in the case when X is strictly contained in Ω as well as in [3, Theorem 4 and Corollary 1; 5, Theorem 4] in the case, when $X \cap \partial\Omega \neq \varnothing$. All these results are formulated in terms of suitable modifications and refinements of the concept of a Nevanlinna domain. One of them is the concept of a *locally Nevanlinna domain* which was introduced as follows (see [5, Definition 2; 7, Definition 4.2]):

Definition 2.3. A bounded simply connected domain Ω is called a locally Nevanlinna domain is there exist a compact set $\Sigma \subset \Omega$ and two functions $u, v \in H^\infty(\Omega \setminus \Sigma)$, $v \not\equiv 0$, such that the equality (2.2) holds on $\partial\Omega$ almost everywhere in the sense of conformal mappings.

It is easy to verify, that a domain bounded by the ellipse which is not a circle is a locally Nevanlinna but not a Nevanlinna domain.

Let us establish now new sufficient approximability condition concerning the question under consideration. For brevity one denotes by ND and LND the classes of all Nevanlinna and locally Nevanlinna domains respectively. Since for a domain $\Omega \in \mathrm{LND}$ there are many different possibilities to determine the desired compact set Σ, let us write $\Omega \sim [\Sigma, u, v]$ to fix which u, v and Σ we are dealing with. Moreover, we may (and shall) always assume that Σ is minimal, which means that the function u/v cannot be meromorphically extended from $\Omega \setminus \Sigma$ to $\Omega \setminus \Sigma_1$ for any proper compact subset Σ_1 of Σ.

Going further and following [5, Definition 1] one says that a closed Jordan arc γ is a *gate* for a bounded open set U, if $\gamma \subset \partial U$ and there exists an open connected set V such that $V \setminus \gamma = V_1 \cup V_2$, where V_1 and V_2 are non empty open connected sets such that $V_1 \subset U$ and $V_2 \subset \mathbb{C} \setminus \overline{U}$. It is clear, that if Ω is a Jordan domain, then any closed subarc $\gamma \subset \partial\Omega$ is a gate for Ω. The following takes place.

Theorem 2.4. *Fix $n \geqslant 3$. Let Ω be a Carathéodory domain in $\mathbb{C}$ such that $\mathbb{C}\setminus\overline{\Omega}$ is connected, $\Omega \in \mathrm{LND}\setminus\mathrm{ND}$, $\Omega \sim [\Sigma, u, v]$, and there exists a rectifiable gate γ for Ω. Let $K \subset \Omega$ be a nowhere dense compact set such that $C(K) = P_n(K)$ and $\Sigma \subset \widehat{K}^\circ$. Assume moreover, that there exists an analytic gate γ_0 for $\widehat{K}^\circ$ such that the Schwarz function S_0 of γ_0 belongs to $A_1(\overline{\Omega})$ and $G_\infty \cap \Sigma \neq \varnothing$, where G_∞ is the unbounded connected component of the set $\mathbb{C} \setminus \overline{(K \setminus \gamma_0)}$. Then $C(K \cup \partial\Omega) = P_n(K \cup \partial\Omega)$.*

Example 4.5 in [7] shows that there exists a compact set Y such that $P_3(Y) = C(Y) \neq P_2(Y)$. Using this Y and taking into account [3, Corollary 4] it is possible to construct the compact set K satisfying all assumptions of Theorem 2.4 and such that $P_3(K) = C(K) \neq P_2(K)$. Furthermore, the situation considered in Theorem 2.4 can not be analyzed using the mentioned above results since $\Sigma \subset \widehat{K}^\circ$.

Proof of Theorem 2.4. Put $K_0 := \overline{K \setminus \gamma_0}$, $X := K \cup \partial\Omega$ and $X_0 := K_0 \cup \partial\Omega$. If $P_n(X) \neq C(X)$ then there exists a measure μ on X such that $\mu \not\equiv 0$ and $\mu \perp P_n(X)$. Define the measure $\mu_0 := \big(\bar{z} - S_0(z)\big)\,\mu$, so that μ_0 is supported on X_0 and $\mu_0 \perp P_{n-1}(X_0)$; it is also clear that $\mu_0 \neq 0$. It follows from [5, Proposition 3] that μ and μ_0, have no atoms on $\partial\Omega$. In the same way as it was done in the proof of [5, Theorem 4] we conclude that $\widehat{\mu}_0 \neq 0$ in $\Omega \setminus \widehat{K}_0$ (where $\widehat{\nu}$ denotes the

Cauchy transform of a measure ν) and that the function u/v must coincides with the function $\widehat{\bar{z}\mu_0}/\widehat{\mu}_0$ in $\Omega\setminus\widehat{\Sigma\cup K_0}$. Therefore, the function u/v can be meromorphically extended to the set $\Omega\setminus\widehat{K}_0$, but $\Omega\setminus\widehat{K}_0\subset G_\infty$ which gives a contradiction because $\Sigma\cap G_\infty$ is not empty and Σ is assumed to be minimal. □

We end this section establishing useful characterizations of Nevanlinna and locally Nevanlinna domains in terms of conformal mappings. For an inner function Θ (we recall, that a function $\Theta\in H^\infty$ is called an inner function, if $|\Theta(\xi)|=1$ for almost all $\xi\in\mathbb{T}$) let us define the space $K_\Theta:=H^2\ominus\Theta H^2$, where H^2 is the standard Hardy space in $\mathbb{D}$. Let now $U\subset\overline{\mathbb{C}}\setminus\overline{\mathbb{D}}$ be an open set such that $\mathbb{T}$ is some component of ∂U.

Definition 2.5. One says that a function $f\in H^\infty$ admits a Nevanlinna-type pseudocontinuation to U if there exist two functions $f_1,f_2\in H^\infty(U)$, $f_2\not\equiv 0$, such that for almost all $\xi\in\mathbb{T}$ the equality $f(\xi)=f_1(\xi)/f_2(\xi)$ holds, where $f_1(\xi)$ and $f_2(\xi)$ are angular boundary values of f_1 and f_2 from U.

One also says, that a function $f\in H^\infty$ admits a *Nevanlinna-type pseudocontinuation* (see [9, Definition 2.1.2; 11, Definition 2]), if f admits the respective pseudocontinuation to $\overline{\mathbb{C}}\setminus\overline{\mathbb{D}}$. The following result takes place.

Theorem 2.6. *Let Ω be a bounded simply connected domain in $\mathbb{C}$ and let φ be some conformal mapping from $\mathbb{D}$ onto Ω.*

(1) $\Omega\in$ ND *if and only if φ admits a Nevanlinna-type pseudocontinuation.*
(2) $\Omega\in$ LND *if and only if φ admits a Nevanlinna-type pseudocontinuation to some open set $U\subset\overline{\mathbb{C}}\setminus\overline{\mathbb{D}}$.*
(3) *If Θ is an inner function, then any bounded univalent function from the space K_Θ maps $\mathbb{D}$ conformally onto some Nevanlinna domain.*
(4) *If $\Omega\in$ ND, then there exists an inner function Θ such that $\varphi\in K_\Theta$.*

The part (1) of Theorem 2.6 has been proved in [7, Proposition 3.1]. The part (2) can be proved analogously, repeating word-for-word the proof of [7, Proposition 3.1] and replacing where it is necessary the functions u and v from the definition of Nevanlinna domain by the respective functions from the definition of a locally Nevanlinna domain.

The statement of the part (3) is exactly the statement of [11, Theorem 1]. In fact, the proof of this result contains also the proof of the statement of the part (4). It is worth to note, that the main ingredients of the respective proofs are the result of the part (1), the result of [9, Theorem 2.2.1], which says that a function $\varphi\in H^\infty$ admits a Nevanlinna-type pseudocontinuation if and only if φ is not a cyclic vector for the backward shift operator $f\mapsto\big(f(z)-f(0)\big)/z$ in H^2, as well as Beurling's classical theorem, in view of which the spaces $K_\Theta\subset H^2$ and only these spaces are invariant subspaces for the backward shift operator in H^2.

3. C^m-approximation by polyanalytic polynomials

For a compact set X in $\mathbb{C}$ and for integers $n\geqslant 2$ and $m\geqslant 1$ we are interesting in the following problem:

Problem 3.1. What conditions on X are necessary and sufficient in order that each function f which is of class C^m in a neighborhood of X and n-analytic on $\mathring{X}$

can be approximated on X by some sequence $(p_k)_{k=1}^{\infty}$ of n-analytic polynomials in such a way, that for all integers $s \geqslant 0$ and $t \geqslant 0$ such that $s+t \leqslant m$ one has

$$\partial_{s,t} p_k \rightrightarrows_X \partial_{s,t} f \tag{3.1}$$

($\rightrightarrows_X$ means the uniform convergence on X and $\partial_{s,t} = \partial^{s+t}/\partial\bar{z}^s\partial z^t$) as $k \to \infty$?

First of all let us note that the convergence (3.1) is in general weaker than the convergence in the classical C^m-Whitney space. Moreover, it is clear, that the condition $f \in \mathcal{O}_n(\mathring{X})$ is a natural necessary approximability condition for a given function f in the problem under consideration. It is appropriate to reformulate Problem 3.1 in terms of the respective classes of functions similarly as in the formulation of Problem 2.1. For an integer $m \geqslant 1$ and for a function f of the class C^m let

$$\nabla^m f := \big(\partial_{m,0} f, \partial_{m-1,1} f, \ldots, \partial_{1,m-1} f, \partial_{0,m} f\big), \qquad \mathcal{D}^m f := \big(f, \nabla f, \ldots, \nabla^m f\big),$$

and $M := (m+1)(m+2)/2$. Let us define the spaces $A_n^m(X)$ and $P_n^m(X)$ to be the $C(X)^M$-*closures* of the subspaces $\{(\mathcal{D}^m f)_{|X} : f \in C^m(\mathbb{C}) \cap \mathcal{O}_n(\mathring{X})\}$ and $\{(\mathcal{D}^m p)_{|X} : p$ is n-analytic polynomial$\}$ respectively. Thus in Problem 3.1 we need to describe such compact sets X for which $A_n^m(X) = P_n^m(X)$.

Problem 3.1 for $m \geqslant n-1$ was solved in [12]. The respective approximability criteria [12, Theorem 2.1] and [12, Theorem 2.4] say that $A_n^{n-1}(X) = P_n^{n-1}(X)$ if and only if the set $\mathbb{C} \setminus X$ is connected, whereas for $m > n-1$ the equality $A_n^m(X) = P_n^m(X)$ holds if and only if $\overline{\mathring{X}} = X$ and the set $\mathbb{C} \setminus X$ is connected.

At the same time, Problem 3.1 remains unsolved when $0 < m < n-1$.

Reverting to the statement of Problem 3.1 one ought to notice that the analogously posed problems of C^1-approximability of functions by harmonic polynomials and by polynomial solutions of general second order homogeneous partial differential elliptic equations with constant complex coefficients on compact sets in $\mathbb{C}$ have been solved earlier in [15, Theorem 1; 16, Theorem 3.2] respectively. In both of these cases a necessary and sufficient approximability condition is also that the set $\mathbb{C} \setminus X$ is connected. Thus [12, Theorem 2.1; 15, Theorem 1; 16, Theorem 3.2] may be considered as natural analogs of the mentioned above classical Mergelyan's theorem.

References

1. M. B. Balk, *Polyanalytic functions*, Math. Res., vol. 63, Akademie-Verlag, Berlin, 1991.
2. A. D. Baranov and K. Yu. Fedorovskiy, *Boundary regularity of Nevanlinna domains and univalent functions in model subspaces*, Mat. Sb. **202** (2011), no. 12, 3–22 (Russian); English transl., Sb. Math. **202** (2011), no. 12, 1723–1740.
3. A. Boivin, P. M. Gauthier, and P. V. Paramonov, *On uniform approximation by n-analytic functions on closed sets in* $\mathbb{C}$, Izv. Ross. Akad. Nauk Ser. Mat. **68** (2004), no. 3, 15–28 (Russian); English transl., Izv. Math. **68** (2004), no. 3, 447–459.
4. J. J. Carmona, *Mergelyan's approximation theorem for rational modules*, J. Approx. Theory **44** (1985), no. 2, 113–126.
5. J. J. Carmona and K. Yu. Fedorovskiy, *Conformal maps and uniform approximation by polyanalytic functions*, Selected Topics in Complex Analysis (V. Ya. Eiderman and M. V. Samokhin, eds.), Oper. Theory Adv. Appl., vol. 158, Birkhäuser, Basel, 2005, pp. 109–130.
6. ———, *On the dependence of conditions for the uniform approximability of functions by polyanalytic polynomials on the order of polyanalyticity*, Mat. Zametki **83** (2008), no. 1, 32–38 (Russian); English transl., Math. Notes **83** (2008), no. 1-2, 31–36.

7. J. J. Carmona, P. V. Paramonov, and K. Yu. Fedorovskiy, *Uniform approximation by polyanalytic polynomials and the Dirichlet problem for bianalytic functions*, Mat. Sb. **193** (2002), no. 10, 75–98 (Russian); English transl., Sb. Math. **193** (2002), no. 9-10, 1469–1492.
8. P. J. Davis, *The Schwarz function and its applications*, Carus Math. Monogr., vol. 17, Math. Assoc. America, Buffalo, NY, 1974.
9. R. G. Douglas, H. S. Shapiro, and A. L. Shields, *Cyclic vectors and invariant subspaces for the backward shift operator*, Ann. Inst. Fourier (Grenoble) **20** (1970), no. 1, 37–76.
10. K. Yu. Fedorovskiy, *On uniform approximations of functions by n-analytic polynomials on rectifiable contours in* **C**, Mat. Zametki **59** (1996), no. 4, 604–610, 640 (Russian); English transl., Math. Notes **59** (1996), no. 3-4, 435–439.
11. ———, *On some properties and examples of Nevanlinna domains*, Tr. Mat. Inst. Steklova **253** (2006), no. Kompleks. Anal. i Prilozh., 204–213 (Russian); English transl., Proc. Steklov Inst. Math. **2 (253)** (2006), 186–194.
12. ———, *C^m-approximation by polyanalytic polynomials on compact subsets of the complex plane*, Complex Anal. Oper. Theory **5** (2011), no. 3, 671–681.
13. M. Ya. Mazalov, *An example of a nonconstant bianalytic function vanishing everywhere on a nowhere analytic boundary*, Mat. Zametki **62** (1997), no. 4, 629–632 (Russian); English transl., Math. Notes **62** (1997), no. 3-4, 524–526 (1998).
14. S. N. Mergelyan, *Uniform approximations of functions of a complex variable*, Uspehi Matem. Nauk (N.S.) **7** (1952), no. 2(48), 31–122 (Russian); English transl., Amer. Math. Soc. Translation **1954** (1954), no. 101, 99.
15. P. V. Paramonov, *Approximations by harmonic polynomials in the C^1-norm on compact sets in* $\mathbf{R}^2$, Izv. Ross. Akad. Nauk Ser. Mat. **57** (1993), no. 2, 113–124 (Russian); English transl., Russian Acad. Sci. Izv. Math. **42** (1994), no. 2, 321–331.
16. P. V. Paramonov and K. Yu. Fedorovskiy, *On uniform and C^1-approximability of functions on compact sets in* $\mathbf{R}^2$ *by solutions of second-order elliptic equations*, Mat. Sb. **190** (1999), no. 2, 123–144 (Russian); English transl., Sb. Math. **190** (1999), no. 1-2, 285–307.

Department of Applied Mathematics, Bauman Moscow State Technical University, 2-nd Baumanskaya, 5, 105005 Moscow, Russia

E-mail address: `kfedorovs@yandex.ru`

Published Titles in This Series

55 **André Boivin and Javad Mashreghi, Editors,** Complex Analysis and Potential Theory, 2012

54 **Daniel Daigle, Richard Ganong, and Mariusz Koras,** Affine Algebraic Geometry, 2011

53 **Bradd Hart, Thomas G. Kucera, Anand Pillay, Philip J. Scott, and Robert A. G. Seely, Editors,** Models, Logics, and Higher-Dimensional Categories, 2011

52 **Dmitry Jakobson, Stéphane Nonnenmacher, and Iosif Polterovich, Editors,** Spectrum and Dynamics, 2010

51 **Javad Mashreghi, Thomas Ransford, and Kristian Seip, Editors,** Hilbert Spaces of Analytic Functions, 2010

50 **P. Robert Kotiuga, Editor,** A Celebration of the Mathematical Legacy of Raoul Bott, 2010

49 **Miguel Abreu, François Lalonde, and Leonid Polterovich, Editors,** New Perspectives and Challenges in Symplectic Field Theory, 2009

48 **David Avis, David Bremner, and Antoine Deza, Editors,** Polyhedral Computation, 2009

47 **John Harnad and Pavel Winternitz, Editors,** Groups and Symmetries, 2009

46 **Jean-Marie De Koninck, Andrew Granville, and Florian Luca, Editors,** Anatomy of Integers, 2008

45 **Panos M. Pardalos and Pierre Hansen, Editors,** Data Mining and Mathematical Programming, 2008

44 **Stanley Alama, Lia Bronsard, and Peter J. Sternberg, Editors,** Singularities in PDE and the Calculus of Variations, 2008

43 **Andrew Granville, Melvyn B. Nathanson, and József Solymosi, Editors,** Additive Combinatorics, 2007

42 **Donald A. Dawson, Vojkan Jakšić, and Boris Vainberg, Editors,** Probability and Mathematical Physics, 2007

41 **André Bandrauk, Michel C. Delfour, and Claude Le Bris, Editors,** High-Dimensional Partial Differential Equations in Science and Engineering, 2007

40 **Vestislav Apostolov, Andrew Dancer, Nigel Hitchin, and McKenzie Wang, Editors,** Perspectives in Riemannian Geometry, 2006

39 **P. Winternitz, D. Gomez-Ullate, A. Iserles, D. Levi, P. J. Olver, R. Quispel, and P. Tempesta, Editors,** Group Theory and Numerical Analysis, 2005

38 **Jacques Hurtubise and Eyal Markman, Editors,** Algebraic Structures and Moduli Spaces, 2004

37 **P. Tempesta, P. Winternitz, J. Harnad, W. Miller, Jr., G. Pogosyan, and M. Rodriguez, Editors,** Superintegrability in Classical and Quantum Systems, 2004

36 **Hershy Kisilevsky and Eyal Z. Goren, Editors,** Number Theory, 2004

35 **H. E. A. Eddy Campbell and David L. Wehlau, Editors,** Invariant Theory in All Characteristics, 2004

34 **P. Winternitz, J. Harnad, C. S. Lam, and J. Patera, Editors,** Symmetry in Physics, 2004

33 **André D. Bandrauk, Michel C. Delfour, and Claude Le Bris, Editors,** Quantum Control: Mathematical and Numerical Challenges, 2003

32 **Vadim B. Kuznetsov, Editor,** The Kowalevski Property, 2002

31 **John Harnad and Alexander Its, Editors,** Isomonodromic Deformations and Applications in Physics, 2002

30 **John McKay and Abdellah Sebbar, Editors,** Proceedings on Moonshine and Related Topics, 2001

29 **Alan Coley, Decio Levi, Robert Milson, Colin Rogers, and Pavel Winternitz, Editors,** Bäcklund and Darboux Transformations. The Geometry of Solitons, 2001

28 **J. C. Taylor, Editor,** Topics in Probability and Lie Groups: Boundary Theory, 2001

PUBLISHED TITLES IN THIS SERIES

27 **I. M. Sigal and C. Sulem, Editors,** Nonlinear Dynamics and Renormalization Group, 2001

26 **J. Harnad, G. Sabidussi, and P. Winternitz, Editors,** Integrable Systems: From Classical to Quantum, 2000

25 **Decio Levi and Orlando Ragnisco, Editors,** SIDE III—Symmetries and Integrability of Difference Equations, 2000

24 **B. Brent Gordon, James D. Lewis, Stefan Müller-Stach, Shuji Saito, and Noriko Yui, Editors,** The Arithmetic and Geometry of Algebraic Cycles, 2000

23 **Pierre Hansen and Odile Marcotte, Editors,** Graph Colouring and Applications, 1999

22 **Jan Felipe van Diejen and Luc Vinet, Editors,** Algebraic Methods and q-Special Functions, 1999

21 **Michel Fortin, Editor,** Plates and Shells, 1999

20 **Katie Coughlin, Editor,** Semi-Analytic Methods for the Navier-Stokes Equations, 1999

19 **Rajiv Gupta and Kenneth S. Williams, Editors,** Number Theory, 1999

18 **Serge Dubuc and Gilles Deslauriers, Editors,** Spline Functions and the Theory of Wavelets, 1999

17 **Olga Kharlampovich, Editor,** Summer School in Group Theory in Banff, 1996, 1999

16 **Alain Vincent, Editor,** Numerical Methods in Fluid Mechanics, 1998

15 **François Lalonde, Editor,** Geometry, Topology, and Dynamics, 1998

14 **John Harnad and Alex Kasman, Editors,** The Bispectral Problem, 1998

13 **Michel C. Delfour, Editor,** Boundaries, Interfaces, and Transitions, 1998

12 **Peter C. Greiner, Victor Ivrii, Luis A. Seco, and Catherine Sulem, Editors,** Partial Differential Equations and Their Applications, 1997

11 **Luc Vinet, Editor,** Advances in Mathematical Sciences: CRM's 25 Years, 1997

10 **Donald E. Knuth,** Stable Marriage and Its Relation to Other Combinatorial Problems, 1997

9 **Decio Levi, Luc Vinet, and Pavel Winternitz, Editors,** Symmetries and Integrability of Difference Equations, 1996

8 **J. Feldman, R. Froese, and L. M. Rosen, Editors,** Mathematical Quantum Theory II: Schrödinger Operators, 1995

7 **J. Feldman, R. Froese, and L. M. Rosen, Editors,** Mathematical Quantum Theory I: Field Theory and Many-Body Theory, 1994

6 **Guido Mislin, Editor,** The Hilton Symposium 1993, 1994

5 **D. A. Dawson, Editor,** Measure-Valued Processes, Stochastic Partial Differential Equations, and Interacting Systems, 1994

4 **Hershy Kisilevsky and M. Ram Murty, Editors,** Elliptic Curves and Related Topics, 1994

3 **Rémi Vaillancourt and Andrei L. Smirnov, Editors,** Asymptotic Methods in Mechanics, 1993

2 **Philip D. Loewen,** Optimal Control via Nonsmooth Analysis, 1993

1 **M. Ram Murty, Editor,** Theta Functions, 1993